1999

Oil Pollution from Ships

International, United Kingdom and
United States Law and Practice

AUSTRALIA AND NEW ZEALAND
The Law Book Company Ltd.
Sydney: Melbourne: Perth

CANADA AND U.S.A.
The Carswell Company Ltd.
Agincourt, Ontario

INDIA
N. M. Tripathi Private Ltd.
Bombay
and
Eastern Law House Private Ltd.
Calcutta and Delhi
M.P.P. House
Bangalore

ISRAEL
Steimatzky's Agency Ltd.
Jerusalem: Tel Aviv: Haifa

MALAYSIA: SINGAPORE: BRUNEI
Malayan Law Journal (Pte.) Ltd.
Singapore

PAKISTAN
Pakistan Law House
Karachi

Oil Pollution from Ships
International, United Kingdom and United States Law and Practice

Second Edition

by

David W. Abecassis   *and*   Richard L. Jarashow
M.A., Ph.D. (Cantab.);   J.D.; LL.M. (London)
of the Middle Temple, *Barrister*   *Partner*, Haight, Gardner,
                                                            Poor & Havens, New York

with

Robert M. Jarvis, J.D., LL.M.
Associate, Haight, Gardner,
Poor & Havens, New York

Malcolm Forster   Victor Sebeck
Legal Counsel   *Secretary*, Advisory Committee
I.U.C.N.   on Pollution of the Sea

*and the following members of the Institute
of Maritime Law, University of Southampton*

Professor R. P. Grime   Professor D. Jackson

Edited by

David W. Abecassis

LONDON
STEVENS & SONS
1985

First Edition 1978
Butterworth & Co. (Publishers) Ltd.

Second Edition 1985
Stevens & Sons Ltd.

Published by
Stevens & Sons Ltd.
11 New Fetter Lane, London
Computerset by Promenade Graphics Ltd., Cheltenham
Printed and bound in Great Britain by
Hazell Watson and Viney Ltd.

British Library Cataloguing in Publication Data

Abecassis, David William
    Oil pollution from ships.—2nd ed.
    1. Oil pollution of the sea—Law and legislation
    I. Title II. Jarashow, R. L.
    341.7'623   K3590.4

ISBN 0–420–47000–X

The sea is his and he made it: and his hands formed the dry land.

Psalm 95–5

And God said, Let us make man in our own image, after our own likeness: and let him have dominion over the fish of the sea . . . and over all the earth . . . .

Genesis 1–26

PREFACE

This work began life as a second edition of my 1978 book, *The Law and Practice Relating to Oil Pollution from Ships*. But as I got into it I realised that, not only had much of the legal picture changed out of all recognition, but that the original structure of the work was inadequate for the needs of the modern reader. As a result, an almost completely new book has been written, with very little indeed of the original text remaining intact. In addition, other authors have joined me in the task. It is therefore appropriate to have called this work by a new name—*Oil Pollution from Ships: International, United Kingdom and United States Law and Practice.*

The new name reflects the new structure: matters of domestic law are gathered together in Parts III and IV, instead of being interspersed among the international law exposition as in the earlier work; all of the United States material is new, as is most of the United Kingdom text. The purpose of these new parts—indeed of the whole work—is not to give an exhaustive and encyclopedic exposition of the law, but rather to outline the structure in reasonable detail, highlight points of particular interest or importance, and enable the reader to find his starting point for the solution of whatever detailed problem he may have before him, if not the solution itself.

Other works deal with aspects of the subject-matter of this work in equal or greater detail but I am not aware of any other which attempts as wide and as comprehensive a survey of the relevant legal principles. In this respect, I hope that this book will appeal to a worldwide readership as, indeed, the original work has done. It is with a worldwide readership in mind that I have unashamedly expanded the international law content, adding three new chapters and an introduction, and considerably adding to the other chapters. I also hope that readers with domestic legal problems on their mind will increasingly find the international law part of use, as more oil pollution legislation around the world comes to be modelled upon international conventions.

But there is also another reason for the prominence of international law in the work, and that is my enduring belief that unilateralism is not the right course for states to take in framing their oil pollution legislation. The better understood international law is, the more hopeful one can be that states will submit to its principles. If this book could contribute in some small measure to the achievement of this goal, then I and my co-authors will have been justly rewarded for our work.

Selbourne, July 1985   David William Abecassis

CONTENTS

Section 2 Salvage and General Average

Section 3 Liability

PART III *SELECTED ISSUES IN UNITED KINGDOM LAW*

TABLE OF CASES AND INCIDENTS

TABLE OF STATUTES

TABLE OF CONVENTIONS

Entry into force:

1954 Convention—26 July 1958

1962 Amendments—18 May and 28 June 1967 (600 UNTS 322; UKTS No. 59 of 1967 (Cmnd. 3354))

1969 Amendments—20 January 1978 (600 UNTS 336; 9 ILMI; IMO Resolution A. 175 (vi))

1971 Amendments—not in force at 31.12.84 (11 ILM 267; IMO Resolutions A. 232 (vii) and A. 246 (vii))

This Convention has been superseded, with effect from 2 October 1983, by the Protocol of 1978 Relating to the International Convention for the Prevention of Pollution from Ships, 1973, as between the States parties thereto.

Contracting States as at 31.12.84: Algeria; Argentina; Australia; Austria; Bahamas; Bangladesh; Belgium; Bulgaria; Canada; Chile; Cyprus; Democratic Yemen; Denmark; Djibouti; Dominican Republic; Egypt; Fiji; Finland; France; German

Entry into force: 1 May 1981

Contracting States as at 31.12.84: Algeria; Argentina; Australia; Bahamas; Barbados; Belgium; Bulgaria; China; Colombia; Czechoslovakia; Denmark; Finland; France; German Democratic Republic; Germany, Federal Republic of; Ghana; Greece; Hungary; Iceland; Ireland; Israel; Italy; Japan; Kuwait; Lebanon; Liberia; Libyan Arab Jamahiriya; Malaysia; Mexico; Netherlands; Nigeria; Norway; Panama; Peru; Poland; Portugal; Republic of Korea; Singapore; South Africa; Spain; Sweden; Switzerland; Tunisia; USSR; United Arab Emirates; United Kingdom; United States; Uruguay; Vanuatu; Yugoslavia.

ABBREVIATIONS

A.C. *or* App.Cas.	Appeal Cases (Law Reports)
ACOPS	Advisory Committee on Pollution of the Sea
AFSONG	Anglo-French Safety of Navigation Group
A.J.I.L.	*American Journal of International Law*
All E.R.	All England Law Reports
A.M.C.	American Maritime Cases
B. & C.	Barnewall and Cresswell's Reports, K.B. 1822–30
BIICL	British Institute of International and Comparative Law
B.L.R.	*Business Law Review*
BYIL	*British Yearbook of International Law*
C.A.	Court of Appeal
C.B.N.S.	Common Bench Reports, N.S., 1856–1865
CBT	Clean Ballast Tanks
CERCLA	The Comprehensive Environmental Response, Compensation and Liability Act 1980
Ch.	Chancery (Law Reports)
C.L.J.	*Cambridge Law Journal*
C.L.R.	Commonwealth Law Reports
C.M.L.R.	Common Market Law Reports
Co.Rep.	Coke's Reports 1572–1616
COW	Crude Oil Washing
Cr.App.R.	Criminal Appeal Reports
Crim.L.R.	*Criminal Law Review*
CRISTAL	Contract Regarding an Interim Supplement to Tanker Liability for Oil Pollution
DAFS	Department of Agriculture and Fisheries for Scotland
D.L.R.	Dominion Law Reports
DPA	The Deepwater Port Act 1975
dwt/DWT	Deadweight. The deadweight tonnage of a vessel is the cargo capacity of that vessel in tons.
E.E.C.	European Economic Community
E.G.	*Estates Gazette*

Esp.	Espinasse's Nisi Prius Reports 1793–1807
F.L.R.	Federal Law Reports
F. 2d.	Federal Reporter, Second Series
F.Supp.	Federal Supplement
Fund Convention	The International Convention on the Establishment of an International Fund for Compensation for Oil Pollution Damage
FWPCA	The Federal Water Pollution Control Act 1948
grt/GRT	Gross registered tons. The grt of a vessel is a measure of the volume of space occupied by the vessel.
H.L.	House of Lords
ICES	International Council for the Exploration of the Sea
I.C.L.Q.	*International and Comparative Law Quarterly*
ILM	*International Legal Materials*
ILO	International Labour Organisation
I.L.T.	Irish Law Times
IMCO	Intergovernmental Maritime Consultative Organisation, now renamed the International Maritime Organisation
IMF	International Monetary Fund
IMO	International Maritime Organisation (formerly the Intergovernmental Maritime Consultative Organisation)
IOPC Fund	International Oil Pollution Compensation Fund
IOPP Certificate	International Oil Pollution Prevention Certificate
ITIA Ltd.	International Tanker Indemnity Association Ltd.
ITOPF	International Tanker Owners' Pollution Federation
J.B.L.	*The Journal of Business Law*
J. Mar. L. & Comm.	*The Journal of Maritime Law and Commerce*
J.P.	Justice of the Peace Reports
J.P.L.	*Journal of Planning and Environmental Law*
Jur.(N.S.)	*The Jurist, New Series, 1855–66*
K.B.	King's Bench (Law Reports)
K.I.R.	Knight's Industrial Reports
Ld.Raym.	Lord Raymond's Reports 1694–1732
L.G.R.	Local Government Reports
Liability Convention	The International Convention on Civil Liability for Oil Pollution Damage
L.J.	*Law Journal Newspaper*

xlviii

L.J.Ch.	*Law Journal Chancery*
L.J.P.	*Law Journal, Probate, Divorce and Admiralty*
L.J.P.C.	*Law Journal, Privy Council*
L.J.Q.B. (or K.B.)	*Law Journal, Queen's Bench or King's Bench*
L.J.R.	*Law Journal Reports*
L.M.C.L.Q.	*Lloyd's Maritime and Commercial Law Quarterly*
LOF	Lloyd's Open Form 1980
LOOP	Louisiana Offshore Oil Port
LNTS	League of Nations Treaty Series
L.R.	Law Reports
L.S.Gaz.	*Law Society's Gazette*
L.T.	Law Times
Ll.L.Rep.	Lloyd's List Reports (before 1951)
Ll.P.C.	Lloyd's Prize Cases
Lloyd's Rep.	Lloyd's List Reports (1951 onwards)
MARPOL 73/78	International Convention for the Prevention of Pollution from Ships 1973, as amended by the 1978 Protocol thereto
MAFF	The Ministry of Agriculture, Fisheries and Food
MEMAC	Marine Emergency Mutual Aid Centre
MOPC Fund	Marine Oil Pollution Compensation Fund
MOPIC	Marine Oil Pollution Insurance Corporation
MPCU	Marine Pollution Control Unit
NCC	Nature Conservancy Council
NDLS	*New Directions in the Law of the Sea*
NERC	National Environmental Research Council
New L.J.	*New Law Journal*
N.I.	Northern Ireland Reports
NRT	National Response Team
NSF	National Strike Force
OBO	Ore/Bulk/Oil Carrier
O.J.	Official Journal of the European Communities
OPC Fund	Offshore Oil Pollution Compensation Fund
OR	*The Official Records* of Conferences sponsored by the IMO are published by the IMO. Page numbers refer to page numbers in the Official Records published by IMO
OSC	On-Scene Co-ordinator

OCSLA	Outer Continental Shelf Lands Act Amendments 1978
P. *or* P.D.	Probate, Divorce and Admiralty (Law Reports)
P.C.	Privy Council
PIOPIC	Protection and Indemnity Oil Pollution Indemnity Clause
PLATO	Pollution Liability Agreement Among Tanker Owners
PWSA	The Ports and Waterways Safety Act 1972
Q.B.	Queen's Bench (Law Reports)
SBT	Segregated Ballast Tanks
S.C.	Session Cases
SDR	Special Drawing Right
S.J. *or* Sol.Jo.	*Solicitors' Journal*
SOLAS 74/78	The International Convention for the Safety of Life at Sea 1974, as amended by the 1978 Protocol thereto
S.L.T.	Scots Law Times
TAP	The Trans-Alaska Pipeline Authorization Act 1973
T.L.R.	Times Law Reports
TOVALOP	Tanker Owners' Voluntary Agreement Concerning Liability for Oil Pollution
TSPP	Tanker Safety and Pollution Prevention Conference
UKOPP Certificate	United Kingdom Oil Pollution Prevention Certificate
UNCLOS III	The Third United Nations Conference on the Law of the Sea
UNEP	United Nations Environment Programme
U.S.	United States Reports
VTMS	Vessel Traffic Management System
VTRS	Vessel Traffic Reporting System
VTS	Vessel Traffic Service System
W.L.R.	Weekly Law Reports
W.N.	Weekly Notes (Law Reports)

PART I
INTRODUCTORY

1. Introduction and Overview

1. OIL POLLUTION FROM SHIPS—A FAST DEVELOPING SUBJECT

1–01   There is no branch of environmental law which progresses at a faster rate than that relating to oil pollution from ships. There seem to be a number of reasons for this. One is that it is a genuine environmental problem perceived by a very wide spectrum of public opinion. Most of us do not live near a nuclear power station, and so dramatic incidents like that at Three Mile Island do not tend to impress themselves as much upon our collective consciousness as the tarballs which stick to our feet when we visit the beach.

1–02   Another is that there is one United Nations body—the International Maritime Organisation, IMO—which has as a major task the development of international instruments relating to protection of the marine environment; another—the United Nations Environment Programme—has as a major task the co-ordination of the activities in this field of other United Nations bodies. Without these two international fora, much of today's international legislation would not exist; and without the international legislation, much of the domestic legislation would not exist. This explains why so much of this book is devoted to international instruments.

1–03   A third reason is that oil pollution from ships is something which governments think they find easy to understand—and it has a more natural appeal than, say, cleaning up one's rivers from land-based discharges or preventing lead discharges from motor vehicle exhausts. The reason for this is that shipowners and oil companies, being inherently international industries, are considered more legitimate targets for regulations which will cost them money than more obviously domestic industry.

1–04   Even if the above suggestions were to prove incorrect, the fact remains that there has been nothing less than an explosion of regulation and other legal development in the field of oil pollution from ships since the first edition of this book in 1978. There has been so much that the subject, once an esoteric branch of international law, is in danger of collapsing under its own weight and complexity. The purpose of this book is to help the reader to come to grips with this complexity, while enabling him to see the subject as a whole, as it were from a distance.

1–05 The United Kingdom material contained in the first edition has been separated into its own Part and largely rewritten by other authors, so that now it forms a discrete adjunct to the international law Part. The United States chapters are all entirely new, and fill a really serious gap in the first edition—a gap made the more important by the fact that as yet the United States has decided more often than not to go its own way in formulating oil pollution laws (or perhaps it would be more accurate to say that it began that way, and has ever since had enormous difficulty in deciding how and when to come into the international fold). The United States is a country where a very large number of oil pollution incidents occur and where damages awards tend to be comparatively high.

2. INTERESTS INVOLVED IN A MAJOR OIL POLLUTION INCIDENT

1–06 One reason why the subject has become so large and complex can be illustrated by summarising the facts of a real case—the loss of the *Tanio* off Brittany, France in March 1980. This ship had been built more than 20 years earlier by a Dutch shipyard. At the time of her loss she was owned by a Swiss leasing company and bareboat chartered to a Panamanian company which subsequently went into liquidation. She was sub-bareboat chartered to a Madagascan company, and sub-chartered again to another Panamanian company, a sister to the first Panamanian company. The ship was then time chartered to a major French oil company, and voyage chartered to a United Kingdom oil trading company. As if this was not complicated enough, she had not long earlier been repaired by an Italian shipyard where the supervision had been done by a French company whose affiliates had contracts for the technical and commercial management of the ship and also contracts relating to the recruitment of certain personnel for the ship. The officers were of French nationality, the crew Madagascan. She was classed by a French classification society. Her third party liability insurance was with a Bermudan mutual insurance company, and the hull cover had been placed with literally hundreds of insurance companies all over the world. The Master, Chief Officer and five others tragically lost their lives in the accident, and one half of the ship sank to a considerable depth taking with it valuable evidence on why the casualty occurred. The only complication not present was a salvage attempt prior to the loss, but that came soon after in that the half which floated had to be towed to shore, and there then followed a long and very expensive operation to pump out the oil remaining in the sunken half of the wreck, after which the company hired for the purpose went into liquidation. The oil released covered miles of the French coastline and had a significant effect on the tourist trade in the immediate vicinity, and it also drifted to the shores of the Channel Islands, under the jurisdiction of the United Kingdom. The resulting damage made it one of the biggest oil pollution incidents ever.

1–07 The *Tanio* case is not atypical in its complexity—other well-known incidents, such as the *Amoco Cadiz*[1] or the most famous of them all, the *Torrey Canyon*—present a similar international patchwork. Hence, when legislation is contemplated, these international complexities must be taken into account. If ever two industries demonstrated the desirability of unifying principles of private law, it is the oil and shipping industries. This need is at its height when oil pollution from ships is being considered.

1–08 Therefore, when a serious casualty occurs to a laden tanker, quite typically a large number of interests become quickly involved. This is particularly so because a large number of commercial interests are involved in almost every voyage, and whenever the coastline is threatened the commercial interests are supplemented by those of the public and the governmental bodies who represent them. First on the scene is often a salvage tug, and right away a hotbed of legal considerations come into play. The terms on which salvage services are offered and accepted are a matter of great importance not only to the salvor, but to those interested in the ship and her cargo. The way in which the expenses involved in the operation will subsequently be divided will depend on whether a no cure-no pay contract is concluded, and, if so, on the success achieved in the salvage operation and on the precise circumstances of the case.[2] Nowadays, the master may be in touch with his owners by radiotelephone the moment the ship gets into difficulties as was the case in the *Amoco Cadiz*, for instance, and so the owner himself may be involved directly from the beginning.

1–09 The coastal governmental authorities, be they coast guard or local authority, are sure to be taking an interest from the earliest moment. Quite apart from initiating their oil pollution response plan (assuming they have one),[3] in a serious case they may be ask-

---

[1] For a detailed study of this case, see paras. 21–61 *et seq.*, below.
[2] Issues related to oil pollution arising out of the salvage of a ship are dealt with in Chap. 8.
[3] Contingency and oil pollution response plans are dealt with in Chaps. 7, 14 and 19.

ing themselves whether they ought not to get involved in the process to the extent of intervening in the drama being played out at sea—for instance by taking direct control of the salvage operation, or by destroying the ship. The question on their minds will be, what rights do they have?[4] And the flag state authorities will be (or should be) taking an interest to see whether an inquiry is necessary or disciplinary action should be taken against those responsible. They may also be concerned to see that the rights of those aboard are not infringed by the coastal state.[5]

1–10 A representative of the Protection and Indemnity Club will probably not be far off. Protection and Indemnity Clubs (usually called P & I Clubs) are to most people rather shadowy and ill-understood bodies. In fact, they are simply insurance companies which are organised on a mutual basis, so that in any one policy year all those receiving cover contribute to the total proven claims of all the assureds.[6] Their interest at the time of the casualty will, therefore, be particularly in liability issues.[7]

1–11 Another person who may be on the scene is the representative of the hull underwriters. Since even in today's market the hull of the ship is likely to be insured for millions of dollars, what those dealing with the oil pollution do to the ship is of as much concern to him as it is to the owner. The hull underwriter also insures the owner's general average expenditure, which will probably include the owner's contribution to a salvage award.[8] Even the cargo underwriter may be there, since the cargo will probably be the most valuable property involved, and by definition the more oil spilled the bigger will be the cargo claim. Also, cargo underwriters will contribute their share to salvage and general average.

1–12 Even the charterers may have a representative around before long. They will be concerned about potential liabilities[9] and possible claims for loss of cargo and, if the ship is time-chartered, loss of bunkers.

---

[4] Intervention by state authorities is dealt with in Chaps. 6 and 18.

[5] The respective rights of coastal states and flag states are dealt with in Chap. 5.

[6] P & I Clubs insure, *inter alia*, their members' third party risks, including the risks that they might become liable for oil pollution damage, or for loss of cargo, or to a fine resulting from a maritime incident. The Clubs therefore protect and insure the third party risks of their members, who are either shipowners or others having an interest in the ship, such as bareboat charterers, mortgagees or, sometimes, ship managers.

[7] See especially Chaps. 9 to 12, 15, and 19 to 21. The international law background to these issues is dealt with in Chap. 2. Liability aspects of salvage are dealt with in Chap. 8.

[8] See Chap. 8.

[9] See especially Chap. 15 for charterers' liabilities and Chap. 9 for their right to limit liability.

6

1–13   Once the accident hits the headlines, people start asking all kinds of questions. Should the crew have been better trained? Should there have been better navigational equipment on board? Was the ship designed to withstand the kind of marine peril she suffered and if not, why not?[10] Meanwhile, the clean-up has to be organised and effected, which in a bad case like the one we are considering may involve a considerable international effort.[11]

1–14   The accident may lead to a public or governmental enquiry, at which the wider issues of pollution control are investigated. Such issues as the rights of states in international law to make regulations to prevent and control pollution in various parts of the seas are likely to come up,[12] as may the potential liability of the state itself.[13] And in subsequent litigation, a host of jurisdictional and procedural issues are almost bound to arise—who can sue whom, and where?[14]

3. OPERATIONAL OIL POLLUTION

1–15   Casualties to ships are the most dramatic, but by far the least frequent part of the overall problem of oil pollution from ships. The tarball on your beach is far more likely to have been caused by an operational discharge of oil from a ship than by oil from a casualty. This aspect of the problem is the least understood by the public, partly because it is not dramatic and newsworthy, partly because it requires an understanding of technical matters relating to ship design and operation. Few people understand why ships have to discharge oil at all; and in fact in many cases now, as we shall see in Chapter 3, they do not have to.[15] There are always those who find it economically, operationally or commercially desirable not to comply with the now very stringent regulations on discharges, and they raise the problem of enforcement of standards.[16]

---

[10] The formulation of legal standards relating to prevention of accidents to oil tankers, and their enforcement, is dealt with in Chaps. 4 and 18.

[11] International co-operation and national contingency plans are dealt with in Chap. 7.

[12] Such issues are dealt with in Chap. 5.

[13] Discussed in Chap. 9.

[14] Such matters are dealt with in Chaps. 5 and 16.

[15] International standards regulating operational oil discharges are the subject of Chap. 3; United Kingdom standards are dealt with in Chap. 14; United States standards are dealt with in Chap. 18.

[16] Jurisdiction generally is the subject of Chap. 5, and specific enforcement provisions relating to operational pollution are dealt with in Chaps. 3, 14 and 18.

1–16 Operational oil pollution is another classic example of the need for uniform international regulation and enforcement—anything less than such uniformity leads to distortions of competition in the international shipping market. While the commercial interests in industry have played a very significant role in the development of solutions to operational oil pollution, it is ultimately governments who have to ensure that the rules are kept. Governments have not always achieved this aim.

4. THE FUTURE

1–17 This book has many messages, but a consistent theme is that the whole problem of oil pollution from ships can best be solved by active international co-operation. Indeed, it is argued in Chapter 2 that states are obliged by international law to do just that. In general, there are now enough international instruments in existence to ensure seas which are tolerably clean from oil pollution if those instruments were universally adhered to—the frailty of man is such that accidents will always happen, and so whatever is written in the regulations, there will always be some oil pollution. Therefore, the task before the international community of states is not to develop yet more and stricter standards, but to ensure that the ones we have now got are universally kept. In this respect, it would help if the Law of the Sea Convention 1982 entered into force, but that would not by itself solve the problem. Only a concerted desire on the part of states and industry alike for cleaner seas, and the money to devote to the task, will achieve it. At present, neither is wholly present.

PART II
INTERNATIONAL LAW

2. The Framework of Legal Obligations in International Law

SUMMARY

2–01   Different strands in the doctrines of customary international law lead to the same broad conclusion, namely that all states are under the same duty to regulate oil pollution from ships under their flag. These doctrines by themselves do not prescribe any particular mode or standard of regulation to which states must adhere, but instead point strongly towards their elaboration by international Convention. This has been the practice of states since 1954, and the Law of the Sea Convention, 1982 endorses the concept of one standard of duty for all states and of the definition of that duty by the process of international agreement. In time the principles of such agreements are capable of affecting the duty itself, and the almost universal acceptance of the provisions of certain Conventions on the prevention of oil pollution from ships, load lines, collision regulations and safety of life at sea means that their fundamental principles, at the least, must now be regarded as having been received into the body of customary international law. This process fills the gaps which the general doctrines were by themselves unable to fill, and the Law of the Sea Convention, 1982 foresees its continuation.

1. INTRODUCTION

2–02   In Section 1 of Part II of this work, we examine the international legal rights and obligations with respect to the prevention, control and clean-up of oil pollution from ships. The most important of these are to be found in the detailed provisions of the Conventions discussed in Chapters 3 to 7. However, these Conventional provisions, with very few exceptions, benefit and bind only the parties to them, so that a complete picture can only be obtained by first examining the general principles of international law which relate to all states. This puts the detailed Conventional law into its proper perspective and provides an explanation for the evolution of the detailed Conventional standards.

2–03 In this Chapter, we examine the general principles of international law which impose obligations on states relating to prevention and clean-up. Rights are best dealt with separately, and so Chapters 5 and 6 each contain an examination of general principles.

2. GENERAL PRINCIPLES

2–04 Environmental problems have not been in the forefront of international relations until the latter half of this century, because it is only recently that the international nature of many pollution problems has been widely appreciated. In fact, oil pollution from ships was one of the earliest to be recognised as requiring an international solution, with attempts being made in 1926 and 1936 to adopt a Convention.[1] Nonetheless, this lack of awareness following the growth of the actual problems has had the effect that international law has had relatively little time to receive and develop applicable general principles. It is not surprising, therefore, to find that such principles as there are were developed either in the context of other aspects of international relations, or were enunciated to meet specific problems submitted to international tribunals. In searching for principles applicable to oil pollution from ships it is therefore necessary to cast a wide net, which leaves ample room for differences of view, with consequent doubt about the validity of the conclusions. The search for such principles is, however, of value because they serve both to highlight the need for, and to form the basis of, the elaboration of effective standards in multilateral Conventions. Although some writers give considerable space to several possible doctrines, including *culpa* and good neighbourliness, there is overlap amongst the theories and doctrines, and only two require examination here.

A. Hazardous Activities

2–05 International law places states under certain duties so to regulate activities under their control that they do not cause harm to other states. Some writers express this doctrine in the phrase *"sic utere tuo ut alienum non laedas"*—use your own so as not to harm

---

[1] For brief accounts of these conferences, see R. M. M'Gonigle and M. M. Zacher, *Pollution, Politics and International Law*, Berkeley, Los Angeles and London, 1979, p. 81; K. Hakapaa, *Marine Pollution in International Law—Material Obligations and Jurisdiction*, Helsinki, 1981, p. 75.

others.[2] However, the starting point of the discussion of this topic is frequently the well-known *Trail Smelter* arbitration between the United States and Canada.[3]

2–06 A smelter in British Columbia, close to the border with the United States, had been emitting fumes which drifted across the border and caused harm to persons on United States territory. In 1935 the parties set up an ad hoc tribunal whose terms of reference provided that the tribunal "shall apply the law and practice followed in dealing with cognate questions in the United States of America as well as international law and practice." In what has become perhaps the most famous statement in international environmental law, the tribunal declared that:

> " . . . under the principles of international law, as well as of the law of the United States, no State has the right to use or permit the use of its territory in such a manner as to cause injury by fumes in or to the territory of another or the property or persons therein, when the case is of serious consequence and the injury is established by clear and convincing evidence."

2–07 Some writers have generalised from this statement to reach a point a long way from the case itself. One, for instance, concludes that:

> " . . . States . . . must take reasonable precautions to prevent persons who are under their territorial jurisdiction as well as ships and other floating devices which are attributable to them from polluting other States' jurisdictional waters. They are finally bound to impose sanctions upon persons who have done so."[4]

Whether this case can, all on its own, be taken that far is open to some doubt.[5] It must be remembered that the statement refers to

---

[2] F. Jimenez de Arichaga, "International Responsibility," in M. Sorensen (ed.), *Manual of Public International Law*, London and New York, 1968, p. 540: "The acting state is in breach of a duty of non-interference established by customary international law, generally stated in the maxim: *sic utere* . . . etc."; K. Hakapaa, *Marine Pollution in International Law—Material Obligations and Jurisdiction*, Helsinki, 1981, p. 136; A. L. Springer, *The International Law of Pollution—Protecting the Global Environment in a World of Sovereign States*, Westport Connecticut and London, 1983, p. 129.

[3] U.N.R.I.A.A., iii, p. 1905.

[4] L. C. Caflisch, "International Law and Ocean Pollution: The Present and The Future," (1972) 8 *Revue Belge de Droit International*, p. 17.

[5] K. Hakapaa, *Marine Pollution in International Law—Material Obligations and Jurisdiction*, Helsinki, 1981, p. 137, concludes that the maxim *sic utere* "can provide only little guidance for the verification of specific state obligations." D. J. Cusine and J. P. Grant, "The Legal Framework" in *The Impact of Marine Pollution* (eds. D. J. Cusine and J. P. Grant), London and Montclair, N.J., 1980, p. 31, say "it is open to doubt how wide the ratio of the Trail

the use of a state's territory, to cases of serious consequence, established by clear and convincing evidence, and that the case was one of atmospheric pollution caused by a continuing state of affairs. On the other hand, it is based on a principle which is clearly capable of forming the basis of a broad duty to protect the international environment.

2–08   The other case of particular relevance here is the *Corfu Channel* case.[6] Between the island of Corfu and the Albanian mainland there is a narrow channel known as the North Corfu Channel. Albania disputed its status as an international strait. In May, 1946, British warships passing through the strait were fired upon by Albanian coastal batteries, and the following October, two British warships passed through the channel within Albanian territorial waters to test Albania's attitude. They struck mines which, the International Court found, had been recently laid. The ships were damaged and members of the crews lost their lives. The court found that there was a right of innocent passage through the strait and that the passage of the ships was innocent. The laying of the minefield could not have been accomplished without the knowledge of the Albanian Government, although it may not have been laid directly on Albanian orders. In holding Albania liable for such damage, the court held that in this situation:

> "The obligations incumbent on the Albanian authorities consisted in notifying, for the benefit of shipping in general, the existence of a minefield in Albanian territorial waters and in warning the approaching British warships of the imminent danger to which the minefield exposed them. Such obligations are based . . . on certain general and well recognised principles, namely, elementary considerations of humanity . . . ; the principle of the freedom of maritime communication; and every State's obligation not to allow knowingly its territory to be used contrary to the rights of other States."[7]

2–09   The reference at the end of that quotation, taken with the reference to the imminent danger posed by the minefield, is significant. In both this case and the *Trail Smelter*, the likelihood that the interests of other states would be affected was extremely high. Had

---

Smelter Arbitration can be extended," and conclude that only a broad construction would be relevant to pollution of the sea by oil. I. Brownlie, *Principles of Public International Law*, (3rd ed.) Oxford, 1979, p. 285, regards the case as having "considerable value in relation to the uses of international rivers, atmospheric pollution and activities in outer space adversely affecting the earth's environment," but he does not mention oil pollution from ships.

[6] [1949] I.C.J. Reports, p. 4.
[7] *Ibid.* at p. 22.

this not been so, both cases might have gone the other way. It would be reasonable to propound the rule from these cases that if an activity or situation within the territory of a state was sufficiently hazardous to the interests of other states to make damage to those interests foreseeable, the state has a duty to prohibit it. Since the control of a state over its territory is at the heart of such a doctrine, there is no logical reason why it should apply only to its territory, but could not be extended to activities under the control of a state.

2–10   The precise content of such a doctrine is, of course, unclear. For instance, it is not clear whether it could apply to preventing accidental emissions as well as operational ones. Nor is it clear what degree of foreseeability attracts the duty to prevent, what level of pollution must be avoided, whether the rule protects only the territory of other states or also includes the high seas and whether the economic and infrastructural resources of the state affect the level of the duty.[8] While the application of such a principle to oil pollution from ships does present problems of definition, the principle can be extrapolated to support the proposition that states are not free simply to sit back and let ships under their flag sail the seas totally unregulated. On the other hand, there is no basis here for concluding that a global ban on operational discharges has become a norm. The truth lies somewhere between the two. The standard will vary with the nature of the activity[9]: oil tankers present a different order of hazard to dry cargo ships, for instance.

B. Abuse of Rights

2–11   While there is general agreement that abuse of rights is a general principle of law capable of forming a source of international law,[10] there is doubt about whether it has been received as a doc-

---

[8] For instance, I. Brownlie, *Principles of Public International Law*, (3rd ed.) Oxford, 1979, p. 476, concludes with others cited therein that there is no absolute duty in international law to contain the effects of ultra-hazardous activities.

[9] See I. Brownlie, "A Survey of International Customary Rules of Environmental Protection," in L. A. Teclaff and A. E. Utton (eds.), *International Environmental Law*, New York and London, 1974, p. 2.

[10] For instance, some writers see it as the basis for the doctrine enunciated in Trail Smelter: D. J. Cusine and J. P. Grant, "The Legal Framework" in *The Impact of Marine Pollution* (eds. D. J. Cusine and J. P. Grant), London and Montclair, N.J., 1980, p. 32; I. Brownlie, *Principles of Public International Law*, (3rd ed.) Oxford, 1979, p. 445.

15

trine of international law, and if so, what is its extent.[11] The relevant rights whose abuse would fall to be considered in the present context are those of the freedoms of the seas. These were succinctly qualified in Article 2 of the Convention on the High Seas, 1958, in the following terms:

> " . . . These freedoms, and others which are recognised by the general principles of international law, shall be exercised by all States with reasonable regard to the interests of other States in their exercise of the freedom of the high seas."

Article 87(2) of the Law of the Sea Convention, 1982 is to similar effect, requiring "due regard" to be had to the interests of other states in the exercise of the freedoms, and Article 300 goes further by specifically endorsing a concept of abuse of rights:

> "States Parties shall fulfil in good faith the obligations assumed under this Convention and shall exercise the rights, jurisdiction and freedoms recognized in this Convention in a manner which would not constitute an abuse of right."

2–12 Provisions such as these do not make it necessary for us to examine further the basis in international law of a doctrine of abuse of rights, because it is clear that international law has accepted as a norm that the freedom of navigation shall be exercised with reasonable regard to the interests of other states. Whatever might be the case with other rights of states in international law, the freedoms of the seas are not unrestricted, but carry with them a burden. This concept of reasonable user suggests that a balance must be struck between the competing state interests.

2–13 As with the doctrines applicable to hazardous activities, there is no guidance given as to precisely how this balance is to be struck in practice. One can therefore conclude that this doctrine leads to very much the same point as the first, and in so doing supports and reaffirms the fundamental duty of states to respond to the problem of oil pollution from ships by making regulations to pre-

---

[11] E. Jimenez de Arichaga, "International Responsibility," in M. Sorensen (ed.), *Manual of Public International Law*, London and New York, 1968, p. 540: "If the prohibition of the abuse of rights has been accepted as a rule of international law . . . ."; I. Brownlie, *Principles of Public International Law*, (3rd ed.) Oxford, 1979, pp. 444–445: " . . . the delimitation of its function is a matter of delicacy . . . . In conclusion it may be said that the doctrine is a useful agent in the progressive development of the law, but that, as a general principle, it does not exist in positive law." K. Hakapaa, *Marine Pollution in International Law—Material Obligations and Jurisdiction*, Helsinki, 1981, p. 138: "The basic difficulty with the doctrine lies in the very question of its functional construction." *Cf.* D. W. Bowett, *The Law of the Sea*, Manchester and Dobbs Ferry, 1967, p. 45: "The 'abuse of rights' involved in the indiscriminate discharge into the seas of waste products has long been recognised in relation to oil." All these works contain copious references to others on the doctrine.

vent it. The principles of balance and reasonableness suggest that the appropriate method is to elaborate standards by international agreement.

3. THE DEVELOPMENT OF INTERNATIONAL STANDARDS

2–14 It is therefore not surprising to find that Article 24 of the Convention on the High Seas, 1958 contains specific mention of this duty to make preventive regulations:

> "Every State shall draw up regulations to prevent pollution of the seas by the discharge of oil from ships . . . taking account of existing treaty provisions on the subject."

At the time the only existing treaty provisions were those of the 1954 Convention, discussed in Chapter 3, but the intent is clearly not to freeze these provisions in time as at 1958. The pollution to be prevented is not limited to operational discharges, but can encompass accidents too. In this connection we may also note Article 10:

> "1. Every State shall take such measures for ships under its flag as are necessary to ensure safety at sea with regard inter alia to:
> (a) The use of signals, the maintenance of communications and the prevention of collisions;
> (b) The manning of ships and labour conditions for crews taking into account the applicable international labour instruments;
> (c) The construction, equipment, and seaworthiness of ships.
> 2. In taking such measures each State is required to conform to generally accepted international standards and to take any steps which may be necessary to ensure their observance."

2–15 These provisions spell out in more detail how the fundamental obligations of states established by general principles of international law and the restrictions to freedoms of the seas enshrined in Article 2 of the Convention are to be elaborated with respect to oil pollution from ships. The process envisaged is clearly one of the negotiation of detailed standards which will apply to all ships sailing the seas. Thereafter, the role of states is seen as one of ensuring compliance. This theme has undergone continuous development since 1958 in a number of instruments, particularly the regional Conventions discussed in Chapter 7.[12]

---

[12] For instance Article 7 of the 1974 Baltic Convention put into effect in advance certain provisions of the International Convention for the Prevention of Pollution from Ships, 1973; Article 6 of the 1976 Mediterranean Convention and Article IV of the 1978 Kuwait Convention place parties under duties to implement applicable international rules.

17

2–16 The protection of the environment obtained a considerable boost at the 1972 Stockholm Conference on the Human Environment, and at this point the idea gained ground that the duties of states might vary according to their economic strength and stage of development. While some of the most important Principles from the oil pollution viewpoint contain no reference to this theme, it finds expression in a number of others, including Principle 23:

> "Without prejudice to such criteria as may be agreed upon by the international community . . . it will be essential in all cases to consider the systems of values prevailing in each country, and the extent of the applicability of standards which are valid for the most advanced countries but which may be inappropriate and of unwarranted social cost for the developing countries."

2–17 The developing countries took the argument to the Law of the Sea Conference, but the objections of developed states to creating differing standards for vessel-source pollution led to a compromise set of provisions whose interpretation is, at first sight, ambiguous. For instance, the general duty of Article 192 of the Law of the Sea Convention to "protect and preserve the marine environment" contains no qualification, nor does the more detailed but still general obligation in Article 194(2):

> "States shall take all measures necessary to ensure that activities under their jurisdiction and control are so conducted as not to cause damage by pollution to other States and their environment . . . ."

Yet the equally important Article 194(1) contains a classic expression of the theme:

> "States shall take . . . all measures consistent with this Convention that are necessary to prevent, reduce and control pollution of the marine environment from any source, using for this purpose the best practicable means at their disposal and in accordance with their capabilities . . . ."

2–18 In so far as standards of duties in relation to the prevention and control of pollution from ships is concerned, it is important to realise that the detailed provisions specifically relating to the nature of the duty—as opposed to those relating to more peripheral matters such as contingency plans, monitoring and enforcement by port states[13]—contain no such qualifications. Articles 194(3)(*b*)

---

[13] Articles 204(1) (monitoring of the risks or effects of pollution), 206 (assessment of potential effects of activities), 218(3) (enforcement by port states) and 219 (measures relating to seaworthiness of vessels to avoid pollution) all contain the phrase "as far as practicable," and Article 199 (contingency plans against pollution) contains the phrase "in accordance with their capabilities."

18

and 211 are universal and make no discrimination.[14] Article 211 is particularly important, because not only does it preserve the notion that the duties on states to prevent and control oil pollution from ships are the same for all states, but it confirms, strengthens and considerably develops the theme that these duties are to be worked out in practice by multilateral agreement:

> "1. States, acting through the competent international organisation or general diplomatic conference, shall establish international rules and standards to prevent, reduce and control pollution of the marine environment from vessels and promote the adoption, in the same manner, wherever appropriate, of routeing systems . . .
> 2. States shall adopt laws and regulations for the prevention, reduction and control of pollution of the marine environment from vessels flying their flag or of their registry. Such laws and regulations shall at least have the same effect as that of generally accepted international rules and standards established through the competent international organisation or general diplomatic conference."

2–19   Equally significant, Articles 94(3), (4) and (5), which are in almost identical (but more detailed) terms to those of Article 10 of the Convention on the High Seas quoted above, support the view of a single standard and relate to "applicable international instruments" and "applicable international regulations."

2–20   It must therefore be concluded that, whatever success the developing states may have achieved over land-based sources[15] and certain peripheral activities connected with pollution caused by ships, it does not spill over into the standards expected of them in relation to the prevention of the polluting discharge itself.

4. THE EFFECT OF SPECIFIC CONVENTIONS ON CUSTOMARY INTERNATIONAL LAW

2–21   International custom is developed by the practice of states. Therefore, if a sufficient number of states accept as law the provisions of an international Convention relating to conduct on the high seas and apply it in their national law, over time its fundamen-

---

[14] Article 194(3): "The measures taken . . . shall include, *inter alia*, those designed to minimise to the fullest possible extent: . . . (*b*) pollution from vessels, in particular measures for preventing accidents and dealing with emergencies, ensuring the safety of operations at sea, preventing intentional and unintentional discharges, and regulating the design, construction, equipment, operation and manning of vessels."

[15] Article 207(4), on land-based pollution, clearly creates differing standards of obligation, for it takes into account "characteristic regional features, the economic capacity of developing States and their need for economic development."

tal principles can be considered as being received into the body of international law which binds all states. This is particularly appropriate in the field of oil pollution from ships, for as we have seen international obligations are to be evolved through the negotiation and implementation of such Conventions. The question arises, therefore, whether any Convention can be considered as having now received such wide acceptance that its principles have passed into the body of customary international law.

2–22   The relevant Convention which, as at December 31, 1984, had received the widest acceptance was, as one might expect, the International Convention on Load Lines, 1966 with 102 parties representing 98 per cent. of the world tonnage, closely followed by two others having general relevance—the Convention on the International Regulations for Preventing Collisions at Sea, 1972 with 88 parties representing 96 per cent. of world tonnage and the International Convention for the Safety of Life at Sea, 1974 which has 78 parties representing 94 per cent. of world tonnage. If the principles of these[16] are not generally accepted, then it is difficult to think what could be.

2–23   Since the International Convention for the Prevention of Pollution of the Sea by Oil, 1954, as amended up to 1969,[17] had 71 parties representing over 90 per cent. of the world's tonnage, that too should qualify. As one writer has put it: " . . . it might be submitted that the discharge provisions laid down in the 1969 Amendments to the 1954 Convention are to the extent recognized that at least their basic features—especially the total discharge prohibition for nearly all tankers within the 50-mile limit—could be considered to have acquired the status of customary law."[18] But the principles of MARPOL 73/78[19] cannot yet be considered to have this status, since at December 31, 1984 it had only 33 parties representing 70 per cent. of the world's tonnage.

2–24   This conclusion means that the problem noted in paragraph 2–10 above, that the level of obligation on states was at some indeterminate point between the two extremes of doing nothing and total prohibition, has to a considerable extent been resolved by the very widespread acceptance by states of certain specific obligations in the 1954 Convention, as amended, and the other Con-

---

[16] See paras. 4–25 *et seq.*
[17] See paras. 3–49 *et seq.*
[18] K. Hakapaa, *Marine Pollution in International Law—Material Obligations and Jurisdiction*, Helsinki, 1981, p. 132.
[19] See paras. 3–52 *et seq.*

ventions just mentioned. These instruments can now be considered to represent the basic framework of international law regarding prevention and control. At one time, they represented the latest state of the art, the highest standards individual states were prepared to accept. It is a measure of the pace at which standards have developed that their principles have passed into the body of customary international law. In time, as the process of implementing the latest standards contained in multilateral Conventions grows and as new standards are evolved, it must be the intention that they will follow the same path.[20]

[20] It would appear that this kind of merger is envisaged in the Law of the Sea Convention, 1982, Article 237(2) of which says: "Specific obligations assumed by States under special Conventions, with respect to the protection and preservation of the marine environment, should be carried out in a manner consistent with the general principles and objectives of this Convention."

3. Operational Discharges

SUMMARY

3–01 This chapter deals with operational discharges from ships—that is, with those discharges which are made deliberately—and with specific provisions relating to the enforcement of those standards. Chapter 4 is concerned with accidental discharges, and Chapter 5 with the general legal and practical issues involved in enforcement of international standards relating to both operational and accidental oil pollution.

3–02 Hence, this chapter begins with a reasonably detailed explanation of why ships have deliberately to discharge oil to sea, and then proceeds to an examination of the international legal standards which have been developed over the years to regulate such discharges. Now that the Protocol of 1978 Relating to the International Convention for the Prevention of Pollution from Ships 1973 (MARPOL 73/78) has entered into force in so far as it relates to oil pollution, it is given particular prominence.

3–03 Since the origin of the problem of operational oil pollution lies in the technicalities of ship operation, it is not surprising to find that industry has played a major role in the development of solutions to the problem. The role of states has been largely to formulate standards and to enforce them. In one respect, however, they have found themselves responsible for an indispensible practical part of the solution to the problem—the provision of adequate reception facilities for oil and oily residues in their ports—but many states have not fulfilled their role in this regard, and this has led to a continuing and unnecessary operational oil pollution problem despite the existence of environmentally acceptable international legal standards.

1. INTRODUCTION: THE PROBLEMS

A. The Crude Oil Tanker

3–04 After the tanker has discharged its cargo, a proportion of it remains in the tanks, caught mainly on the tank walls and horizontal surfaces of the joists which give the tank its structural strength. Such clingage varies largely according to the type of cargo carried, varying from 1.0 per cent. to 0.1 per cent. of the total cargo, the average for crudes being about 0.5 per cent. If the clingage were to remain, a number of consequences would follow. For instance, the heavy fractions would accumulate to impede drainage, to make inspections more difficult and to reduce the cargo capacity of the tank; and the residues may be incompatible with the next cargo to be loaded, thus making it unacceptable to the consignee. Further, a dirty tank can only carry dirty ballast, which gives severe problems of disposal at loading port. The clingage must therefore be regularly removed, and this has traditionally been done by washing the tanks with sea water. The problem this creates is that the water must itself then be returned to the sea and before the early 1960s this was done without separating it from the clingage. The oil was thus deliberately discharged to sea, and an average of 0.5 per cent. of the crude oil transported over the oceans therefore reached the sea.

3–05 Exactly the same problem arose with those tanks filled with ballast. Since discharge of dirty ballast water close to the shore has for some years been prohibited in most jurisdictions, the tanker had to take on ballast into a cleaned tank before arriving at the loading port. The dirty tank washings were therefore discharged out at sea before arrival.

3–06 The tanker shares with other oil-burning vessels the problem of disposal of oily bilges, and of sludge from purification of fuel oil for diesel engines. Both these are described in the next section.

B. Other Ships

3–07 Vessels which burn Bunker C or other heavy fuel oils in diesel engines face the problem of fuel oil purification. During the course of a voyage the purification of the fuel oil and settlement in bunker tanks produces a quantity of sludge which varies with the rate of fuel consumption and quality of fuel. Normally these sludges are kept in sludge tanks, but eventually the contents of these tanks must either be discharged to a shore reception facility, or to sea.

3–08 Another problem is that pumps, tanks and machinery almost inevitably leak small quantities of various types of oil (mainly lubricating oil). These can amount to an appreciable quantity during even a single voyage, and are normally fed via special drainways to separate bilges, or are collected in the normal

23

bilges. There are three possible methods of dealing with these bilges. The oldest and least desirable is to discharge them to sea. A second method is to discharge them to shore reception facilities where they are available. The third is to install an oil/water separator of appropriate capacity; the separated oil may be either stored in a special tank or fed into the fuel oil tanks. The oil content of the water discharged to sea through the separator can be kept within legal limits if the separator is operated properly.[1] However, if the oil which has been separated out is stored rather than burned with the bunkers, it will eventually need to be discharged to a reception facility.

3–09   An additional problem faced by many dry cargo ships is the disposal of oily ballast. This arises on the rare occasions where the vessel has to use any spare bunker tanks for ballast water to ensure stability. The oily ballast is normally discharged only when the vessel is nearing port, but, as with bilges, this may be done via a separator; alternatively shore reception facilities may be used if they are available.

2. PRACTICAL SOLUTIONS

3–10   With the advent and increasing incidence of crude oil washing and segregated ballast for many tankers, the emphasis for tankers has shifted from the earlier solution of Load on Top. However, since the regulations on discharge are best understood in the context of Load on Top and since it is still relevant for many tankers, this section begins with a description of it, and of the limits to its efficacy as a pollution saving procedure.

A. Load on Top

3–11   Following the 1962 Conference to revise the International Convention for the Prevention of Pollution of the Sea by Oil 1954, the oil industry was asked to see if it could not come up with a technical solution to the oil discharge problem, which the 1962 Conference had failed to solve. The industry's solution was to separate the oil and water on board, discharge the water to sea, retaining the oil residues aboard. The next cargo could then, in most situations, be

---

[1] See IMO Resolution A.393(X), adopting the Recommendation on International Performance Specifications for Oily-Water Separating Equipment and Oil Content Meters.

loaded on top of the residues—hence the system was christened Load on Top.

3–12 The Load on Top procedure[2] basically consists of the following steps (see Figure 3.1):
1. After discharge of cargo from the cargo tanks, sea water is taken on as ballast into some of the tanks (Nos. 1, 4, 7, 9 and 11). Other tanks are washed with sea water (Nos. 2, 3, 5, 6, 8 and 10).
2. The cargo tank washings are transferred through "stripping" pipes to a slop tank (No. 12), (which may be just another cargo tank). These wahsings are allowed to settle in the slop tank.
3. (a) Clean ballast water is pumped into the cleaned tanks (Nos. 2, 3, 5, 6, 8 and 10).
   (b) The dirty ballast water taken on at the start has by now partially settled, so the lower layer of water is discharged to sea (from Nos. 1, 4, 7, 9 and 11), and
4. The residual oil and water is stripped to the slop tank (No. 12) to settle.
5. After further settling in the slop tank (and treatment with emulsion breakers, if necessary) the water layer is discharged to sea (from No. 12).
6. A new cargo is loaded on top of the slop tank oil residues (not shown in Figure 3.1).

3–13 By operating Load on Top, a tanker can save nearly all the oil which it would otherwise have discharged to sea, and what it does put over the side can be discharged in a more environmentally acceptable way. The discharge will be close to the side of the ship and so the oil and water mixture will usually be further dispersed by passing through the wake. Some of the oil will remain beneath the surface, but some will rise to form a thin film (from 0.002 mm to 0.005 mm thick) which will normally be degraded within two to three hours. The system is, therefore, an excellent anti-pollution device, and has undoubtedly saved millions of tons of oil pollution since its introduction.

3–14 Since its introduction in 1963 and 1964, adherence to Load on Top quickly reached about 80 per cent. of all tankers.[3] Load on Top was not accepted by 100 per cent. of the tanker industry

---

[2] An operational guide to Load on Top is published by OCIMF/ICS: *Clean Seas Guide for Oil Tankers: The Operation of Load on Top* (1973).

[3] J. H. Kirby, "The Clean Seas Code: a Practical Cure of Operational Pollution," *Proceedings of the International Conference on Oil Pollution of the Sea*, 7–9 October, 1968, Rome, p. 201.

FIGURE 3.1
LOAD ON TOP

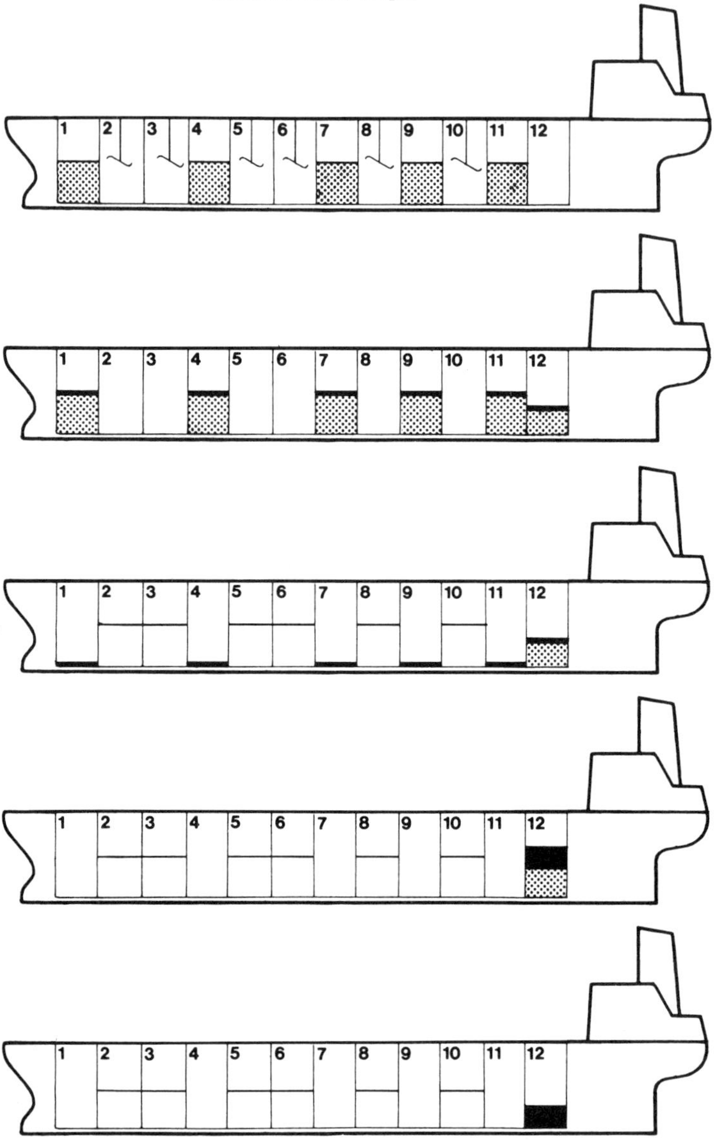

because not all tankers can in practice operate the system—or not all tanker owners for some reason want to. The problems associated with Load on Top which cause this situation are discussed below. Nowadays, Load on Top is being increasingly superseded by crude oil washing, especially those subject to MARPOL 73/78.[4] These requirements grew partly out of the perception that Load on Top was not an environmentally adequate procedure, and partly from the acceptance of the fact that not all tankers could operate it anyway. It is therefore worthwhile to understand why many tankers did not operate Load on Top, particularly as some of these problems are still relevant despite the advent of crude oil washing.

B. Problems associated with Load on Top

3–15 One problem is that the oil retained on board in the slop tank has inevitably got a higher salt content than some refineries can tolerate, and so a tanker whose next cargo is destined for such a refinery will be unable to use the system. Another problem is that the system takes time to operate[5] because of the settling necessary to separate oil and water, and so vessels on short-haul voyages have a problem in using it. A similar problem is encountered by vessels on coastal voyages, whether long or short-haul: because even Load on Top means discharging some oil, there would be a contravention of the total ban on all oil discharges imposed within 50 miles of the coast by most jurisdictions and by the 1954 and 1973 Conventions. Again, if a tanker is to carry as its next cargo a white oil, or some other oil incompatible with the persistent oil just carried, then it too will be unable to operate Load on Top.

3–16 A special problem for the tanker arises before it is due to enter the repair yard: its tanks must be specially well cleaned to remove oily residue and sludge so it can obtain a certificate that the tanks are free of combustible hydrocarbon gases (called a gas-free certificate) which is essential for hot work. Some yards permit tankers to inert the atmosphere in the slop tank,[6] provided the other tanks are gas-free, and so allow the vessel to enter with resi-

---

[4] See below, paras. 3–28 *et seq.* for crude oil washing and paras. 3–65 *et seq.* for MARPOL 73/78 requirements thereon.

[5] The time varies with the weather encountered on the ballast voyage, the size and construction of the tanker, the efficiency of the crew, the type of equipment used on board, and with other factors. The minimum time for a tanker of, say, 65,000–95,000 dwt would be about 50 hours. A more usual time would be 72 hours.

[6] A process of removing all the oxygen in the tank above the oil surface and replacing it with a special, inert gas.

dues on board, but this will not always be the case, and in these circumstances the tanker will be unable to retain any oil on board.

3–17 Combination carriers, such as OBOs and Bulk/Ore carriers,[7] have a special problem connected with their operational peculiarities. When it is desired to change from carrying an oil cargo to carrying a dry cargo (say coal or ore), the tanks must be cleaned of all residues and gas-freed. There is no question of the dry cargo being loaded on top of the slops in the slop tanks! On the ballast voyage before taking on dry cargo, therefore, these vessels have an oil disposal problem.

3–18 There is also a commercial problem. If a charterer wishes to voyage charger a vessel to take a cargo from, say, the Gulf to Rotterdam, he will not necessarily discover whether or not the vessel chartered has slops aboard incompatible with the cargo he wishes to load until inspection is made at the loading terminal. Of course, if there are adequate reception facilities at the loading terminal, there is no operational problem; however, as we shall see,[8] many countries have yet to provide reception facilities in their loading ports adequate for the traffic using them. The result is that an owner chartering out his vessel on the open market may well be reluctant to operate Load on Top on the ballast voyage (during part of which the vessel may be on offer in the market): he will not know where the vessel will be going to for loading until a charter is agreed, and so he cannot be sure that any slops aboard can definitely be discharged ashore. The owner will therefore be under pressure to put the residue over the side.

3–19 Further pressures of a financial nature operate on the owner of such a vessel in the spot market. Where residues have been retained on board, the charterer may well be unprepared to pay freight in respect of the slops or dead freight in respect of the space above them (assuming cargo is not loaded on top of them). If the residues can be discharged to reception facilities at the loading terminal, the charterer may well be unprepared to allow the time taken for such discharge as laytime. There are also similar difficulties concerning the distribution of financial burdens in time charters, both of oil tankers and of combination carriers.[9]

---

[7] The distinction between the two types of vessel for oil pollution purposes is immaterial— both can carry both dry and wet bulk cargoes.

[8] Paras. 3–32 *et seq.*

[9] For these reasons it is important that owners and charterers agree at the time of fixing a charter how they will make provision for the existence of slops on board. See further para. 3–27.

C. Solutions to the Problems Associated with Load on Top

HIGH SALT CONTENT

3–20   There is nothing that the tanker owner can do about this problem except to operate Load on Top as efficiently as he can, thus reducing the salt in the residue which collects in the slop tank. The efficiency which can be achieved will depend on the weather on the ballast voyage (calm weather being best), the care which the officers and crew involved devote to the operation, and, to a certain extent, the availability of equipment on board which aids the operation—for instance, slop tank heating coils.

3–21   Really the optimum solution is to be found in the refining branch of the industry. Any refinery can accept Load on Top cargoes if it has adequate desalting equipment: of course, the installation of such equipment is costly. However, it is fortunate that since the early 1960s more and more refineries have been equipped with desalters, and it is now only a small minority which do not have them.

THE SHORT-HAUL TANKER

3–22   The International Chamber of Shipping and the Oil Companies International Marine Forum have recommended four possible solutions to the problems of the existing short-haul tanker,[10] which are worth attention.

1. Eliminating tank washing and discharging dirty ballast to the loading terminal. The trouble here will often be that the tanks will have to be washed to enable acceptance of the next cargo, and even if such washing is not necessary, the loading terminal may not have adequate dirty ballast reception facilities. This is the first case where the importance of adequate reception facilities at all the world's loading terminals becomes apparent.

2. Washing ballast tanks at the discharge terminal and sailing with clean ballast. This again relies on there being available reception facilities for the washings either at the discharge terminal or at the next loading terminal. In addition, washing ballast tanks at the discharge terminal may in some cases involve the vessel spending more time in port than she can afford to—or than the terminal authorities want her to.

3. Dedicating some cargo tanks for use as permanent ballast tanks, a temporary trading measure. For the tanker on regular short voyages, this normally will be expensive, in that it reduces the

---

[10] *Clean Sea Guide for Oil Tankers: The Operation of Load on Top* (Joint ICS/OCIMF publication, 1973) p. 10.

cargo capacity of the vessel. Particularly in times of recession, ship-owners are sensitive to the great need to make every voyage a commercial success, and so will be reluctant to take such drastic steps. It does, however, represent a real alternative for a tanker which only occasionally, or rarely, makes short voyages.

4. Re-routing the ship or slowing down to allow enough time for proper Load on Top procedures. This is acceptable for the occasional short voyage, but, again in the case of the regular short-haul voyager, the position is different: in times of prosperity such measures would normally prove costly, in terms of lost profit, but in times of recession this may be for some owners an acceptable procedure. The kind of owner who could do this would be the large fleet owner who does not want to lay up the vessels (as a result of decreased demand) but prefers to keep them in service. By slow steaming, the capacity of the fleet to transport oil is reduced down to current demand levels, while allowing for a quick response to increased demand when world trade starts to pick up. In this way losses can be mitigated.

The Coastal Voyage

3–23 The problems of the tanker making coastal voyages, or voyages within the "prohibited zones" described below (*e.g.* within the North Sea) are identical to those of the short-haul tanker, with the exception that it is not possible to slow down so that Load on Top can be operated because Load on Top will involve a discharge of *some* oil, and this is not permitted within the prohibited zone. It is therefore even more important to the alleviation of this problem that adequate reception facilities be provided.

Cargo Incompatibility

3–24 The problem here is not one of time to effect the Load on Top operation, but is that the new cargo cannot be loaded on top of the settled out slops because it is of a different type: for instance, where the tanker has been carrying a waxy crude and next cargo she is to carry a special fuel oil or low cold test crude. The only possible way to retain the oil on board would be to reserve the slop tank space exclusively for the crude slops; this, however, reduces the freight. Earning capacity of the vessel is therefore expensive, particularly for a ship chartered-out on the voyage charter market. The most effective solution is, therefore, again to discharge all the contents of the slop tank to reception facilities, preferably at the loading terminal.

Pre-Maintenance Washing

3–25 Although the burden of this problem can be mitigated by cleaning tanks more intensively as the repair period approaches and by carrying on the last laden voyage a crude which helps to dissolve sludge, such measures do not remove the problem altogether. The slops from washing all tanks have to be discharged to reception facilities somewhere (at the last discharge terminal, at a port between there and the repair yard or at the yard itself), or permission must be sought to enter the repair yard with slops in an inerted tank. Happily, the provision of reception facilities at yards has improved greatly in recent years. The practice of yards in allowing a ship to enter with slops in an inerted tank varies from yard to yard and according to the repairs to be done.

Combination Carriers

3–26 These ships have to clean their entire cargo system when changing from oil to dry bulk cargo. There are only two possibilities for such ships to avoid discharging residues to sea: one is to retain them in the slop tank throughout the dry cargo voyage, and thus be unable to earn freight in respect of the slop tanks, and the other is to discharge them to reception facilities.

Commercial Implications

3–27 The implications of operating Load on Top are that slops remain on board in the slop tank. Owners must therefore ensure in their voyage charters that the charterer agrees that his cargo may be co-mingled with the slops and that the amount of space available for the cargo to be loaded reflects the fact that the slop tank already contains slops. If the charterer is not agreeable, owners are reduced to trying to persuade the charterer to pay freight on the slops and dead freight on the space above the slops in the slop tank. In any event, failure to make clear and adequate provision in the charter is likely to lead to disputes. Commercially, charterers do in fact have some incentive to accommodate owners in this respect, in that out-turn is normally nearer to loaded quantity when the cargo is loaded on top.[11]

---

[11] *e.g.* if a 100,000 dwt ton ship loads 100,000 tons of oil into an absolutely clean ship, with no slops in the slop tank, the out-turn may be, say, 99,500 tons due to clingage. But if the same ship had 500 tons of slops already aboard at time of loading, she will load 99,500 tons only—and still discharge 99,500 at destination.

31

D. Crude Oil Washing[12]

3–28 There is no reason why water has to be used to clean tanks if another substance can be found to do the job efficiently. The only substance which does not present problems of separation after the tank-washing is oil itself. It was found, however, that portable washing machines would pose a safety hazard if used to effect crude oil washing, and so it was not until fixed washing machines became available that crude oil washing became widely used. These machines are supplied with crude from the cargo by permanent piping, and the washing cycle is controlled by individual drive units. Crude oil washing was introduced into some VLCCs in the early 1970s and is now a routine procedure for most tankers; and it is destined to become even more widely used now that MARPOL 73/78 has entered into force.[13]

3–29 Put simply, the process involves part of the cargo being diverted through the fixed tank cleaning system into the tanks as they are being emptied, so that the exposed surfaces where cargo and sludge collects are washed by powerful jets of crude oil. The residues on these surfaces are thus washed down into the cargo as it is being discharged. However, the system is not a cure-all; the process still demands a final water rinse.

3–30 Because less water is used, there is less water and oil separation to be done, and so there is less risk of pollution involved in the discharge of separated water from the ship. Also, the system means that there is less oil remaining on board, for much of the oil which would have clung to the tank and which would have ended up in the slop tank in a Load on Top operation is now discharged with the cargo. The system therefore has economic advantages: cargo out-turn is increased, the cost of routine tank cleaning reduced, corrosion from water washing is reduced, less manual desludging is needed and less salt is discharged to refineries. However, cargo discharge time is increased, and there is an increased workload in port.

3–31 The system not only requires fixed washing machines, but because crude oil washing creates a considerable volume of hydrocarbon vapour in the tank, the ship must also be fitted with an inert gas system in order to reduce the risk of an electrostatic discharge causing an explosion. The oxygen content in the system must be

---

[12] An operational guide to crude oil washing is published by OCIMF/ICS: *Guidelines for Tankwashing with Crude Oil* (1976).
[13] See paras. 3–65 *et seq.*, below.

reduced below 8 per cent. Conversion of existing ships to both fixed washing machines and inert gas systems is expensive, and for this reason the inclusion of crude oil washing in MARPOL 73/78 was for existing ships required progressively rather than immediately.[14]

E. Reception Facilities

3–32 It has been seen above how important the existence of reception facilities is to the solution of certain of the problems associated with the carriage of oil by sea and with the operation of Load on Top. They are also important to all ships—not just tankers and combination carriers—for the handling of oily bilges, sludges from fuel oil purification and dirty ballast. The ports where they are most in need are, of course, where oil is loaded (particularly by short-haul tankers) and where repair yards are sited. Although the need for them was appreciated by the international community as long ago as 1954,[15] progress in providing them has been painfully slow, despite the fact that after the oil price shock of 1973 the value of the recovered oil made it possible for their provision to be an economically viable proposition in some cases. For many years, then, and to some extent even today in a number of locations, reception facilities were non-existent in certain major exporting and transhipment terminals (the Mediterranean Sea area being an example).[16] In particular, the problem is severe at ports where dry cargoes are loaded by combination carriers after they have carried an oil cargo, and generally in relation to facilities for fuel oil purification sludges, dirty ballast and oily bilges from ships of all types. Offshore loading terminals do not have any reception facilities, so that the only solution here is for short-haul tankers to have segregated ballast tanks.[17]

3–33 In view of the way in which the legal requirements of MARPOL 73/78 are framed (discussed below), reception facilities will continue to be needed in particular for:

1. Existing tankers below 40,000 tons dwt in short haul trades;
2. New tankers below 20,000 tons dwt in short haul trades;

---

[14] See paras. 3–71 *et seq.*, below.

[15] The 1954 Convention contains an Article relating to the provision of reception facilities: see para. 3–90 below.

[16] The situation in the Mediterranean is so bad that France had to make a Reservation when ratifying MARPOL 73/78 to the effect that the prohibition on discharges can only be applied to tankers engaged in voyages within the Mediterranean if such tankers are proceeding to a port equipped with the reception facilities required by the Convention.

[17] See paras. 3–36 *et seq.*, below.

3. Tankers which carry fuel oil as cargo (which cannot crude oil wash);
4. Combination carriers switching trades from oil to dry bulk;
5. All ships which have oily bilges;
6. All ships which have sludges from fuel oil purification and settling.

3–34 IMO's Marine Environment Protection Committee has had the matter in hand since 1974, and has produced guidelines on the provision of adequate reception facilities,[18] which are extremely helpful in enabling a state to decide what facilities to provide at its various ports and terminals. The provision of these facilities does not always involve net expenditure: in some cases, carefully planned reception facilities make a substantial profit, but of course this will not always be possible where throughput is relatively low.

3–35 When eventually all large tankers have segregated ballast tanks,[19] the capacity of reception facilities required at terminals used by them will be dramatically reduced. But this does not excuse the failure by many governments to take the necessary action, both in the past and at present, to ensure their provision, despite recommendations by IMO.[20] This failure is in sharp contrast to the efforts made by the oil and tanker industries over the last twenty years to overcome oil pollution. It results, of course, from the fact that reception facilities require tankage and piping which is expensive, and many governments do not like to have to spend the money themselves, or force private interests to do so (although those same governments appear to be prepared to see international Conventions place similar burdens on shipowners).

F. Segregated Ballast Tankers

3–36 If oil and water never mix, there is no problem of having to separate them. This is the principle behind the segregated ballast tanker, which is specially constructed so that in certain tanks only ballast is carried, and in the remainder, only cargo is normally carried—there being an exception in favour of very rough weather, when extra ballast must be taken on to maintain safe navigation. This type of tanker has separate piping arrangements for the cargo and the ballast, so that the ballast tanks are truly segregated.

---

[18] *Guidelines on the Provision of Adequate Reception Facilities in Ports*, Parts I (Oily Wastes) and II (Residues and Mixtures containing Noxious Liquid Substances), (IMO Publication No. 77.02 and 80.03).

[19] See paras. 3–65 *et seq.*, below.

[20] *e.g.* Resolutions A.235 (VII) of 12.10.71 and A.348(IX) of 12.11.75.

34

3–37   The provision of segregated ballast tanks was first made a legal requirement in the International Convention on the Prevention of Pollution from Ships 1973: that provision was limited to new tankers of 70,000 dwt tons and above.[21] Clearly the provision of such tanks has a role to play in the fight against marine pollution. The extension of the mandatory fitting of segregated ballast tanks, with its protective layout, which was achieved in the 1978 Protocol to that Convention forms probably the most important part of it.[22] It is worthwhile briefly summarising here the arguments for and against widespread fitting of segregated ballast tanks, because otherwise the compromise struck at the Conference on Tanker Safety and Pollution Prevention in 1978 (the TSPP Conference) cannot be understood.

3–38   Those who favoured some extension of the 1973 provisions relating to segregated ballast tanks saw their admitted environmental benefits (particularly as they offer a permanent solution requiring little continuing inspection to police) and regarded them as technically and economically feasible. They regarded an extension of the mandatory provisions as being a superior method of achieving a significant reduction of operational pollution to alternatives such as the mandatory operation of crude oil washing. Some states were undoubtedly in favour of such an extension, not so much because of the environmental benefits, but because if the world fleet of existing tankers had to be fitted for segregated ballast, a given quantity of oil would take more ships to carry it than without such an extension. The effect of this would be to stimulate demand in a severely depressed freight market.

3–39   Those against extension to existing tankers (for instance the United Kingdom) took the view that the environmental benefits were too small to justify the enormous costs involved (mainly due to the loss of dead weight and cost of fitting the required piping and pumps).[23] These are suffered not only in having to pay for the retro-fitting, but in time lost during the carrying out of the work. Thereafter, there is the permanent cost associated with the increase in bunkers and other transport costs needed to transport a given quantity of oil. There is also difficulty in seeing that the implementation of the mandatory retro-fitting is fair.

---

[21] See paras. 3–65 *et seq.*, below.

[22] See para. 3–71, below. Protective layout is dealt with in paras. 4–51 and 4–52.

[23] OCIMF estimated $1.5 to $2.5 million for a VLCC, and Italy had an estimate of similar scale (see IMO documents MEPC VI/6 and 6/3). Taking the lowest figure, the world conversion cost for ships over 70,000 dwt tons would alone have been of the order of $1.6 billion.

3–40   One practical alternative to the mandatory fitting of segregated ballast tanks (to either new or existing tankers) is the operation of crude oil washing, which, while still expensive, does not have such serious economic consequences (and indeed, saves oil and increases out-turn). Another alternative for existing tankers is to carry ballast in some of their unaltered cargo tanks which are dedicated exclusively to the carriage of ballast (called "clean ballast tanks"). Clean ballast tanks involve all the economic detriment of segregated ballast tanks apart from the high cost of fitting.

3–41   As we shall see,[24] the 1978 Conference adopted a compromise whereby certain tankers may dedicate ballast tanks or operate crude oil washing as an alternative to fitting segregated ballast tanks.

3. INTERNATIONAL LEGAL STANDARDS

3–42   It was seen in paragraphs 2–14 and 2–15 that international law prescribes that the general obligations of states to protect the marine environment from oil discharges by ships are fulfilled by the negotiation of detailed standards which, in time, will apply to all ships sailing the seas. That this process has not yet resulted in adequate results contributed to calls for wider zones of pollution control now enshrined in the Law of the Sea Convention 1982, discussed in Chapter 5. However, there have since 1954 been the detailed standards referred to, which have been steadily tightened up until now, with MARPOL 73/78, we have standards that are so high and so ambitious that in a number of instances they cannot actually be met in practice! These are examined in some detail in this section.

A. Introduction

3–43   The International Convention for the Prevention of Pollution by Oil 1954 (the 1954 Convention) resulted from a Conference called by the United Kingdom. It has been amended in 1962, 1969 and 1971, each time under the auspices of IMO. As we saw in paragraph 2–23, it attracted widespread support, but by 1973 it was generally felt to require substantial improvement and extension. IMO convened a Conference in that year which adopted what was then a quite revolutionary instrument—the International Conven-

---

[24] Paras. 3–67, 3–71 and 3–76 *et seq.*, below.

tion for the Prevention of Pollution from Ships, known as MARPOL 73—but this was re-enacted in an amended form by the Protocol of 1978 thereto. The Convention, as amended by the Protocol of 1978, is known as MARPOL 73/78, and is the most ambitious treaty of them all. It has the most wide-ranging and strict provisions relating to prevention of pollution from ships ever adopted, and the provisions relating to oil pollution (contained in Annex I) entered into force on October 2, 1983.

3–44 But the development of the law in this field is a relentless and continuing process, so that MARPOL 73/78 is not the last word! As states and industry began to prepare for the implementation both of MARPOL 73 and, later, MARPOL 73/78, it was found that the adopted texts required clarification if they were to be successfully implemented. In this connection it was obviously desirable that the requirements be interpreted and applied in a uniform and consistent manner. With this in view, in 1982 IMO's Marine Environment Protection Committee developed a Unified Interpretation of certain provisions of MARPOL 73/78, and it also developed some proposed amendments thereto, which relate to oil pollution.[25] The Unified Interpretation is not a binding treaty instrument, although it is hoped that states will abide by it. The proposed amendments were adopted in September 1984 by IMO's Marine Environment Protection Committee but are not yet in force. These 1984 amendments are aimed at resolving difficulties and providing practical solutions to problems in the implementation of MARPOL 73/78, created in part by the fact that some of the hardware required is not yet in a stage of development capable of meeting the requirements on a continuing basis. They do not include any matters which would introduce new substantive requirements over those of MARPOL 73/78. They are, therefore, referred to below only in footnotes, where necessary.

B. Discharge Standards for Tankers

THE 1954–62 STANDARDS

3–45 Load on Top was not devised until after 1962, and so the delegates to both the 1954 Conference and to IMO in 1962 proceeded on the basis that oil discharges were inevitable. Reception facilities were much scarcer than they are now, and so

---

[25] IMO Circular Letter No. 968 of January 16, 1984. The MARPOL 73/78, the Unified Interpretation and the 1984 proposed amendments are now published in one document by IMO, Publication No. 525. 82. 19.

the approach was to limit tanker discharges to areas outside prohibited zones. The Convention applied only to discharges of persistent oils and the prohibited zones were:
1. All areas within 50 miles from the nearest land;
2. Any area within 100 miles from the nearest land along the coast of a state which declared such a zone;
3. Certain special areas defined in an Annex to the Convention.

3–46 These standards were virtually useless, for they allowed tankers to discharge oil over the vast majority of the world's seas. Coasts were given marginal protection, but oil residues from tankers not operating Load on Top can travel over 100 miles and remain in the sea for long periods. Further, there were "holes" in the special areas of the Black and Mediterranean Seas into which lawful discharges might be made.

3–47 There was, however, one potentially valuable provision. Article III(c) as introduced in the 1962 amendments prohibited the discharge of oil or oily mixture from new ships of 20,000 gross tons or more, *i.e.* those for which the building contract was placed on or after the entry into force of the 1962 amendments, which occurred in May 1967. However, since there was a worldwide dearth of reception facilities (which were not prescribed by the Convention[26]) the only way in which such a ship could have complied with this requirement was to be fitted with sufficient segregated ballast tankage, but this was not specifically required by the Convention. There was a let-out for ships not so fitted, since an exception was provided for the situation where the master was of the opinion that special circumstances made it neither reasonable nor practical to retain the oil on board. No doubt the master of such a ship would have had this opinion every time he was headed for a port with no reception facilities!

3–48 The problem with these standards was that they were impractical and ineffective because states were not prepared to ensure the provision of adequate reception facilities.[26] It was following the adoption of the 1962 amendments that the oil industry developed Load on Top as a practical solution, but it was not until 1969 that international law caught up with it.

The 1969 Standards

3–49 The 1969 amendments entered into force on January 20, 1978, and introduced a standard which, for the first time, was

---

[26] See paras. 3–90 *et seq.*, below.

based upon Load on Top. Subject to certain exceptions,[27] the amended Article III(*b*) prohibited discharges from tankers[28] of oil or oily mixture except when the following conditions were all satisfied:

  (i) the tanker is proceeding *en route*;
  (ii) the instantaneous rate of discharge of oil content does not exceed 60 litres per mile;
  (iii) the total quantity of oil discharged on a ballast voyage does not exceed 1/15,000 of the total cargo-carrying capacity;
  (iv) the tanker is more than 50 miles from the nearest land.

3–50 These conditions may look rather baffling, but they are not when they are considered in connection with the operation of Load on Top. They are similar to the 1973 standards, and so explanation of them is deferred to paragraphs 3–54 *et seq.* The effect of these standards, when compared to those of 1954–62, is staggering: for a tanker to comply, it had either to operate Load on Top with reasonable efficiency, or it had to retain all oil residues on board for discharge to reception facilities. This placed additional importance on the availability of reception facilities for tankers which, for a legitimate reason, could not operate Load on Top.[29] The standard replaced the old idea of simply keeping discharges off the coastline, to keeping them off the coastline *and* keeping the oil content of the discharge to an acceptable level of dispersion. As was pointed out in paragraph 2–23, above, the basic tenets of this standard can now be considered as representing the minimum level of protection of the environment demanded of states by international law. In 1973, these principles were further refined.

THE 1971 GREAT BARRIER REEF AMENDMENTS[30]

3–51 These amendments, adopted as IMO Resolution A.232(VII), redefined the phrase "nearest land" in Article III so that the Great Barrier Reef was included. Had these amendments ever entered into force, the Reef would have thus regained the status of prohibited zone which it had before the 1969 amendments. It had to wait until the 1973 Convention entered into force before this status was in fact restored.

---

[27] Concerning discharges necessary for the safety of life or the ship or resulting from damage to a ship or its equipment (Article IV).

[28] Article I defined *discharge* as any discharge or escape howsoever caused, and *tanker* as a ship in which the greater part of the cargo space is constructed or adapted for the carriage of liquid cargoes in bulk and which is not, for the time being, carrying a cargo other than oil in that part of its cargo space.

[29] See further paras. 3–20, above.

[30] The 1971 amendments on tank size are dealt with in paras. 4–53 and 4–54, below.

The MARPOL 73/78 Standards[31]

3–52　The 1973 Conference took the opportunity substantially to improve the standards relating to operational oil pollution, and the 1978 Conference made further revolutionary provisions concerning mandatory segregated ballast tanks and crude oil washing which are discussed below.[32] The old concept of specially protected areas is resurrected, and these are called *special areas*; inside them, as we shall see, there is an almost total ban on all discharges. Outside them, discharges are permitted but the standards are improved over the 1969 amendment levels. The new provisions are contained in Annex I to the Convention, and references below are to the Regulations contained therein unless otherwise stated. These standards entered into force on October 2, 1983.

Outside special areas

3–53　The basic standard, which is subject to certain exceptions,[33] is contained in Regulation 9(1)(*a*) and it prohibits any discharge into the sea of oil or oily mixtures except when all the following conditions are satisfied:

(i) the tanker is not within a special area;

(ii) the tanker is more than 50 nautical miles from the nearest land[34];

(iii) the tanker is proceeding *en route*;

(iv) the instantaneous rate of discharge of oil content does not exceed 60 litres per nautical mile;

(v) the total quantity of oil discharged into the sea does not exceed for existing tankers 1/15,000 of the total quantity of the particular cargo of which the residue formed a part—for new tankers the figure is 1/30,000;

(vi) unless exempted,[35] the tanker has in operation an oil discharge monitoring and control system and a slop tank arrangement as required by Regulation 15.

---

[31] Those seeking to implement or apply MARPOL 73/78 should study not only the Convention itself but also the Unified Interpretation thereof published by IMO, and the 1984 amendments to the Convention: see para. 3–44, above. For an academic study of the international law aspects of the Convention, see generally G. J. Timagenis, *International Control of Marine Pollution*, (Dobbs Ferry, 1980), pp. 319 *et seq.*

[32] See paras. 3–65 *et seq.*

[33] These concern certain very small tankers (Regulation 9(2)), discharges necessary for the safety of life or the ship or resulting from damage to the ship (Regulation 11), and certain discharges from machinery space bilges (Regulation 9(1)(*b*)).

[34] *Nearest land* is defined by Regulation 1(9) and incorporates the 1971 amendments definition so that the Great Barrier Reef is protected by measuring the 50 miles from its outer edge.

[35] See para. 3–60.

3–54 It can be readily seen that this list of conditions is very like the 1969 amendments in form; like the 1969 amendments, this standard too is tailored to Load on Top. When a tanker is operating Load on Top it will normally be empty of all cargo and will be proceeding *en route* for a loading port, so Condition (iii) is satisfied. At at least two points in the Load on Top procedure separated sea water is discharged, and this will contain traces of oil.[36] Condition (iv) is designed to keep these discharges to an acceptable minimum concentration. In fact, the concentration will vary according to where the effluent originates—thus the bulk of the settled dirty ballast typically contains about 30 p.p.m. of oil, the bulk of the settled slop tank water some 150 p.p.m. of oil; before MARPOL 73/78 the final discharge from the slop tank may have as much as 1,000 p.p.m. but as we shall see below the system referred to in Condition (vi) should now cut off the discharge before such a concentration can be discharged.[37] Condition (iv) can be met by adjusting the rate of discharge to suit the speed of the ship and the type of effluent being discharged.[38]

3–55 Condition (v) also restricts the manner in which discharges can be made by restricting the total amount of oil which can be discharged. Like Condition (iv), it effectively prohibits the discharge of pure oil. Clingage varies from cargo to cargo and type of ship within a range of about 0.1 per cent. to 1.0 per cent. of cargo volume. If all this was discharged, the limit would, of course, be greatly exceeded. The limit is set so that if Load on Top is practised reasonably efficiently, the discharge will be lawful.[39] The doubling, for new tankers, of this part of the standard over that required by the 1969 amendments[40] can be achieved largely by improvements

---

[36] See paras. 3–12 *et seq.*, above.

[37] Figures from *Clean Seas Guide for Oil Tankers: The Operation of Load on Top* (OCIMF/ICS, 1973).

[38] Regulation 1(11) defines the instantaneous rate of discharge of oil content as "the rate of discharge of oil in litres per hour at any instant divided by the speed of the ship in knots at the same instant." The oil discharge rate in litres per mile equals:

$$\frac{\text{p.p.m. of oil in effluent} \times \text{effluent discharge rate in cu.m/hour}}{\text{ship's speed in knots} \times 1,000}$$

Hence, if the tanker is travelling at 15 knots and the bulk of the separated dirty ballast is to be discharged with an expected oil content of 30 p.p.m., a discharge rate of up to 30,000 cubic metres/hour can be contemplated before condition (iv) is broken. The final discharge from the slop tank, at an expected 1,000 p.p.m., can only be discharged at 900 cubic metres/hour.

[39] If the clingage is 0.1% of the total cargo, then 3.33% of it can be discharged before the limit is exceeded—thus at least 97% of clingage must be saved. Where the oil is heavier, an even greater percentage of the clingage must be saved until with a clingage of 1.0% of total cargo, 99.7% must be retained on board.

[40] New tankers are a species of new ship, which is defined in Regulation 1(6). An existing tanker is one which is not a new tanker.

in shipboard equipment and design.[41] But it is more difficult for smaller tankers to achieve than for the large VLCCs.

3–56 The cargo officer used to face problems with the slop tank water discharge simply because there is no such thing in a slop tank as a definite oil/water interface. Therefore, he had to make sure that when discharging from the slop tank, he turned off the discharge before the concentration of oil was likely to be too high. This created something of a problem with the 1969 amendments in that compliance with them relied quite heavily, in this respect, on the skill with which the slop tank water effluent was discharged. Under MARPOL 73/78, however, Condition (vi) is designed to eliminate this problem, for it requires the ship to have an oil content meter fitted which should have the effect of eliminating reliance on the human element for successful compliance. This is a quite revolutionary change from the position under the 1969 Amendments.

3–57 Regulation 15 requires tankers to be fitted with a prescribed slop tank arrangement and discharge monitoring and control system; existing tankers are given a grace period of three years from the entry into force of the Convention, *i.e.* until October 2, 1986, but for new tankers the requirements are immediate. The purpose is virtually to eliminate reliance on human skill and judgement for compliance with the discharge standard. By Regulation 15(3)(*a*) the discharge monitoring and control system"shall be fitted with a recording device to provide a continuous record of the discharge in litres per nautical mile and total quantity discharged, or the oil content and rate of discharge. This record shall be identifiable as to time and date and shall be kept for at least three years. The oil discharge monitor and control system shall come into operation when there is any discharge of effluent into the sea and shall be such as will ensure that any discharge of oily mixture is automatically stopped when the instantaneous rate of discharge of oil exceeds that permitted by Regulation 9(1)(*a*) . . . ." States must have regard to IMO Specifications for this system.[42]

3–58 In 1973 when this provision was adopted, it was known that

---

[41] For instance, by having more than one slop tank (a requirement which Regulation 15 makes compulsory for new oil tankers over 70,000 dwt tons), by fitting slop tank heating coils and oil/water separators, and by improving the design and siting of discharge pipe ducts in the slop tank. It can also be achieved by operating crude oil washing or recirculatory washing. Recirculatory washing is a variant of Load on Top technique whereby the separated water in the slop tank is used to wash further cargo tanks, rather than being discharged. This improves Load on Top performance, but can only be used by tankers with an inert gas system.

[42] IMO Resolutions A.393(X), A.445(XI) and now A.496(XII) of November 1981.

a great deal of work was needed before the oil content meter required could be developed to an acceptable standard of accuracy, but it was assumed that by the time the Convention entered into force a suitable device would be available. This assumption has, in fact, been proved wrong, and IMO's Assembly was forced to adopt Resolution A543(13) on November 17, 1983, recognising that measurements of oil content in oily water mixtures being discharged may not in all circumstances be accurate, and inviting Governments:

> " . . . to recognise that oil content, instantaneous rate of discharge of oil content and the total quantity of oil discharged, recorded by the existing oil discharge monitoring and control system . . . cannot by itself constitute sufficient evidence of . . . contravention [of Annex I of MARPOL 73/78], but should be taken into account together with other evidence collected, when contemplating any action against ships alleged to have contravened the discharge requirements of the said Annex."

This is still the position to date.

3–59 In addition, regulations concerning the size and arrangement of slop tanks are made by Regulation 15(2). For most ships, these are relatively easy to comply with.[43]

3–60 It was recognised that for certain ships it would be difficult or impossible to comply with Regulations 9(1)(a) and/or 15. Therefore, Regulation 15(5) provides for states to be able to waive the requirements of Regulation 15 for any oil tanker which engages exclusively in short-haul, coastal voyages (those under 72 hours long and within 50 miles of the nearest land); and Regulation 15(6) provides for the waiver of the requirements of Regulation 15(3)(a) for white oil ships if the equipment is, in fact, unavailable. Also, there is now a proposed amendment—Regulation 15(7)—to exempt asphalt carriers and other tankers carrying heavy petroleums which inhibit effective oil/water separation, for whom all residues must be discharged to reception facilities.

3–61 The other feature of this standard worth noting is the definition of *oil*: by Regulation 1(1) oil is "petroleum in any form . . .," and therefore for the first time the definition includes non-persistent oils. This has had immense practical implications for the industry, for there have been very serious difficulties in the development of retention-on-board techniques for this trade (indeed, Load on Top cannot be practised by tankers carrying clean oils), and in the

---

[43] Combination carriers and gas carriers classed for the carriage of naphtha have difficulties, though.

identification of the type and volume of associated reception facilities. Fortunately, the problem is restricted by the fact that with clean oils the clingage is lower and most refineries have ballast facilities for product tankers.

3–62 The standards described above concern the problems that a tanker has with its cargo. Regulations 16 and 17 concern all ships and deal with arrangements to cope with the problems presented by fuel oil sludges and bilges. These are dealt with below, in connection with discharge standards from ships other than tankers.[44] Regulation 18 makes certain requirements as to pumping, piping and discharge arrangements for oil tankers, the most important of which is that, with certain minor exceptions concerning segregated ballast, all discharges shall take place above the waterline—where they can be seen.

Inside special areas

3–63 A special area is defined by Regulation 1(10) as one where— "for recognised technical reasons in relation to its oceanographical and ecological condition and to the particular character of its traffic the adoption of special mandatory methods for the prevention of sea pollution by oil is required." These areas are listed in Regulation 10 and cover the Mediterranean, Baltic, Black and Red Seas and the Gulfs Area. There are no "holes" as there were with the 1954–62 provisions.

3–64 The standard is the highest yet adopted: with certain relatively minor exceptions no ship may discharge any oil whatsoever while in a special area.[45] This makes the provision of adequate reception facilities in special areas absolutely vital,[46] but as was seen in paragraphs 3–32 *et seq.*, the provision of such facilities is in many locations inadequate or non-existent.

Segregated ballast, dedicated ballast and crude oil washing

3–65 The 1973 Convention provided a legal requirement for segregated ballast tanks for the first time: Regulation 13 provided that every new oil tanker of 70,000 dwt tons and above must be provided with segregated ballast tanks of specified capacity, and that in no case should ballast water be carried in the oil tanks of such a

---

[44] See paras. 3–78 *et seq.* below.

[45] The exceptions are highly detailed, particularly in the form they take in the proposed amendments, and there is no space for an account of them here. The main standard is, however, subject also to the Regulation 11 exceptions concerning damage and safety of life and ship.

[46] See paras. 3–93 *et seq.*, below.

vessel except in weather conditions so severe that, in the opinion of the master, it would be necessary to carry additional ballast water therein for the safety of the ship.

3–66   Such provisions were an acceptable addition to the international legal regime because the extra building cost for such large tankers was reasonable, owners were not required to retro-fit existing tankers and the benefits were worthwhile, for there is no doubt that segregated ballast tankers do not need to discharge oil like their non-segregated sisters.

3–67   However, the 1978 Protocol to the Convention replaces these provisions with a completely new and greatly extended set of regulations. References below are therefore to these provisions of MARPOL 73/78 unless otherwise stated. As explained in paras. 3–38 *et seq.*, above, the new provisions adopted at the 1978 Conference form a "package" which represents a delicate compromise between the two main schools of thought represented at the Conference. Nevertheless, three clearly discernible principles are followed in relation to the overall requirements of the Convention:
1. the provisions should be more onerous with increasing size of vessel;
2. the provisions should be more onerous for crude oil carriers than for product carriers;
3. the provisions should be more onerous for new vessels than for existing vessels.

Hence, the most onerous provisions relate to new, crude oil VLCCs.

3–68   A blanket application of these principles would, however, have failed to take account of individual cases, and so would have resulted either in undue hardship or completely unnecessary over-regulation, or both. The Conference therefore increased the complexity of the provisions by including a limited number of exceptions. The resulting Regulations are so detailed that only major provisions can find space here, but it is hoped that this will at least ease the path of those who must have recourse to the text of MARPOL 73/78 itself (along with the Unified Interpretation and proposed amendments).

3–69   **(a) New crude oil tankers.** By Regulation 13(1) every[47] new crude oil tanker[48] of 20,000 dwt tons and above must be pro-

---

[47] Regulation 13(5) provides that states may make separate provision for the segregated ballast in tankers less than 150 metres in length.

[48] Regulation 1(29) defines *crude oil tanker* as an oil tanker engaged in the trade of carrying crude oil.

45

vided with protectively located segregated ballast tanks of specified capacity, and may not carry ballast water in cargo tanks except where necessitated by very severe weather conditions (and if this does happen, the discharge of such ballast must be made in accordance with the discharge standards of the Convention[49] and the cargo tanks used must have been crude oil washed).[50] Thus, the limit for compulsory segregated ballast tanks has been reduced from the 1973 figure of 70,000 dwt tons to 20,000 dwt tons.

3–70 In addition, every new crude oil tanker of 20,000 dwt tons and above must, by Regulation 13(6), be fitted with a cargo tank cleaning system using crude oil washing, which, with limited exceptions to account for practical difficulties, must be operated in accordance with and conform to specifications contained in Regulation 13B.[51] These requirements include the fitting of an inert gas system.[52]

3–71 (b) Existing crude oil tankers. An essential ingedient of the compromise adopted at the 1978 Conference was that, while segregated ballast tanks and crude oil washing could be made concurrent requirements for new crude oil tankers, such a requirement was financially too onerous if applied to existing tankers, and so a third alternative, the operation of dedicated ballast tanks (so-called "clean ballast tanks") was also introduced for them. The requirements are qualified, in the case of existing tankers, by special provisions relating to tankers engaged in specific trades and tankers already having special ballast arrangements.[53]

3–72 Accordingly, Regulations 13(7) and (8) require that every existing crude oil tanker (unless subject to the qualifications mentioned above) of 40,000 dwt tons and above must *either* comply with the segregated ballast tanks requirements *or* comply with the crude oil washing requirements for new crude oil tankers of 20,000 dwt tons and above (described above), from the date when MARPOL 73/78 entered into force (October 2, 1983). A further alternative is provided by Regulation 13(9) for a limited period varying with the

---

[49] See paras. 3–47 *et seq.*, above.

[50] But see the proposed amendments, new Regulation 13(3)(*a*) for a new proposed exception.

[51] See the Revised Specifications for the Design, Operation and Control of Crude Oil Washing Systems as amended, IMO Resolutions A.446(XI) and A.497(XII). Difficulties encountered by ships operating crude oil washing are the subject of Resolution A.498(XII).

[52] As to inert gas systems, see further paras. 4–55 *et seq.*, below.

[53] These qualifications are contained in Regulation 13C and 13D. They are not dealt with further in this work. By Resolution 16 of the Conference, IMO may develop these exemptions. Regulation 13C is the subject of a minor proposed amendment.

deadweight of the tanker, whereby the tanker may operate with dedicated ballast tanks ("clean ballast tanks") in accordance with the specifications of Regulation 13A.[54]

3–73 (c) New product carriers. There can be little environmental justification for the new requirements on segregated ballast and dedicated ballast tanks introduced for product carriers by MARPOL 73/78. It was seen in paragraphs 3–53 *et seq.*, above, that the discharge standards are sufficiently vigorous to ensure that the environmental problem created by non-persistent oils is reduced to a bare minimum. The only environmental merit of these provisions lies in the fact that, being essentially structural in nature, they are easier to police.

3–74 Regulation 13(1) requires every new product carrier[55] of 30,000 dwt tons and above to comply with the same segregated ballast tank requirement as for a new crude oil tanker of 20,000 dwt tons and above (described above), although of course there is no requirement that ballast taken into cargo tanks during very severe weather must be taken into a tank which has been crude oil washed, nor is there any crude oil washing requirement—by definition[55] product carriers do not carry crude oil.

3–75 This means that the limit for compulsory segregated ballast tanks for product carriers has been lowered by MARPOL 73/78 from the old 1973 figure of 70,000 dwt tons to 30,000 dwt tons.

3–76 (d) Existing product carriers. The same principle of alternatives is followed for existing product carriers as it was for existing oil tankers, but of course the crude oil washing alternative is absent. Hence, Regulation 13(10) provides that from the date of MARPOL 73/78 entering into force—October 2, 1983—every existing product carrier of 40,000 dwt tons and above shall comply with the same segregated ballast tanks requirements as for new product carriers (described above), or alternatively it may operate with dedicated ballast tanks in accordance with Regulation 13A.[56] This alternative, unlike that for existing crude oil tankers, is available permanently.

3–77 (e) Impact of the new segregated ballast and crude oil washing requirements. An analysis of how these new provisions will affect the world fleet is beyond the scope of this work, but an

---

[54] See the Revised Specifications for Oil Tankers with Dedicated Clean Ballast Tanks, IMO Resolution A.495(XII).

[55] Regulation 1(30) defines *product carrier* as an oil tanker engaged in the trade of carrying oil, other than crude oil.

[56] See note 54, above.

excellent study has been made by H. P. Drewy (Shipping Consultants) Limited of London for those who need to have this kind of information.[57] Ever since 1975, new tankers over 70,000 dwt tons have very largely been built with segregated ballast tanks which comply with the requirements of MARPOL 73/78, and crude oil washing equipment has been very widely fitted on both new and existing tankers under charter market pressures. For many existing ships, however, owners are faced with alternatives which require careful economic evaluation in the case of each individual ship. One such alternative not mentioned in the Convention, which may be cost effective for certain ships whose size brings them just into a particular category, is for an owner to downrate the dead weight of his ship so that it falls in the category next below, and thereby increases the options open to him or takes the ship out of the provisions altogether. This may be graphically illustrated by the IMO impact chart reproduced (with slight amendments) in Table 3.1.

Table 3.1—MARPOL 73/78 Requirements on ballast tanks and crude oil washing summary at October 2, 1983

New tankers

Product 30,000 cwt tons & above SBT
Crude   20,000 dwt tons & above SBT and COW

Existing tankers

Product 40,000 dwt tons & above SBT or CBT
Crude   40–69,999 dwt tons   SBT or COW or, until 2.10.87, CBT
Crude   70,000 dwt tons & above SBT or COW or, until 2.10.85, CBT

Notes to Table 3.1
SBT   Segregated ballast tanks
COW   Crude oil washing (which may only be carried out by ships fitted with an inert gas system)
CBT   Clean ballast tanks (dedicated ballast tanks)

C. Discharge Standards for Other Ships

3–78 Much of the oil which is discharged from ships other than tankers has not, in the past, been discharged in the environmentally acceptable way in which Load on Top discharges of cargo

---

[57] H. P. Drewy, *The Impact of New Tanker Regulations* (Study No. 94, 1981).

residues are now intended to, and no doubt largely do, take place. The same is true of discharges from tankers of oily bilges (although many tankers can pump oily bilges into the slop tank) and of sludges from fuel oil purification. Discharge standards in relation to ships other than tankers and to tanker bilges are therefore of considerable environmental importance.

THE 1954–62 STANDARDS

3–79 For ballast water carried in fuel tanks, the 1954 Convention applied the standard for tankers[58] to ships as if they were tankers, except that the discharge of oil or of oily mixture from such a ship was not prohibited when the ship is proceeding to a port not provided with reception facilities for ships other than tankers. Since such facilities were at the time very largely lacking in many ports, this standard was completely useless.

3–80 Discharges of oil originating elsewhere in the ship were provided for separately. Discharges of fuel oil purification sludges were exempted from the above standard provided they were made "as far from land as is practicable," and discharges of machinery space oily bilges were exempted if the effluent contained only lubricating oil drained or leaked from the machinery spaces.

3–81 It can therefore be seen that the 1954 standards were environmentally useless, in so far as discharges from ships other than tankers are concerned. The 1962 Conference made only an ineffectual attempt to improve the dirty ballast standard by providing that "carrying ballast in oil fuel tanks shall be avoided if possible." That Conference did, however, adopt Article III(c), which, as we have seen,[59] prohibited oil discharge from large new ships except in special circumstances. However, the effect of this provision on the world's dry cargo fleet was minimal, because so few of such vessels, even today, exceed 20,000 tons g.r.t.

THE 1969 AMENDMENTS

3–82 The 1969 Amendments greatly improved matters: Article III(a) now provides that, subject to certain exceptions and defences:

> "The discharge from a ship to which the present Convention applies, other than a tanker, of oil or oily mixture shall be prohibited except when the following conditions are all satisfied:
> (i) the ship is proceeding *en route*;

[58] See paras. 3–45 *et seq.*, above. This standard prohibits discharges inside prohibited zones.
[59] Para. 3–47, above.

     (ii) the instantaneous rate of discharge of oil content does not exceed 60 litres per mile;

    (iii) the oil content of the discharge is less than 100 parts per 1,000,000 parts of the mixture;

    (iv) the discharge is made as far as practicable from land.''[60]

3–83 Significantly, the old exemptions were deleted, and the discharge of oil from a tanker's machinery space bilges specifically brought within the new Article III(a) standard.[61]

3–84 Conditions (ii) and (iii) of the new Article III(a) mean that oil must be discharged in a more environmentally acceptable way, and in practice this means that ships must discharge oily bilges, dirty ballast and bunker tank washings through a suitable separator, and must retain on board in sludge tanks the sludges from purification of fuel oils. However, as was seen in paragraphs 3–32 *et seq.*, there was and still is a real problem in relation to these sludges due to lack of appropriate reception facilities for them.

The MARPOL 73/78 Standards

3–85 MARPOL 73/78 seeks to improve on these standards. Regulation 9(1)(b) replaces Condition (iv) above with the requirement that the ship be outside a special area and more than 12 nautical miles from the nearest land; and it adds that the ship must have in operation an oil discharge monitoring and control system, oily-water separating equipment, oil filtering system or other installation as required by Regulation 16. The standard for special areas is the same as that for oil tankers within special areas,[62] with certain minor exceptions.

3–86 Regulation 16 requires[63] that any ship of 400 tons g.r.t. and above be fitted with oily-water separating or filtering equipment such as will ensure that any oily mixture discharged to sea after passing through it has an oil content not exceeding 100 p.p.m. In addition, ships of 10,000 tons g.r.t. and above must be fitted *either* with an oil discharge monitoring and control system almost identical to that required for oil tankers,[64] *or* with an oil filtering system which accepts the discharge from the required separating system and reduces the oil content of the effluent to not more than 15 p.p.m.

---

[60] The erroneous omission of the word "nearest" between "from" and "land" was corrected by the 1971 Amendment contained in IMCO Resolution 232(VII).

[61] New Article III(c)(ii).

[62] See paras. 3–63 *et seq.*, above.

[63] See the Recommendation on International Performance Specifications for Oily-Water Separating Equipment and Oil Content Meters, IMO Resolution A.393(X). Special provisions are made in Regulations 16(3) and (4) for ships under 400 tons g.r.t. and for existing ships.

[64] See para. 3–57, above.

3–87   Regulation 14 supplements this by requiring new ships of 4,000 tons g.r.t. and above other than oil tankers, and new oil tankers of 150 tons g.r.t. and above, not to carry any ballast water in any oil fuel tank except in abnormal circumstances. Hitherto, it will be remembered, this was only to have been avoided "if possible." To comply with this Regulation, new ships other than tankers will have to be fitted with segregated ballast tanks, whereas most tankers will comply by using their cargo or, if they have them, segregated ballast tanks for ballast. Existing ships shall comply "as far as reasonable and practicable."

3–88   Perhaps the most significant provision, however, is that of Regulation 17, which requires all ships of 400 g.r.t. tons and above to be provided with sludge tanks of adequate capacity to accept the residues which cannot be dealt with in accordance with the above standards, such as those from the purification of fuel oil. The problem here, though, is what you do with the content of the sludge tank, and as always this relies on the availability of reception facilities for sludges, which are still in too short a supply.[65]

3–89   As with the standards for oil tankers, it can be seen that the MARPOL 73/78 requirements are environmentally superior to anything which had gone before, although it must be questioned whether the new mandatory provisions of Regulation 16 are environmentally cost effective.

D. Reception Facilities

The 1954 Standards

3–90   The approach taken by the 1954 Convention to discharge standards clearly made the provision of adequate reception facilities important to the success of the Convention as a whole. Accordingly, Article VIII provided that states "shall ensure" the provision in each main port of facilities adequate for the reception of residues from oily ballast water and tank washings remaining for disposal by ships other than tankers. While the mandatory language was desirable, it was too much for some states, two of whom entered reservations to the effect that they would not be bound to ensure the provision of such facilities (despite the fact that Article VIII only required reception facilities at main ports, only for ships other than tankers and only for residues rather than sludges or dirty ballast itself).

---

[65] See paras. 3–32 *et seq.*, above.

The 1962 Standards

3–91   The 1962 Conference tried again, amending Article VIII in a number of ways. First, states now had only to take "all appropriate steps to promote the provision of facilities." This provided the escape route required, for it fell short of requiring a state to spend any money, even as a last resort. However, the reception facilities whose provision now had to be promoted were expanded to include all ports, oil-loading terminals and repair ports. Despite this, three states ratifying after these amendments entered into force made reservations about Article VIII.

3–92   No change was made in 1969 or 1971, and it was left to the 1973 Conference to produce some realistic standards. Had adequate reception facilities been available from 1967, when the 1962 amendments entered into force, no ship would have had any excuse for discharging oil into the sea. This failure is that of the international community of states.

The MARPOL 73/78 Standards

3–93   The MARPOL 73/78 standards are contained in Regulation 10(7), for special areas, and Regulation 12, for other areas. The main provision in Regulation 12 is as follows:

> "(1) Subject to the provisions of Regulation 10 of this Annex,[66] the Government of each Party undertakes to ensure the provision at oil loading terminals, repair ports, and in other ports in which ships have oily residues to discharge, of facilities for the reception of such residues and oily mixtures as remain from oil tankers and other ships adequate to meet the needs of the ships using them without causing undue delay to ships."

3–94   Regulations 12(2) and (3) go on to make comprehensive and detailed provisions as to where reception facilities are to be provided and their capacity. Short-haul tankers, coastal tankers, pre-maintenance washing and combination carriers are given special attention, so that the problems associated with the operation of Load on Top are meant to be solved, as are those associated with the disposal of fuel oil purification sludges and oily bilges.

3–95   Even stricter standards are set for special areas by Regulation 10(7), the main difference being that all oil loading terminals and repair ports and all ports handling dry cargo ships must be provided with adequate reception facilities, not just to take residues

---

[66] Dealing with reception facilities in special areas.

left after lawful retention-on-board techniques have been operated, but to take the dirty ballast and tank washings too.

3–96 The MARPOL 73/78 standards are environmentally acceptable. However, as was seen in paragraphs 3–32 *et seq.*, despite the fact that the Convention has entered into force, there is still a dismal failure on the part of some states to comply with the requirements of Regulations 10 and 12. There are, for instance, particular problems in the Mediterranean. In practice, this means for instance that if a crude tanker under 40,000 dwt tons discharges at Genoa and wishes to load at Marsa el Brega, it would have first to call at Fos to discharge dirty ballast and tank washings.

4. ENFORCEMENT OF LEGAL STANDARDS

3–97 It is impossible to say exactly what number of contraventions of international standards there are each year, or exactly what proportion of the world fleet habitually fails to operate anti-pollution procedures, either properly or at all, so as to contravene international standards, but it can be safely concluded that both figures are, relatively speaking, small. Having said that, it is clearly important that the standards are actively enforced, not only to satisfy the demands of the environment and the feelings of those who care for it and want to use it, but to be fair to the great majority of shipowners who do comply.

A. The 1954–62 Position

3–98 The 1954–62 Convention contained two main provisions relating to enforcement, one requiring the completion of oil record books, the other enshrining the traditional doctrine of flag state jurisdiction.

3–99 Article IX provided that every ship which uses fuel oil and every tanker to which the Convention applies shall be provided with a specified oil record book and the book must be completed on specified occasions and signed by the master and one other officer. Basically, such occasions were whenever any act is performed on board which meant that water will or may come into contact with oil: hence ballasting and cleaning of tanks, all discharges of water or oil from tanks, and so on. The idea was that by inspecting a ship's oil record book, it should be possible to tell whether or not an unlawful discharge of oil had been made.

3–100 The competent authorities of any contracting party were given the power to inspect a foreign ship's book while in port, and a true copy made of this had to be made admissible in any judicial proceedings as evidence of the facts stated in the entry. However, no proceedings could be taken by the port state: by Article X, all it could do was to furnish the flag state with particulars in writing of evidence that any provision of the Convention had been contravened by a ship. Upon receipt of such particulars the flag state was obliged to investigate the matter and if satisfied that sufficient evidence was available in the form required by its law, it had to take proceedings as soon as possible against the owner or master[67] of the ship. Article VI demanded that contravention shall be an offence punishable under the law of the flag state, and that the penalties "shall be adequate in severity to discourage such unlawful discharge" and shall not be less than those imposed in respect of the same infringements within the flag state's territorial sea.

3–101 This system was inadequate in many ways. Unless the oil record book by itself betrayed an unlawful discharge, its value for enforcement purposes was much reduced. Avoidance of the use of the oil record book was relatively easy if those responsible for signing it were so minded; and it is of course quite possible to round off figures to hide all but the most blatant breaches. Hence, the effectiveness of the oil record book depended largely on the integrity of the master and officers—and the master or officer who would make an unlawful discharge in the first place can hardly be expected to "own up" to it in the oil record book.

3–102 There was also the most serious defect in that the officials of the port state could inspect the oil record book, but not the contents of the slop tank. Therefore, their first and strongest method of checking the accuracy of what was read in the oil record book was not available to them. However diligent port states were in inspecting the books, the system was entirely dependent for effective enforcement on flag states taking the necessary action. This, in many cases, they manifestly failed to do.

3–103 Lastly, the system was grossly inefficient. Before 1978 the necessary additional evidence was difficult to collect, and in many cases there would be a long gap between detection and prosecution, because the tanker would be far from her flag state. Any good evi-

---

[67] In *Federal Steam Navigation Co. Ltd.* v. *Department of Trade and Industry* [1974] 1 W.L.R. 505 the House of Lords, by a majority of 3 to 2, held that, in the Oil in Navigable Waters Act 1955, s.1(1), the word "or" in this phrase is to be construed conjunctively and not disjunctively, so that both owner and master can be convicted of an offence arising from a single unlawful discharge. See also [1974] C.L.J. 181 (L. Collins).

dence of the offence (*e.g.* a photograph) would normally have to be made available—if it was admissible—and this involved cost, which was further increased if the form of evidence had to conform to a particular type demanded by the flag state's court system (*e.g.* affidavit, signed statement and so on). There has been one case reported in which the prosecution was brought two years after the discharge; this meant that the case could not be handled summarily, but had to proceed on indictment.[68]

B. The 1969 Position

3–104 The 1969 Amendments undoubtedly made proof of an offence less difficult by virtue of the fact that the sighting of a ship discharging oil became much more likely to be evidence of a contravention of the standards. The oil record book was still an important element in the evidence, but whereas up until then it had been the prosecution who needed it, after the 1969 Amendments it was primarily the defendant who wanted to use it. This is because the 1969 Amendments made all discharges *prima facie* unlawful; it was for the defence to show that the discharge was within the lawful exceptions, and the only way this could be done was to plead the operation of Load on Top—as evidenced by the oil record book.

3–105 While the oil record book was improved in form by the 1969 Amendments, there was still no right in the port state to inspect the ship's slop tank. Hence, while prosecutions were easier, there was still much room for improving the system.

C. MARPOL 73/78

3–106 It was the 1973 Convention which really came to grips with the enforcement question. Some of the machinery is familiar: by Article I, parties undertake to give effect to the provisions of the Convention, and, by Article 4(1), flag state jurisdiction is provided for. But Article 4(2) creates a right and duty for parties to enforce the Convention within their jurisdiction, and by Article 9(3) the term "jurisdiction" shall be construed in the light of international law in force at the time of application and interpretation of the Convention. This means that if and when the jurisdictional principles contained in the Law of the Sea Convention, 1982 are received into the body of customary international law, MARPOL

---

[68] *R.* v. *Federal Steam Navigation Co. Ltd.*; *R.* v. *Moran* [1973] 3 All E.R. 849 at p. 850 (C.A.); affd. *sub nom. Federal Steam Navigation Co. Ltd.* v. *Department of Trade and Industry* [1974] 2 All E.R. 97, [1974] 1 W.L.R. 505 (H.L.).

73/78 may be enforced in accordance therewith. As between parties to both Conventions, MARPOL 73/78 may be enforced in accordance with the 1982 Convention before that, when it enters into force.[69]

3–107 MARPOL 73/78 has both standard machinery of enforcement and a number of innovations, both in its Articles and in the Regulations of Annex I. The standard machinery is provided by Regulation 20 and Appendix III of Annex I, where the oil record book provisions were re-enacted with minor changes. The 1978 Protocol made yet further changes, and now IMO's Marine Environment Protection Committee has agreed on a new form of oil record book which will soon be adopted as an Amendment to MARPOL 73/78.[70] The first of the innovations is a set of provisions relating to International Oil Pollution Prevention Certificates. By Regulation 4, all vessels (bar the smallest) shall be subject to regular and complete surveys to ensure that the structure, equipment, fittings, arrangements and materials fully comply with the Convention. A ship which does comply must then be issued with a certificate (Regulations 5 and 6) but, significantly, no certificate shall be issued to a ship entitled to fly the flag of a non-party (Regulation 6(4)). The amendments to Regulation 4 of Annex I effected by the 1978 Protocol, affecting enforcement of design and equipment standards, are discussed in paragraphs 4–58 *et seq.*

3–108 The effect of the certificate is defined in Article 5. It must be accepted by other parties as having the same validity as one issued by themselves; a ship required to hold a certificate is subject to inspection by the officers of a party in its ports or offshore terminals for the purpose of verifying that there is on board a valid certificate, and if there are "clear grounds" for believing that the condition of the ship or its equipment does not correspond substantially with the particulars of the certificate, the inspecting party "shall take such steps as will ensure that the ship shall not sail until it can proceed to sea without presenting an unreasonable threat of harm to the marine environment." In addition, parties shall apply the Convention so that ships of non-parties get no more favourable treatment. It remains to be seen how states will interpret the phrase "clear grounds," but it is least arguable that "unreasonable threat of harm" should be interpreted to mean that a ship must come very close to qualifying for the certificate it lacks (or which it has, but did not at the time of detention substantially correspond with), before it is allowed to sail.

---

[69] See further paras. 5–54 *et seq.*, below.
[70] See M.E.P.C. Circ. 99 of June 30, 1982 and IMO Circular Letter No. 968.

3–109 The second innovation concerns inspection and co-operation. The fact that a ship has a certificate, while indicating that it has the ability to comply with the Convention, does not mean that the discharge standards will actually be complied with. As long as there remains some motive for non-adherence, the need for policing the high seas remains. Recognising this, Article 6 places a duty on parties to co-operate in the detection of violations, and Regulations 9(3) and 10(6) declare that whenever visible traces of oil are observed on or below the water surface in the immediate vicinity of a ship or its wake, governments should "to the extent that they are reasonably able to do so," promptly investigate the situation, and in particular investigate the wind and sea conditions, the track and speed of the ship, other possible sources of the visible traces in the vicinity, and any relevant oil discharge records.

3–110 In addition to the power to inspect for the possession of a certificate and for perusal of the oil record books, Article 6(2) declares that the officers of a port state may inspect a ship "for the purpose of verifying whether the ship has discharged any harmful substances in violation of the provisions of the Regulations," although flag state jurisdiction is still retained for prosecutions in relation to discharges on the high seas (Articles 6(3) to 6(5)). Thus at last the most important element so far lacking in the enforcement provisions of the international conventions is provided.

3–111 The third innovation is that ships must carry an oil discharge monitoring and control system, fitted with a recording device to provide a continuous record of the discharge in litres per nautical mile and total quantity discharged (or the oil content and rate of discharge): see paragraphs 3–57 *et seq.*, above.

3–112 IMO has now elaborated detailed procedures for the exercise of these and the other controls discussed above, including those relating to Crude Oil Washing and Inert Gas Systems: see Resolution A542(13) of November 17, 1983.

3–113 MARPOL 73/78 does, therefore, provide important new machinery which will considerably strengthen the hands of the enforcement agencies. The Conference rejected in 1973 an attempt immediately to move away from the doctrine of flag state jurisdiction and deferred the question to the United Nations Third Conference on the Law of the Sea, and it provided instead that when international law develops away from the doctrine of exclusive flag state jurisdiction, so will the Convention. Those developments have now been embodied in the Law of the Sea Convention 1982,[71] and

---

[71] See paras. 5–54 *et seq.*, below.

they must be seen as the result of a perception by many states that the flag state jurisdiction regime discussed above has failed to achieve adequate protection of the environment.

5. CONCLUSIONS

3–114 It can now be seen that many of the major contributions to the reduction of operational oil pollution have been made by the oil and shipping industries, rather than by the international community of states. Load on Top and crude oil washing were both the result of industry initiative. By contrast, the major contribution which states could have made—the provision of adequate reception facilities—has not been achieved even yet.

3–115 It is also clear that legal standards did not properly address the operational oil pollution problem until 1969, and even then it had to wait until 1973 for the adoption of a proper standard on reception facilities. Of course, part of the problem was that it was not until after 1962 that Load on Top was devised, but once it had been, it took fourteen years for a Load on Top standard to enter into force internationally—on January 20, 1978 (although most of industry and some states had put them into force before that[72]). That is the result of insufficient will on the part of the international community of states.

3–116 Ironically, come 1973 the position was reversed in that states adopted provisions which were so technically ambitious that serious problems have been, and still are being, encountered in their implementation; and in 1978 factors quite unrelated to environmental considerations substantially affected the way in which segregated ballast provisions were built into MARPOL 73/78. Despite these blemishes, MARPOL 73/78 remains a remarkable achievement from the viewpoint of the implementation of the international law duty on states to protect the environment. Its ratification by all states, and the solution of the technical problems it still presents, are matters of priority. As more and more tankers have segregated ballast tanks, the real oil pollution threats will emerge as the disposal of fuel oil purification sludges by all ships and the disposal of dirty ballast and tank washings by tankers unable to operate Load on Top or crude oil washing. Reception facilities to meet these requirements are therefore a major priority.

3–117 But MARPOL 73/78 has clearly not done enough to stave

[72] *e.g.* Canada, Japan, Liberia, Sweden, U.S.S.R. and the United Kingdom.

off calls for wider jurisdiction and powers of enforcement to be given to port and coastal states, which were consummated in the adoption of the provisions of the Law of the Sea Convention 1982 discussed in Chapter 5. The failure of flag state jurisdiction to protect the environment in a manner and to an extent which satisfied the majority of coastal states can only be attributed to poor implementation of the legal regimes which rely for their enforcement on the flag state. If flag states are unhappy with the wider powers of the Law of the Sea Convention, they have only themselves to blame. Whether these wider powers will, when the 1982 Convention enters into force, lead to a significant improvement in adherence to the international standards discussed in this Chapter remains to be seen: that will largely depend on the zeal of the states with the new powers.

4. Accidental Discharges

SUMMARY

4–01 This chapter opens with a brief look at some pollution and casualty statistics, so that the international standards subsequently analysed can be put into context. The available figures suggest that accidental discharges during routine operations are the most important element in the overall problem of accidental pollution, although navigational errors are of much greater importance when the size of the spill is also taken into consideration. All these causes have human error in common, and point to the relevance of personnel standards to accidental oil pollution, while others point to the relevance of navigational and design and equipment standards.

4–02 The current standards in all these fields, which are voluminous and complex, are then identified and examined in varying detail. These relate to training and personnel standards, ships' routeing, regulations on prescribed steering gear, radar and collision avoidance aids, protective layout of segregated ballast tanks, tank size and construction, fire protection and inert gas systems and towing requirements. Specific provisions on enforcement (which is still very largely a question of flag state jurisdiction) contained in the relevant instruments are also examined, but analysis of the extensions in jurisdiction afforded by the Law of the Sea Convention 1982 (when it enters into force) are left for Chapter 5.

4–03 The chapter concludes that there is now no lack of relevant and sufficient international legal standards. The continuing problem of accidental oil pollution from ships is one to which a number of factors contribute, each case being different, but the factors include poor management by shipowners, poor enforcement of standards by governments and, probably, the depression in the shipping industry (particularly the tanker section of it) which helps to perpetuate the commercial life of shipowners who are not safety conscious by encouraging the survival of the lowest cost operator.

1. THE NATURE OF THE PROBLEM

A. Introduction

4–04 Accidental spillage of oil from ships occurs either in association with an operational procedure aboard ship, such as loading or discharging cargo, bunkering or ballasting, or in association with a casualty to the ship, such as a collision, grounding, fire, explosion or structural failure. Frequently, such casualties and the resulting pollution therefrom make international or national news headlines; the less dramatic spills from escapes during operations are rarely newsworthy beyond a column-inch in the local weekly, yet as we shall see they are so much more numerous than the big spills that they form an important part of the overall problem of accidental oil pollution.

4–05 In order to identify what legal norms are relevant to the overall problem, we must find out why accidents happen; and to place the causes in perspective the contribution to the overall pollution problem which each makes must be assessed. Here, an immediate and insuperable difficulty arises. A number of bodies collect data on accidents, but however diligent and professional they are they cannot overcome the fact that inevitably, not every spill will come to their notice. Further, the characterisation of each spill may be different according to who is sorting out the data: for instance, if a ship grounds while taking evasive action to avoid a collision, but still collides, is that to be characterised as a collision or grounding? Both characterisations would have merit, but for the purpose of compiling statistics, only one must be chosen.

4–06 Again, there is the problem of estimating the amount of oil actually spilled in a given case. This is difficult not only in the case of a spillage of a few tons, where the error could easily be of the order of five times, but also in a more dramatic case where, for instance, the calculation of the amount spilled could involve a lengthy and difficult estimate of amounts of oil remaining in a sunken section of the ship or of amounts of oil consumed by fire. Lastly, the amount of oil spilled is only one factor in the importance of a spill in terms of its economic, environmental or legal impact; other such factors obviously include the state of the wind and tide, the temperature of the water, type of oil, distance from shore, type of shoreline polluted, the use made of the waters and shore polluted, and so on.

4–07 There is also a danger that if too great an emphasis is placed on statistics, effects may be mistaken for causes, or symp-

61

toms be mistaken for the disease. For instance, as we shall see, groundings occupy an important position in the picture. The grounding and subsequent pollution are effects, not causes. If you want to find the cause of the pollution, you must look behind the statistic and ask why the grounding occurred. Ships do not hit the rocks by themselves: normally, they are steered there.

4–08   For these reasons, only limited use will be made of statistics here, and the purpose will be simply to illustrate what are the main contributors to the problem, so that the relevant legal instruments may be identified and be placed in context. As will be seen, public perception of the nature of the problem is not always the same as that suggested by either the statistics (despite their qualified accuracy) or by the experience of those most closely involved with the problem.

B. Categories of Casualty

4–09   The London-based International Tanker Owners' Pollution Federation maintains a computerised data collection scheme on oil spills from tankers. The scheme relies on a system of voluntary reporting by the tanker owner members of the Federation and their insurers. These reports are supplemented by information from other sources. Whilst it is probably one of the most comprehensive schemes in existence, its reliance on voluntary reporting inevitably means that the information is not complete, especially for smaller spills. Despite this, it is sufficient for our limited purpose here, and so Tables 4.1 and 4.2 have been compiled from data supplied by the Federation.

Table 4.1—Size and circumstances of world oil spills 1974–1983

	SIZE OF SPILL				
	Under 5 bbl	5–5000 bbl	Over 5000 bbl	Total	%
Routine Operations	2,793	883	12	3,688	75.2
Grounding	34	138	67	239	4.9
Collision	23	121	46	190	3.9
Other	493	227	67	787	16.0
Total	3,343	1,369	192	4,904	100

Table 4.2—Place and circumstances of world oil spills 1974–1979

| | PLACE OF SPILL | | | |
	In Port	At Sea	Total	%
Routine Operations	3,330	235	3,565	86.0
Grounding	147	53	200	4.9
Collision	87	27	114	2.8
Other	208	27	235	5.7
Total	3,772	342	4,114	100

Notes to Tables 4.1 and 4.2

(a) *Routine Operations* include ballasting, deballasting, loading and discharging cargo, tank cleaning, bunkering, bilge pumping and internal transfers of cargo.

(b) *Other* includes all other circumstances, including cases where fire, explosion or structural failure are the *primary* cause; since categorisation is by *primary* cause of spill, cases involving fire, explosion and structural failure are also present amongst the figures given in other categories because they have followed upon the primary cause of the spill.

(c) *In Port* means at or close to a port; *At Sea* means spills other than those *In Port*.

(d) Table 4.2 only goes up to 1979 because after that, difficulties in characterising spills as being in port or at sea led to the Federation discontinuing an analysis of the data in this way.

***

4–10 From Table 4.1, it can be seen how important accidental spills from routine operations are, at any rate in terms of numbers of incidents. These spills are almost always in a port or at a terminal, as Table 4.2 confirms, and are nearly all in the medium or small category. Thus, while they are important, they are more likely to be relatively manageable since means of spill control and clean-up are more likely to be within reach. Legal problems of a jurisdictional nature rarely arise, although sometimes clean-up costs are more expensive than might be expected precisely because the incident takes place in port, where other amenities are easily threatened. The most significant cause of these accidental spills is human error in the conduct of the operations: mismanipulation of valves, making poor hose connections and so on. This type of accidental spill can therefore be expected to vary with standards of shipboard personnel.

4–11 Although in relation solely to the numbers of incidents, collisions, groundings and other circumstances pale into insignificance beside routine operations, they form the overwhelming majority of the large spills. As so many of these cases have shown, they tend to create more political pressure due to their size, intensity and public attention than other types of spill. Repeated press coverage of ships streaming oil from positions stranded on rocks or reefs creates the public impression that these cases occur largely at sea, but the figures in Table 4.2 suggest otherwise. This is to be expected. Crowded waters, fog, shallow water and obstacles (whether marked on charts or not), all of which contribute to collisions and groundings, tend to occur close to the shore and within port limits rather than at sea.

4–12 Collisions and groundings are essentially navigational in character, and so can be expected to be sensitive to standards of navigation exhibited by the shipboard personnel as well as to other factors such as the navigational aids available on board and the existence of traffic separation in crowded waters.

4–13 One important factor which from experience of major casualties is known to be relevant is the structural integrity of the ship. Not only is this relevant to a ship's ability to withstand a collision, grounding or other casualty, but it is also important to its ability to withstand heavy weather. A classic recent example is the *Tanio* case off Brittany, France in 1980, when some 12,500 tons of fuel oil cargo caused widespread pollution following the breaking up of the ship in heavy weather.

4–14 There is a popular feeling that fire and explosion aboard tankers must be serious pollution problems. Cases such as the *Betelgeuse*, in Bantry Bay, Ireland, in 1979 and the *Castillo del Bellver* off Cape Town, South Africa, in 1983 have contributed to this. Table 4.1 does not, unfortunately, distinguish between fire and explosion and other categories; but figures for serious casualties to oil and chemical tankers from 1968 to 1983 produced by IMO[1] show that 28 per cent. of the serious casualties were occasioned by fire or explosion involving cargo, machinery spaces or other part of the ship. There is, therefore, some evidence that fire and explosion are important features of the casualty statistics, and so they do present a significant pollution threat.

---

[1] See MSC 50/14. Figures for 1968–1982 are more widely available in *Analysis of Serious Casualties to Sea-going Tankers, 1968–1982* (IMO, 1983). *Serious Casualty* is widely defined to include: (i) a fire, explosion, collision, grounding, contact, heavy weather damage, ice damage, hull-cracking or suspected defect resulting in structural damage rendering the ship unseaworthy or loss of life or pollution (regardless of quantity); (ii) breakdown necessitating towage or shore assistance and (iii) total loss.

4–15 One of the major risks of explosion in a tanker is associated with tank washing, when it is known that electrostatic charges can build up in the tank and cause an explosion if a flammable gas mixture is also present. The tanker is normally empty of oil at the time when tank washing is in progress. This is not so in the case of crude oil washing, and so the provisions requiring inert gas systems for crude oil washing are relevant to pollution control.

4–16 All these considerations point towards the international regime of standards of training, of navigation and navigational aids as having the most direct relevance to the prevention of accidental oil pollution. In this respect, pollution is simply part of the overall problem of maritime safety. Since our concern here is with oil pollution from ships, rather than with the much wider question of maritime safety, discussion of the relevant instruments must necessarily be limited and their impact on oil pollution kept firmly in mind.

4–17 One other set of provisions falls to be considered below. It stands to reason that the way in which a ship is designed can have a dramatic effect on its ability to retain its oil on board after a casualty and on its ability generally to withstand the effects of the casualty. It will, therefore, be necessary to summarise provisions designed to limit the amount of oil which can escape, even from a very large tanker, in the event of a serious casualty.

2. TRAINING AND PERSONNEL STANDARDS

4–18 The earliest work on this subject was done by the International Labour Organisation, although the primary motivation for the work of ILO is not environmental. As early as 1936, ILO adopted the Officers' Competency Certificates Convention (ILO Convention No. 53), concerning the minimum requirement of professional capacity for masters and officers on board merchant ships. This was followed by a similar Convention in 1946 (No. 74) concerning Able Seamen. In 1958 Convention No. 109, revising earlier Conventions,[2] was adopted, and this concerns wages, hours of work and manning at sea. The latest relevant ILO Convention is the Merchant Shipping (Minimum Standards) Convention, 1976 (No. 147 of 1978), which came into force in November 1981. The wording of the Convention is very general; for instance, parties undertake "to ensure that seafarers on ships registered in its territory are

---

[2] No. 57 of 1936; No. 76 of 1946; and No. 93 of 1949.

properly qualified or trained for the duties for which they are engaged."

4–19   IMO also interested itself in the problem of shipboard standards, but for many years confined itself to passing non-binding resolutions and issuing guidelines. This work culminated in the adoption of the International Convention on Standards of Training, Certification and Watchkeeping for Seafarers, 1978, which entered into force on April 28, 1984 (although since then IMO has had to pass further Resolutions and guidelines for the application of the Convention[3]). As a response to a Resolution passed by the International Conference on Tanker Safety and Pollution Prevention four months earlier,[4] this Convention contains several provisions dealing specifically with tankers. Furthermore, protection of the environment has been incorporated into the Convention as an objective additional to safety of life and property.

4–20   Broadly speaking, the Convention lays down basic standards of training, certification and watchkeeping for responsible officers in all departments aboard sea-going ships. In addition, Chapter V contains special mandatory minimum requirements for the training and qualification of masters, officers and ratings of oil tankers. However, as with the other provisions of the Convention, the drafting is very general and interpretation and implementation is left to the flag state. An example of this is seen in Regulation V/1, which provides as follows:

> "1. Officers and ratings who are to have specific duties . . . in connexion with cargo and cargo equipment on oil tankers and who have not served on board an oil tanker as part of the regular complement, before carrying out such duties shall have completed an appropriate shore-based fire-fighting course; and
>
> > (*a*) an appropriate period of supervised shipboard service in order to acquire adequate knowledge of safe operational practices; or
> >
> > (*b*) an approved oil tanker familiarisation course which includes basic safety and pollution prevention precautions and procedures, . . .
>
> 2. Masters [and other relevant officers and] . . . any person with the immediate responsibility for loading, discharging and care in

---

[3] See, *e.g* Resolutions A.285(VIII) of 20.11.73, A.337(IX) of 12.11.75 on basic principles of watchkeeping and IMO's Document for Guidance on personnel training, first adopted as Resolution A.89(IV) in September 1965 and updated until 1975. Most recently, in November 1981, IMO adopted Resolutions A.481(XII) to A.486(XII) on principles of safe manning, training in the use of radar and the training and use of pilots.

[4] Resolution No. 8. As to the T.S.P.P. Conference see further paras. 3–37 to 3–41, above.

transit or handling of cargo, in addition to the provisions of paragraph 1, shall have:

    (*a*) relevant experience appropriate to their duties on oil tankers; and

    (*b*) completed a specialized training programme appropriate to their duties, including oil tanker safety, fire safety measures and systems, pollution prevention and control, operational practice and obligations under applicable laws and regulations."

4–21   Hence, flag states are free to determine what is "an appropriate period" under paragraph 1(*a*) and what constitutes "relevant experience" under paragraph 2(*a*). The content of the training to be given is elaborated in more detail in Resolution 10 of the Conference, but of course that Resolution is not binding. Therefore the implementation of these very basic standards depends entirely on the flag state. Since the training capabilities and general standards vary from state to state, one can still expect a wide range of shipboard standards from states party to the Convention. This is true not only of Chapter V, but of all the provisions in the Convention. Resolution 16 seems to recognise this in urging states to provide technical assistance to other states who request it.

4–22   The Convention provides for a measure of port state control, but this is limited to verification of certificates and, if an incident has given grounds for believing that standards are not being maintained, assessment of the ability of the seafarers to maintain watchkeeping standards. The only two grounds for detaining a ship are, under Regulation I/4 and Article X(3), failure to correct deficiencies in proper certification or in proper watch arrangements— and even then, this must pose a danger to persons, property or the environment.

4–23   Article X(5) provides that this power of control shall be applied so that no more favourable treatment is given to ships of non-parties to the Convention. Article X also contains a two-edged sword; on the one hand, a port state *must* detain a ship in the prescribed circumstances, and on the other a ship unduly detained is entitled to compensation, so that the limited power to detain is likely to be used sparingly. The Convention does not grant a power to port states to prosecute individual offenders.

4–24   It would be easy to conclude that the 1978 Convention and the ILO Conventions are of little use, but this would be too hasty. Rather, they should be considered as a start. The 1978 IMO Convention in particular has useful elements in it, and it can be used as a set of standards on which to build internationally. It must also be remembered that it is impossible to legislate away human frailty. However strictly legal standards are set, they cannot eradicate negligence, nor can they ensure that a shipping company's collective management philosophy is fundamentally safety-oriented.

3. NAVIGATIONAL STANDARDS

A. International Conventions

4–25   The earliest international agreement on safety of life at sea was the International Convention for the Safety of Life at Sea 1948, with Collision Regulations attached, which was superseded by the International Convention for the Safety of Life at Sea 1960 (SOLAS 60), and the International Regulations for Preventing Collisions at Sea 1960. SOLAS 1960 was amended regularly since its entry into force, but by 1974 (and, indeed, even today) none of these amendments had gained sufficient acceptances to enter into force. Partly with the hope of bringing the provisions of these amendments into force, they were readopted with other changes in the International Convention for the Safety of Life at Sea 1974 (SOLAS 74) which entered into force on May 25, 1980. The 1960 Collision Regulations were also revised by the Convention on the International Regulations for Preventing Collisions at Sea 1972, which entered into force on July 15, 1977.

4–26   The task of revision did not, however, end there. SOLAS 74 was amended in 1978 by a Protocol adopted at the International Conference on Tanker Safety and Pollution Prevention (the TSPP Conference),[5] and this Protocol entered into force on May 1, 1981. In 1981 there were further Amendments adopted by IMO's Maritime Safety Committee, which entered into force on September 1, 1984, and in 1983 it adopted yet another set of Amendments which are set for entry into force on January 1, 1986. The 1972 Regulations have undergone a similar process, with relatively minor Amendments adopted by IMO's Assembly in 1981 entering into force on June 1, 1983; further amendments are under consideration by the Maritime Safety Committee.

4–27   The two basic instruments, SOLAS 74 and the 1972 Collision Regulations, have received such widespread ratification that,

---

[5] See further paras. 3–37 *et seq.*, 3–67, 3–71 and 3–76 *et seq.*, above.

as was pointed out in paragraph 2–22, above, their fundamental principles must be considered to have passed into customary international law.

4–28 SOLAS 74 and the 1972 Regulations, as amended, between them lay down basic international standards relating to the navigation of all ships on the high seas and, although most of their provisions have as their prime object the saving of life, it is clear that they are directly relevant to the prevention of collisions and stranding, and so are environmentally important. Almost every aspect of navigation safety is dealt with, including lights and shapes, behaviour in restricted and unrestricted visibility, speed of vessels, equipment to be carried on board (such as radar), manning, danger messages and even the Ice Patrol. By far the most important navigational provisions, from the oil pollution point of view, are those concerning ships' routeing.

SHIPS' ROUTEING

4–29 Ship's routeing has a very long history, dating back to the nineteenth century.[6] We can pick up the story in 1964, when an investigation by the UK Institute of Navigation discovered that between 87 per cent. and 100 per cent. of mariners (depending on the country from which they came) favoured routeing, specifically in the Dover Strait.[7] The idea having thus received considerable impetus, IMO began devising traffic separation schemes and incorporating them into Resolutions (compliance being voluntary), of which a large body now exists. They recommend predetermined routes, deep water routes and areas to be avoided in certain parts of the Baltic Sea, Western European Waters including the English Channel, the Mediterranean, the Arabian Gulf and the Atlantic and Pacific coasts of North America.[8]

4–30 Then in 1971 the IMO Assembly adopted an amendment to Regulation 8 of Chapter V of SOLAS 60[9] which included the recognition of IMO as "the only international body for establishing and adopting measures on an international level concerning routeing," although the selection of routes was left "primarily the

---

[6] See J. H. Beattie, "Traffic Routeing at Sea 1857–1977," (1978) 31 *Journal of Navigation* 167.
[7] *The Separation of Traffic at Sea*, Report of a Working Group, Institute of Navigation, London, 1964. A brief history of routeing is contained in L. Oudet, "The Economics of Traffic Circulation," (1972) 25 *Journal of Navigation* 60.
[8] For IMO's General Provisions on Ships' Routeing, see Resolution A.378(X), as amended by Resolutions A.428(XI) and A.527(13). For complete and up to date descriptions of each scheme, see *Ships' Routeing* (5th ed., 1984 IMO).
[9] Annex VI of Resolution A.205(VII).

responsibility of the Governments concerned." For the first time, an element of compulsion was enacted: "where the Organisation (IMO) has adopted traffic separation schemes which specify one-way traffic lanes, ships using these lanes *shall proceed in the specified direction* of traffic flow" (author's italics). Aware that these words were meaningless without some form of sanction, the IMO Assembly also passed Resolution A228(VII), which recommended that enforcement in such cases shall be executed by governments making it an offence for ships of their flag to contravene.

4–31 The mandatory wording of this amendment was adopted in Rule 10 of the 1972 Collision Regulations as well, so that when they entered into force on July 15, 1977[10] it became binding on the ships registered in states party to these Regulations to comply with the schemes adopted by IMO when passing through the waters covered by them.[11] However, there is no obligation on a ship to use such a scheme: if physically possible, the ship can avoid the scheme and take another route.

4–32 The oldest scheme is the one operating in the Dover Strait, where traffic has been well monitored. Its effect on the collision rate by 1975 was summed up by Captain R. K. Emden of H.M. Coast-guard's Dover Strait Operations Centre in these words: "Traffic separation began in the early 60s and the annual statistics for collisions and groundings show a gradual decrease from this date, and particularly since the multiple collision of 1971, and this over a period during which the world-wide figures have been showing a disturbing increase."[12]

4–33 On a wider scale, Captain A. N. Cockcroft reported in 1976 that, "most of the IMO designated traffic separation areas are located in the areas in which there has been a significant decrease [in the incidence of collisions]. Collisions in the vicinity of what are

---

[10] Rule 10 now has a minor exception enacted in the 1981 Amendments, which does not have any oil pollution impact.

[11] Somewhat anomalously, however, at the 1974 Conference on the Safety of Life at Sea, the 1971 Amendment to SOLAS 60 (which formed one of the reference papers upon which the Conference proceeded) was not fully incorporated in SOLAS 74. While the recognition of IMO as the only international body for adopting routeing schemes was retained, the old, ambiguous formula was reverted to and Chapter V, Regulation 8 of SOLAS 74 now reads as follows: " . . . (d) Contracting Governments will use their influence to secure the appropriate use of adopted routes and will do everything in their power to ensure adherence to the measures adopted by (IMO) in connection with the routeing of ships." This is only better than the 1960 formula in that it omits the words, "so far as circumstances will permit."

[12] "The Dover Strait Information Service," (1975) 28 *Journal of Navigation* 129 at p. 136. A further article by Captain Emden is "The Dover Strait Information Service: Recent Progress" (1976) 29 *Journal of Navigation* 263.

now designated as traffic separation areas accounted for almost 50 per cent. of the total number of sea collisions during the 8-year period 1959–66, and for approximately 40 per cent. in the following 8 years. The proportion had decreased to less than 25 per cent. for the 4-year period 1971–74."[13]

4–34 In some cases, however, coastal states have felt it necessary to do more than simply designate a traffic separation scheme, and have instead prescribed (within their territorial sea and/or internal waters) a greater degree of control over ships, varying from a requirement that the ship report to shore stations on arrival at a certain point,[14] to almost full-scale control akin to an Air Traffic Control system. One writer has distinguished these more sophisticated schemes as Vessel Traffic Reporting Systems (VTRS) and Vessel Traffic Management Systems (VTMS), to distinguish them from the ship routeing schemes and traffic separation schemes discussed above.[15] These more sophisticated schemes are usually only justified in heavily congested port approaches or confined waters (such as the St. Lawrence Seaway), although in at least one case— the Eastern Canadian Traffic Regulations System (ECAREG)—it has been extended to the entire twelve mile territorial sea. The proliferation of this kind of reporting requirement has led the IMO Assembly on November 17, 1983 to pass Resolution A531(13) recommending a standard form of report for use in all of them, and the adoption of certain principles, including keeping the number of reports required to the minimum.

SETTING UP A SHIP ROUTEING SCHEME

4–35 Even the simplest of ship routeing schemes is usually expensive, and the more sophisticated they get, the more they cost. Before a recommended route can be adopted, a new hydrographic survey may need to be done to determine exact bottom depths and the direction of natural channels, so that the choice of deep water routes, sea lanes and areas to be avoided can be made on an informal basis. Once the scheme is adopted, the routes must be lit and buoyed, and perhaps also dredged and further surveyed. Unless no form of enforcement is envisaged at all, radar and other communi-

---

[13] "Statistics of Collisions at Sea," (1976) 29 *Journal of Navigation* 215 at p. 220.
[14] An example being the European Economic Community reporting scheme under Council Directive 79/116/EEC of 21.12.78, as amended by Council Directive 79/1034/EEC of 6.12.79, under which all tankers of 1,600 g.r.t. and above must, before entering or leaving the seaport of a Member State make a report in prescibed form to the competent authorities.
[15] E. Gold, "Vessel Traffic Regulation: The Interface of Maritime Safety and Operational Freedom" (1983) 14 J. Mar. Law & Comm., 1.

cations stations appropriate to the type of scheme must be set up on shore, and permanent staff employed to man these shore installations. Provision should be made for the removal and marking of wrecks, and for the accurate identification of offenders, which may mean arranging for helicopter or fixed wing surveillance, and/or patrol craft (naval or civil); all of this entails the permanent or part-time employment of further manpower or existing services.[16]

4–36   Another problem, of course, is who is to pay for all this? Since ultimately all nations benefit from an internationally adopted scheme, all nations would, in an ideal world, pay. However, few nations would be prepared to contribute to the cost of a scheme in anything more than the proportion which they conceived they benefited from it: naturally, doing such a sum would be impossible. So in practice it falls to the coastal state or states concerned to do all the work and pay for it. Sometimes, this cost is directly recovered from shipowners in increased port charges, but in cases where coastal traffic is affected, this is somewhat arbitrary because many of the ships using the scheme will not enter a port in the coastal state concerned, and because at least some ships which do enter such a port will not use the scheme.

4–37   Nonetheless there is a clear trend towards greater shore control of coastal and near-coastal shipping, motivated in many cases by a desire to reduce the risk of pollution.

B. Enforcement of Navigational Standards

4–38   The main problem with the Collision Regulations generally and with traffic separation schemes in particular is the identification of offenders. While a radar trace can indicate to the shore station that someone is contravening the rules, and this trace can be recorded for subsequent production in court (subject to applicable rules of evidence), it needs a visual identification before any action can be taken. In the case of a scheme covering the high seas, this will be limited to notifying the flag state so that it can take appropriate action.[17] Such identification is next to impossible at

---

[16] See generally L. Oudet, "The Economics of Traffic Circulation" *Journal of Navigation* (1972) 25, 60 and the 1975 article by Captain Emden, "The Dover Strait Information Service" *Journal of Navigation* (1975) 28, at p. 129, and C. Warbick, "The Regulation of Navigation" in Churchill, Simmonds and Welch (eds.) *New Directions in the Law of the Sea* (Vol. III) 137.

[17] Hence, IMO Resolution A.432(XI) of 19.2.80 laments "the large number of contraventions of the 1972 Collision Regulations," but is reduced to urging member governments to submit reports and sufficient evidence of infringements to flag states, and to urging flag states to take "appropriate action" when receiving such a report.

night, and is out of the question in fog. Even during good daytime visibility the wrong ship's name and flag may be taken, which is embarrassing for the coastal state. The other main problem is that unless an even greater amount of money is spent, it is impossible to identify all those vessels which the radar indicates are offenders.

4–39 In a scheme covering the high seas, all the enforcement work done by the coastal state will be thwarted if the flag state does not take sufficiently punitive action upon receiving notification of an observed infringement,[18] or if action is taken too slowly; the need for taking prompt action is evident from the fact that the master may have left the employ of the shipowner soon after the offence, and have all but disappeared. Such difficulties are inherent in the concept of flag-state jurisdiction, and have undoubtedly contributed to the development in the Law of the Sea Convention 1982 of much wider zones of coastal state jurisdiction.[19]

4–40 But even when this new regime enters into force, not all of the practical problems will be over. A scheme, however sophisticated, must still be enforced if it is to have real effect, and for this a certain co-operation by the ships affected is required, from reporting its arrival in the vicinity of the scheme to complying with its requirements thereafter. Enforcement is, therefore, a process of *ex post facto* prosecution of offenders. This is possible fairly speedily if and when the ship reaches a port in the coastal state (if the offence occurred within the jurisdiction of the coastal state). Otherwise, it is a question of relying on prosecution by the flag state itself.[20]

4–41 In this connection, an international instrument[21] which may well become a model for others is The Memorandum of Understanding on Port State Control, done at Paris on January 26, 1982, between the maritime authorities of Western Europe and Scandinavia. This is designed to supersede the Memorandum of Understanding between Certain Maritime Authorities on the

---

[18] IMO Resolution A.499(XII) of 19.11.81, expressing concern that existing penalties "in many cases do not effectively discourage violations," urges states party to MARPOL 73/78 to ensure that in future they do. *Cf.* Article 217(7), Law of the Sea Convention: "Penalties provided for by the laws and regulations of States for vessels flying their flag shall be adequate in severity to discourage violations wherever they occur."

[19] See further Chapter 5.

[20] The 8th Report of the Royal Commission on Environmental Pollution (Cmnd. 8358) reports in para. 7.104 that in 1980 there are 1,221 identified contraventions of Rule 10 in the Dover Strait scheme, 50 by tankers (including gas carriers). Of the non-UK ships involved, 259 were reported to their flag state, but this resulted in only 32 fines and 11 warnings.

[21] The instrument is not a treaty because it not between states: it is between maritime authorities.

Maintenance of Standards on Merchant Ships, signed at The Hague on March 2, 1978. Section 5 of the Paris Memorandum states that:

> "The Authorities will upon the request of another Authority, endeavour to secure evidence relating to suspected violations of the requirements on operational matters of Rule 10 of the International Regulations for Preventing Collisions at Sea, 1972 . . . ."[22]

Although this is not an agreement that another port state shall have jurisdiction over the offence, it is a considerable help to flag states who do want to control their fleets.

4. DESIGN STANDARDS

A. International Conventions

4–42 Here again, it is beyond the scope of this study to attempt an evaluation of the complicated and highly technical provisions relating to the design of ships. The appropriate standards will be merely mentioned, so that it can be seen roughly what role the law plays in this field.

4–43 The major provisions are contained in SOLAS 74, Chapters I, II and V, as amended in 1978, 1981 and 1983, which deal with subdivision, stability, machinery, electrical installations and fire precautions. Further, the International Convention on Load Lines 1966 regulates the assignment to a ship of her various freeboards.[23] Most of these provisions deal with the design of ships in general and while they are obviously relevant to pollution prevention, they are more specifically aimed at safety of life at sea and ship safety generally. However, there are certain provisions which, because of their particular effect on pollution prevention, require special mention here.

STEERING GEAR

4–44 It is self-evident that if the steering gear of a vessel fails, the vessel immediately becomes in need of salvage assistance. The pollution danger of a tanker whose steering gear has failed has long been appreciated, and was most sadly underlined recently by the

---

[22] For an analysis of the Paris Memorandum see A. V. Lowe, "A Move Against Substandard Shipping," [1982] *Marine Policy* 326.

[23] This Convention has been amended several times by IMO's Assembly: see Resolutions A.231(VII) of 1971, A.319(IX) of 1975 and A.411(XI) of 1979.

largest pollution disaster to date, the stranding on March 16, 1978 off Portsall, France, of the *Amoco Cadiz*, following a failure of her steering gear.

4–45 In fact, this particular subject has a long history. As early as 1971 the IMO Assembly adopted a Resolution—A210(VII)—which recommended the provision of certain steering gear on large ships; but SOLAS 74 failed to include such recommendations, enacting relatively modest requirements instead, but referring in Regulations 29 and 30 of Chapter II–1 to Resolution A210(VII). In 1975 the IMO Assembly adopted another Resolution—A325(IX)—on improved steering arrangements for new ships of various tonnages, but it was not until 1978 that SOLAS got altered, when the TSPP Conference adopted amendments to Regulation 29 of Chapter II–1 by making special extra provisions for tankers. These provisions underwent yet further, and fairly radical, change in the 1981 Amendments to SOLAS 74/78, which entered into force on September 1, 1984. Regulation 29 now has very elaborate requirements, of which but two require mention here. All ships must now have a main and an auxiliary steering gear. Additionally, in new and existing tankers of 10,000 tons gross tonnage and above (amongst other ships), the main steering gear must comprise two or more identical power units which shall be so arranged that if steering capability is lost due to a single failure in any part of one power actuating system of the main steering gear, it can be regained in no more than 45 seconds. There are transitional provisions for existing tankers.

4–46 These new requirements do not present technical difficulties in the construction of new tankers, and many well designed existing tankers already comply: but of course they will add to the cost of a new tanker and will engender conversion costs for existing tankers not already so fitted. Nonetheless they are clearly an important and welcome addition to international legal standards on accidental pollution prevention.

Radar and Collision Avoidance Aids

4–47 SOLAS 60 contained no requirement that ships should carry radar, although Regulation 12 of Chapter V did require all ships of 1,600 tons gross tonnage and above to be fitted with a specified direction-finding apparatus quite different from radar. It was not until SOLAS 74 that there was a requirement that all such ships carry radar of a type approved by the ship's flag state. At the same time SOLAS 74 made provision for the mandatory fitting of

gyro-compass, echo-sounder and a radiotelephone distress fre-
quency homing device.

4–48 However, a radar device can always fail, and the 1978
TSPP Conference adopted an amendment to Regulation 12 of
Chapter V so that all ships of 10,000 tons gross tonnage and above
should be fitted with two radars capable of operating independently of each other. Partly in response to Resolution 13 of the
TSPP Conference, this Regulation underwent yet further change in
the 1981 Amendments, so that it now provides not only for all of the
above equipment, but also, *inter alia*, for the following:

(a) an automatic radar plotting aid and a device to indicate
speed and distance through the water for all new and existing
tankers of 10,000 tons gross tonnage and above (with transitional
provisions for existing ships);

(b) for new ships of 100,000 tons gross tonnage and above, a rate-
of-turn indicator;

(c) for new and existing ships of 1,600 tons gross tonnage and
above, indicators showing rudder angle and rate of revolution of
each propellor.

4–49 Some of these new provisions are already complied with by
well-equipped tankers but there has been controversy over the wis-
dom and cost-effectiveness of some of the requirements, notably the
automatic radar plotting aid.

DOUBLE BOTTOMS AND PROTECTIVE LAYOUT OF SEGREGATED
BALLAST TANKS

4–50 The United States had wanted the 1978 TSPP Conference
to make double bottoms mandatory for new oil tankers of 20,000
tons dwt and above.[24] The argument in favour of double bottoms is
simply that, if a tanker is involved in a collision or stranding, the
outer skin may be pierced but the inner one, inside which is the oil,
has a good chance of remaining intact.[25] The arguments against
such a proposal,[26] again put simply, are that double bottoms create
a high risk of explosion, they cause salvage difficulties because they
lead to a loss of buoyancy, and they are difficult to inspect, drain
and ventilate. Perhaps the most effective argument against them is
that in many cases they still will not be effective to prevent an

[24] See Conference Documents TSPP II/2, paras. 11–16, 31–34 and Annex VI; TSPP III/8,
paras. 14–20, 28–34 and Annex III; MSC/MEPC/10, para. 70 and Annex XI.
[25] Detailed and technical argument and analysis is contained in J. C. Card, "Effectiveness of
Double Bottoms in Preventing Oil Outflow from Tanker Bottom Damage Incidents"
(1975) 12 *Marine Technology* 60.
[26] See TSPP II/2, para. 14 and TSPP III/3/2.

escape of oil due to the seriousness of the casualty, and hence they are not cost-effective.[27]

4–51 An alternative to the double bottom is to place the segregated ballast tanks which a tanker has (whether because such a requirement is mandatory[28] or by choice of the shipowner) protectively around the ship, so that if the hull is breached, the chances of the breach occurring in a tank containing ballast water or nothing at all are greatly increased. After considerable discussion at the 1978 Conference, it was this approach which was adopted.

4–52 The 1978 Protocol to the International Convention on the Prevention of Pollution from Ships 1973 (MARPOL 73/78) adds to Annex I thereof a new Regulation 13E, which requires every new crude oil tanker of 20,000 tons dwt and above and every new product carrier of 30,000 tons dwt and above to have their mandatory segregated ballast tanks arranged in a specified way. This specification is not linked, as the United States had wished, to the oil outflow requirements described below, but in deference to those who supported this approach the TSPP Conference adopted Resolution 17 recommending further work to be done by IMO.

TANK SIZE AND CONSTRUCTION

4–53 The first set of international design provisions which were specifically pollution-oriented were the 1971 Tanks Amendments to the International Convention for the Prevention of Pollution of the Seas by Oil 1954.[29] They were evolved by the IMO Sub-Committee on Design and Equipment, and they aimed to avoid or limit the escape of oil cargo from a tanker in the event of stranding or collision. The basic formula was that tankers ordered after 1972 had to have cargo tanks so constructed and arranged that if certain assumed side or bottom damage is sustained, the "hypothetical outflow of oil" would not exceed a figure varying with the tanker's deadweight, but which would not for any tanker, however large, exceed 40,000 cubic metres. In addition the volume of centre and wing cargo tanks was limited. Although these Amendments never entered into force, most tankers ordered since January 1, 1972 have, in fact, been built in accordance therewith.[30]

---

[27] For a full analysis of the pros and cons of double bottoms and protective layout of segregated ballast tanks, see W. O. Gray, "Accidental Spills from Tankers and Other Vessels," in J. Wardley-Smith (ed.), *The Prevention of Oil Pollution* (1979, London) p. 79.

[28] As to the legal requirements for segregated ballast, see paras. 3–66 *et seq.*, above.

[29] IMO Resolution A.246(VII).

[30] See Resolution 11 of the 1973 Conference.

4–54 The current set of provisions are contained in the International Convention for the Prevention of Pollution from Ships 1973, and they have not undergone any revision since then. Regulations 22–25 of Annex I re-enact the basic formula of the 1971 Amendments (size and arrangements of tanks) just described, but make two important additions for new tankers: (1) when a cargo transfer system interconnects two or more cargo tanks (as is invariably the case) valves for separating the tanks from each other must be provided, and must be kept closed while the tanker is at sea, and similar provisions are made for piping which runs near the side or bottom of the ship—see Regulation 24(5) and (6); (2) tankers must be so constructed as to comply with the subdivision and damage stability criteria specified in Regulation 25, the object here being to stop a damaged tanker flooding too much in the event of the assumed damage.

FIRE PROTECTION AND INERT GAS SYSTEMS

4–55 Fire aboard ship is one of the mariner's greatest fears. It is not surprising, then, to find that there have long been international regulations on the subject. Chapter II–2 of SOLAS 60 dealt with fire protection, detection and extinction, and this has been revised and updated at every opportunity. The current provisions are to be found in SOLAS 74 as amended in 1981 and 1983, which contains a special part (Part D) on fire safety measures for tankers.

4–56 Here is not the place for a detailed account of the extensive and complicated requirements of Chapter II–2. It is sufficient to single out for mention the current requirements of Regulations 60 to 62 concerning inert gas systems, which originated with SOLAS 74 and have been progressively widened in scope since then. These are of special interest from the oil pollution viewpoint, because, as Regulation 60(6) provides, they are a pre-requisite for any tanker operating crude oil washing.[31] Hence Regulation 60(1) of SOLAS 74, as amended by the 1981 Amendments, now provides that all tankers of 20,000 tonnes dwt and above must have an inert gas system complying with Regulation 62 (or its equivalent).[32] There are grace periods for different types of tanker, varying with size and age.

---

[31] For the crude oil washing requirements, see paras. 3–65 *et seq.*, above; a summary is at para. 3–77.

[32] For details, see IMO Publication "Guidelines for Inert Gas Systems," No. 860.83.15 (1983).

Towing Requirements

4–57 There was some evidence at the Liberian Inquiry into the loss of the *Amoco Cadiz* that the tow rope may have broken as a result of the design of the tanker's fairlead, and in subsequent litigation in the United States certain parties unsuccessfully alleged that it had been poorly designed.[33] However, the incident prompted the adoption by IMO's Assembly of Resolution A535(13) on November 17, 1983, which recommends that all new tankers over 50,000 tonnes dwt, and, at their next scheduled dry-docking, all existing tankers over 100,000 tonnes dwt, be fitted with emergency towing arrangements of the specified type at the bow and stern. The purpose of this recommended harmonisation is clearly stated as being "primarily to reduce the risk of pollution."

B. Enforcement of Design and Equipment Standards

4–58 Although still based largely on flag state jurisdiction, the enforcement of design and equipment standards does not present quite the same problems which the enforcement of navigational or operational discharge standards presents. A flag state can ensure by inspection that no ship which fails to comply with the relevant standards is registered in that state, and that no such ship is issued with the appropriate certificate of compliance; port states can inspect the certificate. For instance, Regulation 6 of Chapter I of SOLAS 74/78 provides for the survey and certification of ships, including unscheduled inspection during the validity of a certificate.[34] Regulation 19 of Chapter I provides for inspection for certificates by port states and a limited power to detain the ship until she can proceed without danger to the ship or persons on board.[35] The Load Lines Convention 1966 has similar provisions,[36] as does MARPOL 73/78.[37]

4–59 However, the large number of sub-standard ships still sailing the high seas is a witness to the fact that much is yet to be desired concerning inspection and certification. Part of this con-

---

[33] See In re Oil Spill by the "Amoco Cadiz" off the Coast of France, [1984] 2 Lloyd's Rep. 304, para. 132.

[34] See IMO Guidelines on Surveys, Resolutions A.413(XI) and A.465(XII); IMO Publication No. 857.82.06 (1982).

[35] See IMO Resolution A.466(XII) of 19.11.81 giving detailed guidance on procedures for the control of ships, especially by port states, under SOLAS 74 and the Load Lines Convention. IMO Publication No. 819.82.13 (1982).

[36] Articles 12 to 21.

[37] Article 5, and Regulations 4 to 8 and Appendix II of Annex I. IMO guidelines for control procedures under MARPOL 73/78 are contained in Resolution A.542(13) of 17.11.83.

cerns the will of both flag and port states, but part also relates to the provisions of the International Conventions themselves. Only diplomatic pressure and public opinion can do much to improve the former; the 1978 TSPP Conference tried to do something about the latter by adopting modifications to two of them: the SOLAS 74 and MARPOL 73.

4–60   The 1978 Protocols to those two instruments amend the relevant provisions, *inter alia* by introducing mandatory unscheduled inspection by the flag state (or annual scheduled inspections instead), and by tightening up the regulations relating to the duration of certificates (so that their life is in fact shortened). As before, ships can be prevented from proceeding to sea if they do not comply with the relevant standards. Additionally, the new Regulation 4(4)(*c*) of Annex I of MARPOL 73/78 creates an obligation on the master or owner, whenever an accident occurs to a ship or a defect is discovered which substantially affects the integrity of the ship or the efficiency or completeness of the equipment required by Annex I, to report it to the flag state or person responsible for the issue of the relevant Pollution Prevention Certificate; this will enable inquiries to be initiated to see if a survey is needed. If the ship is in a port of a state party to the Protocol, the appropriate authorities of that state must also be informed. Prior to these provisions, the duty to report had been limited to arising when a spill had occurred or probably occurred.[38]

4–61   Perhaps more significantly, Article II(3) of the 1978 Protocol to SOLAS 74 brings that Convention into line with Article 5(4) of MARPOL 73 and Article X(5) of the 1978 Training and Watchkeeping Convention discussed above,[39] by providing that states shall apply the requirements of SOLAS 74 and its Protocol so that no more favourable treatment is given to the ships of non-parties. Resolution 6 of the 1978 Conference invited IMO further to develop procedures and guidelines for the operation of the enforcement provisions of SOLAS 74 and the 1973 Convention, and these are now contained in IMO Resolutions A466(XII) of November 19, 1981 (for SOLAS 74) and A542(13) (for MARPOL 73/78).

4–62   States can enhance the effectiveness of these provisions by agreeing to co-ordinate their inspection activities on a regional basis, as for instance was done in the Paris Memorandum of Understanding discussed in paragraph 4–41, above, or as envisaged in the many Regional Conventions discussed in Chapter 7.

---

[38] Article 8 and Protocol 1 of MARPOL 73.
[39] See para. 3–108 (MARPOL) and 4–23 (STW).

5. CONCLUSIONS

4–63 It can now be seen that the existence of accidental oil pollution is not the result of any lack of international standards relating to the protection of the environment. The detailed rules and regulations mentioned briefly in this chapter run to several hundred pages, and the task of reviewing and updating them continues at IMO and in other international fora at a quite bewildering rate.

4–64 Yet the explosion in international regulations aimed generally at prevention of shipping casualties and specifically at prevention of accidental oil pollution from ships has not led to any very noticeable drop in the worldwide casualty rate. There has been a drop in the last two years for which statistics are available, which is hopeful, but too much store cannot be placed on that unless it is maintained. The serious casualty rate per 100 tankers dropped below the 16-year average of 2.29 to 1.84 in 1982 and 1.87 in 1983.[40] It is too early to say whether this represents the start of a trend, particularly as similar two-year drops were recorded in 1970 and 1971, and again in 1973 and 1974, only to be followed by rises. It is still more difficult to say to what extent, if any, the results of the last two years have been achieved because of the implementation of some of the standards discussed in this chapter. It is likely, for instance, that the last two years have seen a drop in the casualty rate at least in part because many tankers at risk have spent prolonged periods idle, waiting for cargoes in a depressed market, and because there has been widespread slow steaming. Nonetheless, a contribution from inert gas systems, ships' routeing and indeed all the other standards mentioned above cannot be ruled out, even if it cannot be quantified.

4–65 But the current absence of a long-term downward trend must be evidence that legal and operational rules are only part of the remedy, and that whatever sophisticated new ideas might be tried—such as compulsory deep sea pilotage, more ship reporting, more stringent design standards—they will only have a marginal effect on the overall accidental oil pollution picture. At the end of the day, accidents are usually caused by people making mistakes, and hence there must be a strong argument that the statistics will be more sensitive to the collective management philosophies of shipping companies and seafarers than to the provisions of International Conventions. There is, for instance, some evidence that the accident rate for independently owned tankers is substantially

---

[40] The latest figures referred to are those compiled by IMO in MSC 50/14 of June 26, 1984.

higher than that for tankers owned by oil companies.[41] This is not to suggest that the elaboration and enforcement of standards is unimportant—far from it, and the fact that ships of certain flags consistently suffer higher casualty rates than others bears witness to that.[42] Nor is it to suggest that elaboration of tighter standards is not worthwhile, or does not affect a substantial proportion of ship-owners. But it is to suggest that voluntary compliance with high standards, and a safety-conscious approach to the job of ship management, is worth a hundred inspectors and pages of detailed regulations.

4–66 Another factor in the picture is probably the collapse of the tanker freight market which really began in 1973 and has continued since then. In a low freight market, the cheapest operator is the one who survives. Such a market therefore assists those whose attitude to safety leaves much to be desired to survive, at the expense of those many shipowners who are naturally safety-conscious.

4–67 The attitude of governments to effective enforcement is crucial, and is also a most important element in the picture. At present, there is adequate machinery, even without the Law of the Sea Convention's wider zones of coastal state control, for the effective enforcement of international standards, but this is left very largely to flag states, and port states play a secondary role. What is lacking in some states is the will to use what has been provided, and it is this which has led to the development of the wider powers of port and coastal states in the Law of the Sea Convention 1982.[43]

[41] F. M. Van Poelgeest, "Sub-standard Tankers," (1978, Netherlands Maritime Institute), quoted in the 8th Report of the Royal Commission on Environmental Pollution, (1981, Cmnd. 8358), para. 7.51.
[42] See E. Osieke, "Flags of Convenience Vessels: Recent Developments," (1978) 73 A.J.I.L. 123; F. M. Van Poelgeest, note 41 above; and Lloyd's Register Statistical Tables (published annually).
[43] See generally Chapter 5.

5. Jurisdiction and Enforcement of Standards

SUMMARY

5–01   In this chapter the current regime of the seas relating to the jurisdiction of states over oil pollution from ships is discussed, together with the regime envisaged under the Law of the Sea Convention 1982. The power of a state to intervene following a pollution incident is considered in Chapter 6. The main feature of the current jurisdictional regime is that the flag state is the most important state when it comes to both enacting and enforcing standards, although coastal states have a certain competence over pollution control zones close to their shores. In 1973, an attempt was made to depart from this regime to wider jurisdiction for port and coastal states in MARPOL 73, but the result was a postponement of the issue to the Third United Nations Conference on the Law of the Sea, which began almost immediately after.

5–02   The Convention which that Conference eventually adopted in 1982 is the single most important instrument relating to the international law of the sea ever to be adopted. It achieves the departure from flag state jurisdiction which was demanded by many states, who saw the current regime of virtually exclusive flag state jurisdiction as essentially a failure in so far as pollution was concerned. Its provisions have yet to enter into force, but this does not diminish its present significance, since the Conference, which was the largest ever to take place, proceeded on the pollution issues (if not on all issues) by consensus.

1. INTRODUCTION

5–03   In Chapter 2 we saw that international law places upon states certain fundamental obligations to protect and preserve the marine environment, and to define them more specifically by the process of international agreement. In Chapters 3 and 4 the relevant agreements already adopted which achieve just such a definition were examined, some in considerable detail. In this chapter and in Chapter 6, we turn attention to the rights of states regarding the protection of the marine environment from oil pollution by ships. The division between obligations and rights is one with only a limited value, since many rights carry with them corresponding obligations, and the division ignores the important legal concept of immunity. However, it is useful here to draw together the provisions on international law which relate to the jurisdiction of states to prescribe standards relating to oil pollution from ships, and to enforce those standards or standards prescribed by others; and these are generally thought of in the context of rights.

5–04 In Chapters 3 and 4 specific provisions of international conventions relating to enforcement of standards were examined.[1] In this Chapter, the general regime of jurisdiction and enforcement, which forms the background to those specific provisions, will be examined. This regime, with most of the specific provisions referred to, has been found by many states to be inadequate to protect the marine environment from oil pollution by ships (and other forms of pollution), and so the new regime envisaged in the Law of the Sea Convention 1982, which was evolved in response to this situation, is also examined in some detail. Chapter 6 is devoted to the individual topic of intervention by states following a maritime casualty, although it is, of course, an integral part of the general subject of state jurisdiction relating to pollution control.

5–05 This work does not set out to be a general treatise on the law of the sea: for that, reference must be made to other works. However, the particular position of jurisdiction over oil pollution from ships cannot be understood without referring to certain fundamental concepts of general application; indeed, until the Law of the Sea Convention 1982 enters into force, these general concepts are the only ones to form the backcloth to the specific treaty provisions examined in Chapters 3 and 4.

2. STATE JURISDICTION OVER OIL POLLUTION FROM SHIPS: THE CURRENT REGIME OF THE SEAS

5–06 A distinction must be made not only between the powers of states to make rules of law prescribing conduct (regulatory jurisdiction) and to enforce such laws, but also between those powers, on the one hand, and the civil jurisdiction of states, which relates to their power to apportion the financial consequences of oil spills and to hear such cases, on the other. In connection with the former type of jurisdiction, the current regime of the seas is divided into geogra-

---

[1] See paras. 3–97 *et seq.*, 4–22 *et seq.*, 4–38 *et seq.*, and 4–58 *et seq.*, above.

phical areas, each of which may be regarded as a pollution control zone. The jurisdiction of states in each zone is different.

A. Regulatory and Penal Jurisdiction

INTERNAL WATERS

5–07   The sovereignty of a state extends not only to its land territory but also to its internal waters, the outer limits of which have been a matter of controversy. Much of this controversy appeared to have been settled with the adoption of the Convention on the Territorial Sea and the Contiguous Zone 1958 ("the Territorial Sea Convention"),[2] but delimitation problems still lingered and it was not until the Law of the Sea Convention 1982 was adopted that a comprehensive text achieved a real measure of consensus.[3]

5–08   In its internal waters a state therefore has freedom to regulate navigation, pollution discharges, manning, design and construction criteria and other matters directly relating to control of oil pollution from ships, limited only by the following:
   (i) constraints imposed by international law generally upon a state's treatment of aliens within its territory;
  (ii) certain rights of innocent passage through waters which, before the application of a straight baseline in accordance with Article 4 of the Territorial Sea Convention, had been considered as part of the territorial sea or high seas[4];
 (iii) certain possible constraints upon the coastal or port state's jurisdiction over criminal offences committed on board the ship which do not affect the interests of the state[5];
 (iv) constraints agreed to by the state in a treaty.

5–09   By far the most important of these is the last, since, as we have seen,[6] states have made specific rules about enforcement which bind them in their treatment of ships flying the flag of other state parties while such ships are in their internal waters. Therefore, if for instance a state wishes to make it a criminal offence for

---

[2] This Convention entered into force on September 10, 1964. Where a provision of it is quoted in this chapter, the author is of the opinion that it is generally declaratory of international law, unless otherwise stated.

[3] Discussion of the delineation of internal waters is outside the scope of this book.

[4] Article 5(2) of The Territorial Sea Convention.

[5] See *e.g.* J. L. Brierly, *The Law of Nations*, (6th ed.,) (ed. Sir H. Waldock), 1963 London, p. 223; M. Sahovic and W.M. Bishop, "The Authority of the State: its Range with Respect to Persons and Places," in M. Sorensen (ed.), *Manual of Public International Law*, 1966 London, p. 335. Both these writers concur with earlier ones that this is more a matter of practice than a binding obligation.

[6] See paras. 3–97 *et seq.*, 4–22 *et seq.*, 4–38 *et seq.*, 4–58 *et seq.*

oil to escape from a ship into its internal waters even where the escape results from a casualty, it would be constrained by Annex I, Regulation 11 of MARPOL 73/78[7] if that state is a party thereto. In fact a state which is not a party to MARPOL 73/78 might be constrained by the principle in Article IV of the International Convention for the Prevention of Pollution of the Sea by Oil, 1954, which is to similar effect, even if not a party thereto because, as it was argued in paragraph 2–23, the principles of the 1954 Convention have now passed into customary international law. Rule-making constraints are also placed upon states party by all the Conventions discussed in Chapters 3 and 4.

5–10 It nonetheless remains true that the internal waters of a state constitute the zone where the state's powers to control pollution are the least restricted. As we shall see, these restrictions increase the further away one goes from internal waters.

THE TERRITORIAL SEA

5–11 The regime of the territorial sea has given rise to more controversy than either the regime of internal waters or the delimitation of the baselines which divide them. Perhaps the most thorny issue has been the breadth of the territorial sea, which both the 1958 and 1960 United Nations Conferences on the Law of the Sea were unable to resolve, and which only reached a consensus at the 1982 Conference. When the 1982 Convention enters into force, the issue will be resolved (at twelve nautical miles).[8] For our present purpose, it is enough to note that currently, no state claims a territorial sea of less than three miles and many claim more, and that claims over twelve miles are likely not to be accepted internationally.

5–12 The sovereignty of a state extends over its territorial sea, but, as with internal waters, is subject to certain restrictions.[9] These are the same as those for internal waters, but with one very important addition: subject to the provisions of the Territorial Sea Convention and other rules of international law, ships of all states

---

[7] The relevant part of Regulation 11 of Annex I provides that the operative prohibition provisions do not apply to discharges for the purpose of securing the safety of a ship or saving life at sea, or to discharges resulting from damage to a ship or its equipment provided reasonable preventive precautions are taken after the discharge (unless the owner or master acted with intent or recklessness regarding the causing of damage).

[8] See para. 5–57, below.

[9] Article 1, Territorial Sea Convention. Claims to a wider jurisdiction in the territorial sea than is described below are therefore contrary to international law.

enjoy a right of innocent passage through the territorial sea.[10] The character of innocent passage is described in Article 14 and the nature of the limits to the coastal state's sovereignty incidental to the right of innocent passage is set out in Articles 15 to 23. The most important points to note from the pollution control viewpoint, are as follows.

5–13   A ship's passage is innocent "so long as it is not prejudicial to the peace, good order or security of the coastal state." This may be taken to put the burden of proof that passage is not innocent on the coastal state. A seaworthy tanker, fully laden with oil, cannot be said to prejudice the peace, good order or security of the coastal state merely on account of its physical characteristics of size and being laden with oil. However, the question arises as to whether there are any circumstances in which the condition of the ship would affect the character of her passage. For instance, if the ship is known to present, or is reasonably suspected of presenting, a pollution threat, is its passage innocent? The more serious the leak or pollution threat, the more difficult the case becomes; but in any event there comes a point when the doctrine of necessity[11] would take over and empower the state to intervene.

5–14   There is, however, an argument that the condition of the ship cannot alter the character of *the passage*; and that this can only be affected by its *purpose*. This is the approach taken by the Law of the Sea Convention 1982, Article 19 of which has a long list of acts which render passage not innocent; all of these address the purpose of the passage and specifically "any act of wilful and serious pollution" is mentioned. The purpose approach is probably the better one for the 1958 Convention as well, so that a state would not be entitled to interfere with a tanker's innocent passage merely because it is a tanker, or because it poses a less than serious threat of pollution.

5–15   The right of unhampered[12] innocent passage, although powerful, is not an absolute right for ships to sail where they please or in the manner they please. Thus, by Article 16(2), coastal states may take the necessary steps to prevent any breach in the territorial sea by ships proceeding to its internal waters of the conditions to which admission of those ships to those waters is subject. A coastal

---

[10] Article 14, Territorial Sea Convention. By Article 21, the regime applies equally to merchant ships and government ships operated for commercial purposes; Articles 22 and 23 provide for certain immunities for other government ships and warships.

[11] See para. 6–06 *et seq.* below.

[12] See Article 15(1), Territorial Sea Convention.

state may prohibit the entry of any ship to its internal waters, for instance on the grounds that it is in the opinion of the local authorities a pollution threat, and turn it away *at the edge of the territorial sea* if the ship is heading for internal waters—but not, as we have just seen, if the ship is passing through the territorial sea for another destination.

5–16   Further, by Article 17:

> "Foreign ships exercising the right of innocent passage shall comply with the laws and regulations enacted by the coastal State in conformity with these Articles and other rules of international law and, in particular, with such laws and regulations relating to transport and navigation."

This means, for instance, that states may make and enforce laws to apply to all ships in their territorial sea which prescribe routes to be taken, require ships to report their position on arrival and thereafter during the passage, and to comply with reasonable routeing directions from the competent authorities of the state; and they may also require ships to have on board a prescribed certificate of financial responsibility—all so long as innocent passage is not hampered. But a provision which, for instance, required that ships in passage through the territorial sea destined for the high seas must be manned, or built to, standards in excess of those prescribed by international law[13] would be unlawful. Furthermore, by Article 18 no charge may be levied on foreign ships by reason only of their passage through the territorial sea (but a charge may be made for specific, compulsory services rendered to all ships, such as pilotage or a vessel traffic management system charge).

5–17   Most states prescribe that unlawful oil discharges will be punishable by criminal proceedings, although some, like the United States,[14] also prescribe so-called civil penalties as well. The application of such criminal jurisdiction to ships in innocent passage is lawful under Article 19(1)(a) because the offence is one whose consequences extend to the coastal state, although in such a case Article 19(3) requires the state to advise the consular authorities of the flag state before taking any steps (if the captain so requests), and to facilitate contact between such authorities and the crew. In emergencies, this may be done while the steps are being taken.

5–18   Article 19(5) maintains a fundamental principle designed

---

[13] See paras. 2–14 *et seq.* and 2–21 *et seq.*
[14] See paras. 18–25, 18–67, 19–10 and 20–39 *et seq.*

to protect the doctrine of flag state jurisdiction on the high seas discussed below: namely that the coastal state may not take any steps in connection with any offence committed *before* the foreign ship entered the territorial sea "if the ship, proceeding from a foreign port, is only passing though the territorial sea without entering internal waters." It might seem, from this, that if the ship *is not* so proceeding, and the oil discharge took place on the high seas, such action would be lawful, but this would be inconsistent both with Article 11 of the High Seas Convention and with the provisions of the International Convention for the Unification of Certain Rules Relating to Penal Jurisdiction in Matters of Collision or Other Incidents of Navigation, 1952, which have the effect of nullifying the effect of the decision of the Permanent Court of International Justice in the *Lotus* collision case in 1927.[15] Article 19(5) must be interpreted accordingly.

5–19 The right of innocent passage gives foreign ships only limited protection from civil jurisdiction in the course of that passage. Article 20 of the Territorial Sea Convention makes a distinction between ships *passing through* the territorial sea, and those *lying in* it. Ships passing through may not be stopped or directed in order to exercise civil jurisdiction in relation to a person on board unless the proceedings relate to a liability incurred by the ship itself in the course of its voyage through the waters of the coastal state. Hence, this does not protect against stoppage or arrest in civil proceedings against persons on board in connection with an oil discharge during passage which may have caused pollution damage. A ship lying in the territorial sea, on the other hand, may be arrested, or judgment may be executed against it, irrespective of where the liability was incurred.

THE CONTIGUOUS ZONE

5–20 Article 24 of the Territorial Sea Convention provides for a further belt of sea contiguous to the territorial sea, whose breadth may not extend beyond 12 miles from the baseline from which the breadth of the territorial sea is measured, called the contiguous zone. This zone has the character of high seas—discussed below— and so is not under the sovereignty of the coastal, or any other, state; but the coastal state may in the contiguous zone exercise the control necessary to prevent infringement of its customs, fiscal, immigration or sanitary regulations within its territory or territor-

---

[15] PCIJ, Series A, No.10. In that case a collision on the high seas led to Turkey bringing to trial the officers of the French ship involved (as well as of the Turkish ship).

ial sea, and to punish infringements of such regulations committed within its territory or territorial sea.[16] The question therefore arises whether oil pollution discharge regulations could be considered to be "sanitary" regulations. Although the matter is not entirely free from doubt, the better view is that they cannot.[17] Certainly, manning, design and construction standards cannot be so considered, nor can navigational standards.

THE HIGH SEAS

5–21   At the 1958 Conference, the only one of the four Conventions adopted which was recognised as being generally declaratory of established principles of international law is the High Seas Convention,[18] which was adopted pursuant to a stated desire to codify the rules of international law relating to the high seas. The high seas are defined in Article 1 as being all parts of the sea that are not included in the territorial sea or in the internal waters of a state; thus they constitute the considerable majority of the earth's surface.

5–22   Jurisdiction over the high seas is of at least equal importance to oil pollution control as jurisdiction over the territorial sea and internal waters. This might seem surprising when it is remembered that most oil discharges take place in port (*i.e.* in internal waters) or in confined waters, many of which are in the territorial sea.[19] However, many routine operations relating to load on top can take place on the high seas, and most of the time of deep sea ships is spent there. The collisions and groundings leading to oil pollution which have occurred on the high seas are too numerous to mention. Perhaps most important of all, the flag state is responsible in international law for manning, design and construction standards, and as we shall see the high seas is characterised by this regime of flag state jurisdiction.

5–23   Article 2 of the High Seas Convention provides that, "the high seas being open to all nations, no State may validly purport to subject any part of them to its sovereignty." The high seas are, therefore, *res communis omnium*. But anarchy must not prevail, and so the freedoms of the high seas, which of course include the freedom of navigation, must be exercised in accordance with international law and with reasonable regard to the interests of other states in the

---

[16] Article 24(1).
[17] See para. 6–12 and references cited there. For the United States policy, see para. 19–05.
[18] Entered into force on September 30, 1962.
[19] See Tables 4.1 and 4.2 and para. 4–09.

exercise of their freedoms. In order to implement this, Article 5 provides that each state shall fix the conditions for the grant of its nationality to ships, for the registration of ships in its territory and for the right to fly its flag. It must also "effectively exercise its jurisdiction and control in administrative, technical and social matters over ships flying its flag." In so doing, it must under Article 10 take measures which conform to international standards to ensure safety at sea with regard to construction, equipment, seaworthiness, manning and prevention of collisions.

5–24 Article 6 enshrines the heart of the doctrine of flag state jurisdiction:

> "1. Ships shall sail under the flag of one State only and, save in exceptional cases provided for in international treaties or in these Articles,[20] shall be subject to its exclusive jurisdiction on the high seas . . . ."

This doctrine makes unlawful any claim by a coastal state to jurisdiction or control over navigational matters or oil discharges on the high seas, except in relation to ships of its own flag or as provided by international treaty. Article 11(3) backs this up by providing that no arrest or detention of a ship following a collision or other incident of navigation concerning a ship on the high seas, even as a measure of investigation, shall be ordered by any authorities other than those of the flag state.[21] Taken with Article 6(1), an oil discharge at sea which is accidental would clearly be an "other incident of navigation"; a deliberate discharge is not so clearly within that description, but to conclude that it is not would be to place such discharges in the same category of exceptions to the doctrine of flag state jurisdiction as piracy and the slave trade, and so the better view is that they, too, are covered by Article 11(3).

5–25 The doctrine of flag state jurisdiction also means that coastal states are unable to establish any form of vessel traffic management scheme[22] on the high seas off their shores without securing international agreement. A further result is that a coastal state

---

[20] *e.g.* amongst others the right of hot pursuit dealt with in Article 23, jurisdiction over ships in the contiguous zone described in para. 5–20, above and jurisdiction over piracy. The exceptions generally do not apply to warships or ships owned or operated by a State and used only on government non-commercial service, which on the high seas have complete immunity from the jurisdiction of any State other than the flag state: see Articles 8, 9 and 16. *Cf.* paras. 9–10 *et seq.* below.

[21] Article 2 of the International Convention for the Unification of Certain Rules Relating to Penal Jurisdiction in Matters of Collision or Other Incidents of Navigation, 1952, is to identical effect.

[22] See paras. 4–29 *et seq.* above.

injured by an oil discharge made on the high seas by a foreign ship is powerless in international law to take any action to arrest or detain the ship unless the doctrine of necessity applies[23] or a treaty with the foreign state so empowers it,[24] *even if subsequently the ship enters its territorial sea or internal waters.* Of course, the coastal state may refuse it subsequent entry to its internal waters, but despite the fact that this has occurred in a number of cases, it may be thought cold comfort.[25] It may also institute a civil action for damages.[26]

5–26 It is therefore often reduced to gathering evidence of the discharge, notifying the flag state, and requesting the flag state to take enforcement action in accordance with its law. In so far as criminal proceedings against those on board are concerned, it has, in theory, an alternative course, if those on board are of a different nationality to the flag state, of notifying the state of such nationality. Under Article 11(1) of the High Seas Convention,[27] that state has concurrent penal and disciplinary jurisdiction over such persons. Since a very large number of oil tankers and other ships are crewed by nationals of a state other than the flag state, this is not a useless option. The problem in many cases is getting hold of the necessary information, or evidence that can be used in penal or disciplinary hearings in a form which conforms to the foreign state's legal requirements.

5–27 The other possibility—concluding a treaty which would allow the coastal or port state to take its own criminal proceedings—had, prior to 1973, been resolutely resisted by many states, notably those with large fleets. But at the 1973 Conference which adopted MARPOL 73, the question of jurisdiction was a key item on the agenda, and was keenly fought. At that time the Third United Nations Conference on the Law of the Sea (UNCLOS III) was just about to start its first session, and states were jockeying for position in that forum.[28] The 1973 Conference considered proposals which widened the competence of individual states to prescribe standards to apply within its own jurisdiction, and also

---

[23] See para. 6–06 *et seq.* below.

[24] *e.g.* the 1969 Intervention Convention: see paras. 6–14 *et seq.*, below.

[25] *Cf.* the sweeping powers given to the United States Coast Guard in this regard: paras. 18–21 *et seq.* The right to refuse entry has given rise to the problem of the "flying dutchman," such as the *Andros Patria* in 1978, discussed in para. 8–35 below.

[26] See paras. 5–30 *et seq.* below.

[27] And Article 3 of the International Convention for the Unification of Certain Rules Relating to Penal Jurisdiction in Matters of Collision or Other Incidents of Navigation, 1952.

[28] See G. J. Timagenis, *International Control of Marine Pollution*, 1980, Dobbs Ferry, Chapter 14, for a full account of the proceedings of the 1973 Conference on the jurisdictional issue.

proposals which enabled port and coastal states to enforce the standards to be enshrined in the Convention by instituting their own proceedings. The former failed to secure the necessary two-thirds majority in Plenary.[29]

5–28 The latter were not adopted in their original form,[30] but a compromise position was finally reached[31] which paved the way for the discussion at UNCLOS III. Articles 4(2) and 9(3) of MARPOL 73/78 (these Articles not being amended at the 1978 TSPP Conference) now provide as follows:

> Article 4(2): "Any violation of the requirements of the present Convention within the jurisdiction of any Party to the Convention shall be prohibited and sanctions shall be established therefor under the law of that Party. Whenever such a violation occurs, that Party shall either:
>   (a) cause proceedings to be taken in accordance with its law; or
>   (b) furnish to the administration of the ship such information and evidence as may be in its possession that a violation has occurred."
> Article 9(3): "The term 'jurisdiction' in the present Convention shall be construed in the light of international law in force at the time of application or interpretation of the present Convention."

5–29 "Any violation" in Article 4(2) can only mean what it says, and so would apply to the operational and discharge standards, as well as the design and equipment standards of the Convention. As we have seen, these underwent considerable expansion at the 1978 TSPP Conference.[32] Article 9(2) and Resolution 23 of the 1973 Conference expressly state that these provisions do not prejudice the deliberations of UNCLOS III. The effect of Articles 4(2) and 9(3) is that when international law receives a different basis for jurisdiction to enforce pollution standards against ships and those on board, MARPOL 73/78 may be enforced in accordance therewith. In practice, it is possible that when the Law of the Sea Convention 1982 enters into force, MARPOL 73/78 will be enforced in accordance with its terms, even if at that time that Convention cannot be said to represent international law.

[29] See MP/CONF/SR.12, the vote on what was then Article 9, MP/CONF/WP.17.
[30] See, *e.g.* MP/CONF/C.1/WP.25 and WP.24 (port state jurisdiction over discharges wherever the violation occurs) and MP/CONF/4, Alternatives I and II of Article 4 (coastal state jurisdiction over violations in the territorial sea).
[31] See MP/CONF/SR.10 where the vote was 49 in favour, 3 against and 5 abstentions.
[32] See paras. 3–65 *et seq.*, above.

B. Civil Jurisdiction

5–30 The above discussion has centred on questions relating essentially to matters of penal jurisdiction—rights to prescribe standards of conduct, rights to take action to enforce them—in relation to polluting discharges, design and construction criteria, navigation and manning. In relation to civil jurisdiction, as with penal, the discussion starts with the proposition that as a fundamental rule of customary international law a state has sovereignty over its land territory and its internal waters. Therefore, if a foreign ship or foreign national pollutes the shores of a state while in its internal waters or territorial sea, that state's courts do have jurisdiction in international law to entertain a civil action for damages. If, however, the defendant is abroad, such as a foreign company not within the jurisdiction, a number of questions arise, such as whether the court would exercise jurisdiction over the defendant and, if the court of the state where the defendant is will exercise jurisdiction, whether the defendant has a defence that the act or omission occurred in another state.

5–31 International law had not, prior to the 1969 Liability Convention discussed in Chapter 10, developed any special rules on such subjects to deal with oil pollution claims, although there are some treaties, discussed below, which deal with certain limited aspects such as collision cases and actions *in rem* brought against the ship itself. In fact, aside from such treaties, international law has left the generality of the cases very much to individual state discretion. There have, therefore, been widely varying state practices, none of which have acquired the status of custom. An examination of the law and practice of any one state on such matters is outside the scope of this chapter, but the United Kingdom approach is discussed in Chapter 16 below.

5–32 As with limitation of liability,[33] the scope of the treaties on the unification of private law was not, until 1969, specifically pollution-oriented and so it is necessary to look at instruments drawn up at a time when pollution was not perceived as a widespread problem or one of special importance. The treaties therefore deal with more general types of maritime claim, and so the solutions to the problems an oil pollution claimant might face are essentially somewhat haphazard.

5–33 Hence, the International Convention for the Unification of

[33] See paras. 9–29 *et seq.*

Certain Rules Relating to the Arrest of Seagoing Ships, 1952 ("the Arrest Convention") provides in general terms that a ship flying the flag of a contracting state may be arrested in the jurisdiction of any contracting state, in respect of any maritime claim against that ship, but in respect of no other claim.[34] "Maritime claim" includes damage caused by any ship "in collision or otherwise," and so includes oil pollution damage.[35] Under Article 3, a sister ship may be arrested instead of the ship giving rise to the claim, and the Convention goes on to make detailed rules concerning release and bail. "Arrest" is defined in Article 1(2) to mean detention of the ship as a pre-judgment remedy and to exclude the seizure of a ship in execution or satisfaction of a judgment.

5–34 But the Convention does not attempt to define what is the geographical scope of a state's jurisdiction, which therefore must be determined and interpreted in the light of the customary and conventional international law discussed above. In particular the rule in Article 20(2) of the Territorial Sea Convention 1958[36] would limit a state's right to arrest a ship while it is actually passing through its territorial sea.

5–35 So far, then, the Arrest Convention establishes a rule that a court may order the arrest of a ship when it is within its jurisdiction. An oil pollution claimant may therefore secure the arrest of the ship, or of its sister ship, outside the state where the damage was suffered. But this does not address the issue of whether that court may hear the merits of the claim. Article 7 provides that it does have power to hear the merits, in certain cases where the claim has a stated connection with that court's jurisdiction, the ones relevant to oil pollution being:
  (a) if the claimant has his habitual residence or principal place of business in the state of arrest;
  (b) if the claim arose in the country in which the arrest was made;
  (c) if the claim concerns the voyage of the ship during which the arrest was made;
  (d) if the claim arose out of a collision, and
  (e) if the claim is for salvage.

5–36 If the claim is not within one of these, Article 7 leaves the question to the domestic law of the court where the arrest was

---

[34] Article 2, which goes on to preserve any right or power vested in any government or public authority under their existing domestic law to arrest, detain or otherwise prevent the sailing of vessels within their jurisdiction.

[35] Article 1(1). As to whether oil pollution damage is damage done by a ship, see paras. 16–06 *et seq.*, below.

[36] See para. 5–19, above.

made. Hence, if a ship goes aground and pollutes state A while on a voyage on the high seas, puts into a port in state B, and is arrested on her next voyage in state C, Article 7 leaves it to the domestic law of state C to determine whether state C, or some other state, has jurisdiction on the merits; and this would also be so if the grounding had occurred in the territorial or internal waters of state A.

5–37 The Arrest Convention does not deal with actions which are not based upon arrest of a ship. In cases of collision, the International Convention on Certain Rules Concerning Civil Jurisdiction in Matters of Collision, 1952 ("the Collision Convention") does deal with such actions, but it has a limited amount to say on the subject. This Convention deals with actions "for collision" and certain similar actions where those on board another ship have suffered damage even though there has been no actual collision, and so it would not be applied to the normal oil pollution claim; but it would apply where two ships, A and B, have collided, and A sues B for an indemnity for any oil pollution liability A may have incurred. Also, in some states such as the United Kingdom[37] the Convention has been extended to claims for damage "arising out of " collision, and so it may be of wider oil pollution interest in such states.

5–38 The rule in the Collision Convention may be summarised thus: an action for collision can *only* be introduced before the Court:
  (a) where the defendant has his habitual residence or a place of business;
  (b) where the defendant ship or its sister ship was arrested or security has been furnished;
  (c) where the collision occurred (if within inland waters).
This Convention again requires the establishment of a link between the court and the claim. The significant point is that really the Convention operates as a restriction to a plaintiff, and in a case governed by it the plaintiff cannot sue in the state where the damage occurred (which is most likely to be his own state) unless that happens to be one of the three named, which it may very well not.

5–39 Neither the Arrest nor the Collision Convention deals in a comprehensive way with civil jurisdiction over oil pollution claims. While inevitably the Arrest Convention facilitates enforcement of a judgment given in the action based on the arrest—in that the ship or bail would normally be available in satisfaction of such a judgment—recognition and enforcement of judgments is not dealt with

---

[37] See further paras. 16–21 *et seq.*, below.

generally, and plaintiffs would be left to the rules created under *ad hoc*, mainly bilateral, treaties.[38]

5–40 In this connection, although limited to states which are members of the European Economic Community, the Brussels Convention on Jurisdiction and the Enforcement of Judgments in Civil and Commercial Matters, 1968 ("the 1968 Brussels Convention") is of great importance, because it deals comprehensively with civil jurisdiction, recognition and enforcement within the EEC states. The basic rule on jurisdiction, under Article 2, is that persons shall be sued in the courts of the contracting state in which they are domiciled; but, of great significance in oil pollution cases, a person may be sued under Article 5(3) in matters relating to tort, delict or quasi-delict in the courts for the place where the harmful event occurred. Further, under Article 6(1), a person domiciled in a contracting state may also be sued, where he is one of a number of defendants, in the courts for the place where any one of them is domiciled. Lastly, under Article 6A a court having jurisdiction under the Convention in actions relating to liability arising from the use or operation of a ship also has jurisdiction over claims for limitation of that liability. The Convention goes on to provide a comprehensive regime for the mutual recognition and enforcement of judgments by contracting states.

5–41 The only real problem with the 1968 Brussels Convention from the oil pollution viewpoint is that membership is limited to states within the EEC although it is not limited to citizens of such states, nor to acts occurring therein. But its existence serves to highlight the need for a global regime, and that is why the provisions of the 1969 Liability Convention discussed in Chapter 10[39] were so necessary: oil pollution in Europe is caused not only by those domiciled in the EEC!

5–42 None of the Conventions discussed above deals with two other problems of interest to the oil pollution claimant. One is the question of the law which governs his claim. There is no international regime on this subject, and so, again, the adoption of the 1969 Liability Convention provided the first unification of law on that issue by providing the same rules on claims thereunder for all contracting states thereto. The other is the question of the geographical scope of damage which may be claimed. This was, as we shall see,[40] a controversial issue at the 1984 Conference to amend the 1969 Liability Convention.

---

[38] See, *e.g.* the long list of treaties in Article 55 of the 1968 Brussels Convention.
[39] See paras. 10–84 *et seq.*, below.
[40] Paras. 10–136, below.

5–43 This problem manifests itself in two forms. The first is whether a state can allow a claim for damage suffered within the jurisdiction of another state. This will very often be resolved affirmatively if the state seized with the case has jurisdiction on the merits. The second is whether a state can allow a claim for damage suffered on the high seas.

5–44 Article 2 of the High Seas Convention 1958, embodying customary international law, provides that no state may validly purport to subject any part of the high seas to its sovereignty. But this has never been taken by states to preclude them from hearing cases for collisions taking place on the high seas and there is no conceptual reason why, if collision damage suffered on the high seas can be claimed in a collision case, oil pollution damage suffered on the high seas resulting from an incident of navigation should not be claimed. Hence, it is perhaps a little surprising that at the 1969 Conference which adopted the Liability Convention, the right to recover thereunder was limited to pollution damage suffered on the territory or in the territorial sea of a contracting state and to measures, wherever taken, to prevent such damage.[41] It is equally surprising that when it came in 1984 to extending this rule to the exclusive economic zone, the battle was so keenly fought.[42] It may be surmised that what states were really worried about in 1969 was validating claims for fishing damage too far out to sea, thereby increasing the exposure of owners; and that, in addition to this, in 1984 they were worried about contributing to the status of the exclusive economic zone as a concept known to international law. These fears, and the form of the 1969 and 1984 Liability Conventions, must not be taken to deny the state the appropriate civil jurisdiction over damage suffered on the high seas, where it has otherwise got jurisdiction on the merits.

C. Conclusion on the Current Regime

5–45 The enforcement machinery adopted in the Conventions discussed in Chapters 3 and 4 provides for inspection of certificates, inspection to see if ships are in the proper condition, and for limited powers of detention if the certificates are not present or if the ship is not in a proper condition.[43] Under MARPOL 73/78, there is scope

---

[41] See paras. 10–34 *et seq.*, below.
[42] See paras. 10–31 *et seq.*, below.
[43] See, *e.g.* paras. 3–97 *et seq.*, for the oil pollution Conventions, 4–22 *et seq.*, for the 1978 Standards of Training and Watchkeeping Convention, paras. 4–38 *et seq.*, for the Collision Regulations and paras. 4–58 *et seq.*, for SOLAS 74/78 design and equipment standards.

for a wider role for port and coastal states once international law alters the basis of jurisdiction generally. But in 1973 (and earlier) many states had begun to feel that the system of flag state jurisdiction, with limited powers for coastal states in their territorial seas, was inadequate for the purpose of protecting their coasts from pollution from ships, particularly from oil pollution.[44] This might not have been so if all flag states had pursued a more rigorous policy of enforcing and implementing the now very complex, detailed and numerous rules in the relevant Conventions. Given this situation, it is hardly surprising that when the largest number of states ever to attend an international Conference sat down at UNCLOS III in New York on December 3, 1973, with the purpose of rewriting the law of the sea which in 1958 had been negotiated in a very different political environment, many had as one of their objectives the development of wider zones of pollution control for port and coastal states and a substantial increase in their powers of control. As we will see, this was achieved, by consensus, in 1982.

3. THE EMERGENT REGIME UNDER THE LAW OF THE SEA CONVENTION 1982

A. Introduction

5–46   The Law of the Sea Convention 1982 has a special importance even today, before it has entered into force, because the Conference proceeded by consensus for almost its entire length, and would probably have adopted the final text by consensus if the United States had not caused the breakdown in that principle.[45] Furthermore, it was not the regime relating to protection of the marine environment generally or to pollution in particular which was the cause of the need finally to resort to voting, but rather the deep sea-bed mining regime.

5–47   The Convention's present importance is enhanced by a

---

[44] Some might argue that the so-called concern of some of these states was simply to bolster their argument for an exclusive economic zone, and that their real level of concern for the protection and preservation of the marine environment from oil pollution can be discerned from their record in providing the necessary reception facilities and enforcing international standards for ships under their flag.

[45] The Conference and Convention have already given rise to a voluminous literature which does not need to be cited here. A useful summary of the Convention and its proceedings is contained in the United Nations Publication, *The Law of the Sea* (1983, Sales No. E.83.V.5) which includes the Final Act of the Conference, an introduction by B. Zuleta and some remarks by T. B. Koh, the second President of the Conference.

number of other facts. The Conference was attended by the greatest number of states ever, many of whom had not been in existence as independent sovereign states at the First and Second Conferences in 1958 and 1960. Therefore, it can be said that virtually the entire family of nations had the opportunity to present their view and to advance their individual interests. The "package" nature of the final text, in the context of the comprehensive breadth of scope of the Convention, renders each individual part of it of greater importance than it would otherwise have been—all the parts are inter-related in political importance, and in general by Article 309 no reservations or exceptions may be made on ratification or accession. This enhances the significance of the fact that the Convention has already been signed by over 130 states, which, under Article 18 of the Vienna Convention on the Law of Treaties 1969, are obliged to refrain from acts which would defeat the object and purpose of the treaty until in any particular state's case its intention is made clear that it will not become a party.

5–48 The purpose and object of the Convention, in the words of its Preamble, is none other than to settle all issues relating to the law of the sea and to codify and progressively develop that law to that end. Therefore the Convention is designed to replace the 1958 Geneva Conventions entirely.[46] In practice, it is likely that the Convention will stand for a generation. With such a purpose, it is clear that the adopted text contains some rules which codify and clarify existing rules of international law, and others which are entirely new. This is particularly so in relation to the jurisdictional regime of the seas, and to the place of the protection and preservation of the marine environment within it. But the new rules are important even today because they point in the direction in which international law is likely to develop—even if the Convention were to be superseded by a new, renegotiated one or if it failed to enter into force.

5–49 But from the oil pollution control viewpoint, probably the greatest importance of the Convention lies in the very fact that in 1958, pollution was not perceived as an important subject and, as we have seen, very little attention was given to it in the Geneva Conventions. In the Law of the Sea Convention, however, pollution control and related navigational restrictions and controls are a major subject, and an entire part—Part XII—is devoted to the protection and preservation of the marine environment.

5–50 Given the history of the evolution of the text (essentially a

---

[46] See Article 311.

series of negotiating texts, each successively felt to be more likely to attract consensus than the last), the comprehensive scope of the discussions, the new ground being broken and the delicacy of the negotiating process at the Conference, it would be foolish to look for complete precision in the text. The Convention is both a Bill of Rights and a framework in which more detailed rules can be worked out. Hence, rather than attempt a detailed analysis of each relevant Article, our purpose in this chapter will be to outline the scheme of the Convention relating to state jurisdiction over oil pollution from ships, particularly in relation to competence to develop standards and competence to enforce them, so that a general picture emerges of how the Convention envisages the new international legal regime on these points. The contrast with the existing regime will then become sufficiently apparent.

B. Terminology: Port State, Coastal State, Flag State, International Standards

5–51 The Convention provides for an entirely new set of pollution control zones, both in terms of prescribing standards and enforcing them. In particular, the enforcement regime envisages concurrent jurisdiction in many zones. In this connection, "port state" will be used here to refer to the state in whose port or at whose offshore terminal a ship is voluntarily present; "coastal state" will be used to refer to a state having jurisdiction over events occurring in a zone of jurisdiction off its coast; and "flag state" will be used to refer to a state in relation to ships flying its flag or of its registry. "Port state jurisdiction", "coastal state jurisdiction" and "flag state jurisdiction" may be understood accordingly.

5–52 The Convention uses a number of different concepts having similar meaning. Examples are:

> "international rules and standards to prevent, reduce and control pollution of the marine environment from vessels" (Article 211(1)).
> "generally accepted international rules and standards established through the competent international organisation or general diplomatic conference" (Article 211(5)).
> "applicable international rules and standards, established through the competent international organisation or general diplomatic conference" (Article 218(1)).
> "laws and regulations adopted in accordance with this Convention or applicable international rules and standards for the prevention, reduction and control of pollution from vessels" (Article 220(1)).

5–53 The difference between these formulations is precisely the sort of question that space prohibits us from examining here. The point that needs highlighting is that, although the words used to express the main concept vary from Article to Article, the general emphasis is on internationally agreed and evolved standards (as opposed to ones which are unilaterally evolved) and it is in respect of these that the Convention envisages new powers of enforcement. In the text below, to avoid making the exposition too turgid, the distinction between each of the above formulations has not normally been highlighted and the phrase "international standards" has been used instead to cover them all.

C. Regulatory and Penal Jurisdiction

INTERNAL WATERS

5–54 In so far as a state's jurisdiction to prescribe standards is concerned, the regime envisaged in the Convention is virtually the same as the current one,[47] so that internal waters remain the zone in which a state has the widest powers and the main limitations are likely to be those voluntarily agreed to in the treaties.

5–55 There is, however, a dramatic change in the enforcement regime relating to internal waters. We saw that at present a state whose coastline and internal waters are threatened or damaged by a discharge violation on the high seas is powerless to take penal proceedings in respect of that violation.[48] Under the Convention, however, the state may make investigations and institute proceedings itself for a violation of international standards if the ship subsequently enters one of its ports or offshore terminals: under Article 220(1) if the violation occurred within its territorial sea or exclusive economic zone, and under Article 218(1) if the violation occurred outside those zones. If the state is not so lucky as to find the ship within its ports or offshore terminals, it may under Article 218(2) request another port state to take the proceedings, and under Article 218(3) that state must at least investigate the violation as far as practicable. Any such proceeding must be suspended on request of the coastal state where the violation occurred within its internal waters, territorial sea or exclusive economic zone.

5–56 Those provisions only apply to discharges: the regime relating to the condition of the ship is different. In so far as seaworthiness of the ship is concerned, Article 219 provides that if a ship

---

[47] See paras. 5–07 *et seq.*, above. Article 8 is very similar to Article 5 of the Territorial Sea Convention, the difference in substance being that archipelagic states are taken into account; and see also Article 211(3) for a small, new burden of publicity in certain cases.

[48] Para. 5–25, above.

within a port or offshore terminal is in violation of international standards, so as to threaten damage to the marine environment, it not only may, but it must, take measures to prevent it from sailing (as far as is practicable). Of course, not all requirements relating to the design, construction, equipment and manning of ships affect their seaworthiness (for instance, an oil tanker is not unseaworthy if it is not fitted with the required segregated ballast tanks or crude oil washing equipment): enforcement of such standards is generally reserved to the flag state (under Article 217(2)) or to the detailed enforcement provisions of the relevant Conventions like MARPOL 73/78 (see Article 237).

THE TERRITORIAL SEA AND RELATED CONCEPTS

5–57   The Convention settles many, if not all, of the uncertainties surrounding the current regime of the territorial sea, including the most difficult of them all: the breadth will be a maximum of twelve nautical miles. It develops the concept of straits used for international navigation which was the subject of Article 16(4) of the Territorial Sea Convention 1958 (herebelow "international straits") and creates a new right, called transit passage, in connection with certain of them. It also creates an entirely new concept called archipelagic waters, which have some of the characteristics of territorial sea, but which lie inside the territorial sea's baselines.

The Territorial Sea

5–58   As at present, subject to the provisions of the Convention, the sovereignty of a state extends to its territorial sea,[49] but its maximum breadth is settled at 12 miles.[50] Ships of all states enjoy the right of unhampered innocent passage through the territorial sea,[51] and "innocent", while still defined as being not prejudicial to the peace, good order or security of the state,[52] is defined in more detail in Article 19(2). In particular, passage is not innocent if the ship engages in the territorial sea in "(h) any act of wilful and serious pollution contrary to this Convention."

5–59   Although the list in Article 19(2) is not exclusive, this resolves most of the uncertainty noted above[53] about the relationship between innocent passage and acts of pollution or threatened pollution in the territorial sea. All the items listed involve some

---

[49] Article 2.
[50] Article 3.
[51] Articles 17 and 24.
[52] Article 19(1).
[53] See paras. 5–13 *et seq.*, above.

deliberate act, and hence intent becomes the distinguishing feature of passage which is not innocent.

5–60 Concerning the competence to make standards, by Article 21(1), the coastal state may adopt laws and regulations, in conformity with the provisions of the Convention and other rules of international law, to regulate innocent passage in respect of:
  (a) the safety of navigation and the regulation of maritime traffic, and
  (f) the preservation of the environment of the coastal state and the prevention, reduction and control of pollution thereof.
Most importantly, by Article 21(2) these shall not apply to the design, construction, manning or equipment of foreign ships "unless they are giving effect to generally accepted international rules or standards."[54] Article 211(4) is to similar effect as Article 21(1)(f), merely stating that the laws must not hamper innocent passage of foreign vessels,[55] and so unlike the present there might be here a basis for making discharge regulations stricter than international standards (subject to treaty obligations a state may have undertaken).

5–61 Article 22 supplements Article 21(1)(a) by going into some detail about how sea lanes and separation schemes are to be declared, and in particular the coastal state must take into account the recommendations of the competent international organisation.[56] As at present, if a service is rendered, a charge may be levied without discrimination, but not otherwise.[57]

5–62 Therefore, except possibly in relation to discharge standards where the door might be said to be left open to making stricter standards than international ones (so long as innocent passage is not hampered), the rule-making competence of states in their territorial sea is basically unchanged from the present. But, as with internal waters, there is a fundamental change in the powers of enforcement.

5–63 Articles 27(1)(a) and 28 re-enact Articles 19(1)(a) and 20 of the Territorial Sea Convention discussed above,[58] and Article 25(2) re-enacts Article 16(2) thereof,[59] so that the underlying enforcement regime remains as at present. But Article 27(5) pro-

---

[54] By Article 24(1), foreign ships must comply. The immunities of warships and other government ships operated for non-commercial purposes are the subject of Articles 29–32.
[55] And see Article 24.
[56] Note the special status given to IMO in SOLAS 74/78: see paras. 4–30 *et seq.*, above.
[57] Article 26.
[58] Paras. 5–17 and 5–19.
[59] Para. 5–15.

vides an exception to the rule that coastal states may not take any steps to board a foreign ship passing through the territorial sea in connection with an offence before the ship entered the territorial sea, in favour of the provisions of Part XII and of violations of laws adopted in accordance with Part V (on the exclusive economic zone).

5–64   Part XII makes some radical departures. The coastal state is given in respect of discharge and other violations of international standards exactly the same rights under Articles 220(1) and 218(1) to (3) as are enjoyed in respect of its internal waters.[60] Further, by Article 220(2) the coastal state is given powers to undertake physical inspection of a ship navigating in its territorial sea in respect of violations of international standards taking place therein, and it may also institute proceedings (including detention of the ship) where the evidence so warrants.

5–65   However, powers over ships navigating in the territorial sea in respect of such violations occurring in the exclusive economic zone are limited by Article 220(3) to requesting information, unless the violation resulted in a substantial discharge causing or threatening significant pollution of the marine environment, in which case Article 220(5) would permit physical inspection where the information is refused or looks untrue; and if the violation in the exclusive economic zone resulted in a discharge causing major damage or threat of major damage to the coastal state including its territorial sea and exclusive economic zone, the coastal state is given power to detain the ship and institute proceedings under Article 220(6). The provisions of Articles 220(5) and (6) are, in practice, likely only to be rarely used.

5–66   As with internal waters, then, the territorial sea is a zone both in respect of which, and in which, the coastal state has increased powers of enforcement.

Straits used for international navigation

5–67   Articles 37 to 45 in Part III of the Convention create a special legal regime in straits used for international navigation, namely those which are used for international navigation between one part of the high seas or an exclusive economic zone to another part of the high seas or an exclusive economic zone.[61] These straits are, for ease of reference, referred to below as "international straits." The concept is not new, since it has been developed from

---

[60] See para. 5–54, above.
[61] Article 37.

one contained in Article 16(4) of the Territorial Sea Convention 1958, but a new right called transit passage through international straits is created which is similar to, but distinct from, the right of innocent passage through the territorial sea.

5–68 With certain exceptions, Article 38(1) creates for all ships a right of unimpeded transit passage through international straits; transit passage is further defined in Article 38(2) to be continuous and expeditious passage in accordance with Part III from one part of the high seas or an exclusive economic zone to another part of the high seas or an exclusive economic zone. Only transit passage is specially protected in international straits, although passage which is not transit passage might still be protected as innocent passage (e.g. if the ship is heading for internal waters).[62]

5–69 Ships in transit passage must, by Article 39(2), comply with international standards on safety at sea and the Collision Regulations and on prevention and control of pollution from ships. Article 41 complements this by empowering states bordering straits to designate sea lanes and prescribe traffic separation schemes in straits where necessary to promote safe passage, after adoption by the competent international organisation.[63] Article 42(1)(a) and (b) gives powers to such states to adopt laws and regulations giving effect to the standards of Article 41 and to international standards on oil discharges, which must not have the effect of hampering transit passage to any ship. Article 43 provides that user states and states bordering a strait "should" co-operate by agreement in establishing and maintaining the necessary navigational safety aids, and for preventing pollution from ships.

5–70 The message, then, on competence to evolve standards relating to transit passage is clear: only internationally accepted standards are to be implemented, not higher, national ones, and transit passage must not be hampered.

5–71 As to enforcement, however, the Convention is not so clear. Article 233 says that the port, coastal and flag state jurisdiction regimes mentioned above do not affect the legal regime of straits used for international navigation, unless the ship has committed a violation of the Article 42(1)(a) and (b) laws and has caused or threatened major damage to the marine environment of the straits, in which case states bordering the straits may take "appropriate measures."

---

[62] Articles 38(3) and 45.
[63] Which is IMO: see para. 4–30, above.

5–72   Article 233 is ambiguously drafted, for it is not clear what it means when it says nothing, in Part XII, Sections 5 to 7, "affects" the legal regime of international straits. A plain reading would suggest that, when taken as a whole, a ship remains entitled to unimpeded transit passage even when in violation of international standards, unless the violation and its consequences are those specifically mentioned. Thus, arrest and detention under Article 220(2) would not normally apply to such a ship.[64] However, there is the alternative reading, which is that since transit passage is defined in Article 38(2) as "the exercise (in accordance with this Part) of the freedom of navigation, etc.," and since the requirements of this Part include Article 42(4) which demands compliance with Article 42(1)(*a*) and (*b*), such a ship is not in transit passage when in violation of any requirement of Part III and so the first sentence of Article 233 can offer it no protection.

5–73   The problem with this alternative view is that, although it follows the strict drafting, it has the effect of making transit passage and innocent passage virtually indistinguishable in practice, except that the former may never be altogether suspended: Article 44. An entire Part of the Convention is hardly needed just for that.

5–74   The best way, then, to resolve the ambiguity of Article 233 is to interpret it to mean that ships in transit passage are, indeed, not to be hampered *while they are actually in passage*, unless the second sentence applies, but that once the passage has ended, and the ship is no longer in transit passage, Part XII may be used. In this way, it could be rightly said that the legal regime of straits used for international navigation had not been affected.

Archipelagic waters

5–75   Part IV of the Convention deals with archipelagic States. Archipelagic waters are those enclosed by archipelagic baselines drawn by an archipelagic State, and, subject to the provisions of Part IV, the archipelagic State has sovereignty over such waters,[65] The territorial sea of such a state lies outside these baselines, and its internal waters lie between its land territory and closing lines drawn in accordance with Articles 9, 10 and 11. Archipelagic waters are a hybrid legal concept, a sort of cross between territorial sea and international straits.

5–76   Hence, by Article 52, ships of all states have the right of

---

[64] See para. 5–64, above.
[65] Articles 49(1) and 2(1).

innocent passage[66] through archipelagic waters. In addition, a new right of archipelagic sea lanes passage is created by Article 53, which is almost identical to transit passage.[67] However, whereas transit passage is specifically stated to be a right which shall not be impeded, there is a conspicuous absence of such protection from Article 53(2) in respect of archipelagic sea lanes passage. In addition to designating the centre lines of the sea lanes through which such passage is to take place, the archipelagic State may prescribe traffic separation schemes for safe passage through narrow channels of the sea lanes which conform to international standards, but only after adoption by the competent international organisation.[68]

5–77 Articles 39, 40, 42 and 44, the relevant provisions of which were discussed above,[69] relate to the duties of ships during transit passage and the law-making competence of states bordering international straits relating to transit passage; by Article 54 they are applied *mutatis mutandis* to archipelagic sea lanes passage, so that in this respect archipelagic waters are treated as if they were international straits and archipelagic sea lanes passage a right of transit passage through them.

5–78 As to enforcement, the situation is unfortunately no clearer than it is for transit passage. Part IV has no enforcement provisions, and so recourse must be had to Part XII. Article 233 does not mention archipelagic waters, although it does refer to transit passage and in particular to Article 42 which, as we have just seen, is applied *mutatis mutandis* to archipelagic sea lanes passage.

5–79 This is not the only area for interpretation. While Article 217 on flag state jurisdiction expressly covers archipelagic waters, it is unclear how Article 218(2) and (3) and Article 220 are to apply with respect to such waters.

5–80 A logical approach (which may not be the right one) would be to assimilate archipelagic waters to the territorial sea and archipelagic sea lanes passage to transit passage, for purposes of interpreting Part XII. This would give significance to the distinction made in Part IV between archipelagic sea lanes passage and innocent passage, and to the application in Article 54 of transit passage concepts to archipelagic sea lanes passage.

[66] See paras. 5–58 *et seq.*, above.
[67] See paras. 5–68 *et seq.*, above.
[68] IMO: see para. 4–30 above.
[69] Paras. 5–69 *et seq.*, above.

THE CONTIGUOUS ZONE

5–81 The Convention retains this concept in Article 33, but does not elaborate it. Article 33(1) is identical to Article 24(1) of the Territorial Sea Convention,[70] and the conclusion that "sanitary" does not cover oil pollution is here strengthened by the prolific mention of pollution elsewhere in the Convention.

THE EXCLUSIVE ECONOMIC ZONE

5–82 Probably the most revolutionary jurisdictional concept in the Convention (with the possible exception of port and coastal state jurisdiction) is that of the exclusive economic zone, the subject of Part V. The exclusive economic zone is entirely new and cannot even be considered a development of concepts already present in the 1958 Conventions. Its inspiration is not the desire for a wider area of pollution control, but is instead economic. Pollution control was, however, seen at an early stage as an appropriate added jurisdiction within the exclusive economic zone. The exclusive economic zone is a concept *sui generis*, being neither territorial sea (in which the coastal state has sovereignty) nor high seas (which are open to all states), nor any hybrid concept such as archipelagic waters.

5–83 The exclusive economic zone is an area beyond and adjacent to the territorial sea, which by Article 57 shall not extend beyond 200 nautical miles from the territorial sea baselines. The coastal state does not enjoy sovereignty over its exclusive economic zone: rather, under Article 56(1) it has certain sovereign rights for the purpose of exploring and exploiting the natural resources of the sea-bed, its subsoil and the superadjacent waters, and it has certain limited jurisdictional rights, *inter alia* with regard to the protection and preservation of the marine environment.

5–84 The pollution control jurisdiction starts from the proposition in Article 58 that all states enjoy freedom of navigation in the exclusive economic zone, subject to the restrictions provided in the Convention, and that the provisions of Articles 88 to 115—the general provisions relating to the high seas—apply to the exclusive economic zone unless incompatible with a provision of Part V. One can, therefore, see the pollution regime as one which essentially provides exceptions to the high seas regime to be discussed below.

5–85 Concerning rule-making competence, Article 211(5) gives coastal states a right to adopt laws and regulations for pollution

[70] See paras. 5–20 *et seq.*, above.

control conforming to and giving effect to international standards. Article 211(6) gives in addition an elaborately formulated right to get the competent international organisation to designate a particular, clearly-defined area where special measures are needed to protect the marine environment. Under this Article a coastal state may prescribe measures stricter than international standards, but since even then they have to conform to the international standards that the competent organisation makes applicable to special areas, this can hardly be regarded as an exception to the general rule. However, in respect of ice-covered areas within the exclusive economic zone, there is a genuine exception and the coastal state is given *carte blanche* if pollution of the marine environment could cause major harm to, or irreversible disturbance of, the ecological balance.

5–86 In relation to ice-covered areas the coastal state is also given *carte blanche* to enforce the laws adopted under Article 234, but except in relation thereto, the regime of enforcement follows a similar pattern to the one we have seen emerging in respect of other zones, only the powers are a little less strong.[71] At all times, the coastal state must exercise its rights in the exclusive economic zone with due regard to the rights and duties of other states.[72]

5–87 The regime of port state jurisdiction under Articles 218(1) and (2) applies equally to discharge violations in the exclusive economic zone as it does to internal waters or the territorial sea,[73] and Article 220(1) gives the coastal state a port state jurisdiction in respect of violations within its own exclusive economic zone.

5–88 However, the power of inspection and detention of Article 220(2) does *not* apply to vessels navigating in the exclusive economic zone, and so in this respect ships have a greater freedom in the zone than in those closer to shore. Instead, the coastal state's powers are related to the gravity of the situation. Article 220(3) gives a carefully defined power to a coastal state to request information in respect of violations of international standards in the exclusive economic zone from ships navigating therein or in its territorial sea; and in a blatant case where such a violation has caused or threatened significant pollution, the coastal state may under Article 220(5) inspect the ship if true information has not been

---

[71] Note that Article 73, which gives powerful rights of arrest and detention in the exclusive economic zone, only applies where this is in the exercise of the coastal state's sovereign rights to explore, exploit, conserve and manage the living resources of the zone. In view of the other elaborate and specifically pollution oriented provisions in the Convention, Article 73 should not be taken to cover enforcement of standards relating to oil pollution from ships.

[72] Article 56(2).

[73] See paras. 5–55 and 5–64 above.

forthcoming. Only where the violation results in major damage to the coastline or related interests of the coastal state, may the state detain the ship (Article 220(6)). These provisions apply also to the violations of the national laws for special areas established in accordance with Article 211(6).[74]

THE HIGH SEAS—FLAG STATE JURISDICTION

5–89 The high seas, which are the subject of Part VII of the Convention, are envisaged by Article 86 as all parts of the sea not included in the exclusive economic zone, the territorial sea, the internal waters or the archipelagic waters of any state. They are less extensive in area than under the present regime due to the new exclusive ecomomic zone, but are not therefore to be regarded as less important from a pollution control viewpoint. The general regime of the high seas remains very much the same as at present, although, as we shall see, there is a new departure regarding enforcement of standards.

5–90 Thus, under Article 89 as at present no state may validly purport to subject any part of the high seas to its sovereignty—they are still *res communis omnium*. Article 87 lists the freedoms of the high seas, which must be exercised by all states with due regard for the interests of other states. The cornerstone both of the rule-making jurisdiction and of the enforcement jurisdiction remains the flag state. Article 92 provides that ships shall sail under the flag of one state only and, except where this Convention or international treaties expressly provide otherwise, shall be subject to its exclusive jurisdiction on the high seas. Those express provisions otherwise are mainly familiar too: piracy; the transport of slaves, hot pursuit; but unauthorised broadcasting is a newcomer.[75]

5–91 Rule-making competence regarding administrative and technical matters including safety, construction, manning, equipment, seaworthiness and pollution is made the right and duty of the flag state under Article 94, in familiar but more detailed terms to those of Article 10 of the High Seas Convention 1958.[76] However, a significant departure is that under Article 94(5) flag states are clearly required in taking such measures to conform to international standards.[77] Another new and valuable provision is that

---

[74] Article 22(8). And see para. 5–85, above.
[75] Articles 99–111. And see Article 218(1), discussed below.
[76] See paras. 2–14 and 5–23, above.
[77] But this does not prevent flag states from prescribing higher standards, since other states do not have concurrent rights relating thereto. This is made clear specifically in relation to pollution standards by Article 211(2) which uses the phrase "at least."

flag states are required by Article 94(7) to have an enquiry whenever an incident of navigation on the high seas causes serious
damage to the marine environment. Article 211(2) backs up these
provisions with a specific duty on flag states to enact laws which
have at least the same effect as international standards.

5–92 As to enforcement, Article 97 re-enacts word-for-word
Article 11 of the High Seas Convention 1958.[78] But this must now
be interpreted also in accordance with Part XII,[79] so that effectively it does not prescribe a completely exclusive flag state jurisdiction regime. Article 218(1) entitles port states to exercise penal
jurisdiction against those on board in respect of discharges on the
high seas in violation of international standards. This is, however,
the only exception to the general rule, since Article 220 does not
apply to violations on the high seas.

5–93 Article 217 reinforces the generality of Article 97 by providing for a right and duty on flag states to ensure compliance by ships
flying their flag with international pollution standards, *irrespective of
where a violation occurs.* Hence, flag state jurisdiction is global in
scope and, in relation to the high seas, exclusive except for Article
218(1). Further, Article 217(4) adds to Article 94(7) by requiring
that the flag state must provide for immediate investigation of
alleged violations irrespective of where they occur. Under Article
217(3) it must ensure that its ships are properly certificated and
periodically inspected and by Article 217(2) it must take appropriate measures to ensure its ships are prohibited from sailing until
they can proceed to sea in compliance with international standards.

5–94 The result of these provisions is that the regime of the high
seas itself has undergone a certain shift in emphasis towards international standards being the normative rule, and a lessening of the
exclusivity of flag state jurisdiction by the introduction of a limited
but significant right of port state jurisdiction. Otherwise, there is
less change from the present than in respect of other zones of pollution control. It still remains true that flag state jurisdiction becomes
more important the further away from land one goes.

SAFEGUARDS

5–95 With the concurrent regimes of flag, port and coastal state
jurisdiction envisaged in the Convention, it was extremely important to enact safeguards both against abuse of jurisdiction and

---

[78] See paras. 5–24 to 26 above.
[79] As indicated by Article 92(1).

against double jeopardy. Accordingly, in addition to the kind of specific safeguard in Articles 218(2), 218(4) and 233 noted already, Section 7 of Part XII contains extensive and detailed safeguards applying to the powers in Part XII. Some of the powers contained in Part XII are made expressly subject to Section 7 safeguards (*e.g.* those under Article 220(1),(2) and (6)) but others, such as those in Article 218(1), are by the express drafting of the Section 7 provisions also subject to safeguards.

5–96 Among the most important of these are the duty not to endanger safety of navigation or otherwise create any hazard to a ship or to the marine environment,[80] nor to unduly delay a ship in exercising powers of enforcement.[81] Only monetary penalties may be imposed with respect to violations of international standards by foreign ships, except in a case of a wilful and serious act of pollution in the territorial sea.[82] States are liable for unlawful measures or those which are reasonably required in the light of available information.[83] However, by far the most significant, from the jurisdictional viewpoint, are the provisions which enable the flag state to pre-empt proceedings in many cases.

5–97 Article 231 places states under a duty promptly to notify the flag state of any measures it has taken against a foreign ship, and to submit to it all official reports relating thereto (although where the violation was committed in the territorial sea, this applies only to proceedings). This is designed to ensure that the flag state knows what is going on and takes action itself. Article 228 enables the flag state to take corresponding proceedings within six months, where the violation was committed beyond the territorial sea of the coastal state which has instituted proceedings, in which case the coastal state's proceedings shall be suspended. There is, however, an exception where the case involves major damage to the coastal state[84] and also where "the flag State in question has repeatedly disregarded its obligation to enforce effectively the applicable international rules and standards in respect of violations committed by its vessels." The former exception is unlikely to be used much in practice; the latter reflects the whole philosophy of Part XII—that flag state jurisdiction has in some cases failed adequately to protect the interests of coastal states.

---

[80] Article 225.
[81] Article 226.
[82] Article 230.
[83] Article 232.
[84] *Cf.* Article 220(6); para. 5–65, above.

D. Civil Jurisdiction

5–98 As with the 1958 Conventions, the Law of the Sea Convention 1982 does not really create any framework for civil jurisdiction. Article 229 simply says that nothing in the Convention affects civil proceedings in respect of any claim for loss or damage resulting from pollution of the marine environment, and Article 28 re-enacts word-for-word the provisions of Article 20 of the Territorial Sea Convention 1958.[85] Hence, there is really no change from the present position.[86]

E. Conclusion

5–99 The sometimes vague and ambiguous drafting of the Convention, and its detailed rules and exceptions, reflect the process of consensus and compromise that went on from 1973 to 1982. Hence, much detail is left to states in their implementing legislation. Despite this, the Convention works a revolution in the international legal regime relating to jurisdiction over pollution offences by ships. Its twin philosophies are, on the one hand, to emphasise and strengthen standards evolved by international agreement, and on the other hand, to provide adequate exceptions to the traditional principle of flag state jurisdiction outside seas subject to the sovereignty of a state to satisfy those who feel that the principle has failed. It remains to be seen to what extent these philosophies will be implemented, and exactly how states will resolve the ambiguities.

---

[85] See para. 5–19 above.
[86] See paras. 5–30 et seq., above.

6. Intervention by a Coastal State

SUMMARY

6–01   In this chapter a brief analysis is made of the basis on which
a coastal state may intervene in the case of an oil pollution casualty
and direct the action to be taken with respect to private property
and interests. After dealing with the position under customary
international law, two multilateral treaties are examined: the
Geneva Convention on the Territorial Sea and the Contiguous
Zone 1958, and the International Convention Relating to Intervention on the High Seas in Cases of Oil Pollution Casualties 1969.
The 1969 Convention leads to a dichotomy of international legal
regime between the territorial sea and the high seas (although this
does not amount to very much in practice) and succeeds mainly in
codifying the international law doctrine of necessity. Developments
in the Law of the Sea Convention 1982 relating to wider zones of
pollution control were discussed in Chapter 5.

1. INTRODUCTION

6–02   Ever since the *Torrey Canyon* grounded on Seven Stones reef
on March 18, 1967, states have regarded as important the question
of whether or not a coastal state threatened with oil pollution
damage may take control of the situation, even against the will of
other interested states or persons (who may include the shipowner,
charterer, cargo owner, insurers, salvors and the state of registry).
In the case of a serious casualty, where the state is most likely to
want to intervene, this question arises nowadays, almost invariably, in the context that experts employed by or acting on behalf of
the private interests (notably hull and P & I insurers and salvors)
will be involved from the earliest stage.

6–03   Since 1967, techniques of tanker salvage and the capabilities of petroleum-safe pumps have developed considerably, so that
wherever possible the vessel will be emptied of her cargo (or such
part of it at risk of leaking) into a relief tanker, and this can often be
arranged as well (if not better) by the private interests as by the
coastal state. Familiarity with the ship herself is always important,

and this is an advantage enjoyed by the private interests over the civil servants of the coastal state. Only in the more unusual cases will the operation be beyond the resources of the private interests, or will the coastal state feel it important to become involved in, or to direct, the operations being performed by the private interests.

6–04 But those unusual cases do sometimes occur, and in addition there are sometimes more run-of-the-mill cases where, despite the fact that the private interests may be doing all that can be done as well as it can be done, the state will wish to intervene itself—perhaps in order to respond to strong public pressure.

6–05 The power of intervention is, of course, open to abuse, or to well-intentioned misuse. Its extent is therefore a matter of concern to shipowners, cargo owners, their insurers and others. The exercise of such a power on the high seas is prima facie an interference with the freedom of navigation, and so it requires justification. This may be accorded by a doctrine of customary international law, or by treaty. The powers under customary international law discussed below are of more than academic interest, since it still applies in all cases except where both the coastal and flag states concerned are parties to a relevant treaty—for instance, when the United Kingdom took control of the salvage of the Greek flag tanker *Christos Bitas*, which grounded off South Wales in October 1978, and ordered that she be towed out to sea and sunk, the action could be justified pursuant to such powers, as Greece and the United Kingdom were not both party to the 1969 Intervention Convention.

2. CUSTOMARY INTERNATIONAL LAW

6–06 Under customary international law, there are two doctrines which might form the basis of a justification of intervention on the high seas by the coastal state: self-defence and necessity. These two have been distinguished by one author[1] as follows:

> "The essence of self-defence is a wrong done, a breach of a legal duty owed to the state acting in self-defence. . . . The breach of duty violates a substantive right, for example the right of territorial integrity, and gives rise to the right of self-defence. It is this precondition of delictual conduct which distinguishes self-defence from the 'right' of necessity."[2]

---

[1] D. W. Bowett, *Self-Defence in International Law* (Manchester U.P., 1958) p. 9.
[2] See note 1, above. See further R.Y. Jennings, "The Caroline and McLeod Cases" (1938) 32 A.J.I.L. 82 at p. 91 for the distinction between self-defence and self-preservation.

6–07 It is very doubtful whether a shipowner owes an international legal duty to a coastal state to take reasonable care to prevent oil pollution of its shores. As to the flag state, it was seen in Chapter 2 that the extent of its duties probably goes no further than the making of appropriate legal rules and exercising reasonable efforts to enforce them. In most cases of serious casualty, this duty will not have been breached (or its breach will only be discoverable long after the decision to intervene or not must be taken). Further, it is doubtful whether the injury suffered by the coastal state constitutes violation of a sufficiently substantive right to justify self-defence.[3] On these grounds, then, it can be concluded with reasonable surety that self-defence cannot support intervention in most, if not all, cases of actual or threatened oil pollution.

6–08 On the other hand, there may be some grounds for thinking that the doctrine of necessity could justify intervention in an oil pollution case, if the degree of necessity is evaluated by balancing the interests of the coastal state with those of the ship's state of registry.[4] The stringent conditions which must be met are well exemplified by one learned writer in this way:

> "If an imminent violation . . . can be prevented and redressed otherwise than by a violation of another state on the part of the endangered state, this latter violation is not necessary, and therefore not excused and justified. When, to give an example, a state is informed that a body of armed men is being organised on neighbouring territory for the purpose of a raid into its territory, and when the danger can be removed through an appeal to the authorities of the neighbouring country, no case of necessity has arisen. But if such an appeal is fruitless or not possible, or if there is danger in delay, a case of necessity arises, and the threatened state is justified in invading the neighbouring country and disarming the intending raiders."[5]

6–09 The same learned writer was further of opinion that necessity applied to dangers threatened by the work of nature.[6] Thus if the doctrine of necessity is to excuse an intervention, it would appear likely that the danger must be imminent, the rights or interests threatened substantial, and the sole remedy the act sought to

---

[3] L.C. Caflisch, "International Law and Ocean Pollution" [1972] *Revue Belgique de Droit International* 7 at p. 20 concludes that only an armed attack can justify self-defence, and so the doctrine cannot justify intervention in an oil pollution case.

[4] This argument is adopted by E.D. Brown, *The Legal Regime of Hydrospace* (London, 1971), pp. 144–5, following Professor B. Cheng and the authority of three arbitral awards: the *Neptune* case, J.B. Moore (1979) 4 *International Adjudications Modern Series* 372; the *Russian Indemnity* case (1912), J.B. Scott (1916) *1 Hague Court Reports* 532; and the *Oscar Chinn* case (1934) P.C.I.J. Series A/B No. 63.

[5] L. Oppenheim, *International Law* (8th ed.), (H. Lauterpacht, ed.) London, 1955, p. 298.

[6] *Ibid.* at p. 298, n. 3.

be justified; further, inherent in the concept of necessity, the measures taken must be proportionate to the importance of the threatened rights.[7]

3. THE GENEVA CONVENTION ON THE TERRITORIAL SEA AND CONTIGUOUS ZONE 1958

6–10 Prior to 1969 the only multilateral treaty under which such justification might have been sought was the Geneva Convention on the Territorial Sea and Contiguous Zone, 1958, which was discussed above in paragraphs 5–07 *et seq.*, and which entered into force on September 10, 1964. This Convention enshrines the concept of the territorial sea by providing, in Article 1(1), that "the sovereignty of a State extends, beyond its land territory and its internal waters, to a belt of sea adjacent to its coast, described as the territorial sea." But this sovereignty is not unrestricted since, by Article 1(2), it "is exercised subject to the provisions of these Articles and to other rules of international law." The right of innocent passage (Articles 14 to 23) is the most important restriction on the coastal state's sovereignty in the territorial sea, but while the coastal state must not hamper innocent passage, the right does not grant an unlimited immunity to the coastal state's laws. Article 17 provides that:

> "Foreign ships exercising the right of innocent passage shall comply with the laws and regulations enacted by the coastal state in conformity with these Articles and other rules of international law and, in particular, with such laws and regulations relating to transport and navigation."

6–11 Such laws clearly include those regulating the prevention and control of pollution, but the question still remains as to whether a law justifying intervention by the state against the wishes of the private interests involved would be "in conformity with . . . international law." Clearly, one which justified intervention in the same circumstances that would be justified by the doctrine of

---

[7] Questions which have been examined by others but which cannot find space here include: (1) whether there is merely an immunity for what is an unlawful intervention (as opposed to a *right* to intervene); (2) whether this right (or immunity) extends to taking action inside the territorial sea of another state, or whether it merely extends to action taken on the high seas; and (3) the extent to which the sovereignty of a state exists over its own territorial sea. See, *e.g.* E.D. Brown, *The Legal Regime of Hydrospace* 1971, Stevens, London, pp. 139–146; L.C. Caflisch, "International Law and Ocean Pollution" [1972] *Revue Belgique de Droit Internationale* 7; L.C. Caflisch, "Some Aspects of Oil Pollution from Merchant Ships" [1973] *4 Annales D'Etudes Internationales* 213. All three contain copious references to earlier work. See also E. D. Brown (1968) 21 *Current Legal Problems* 113–136.

necessity would be hallowed by Articles 1 and 17, but the drafting of Article 17 does not seem adventurous enough to allow of much wider an interpretation. The result appears to be that the coastal state probably does not enjoy in its territorial sea any wider rights under the Convention than are accorded by the doctrine of necessity for the high seas.

6–12   The question of whether there is a wider right to intervene in the contiguous zone beyond the territorial sea is equally doubtful. Article 24(1) provides as follows:

> "In a zone of the high seas contiguous to its territorial sea, the coastal state may exercise the control necessary to:
> (a) Prevent infringement of its customs, fiscal, immigration or sanitary regulations within its territory or territorial sea;
> (b) Punish infringement of the above regulations committed within its territory or territorial sea."

The question here, then, is simply whether oil pollution is a matter within "sanitary regulations." This has been considered by other writers.[8] Probably the better view is that "sanitary" does not cover pollution regulations justifying intervention in cases other than those justified by the doctrine of necessity, because there is insufficient to suggest that it is intended to overcome the freedom of the high seas in such cases. If it were otherwise, there might be a wider right to intervene in the contiguous zone than in the territorial sea!

6–13   The conclusion to which this reasoning leads is that the 1958 Convention does not really add anything to the doctrine of necessity, either concerning the territorial sea or concerning the contiguous zone; or at any rate it does not clearly do so. Therefore the position in 1967 at the time of the wreck of the *Torrey Canyon*, which brought all this into sharp relief, was unsatisfactory because the right to intervene in oil pollution cases was incapable of formulation with the kind of precision and certainty which states and the maritime community would like. In particular, it was difficult to say *before* intervention whether or not the intervention would be

---

[8] E. D. Brown, *The Legal Regime of Hydrospace* (1971, London) 146, suggests that action taken after an accidental spill would be within the phrase, but N. A. Wulf, "Contiguous Zones for Pollution Control" (1971) *Univ. of Miami Sea Grant Program* 133–155, points out that from the deliberations of the International Law Commission, "sanitary" was intended to be limited to disease, and that the Conference cannot be taken from the official records to have altered such intention. The same author, in "Contiguous Zones for Pollution Control" (1972) 3 J. Mar. Law. and Comm., 587, suggests that because there is a necessity to accommodate other users of the contiguous zone who have an interest in navigation, the rights of the coastal state in that zone are "limited to that which is reasonable to protect coastal state interests."

lawful. Accordingly, after a request to IMO by the United Kingdom, a draft convention on the subject was prepared.

4. THE INTERVENTION CONVENTION 1969

A. The Right to Intervene

6–14   The International Convention Relating to Intervention on the High Seas in Cases of Oil Pollution Casualties 1969 was adopted on November 28, 1969 and entered into force on May 6, 1975. The Preamble indicates that the basis of the Convention is none other than necessity:

> "The States Parties to the present Convention, CONSCIOUS of the *need* to protect the interests of their peoples against the grave consequences of a maritime casualty resulting in danger of oil pollution of sea and coastlines, CONVINCED that under these circumstances measures of an exceptional character to protect such interests might be *necessary* on the high seas and that these measures do not affect the principle of freedom of the high seas . . . ." (emphasis supplied).

6–15   The core of the Convention is contained in Article I:

> "(1) Parties to the present Convention may take such measures on the high seas as may be necessary to prevent, mitigate or eliminate grave and imminent danger to their coastline or related interests from pollution or threat of pollution of the sea by oil, following upon a maritime casualty or acts related to such a casualty, which may reasonably be expected to result in major harmful consequences."

It may be immediately seen that this provision is very close to the doctrine of necessity, although of course formulated with greater precision. The most important feature is that no power to intervene arises unless there is grave and imminent danger. Thus where the tanker grounds a very long way offshore, there will inevitably arise the problem of whether the danger is grave and imminent. It may become so after a while, of course, and this can only be a question of fact for each case: but the Convention gives no guidance itself as to precisely where the line is to be drawn.

6–16   The interest threatened is expressed widely by virtue of the definition of "related interests" in Article II(4)—they include fishing activities, tourism and "the well-being of living marine resources and of wildlife," although these interests must be "directly affected or threatened by the maritime casualty." Hence there is a power to intervene based on purely environmental grounds, and it is not necessary to show that any proprietary or pecuniary interest is threatened.

6–17   The operation of the Convention is limited to cases where a "maritime casualty" has occurred. This phrase is defined in Article II(1) to mean an incident of navigation limited to ships, and by Article II(2), "ship" excludes an installation or device engaged in the exploration and exploitation of the resources of the sea-bed and the ocean floor and the subsoil thereof. By Article I(2), warships and ships owned or operated by a state and used, for the time being, only on government non-commercial service are also excluded. However, there is nothing in the definition of ship to limit the Convention to tankers: apart from the noted exceptions, it applies to any sea-going vessel of any type whatsoever and any floating craft.

B. Geographical Scope

6–18   One of the most important limitations of the Convention is that it only applies to measures taken on the high seas. The question of whether or not an extension to cover the territorial sea of parties should be made was one which occupied considerable time at the Conference.[9] Those who opposed such an extension, led by Canada, argued that it would erode the sovereignty which under international law the coastal state enjoyed over its territorial sea, and that this would reduce the support which nations gave the Convention. Those who supported the extension, one of the most vehement being the United Kingdom, argued that it was illogical to have two separate regimes, one on the high seas where powers of intervention were to be limited, the other on the territorial sea where a state's power to intervene was virtually unhampered.[10] The proposal that the Convention should extend to a foreign ship using its right of innocent passage through a party's territorial sea[11] was heavily defeated in Committee on a roll-call vote by 24 votes to 8, with five abstentions, and there was no attempt in plenary session to reverse that decision.[12]

[9] See LEG/CONF/3, *OR* 190 (Canada), 191 (Norway), 192 (U.K.) and LEG/CONF/3/ Add. 1, *OR* 240 (Liberia); and LEG/CONF/C.1/SR.3 and 4, *OR* 285 to 296, and LEG/ CONF/C.1/SR.19, *OR* 408–410.
[10] Clearly the United Kingdom did not agree with the analysis in paras. 6–06 *et seq.*, above, where it is seen that the coastal states' rights are rather hampered by the limited nature of the doctrine of necessity.
[11] LEG/CONF/C.1/WP.2, *OR* 249.
[12] LEG/CONF/SR.5, *OR* 93, where Article I was adopted by 41 votes to none with 4 abstentions.

6–19 The illogicality remains, but whether states will take advantage of it by prescribing different sets of regulations for measures taken inside the territorial sea and those taken on the high seas is not a foregone conclusion. At the Conference the concern of those opposing the extension to the territorial sea did not appear to be so much with the practicalities of control in a real situation as with the preservation of what they saw as a cornerstone of international law, which preservation would provide them with an advantageous position in other fora (*e.g.* the United Nations Third Law of the Sea Conference) and in other situations (*e.g.* in justifying, *ex post facto*, excessive action taken in a particular case within the territorial sea). It is interesting to note that the United Kingdom, for instance, does have two sets of criteria for intervention,[13] and that within the territorial sea there is no need for the danger to be "grave and imminent"; but in practice the real difference is only likely to be brought out in a relatively minor case, where the chances of the state wanting to intervene are reduced.[14]

C. Definition of Oil

6–20 The final important limitation on the operation of the Convention is that it applies only to pollution (or a threat of pollution) "by oil": oil is defined by Article II(3) as "crude oil, fuel oil, diesel oil and lubricating oil." In the original draft Convention before the Conference, the words "by oil" were in square brackets[15] and it was an open question whether the Convention should apply to all pollution. It was the first issue to be discussed in Committee,[16] and

---

[13] Prevention of Oil Pollution Act 1971, ss. 12–16; and the Oil in Navigable Waters (Shipping Casualties) Order 1971 (S.I. 1971 No. 1736). If any ship inside U.K. territorial waters or any British ship outside thereof suffers an accident the Secretary of State may give directions, or if that is inadequate, take action himself, where in his opinion "oil from the ship will or may cause pollution on a large scale" in the United Kingdom or her territorial waters; whereas these powers may only be exercised in respect of a foreign ship outside U.K. territorial waters "in any case in which the Secretary of State is satisfied that there is a need to protect the coast of the United Kingdom or the waters in or adjacent to the United Kingdom up to the seaward limits of territorial waters against grave and imminent danger of oil pollution."

[14] A more marked distinction is provided by Spain. She has enacted the Convention verbatim into her municipal law (Law of February 28, 1976) so that the high seas test is whether there is a grave and imminent danger. Her pre-existing law, The Law of Ports 1928, only applied to the territorial sea, and so it remains unamended by the ratifying Law of February 26, 1976. The powers granted to the Comandante de Marina are very wide, and exist when a vessel is considered to be a pollution risk. The distinction between the regimes on the high seas and the territorial sea could be inadvertently created by a state adopting the Convention verbatim without considering amendment of existing law (although there is no evidence that this is the case in Spain).

[15] LEG/CONF/3, *OR* 195.

[16] LEG/CONF/C.1/SR.1–3, *OR* 275–289.

opposition to covering substances other than oil was forthcoming on a number of grounds, *e.g.* that it was more realistic and practical to limit the Convention to oil, that this would assist wide acceptance, that insufficient data on the effect of substances other than oil was available and that drawing up a list (or an exhaustive list) of such substances was impossible. As a compromise, it was agreed to limit the Convention to oil but to investigate the possibility of an additional Protocol on other substances.[17] A Working Group produced such an instrument (but called it an Additional Act) and a draft Resolution,[18] but after a debate only the latter was adopted.[19] The Resolution on International Co-operation Concerning Pollutants Other than Oil[20] recommended that IMO intensify its work on substances other than oil, and this led to the eventual adoption of a Protocol to the Convention at the 1973 Conference.[21]

6–21 The decision of the Conference to confine the Convention to oil called for a definition of oil, and this, proposed by Sweden,[22] was adopted in Committee.[23] Sweden had deliberately omitted the word "persistent" because it was believed that non-persistent oils might cause considerable damage during the time needed to remove them,[24] and for much the same reason the qualification of "heavy" for "diesel oil" was also omitted.[25] However, crude oil, fuel oil, diesel oil and lubricating oil are all normally regarded as persistent, and so the wording chosen seems to have defeated the intent to include non-persistent oils (such as gasolene).[26]

D. Exercise of the Right to Intervene

6–22 While Article I describes when the right of intervention arises, Articles III and V describe how that right shall be lawfully exercised. Article III provides that the measures taken shall be preceeded by due consultation with states or persons whose interests

---

[17] LEG/CONF/C.1/SR.3, *OR* voting on LEG/CONF/C.1/WP.1, *OR* 248.
[18] LEG/CONF/C.1/WP.8, *OR* 252.
[19] LEG/CONF/C.1/SR.18 to 19, *OR* 401 to 407.
[20] *OR* 184.
[21] Adopted on November 2, 1973, entered into force March 30, 1983. The Protocol extends the Convention to "substances other than oil," defined in Article I(2) as "(*a*) those substances enumerated in a list which shall be established by an appropriate body designated by IMO and which shall be annexed to the present Protocol, and (*b*) those other substances which are liable to create hazards to human health, to harm living resources and marine life, to damage amenities or to interfere with other legitimate uses of the sea."
[22] LEG/CONF/C.1/WP.5, *OR* 250.
[23] LEG/CONF/C.1/SR.8, *OR* 324–5 (debate); LEG/CONF/C.1/SR.19, *OR* 407–8 (vote).
[24] LEG/CONF/C.1/SR.8, *OR* 408.
[25] LEG/CONF/C.1/SR.19, *OR* 408.
[26] As to the meaning of persistent, see paras. 10–12 *et seq.*, below.

are affected, except in cases of extreme urgency. This is an important safeguard against abuse.[27] Article III also provides, as a partial
safeguard against misuse of the powers, that states "may" consult
with independent experts chosen from a list maintained by IMO
under Article IV (this list is now in existence).

6–23 More significantly, Article V provides for the degree of the
measures taken. By Article V(1), they shall be "proportionate to
the damage actual or threatened"—the so-called proportionality
principle. Hence, by Article V(2):

> "Such measures shall not go beyond what is reasonably necess-
> ary to achieve the end mentioned in Article I[28] and shall cease
> as soon as that end has been achieved; they shall not unneces-
> sarily interfere with the rights and interests of the flag state,
> third states and of any persons, physical or corporate, con-
> cerned."

Although this principle is inherent in the customary doctrine of
necessity, the Convention improves considerably thereon by enu-
merating (in Article V(3)) the considerations to be taken into
account when deciding what measures are proportionate to the
damage threatened or suffered: "(a) the extent and probability of
imminent damage if those measures are not taken; and (b) the likelihood of those measures being effective; and (c) the extent of the
damage which may be caused by such measures."

E. Compensation

6–24 Article VI makes important provision for compensation
where the measures taken are in contravention of the Convention.
This element too is probably present in the concept of necessity in
customary international law[29] and so it is not surprising to see it
included in the Convention. However, it had a lengthy consideration at the Conference and aroused considerable disagreement.[30]
An attempt by Canada[31] to link the Civil Liability Convention and

---

[27] The value of negotiations by the coastal state is illustrated by the *Wafra* case in February
1971 off South Africa; see W.K. Bissell, "Intervention on the High Seas" (1976) 7 J. Mar.
L. & Comm., 718 at p. 727.
[28] *i.e.* the prevention, mitigation or elimination of the danger.
[29] "It is a fact that in certain cases violations committed in self-preservation are not prohibited by the Law of Nations. But, nevertheless, they remain violations, may therefore be
repelled, and indemnities may be demanded for damage done" *per* L. Oppenheim (H.
Lauterpacht, ed.) *International Law* (8th ed., London, 1955) p. 298.
[30] LEG/CONF/C.1/SR.11–12, *OR* 347–358; LEG/CONF/C.1/SR.16–18., *OR* 380–401.
[31] LEG/CONF/3, *OR* 212–4.

this one failed on a roll call vote[32] by 24 votes to 10 with 4 abstentions, but unfortunately another proposal by Canada[33] that compensation be recovered by direct action in the courts rather than by inter-party conciliation and arbitration, was also rejected. The result is that the Convention contains a lengthy Annex on conciliation and arbitration and the only way for a private party who seeks compensation under Article VI to succeed is if the State of which he is a national takes up the case under Article VIII (unless, of course, the coastal state voluntarily agrees to negotiate and settle direct with such a claimant).

F. Conclusion

6–25 The Convention has relatively few shortcomings, and is much to be preferred to the customary international law, which it may be regarded as it codifying,[34] simply because it puts matters on a more certain footing: shipowners and governments alike need to know where they stand, especially when a serious casualty requires urgent measures. It is therefore to be hoped that more states enjoying long and exposed coastlines will ratify or accede to it.[35]

---

[32] LEG/CONF/C.1/SR.17, *OR* 394.

[33] Introduced at LEG/CONF/C.1/SR.18, *OR* 401, and rejected by 29 votes to 4 with 5 abstentions.

[34] See L.F.E. Goldie, "International Principles of Responsibility for Pollution" (1970) *9 Columbia J. of Transnational Law* 283 at pp. 301–303, concurring that the Convention effects a codification.

[35] A measure of the Convention's success is that the Law of the Sea Convention 1982 addresses intervention on the high seas only in Article 221, only to say that nothing in Part XII (Protection and Preservation of the Marine Environment) prejudices states, exercising customary and conventional rights.

7. Oil Pollution Response and Regional Co-operation

SUMMARY

7–01   This Chapter discusses the need for states to prepare contingency plans against the possibility of oil spills, particularly major ones, and for states both to co-operate in times of need and to harmonise their oil pollution response policies and procedures. The international legal framework is then examined and the conclusion reached that a state has a duty at the very least to warn other states of a spill which has come to its knowledge. States have in fact developed and refined this basic duty in the adoption over recent years of a large number of regional conventions dealing with pollution and pollution response, a process specifically envisaged in the Law of the Sea Convention 1982 and now greatly encouraged and assisted by the United Nations Environment Programme.

1. INTRODUCTION

THE NEED FOR CONTINGENCY PLANNING

7–02   The degree of a state's preparedness for an oil pollution incident will clearly affect the efficiency and success of the clean-up operation. For instance, if the sort of equipment, and the people with the right skill to operate it, are not readily available at the time of a spill, the oil is more likely to reach the shorelines and once there is more likely to stay longer and inflict more damage than if a well trained response team goes into action at the start. As with so many initiatives in the oil pollution field, it was the *Torrey Canyon* incident in 1967 which first alerted the international community to the dangers of exacerbating the effects of a major spill by poor response technique.[1] Since then, the implementation of national

---

[1] The General Assembly of IMO, evidencing a generally increased awareness after the *Torrey Canyon* incident, passed a number of resolutions in 1968 and 1969 recommending various forms of action on the part of governments, covering the reporting of accidents involving significant spillages of oil (Resolution A.147(ES.IV)), national arrangements for dealing with them (Resolution A.148(ES.IV)), regional co-operation (Resolution A.149(ES.IV)) and exchange of information concerning pollution incidents (Resolution A.189(VI)).

contingency plans has proceeded in a less than uniform manner. Some countries (such as the United States, whose plan is outlined in paragraph 19–14, and the United Kingdom, whose response arrangements are outlined in Chapter 14, have developed plans appropriate to the threat they face, while others are still very exposed through lack of adequate planning.

7–03 Here is not the place to detail the general form and content of a model contingency plan, or to discuss how it may be implemented in any one country.[2] However, a few basic features are worthy of note so that the relevant legal instruments discussed below can be set in context. The most important basis for any plan is to have the right information in the right place at the right time. Hence, it is vital that an accident involving the escape of oil (or the possibility thereof), or the existence of a slick of unknown origin, be known about as soon as possible by those likely to be affected by it, and so reporting, alerting and communications functions are an essential element in any plan. The response must, of course, be co-ordinated: so often experience has been that once oil comes ashore there are too many local groups and authorities involved each of whom is not answerable to any other, and whose efforts are unco-ordinated. Sometimes a chaos of duplication and omission results. Co-ordination and administration are therefore also essential parts of any plan.

7–04 Again, experience has shown that the efforts of untrained and ill-informed personnel are very often counter-productive, if not merely ineffective, and so education and training are also essential.[3] But even the best trained personnel need the right equipment, available in the right place at the right time, and so stockpiling, equipment and logistics are also an essential element. Lastly, all these things cost money, and so funding is also of integral importance.

THE NEED FOR INTERNATIONAL CO-OPERATION AND THE ROLE OF UNEP

7–05 Where a stretch of coastline contains an international boundary, or where an expanse of sea is bordered by a number of

[2] A great deal of work on this has been done by various bodies, notably IMO's Marine Environment Protection Committee and the United Nations Environment Programme in its Regional Seas Programme. See, *e.g.* IMO Document MEPC/Circ.122 of February 14, 1984, "Guidelines for International Marine Oil Spill Contingency Plans"; D.Cormack, *Response to Oil and Chemical Marine Pollution*, (Applied Science Publishers, 1983); R. W. Hann & H. N. Young, "International Oil Spill Control Training Programme," *Proceedings of the 1981 Oil Spill Conference 2–5 March, Atlanta Georgia*.

[3] IMO has a technical assistance programme for developing countries in contingency planning and training, and interested states should apply to IMO for detailed information about this programme.

states, a single pollution incident is capable of threatening more than one state, and so there comes a need for international co-operation. In fact, such a need is not limited to just this situation: it makes eminent good sense that neighbouring states should co-operate in appropriate cases where only one state is threatened, so that the maximum resources can be brought to bear at any one time. From this need have arisen a large number of bilateral and regional Conventions on oil pollution response, the basic features and trends of which are outlined below.

7–06 The United Nations Environment Programme, UNEP, which is the United Nations' specialised agency whose role is to co-ordinate the work of the United Nations in environmental matters, has developed a Regional Seas Programme whose overall strategy is to further co-operation among governments in designated regions and to formulate and assist the adoption by governments of Action Plans. Each UNEP Action Plan has a number of components— environmental assessment, environmental management, legal agreements on co-operation and institutional and financial arrangements—and each Action Plan covers all sources of marine pollution within its sea area. As a result of this programme, a number of regional Conventions building upon earlier models have been adopted, the oil pollution response aspects of which are discussed below in paragraphs 7–16 *et seq.* Further instruments adopted under the auspices of UNEP can be expected in future.

THE ROLE OF INDUSTRY

7–07 However, it is important to note that, as elsewhere in oil pollution control, neither a national contingency plan nor a regional Convention need rely exclusively on governmental agencies or departments for their implementation. Very often, oil companies have available the very resources, in terms of skilled manpower and equipment, which are required. Further, there are two examples known to the author where oil companies formalised their own co-operation arrangements long before the governments of the same region began to take action to implement international co-operation. These are the Gulf Area Oil Companies Mutual Aid Organisation, which covers the southern part of the Arabian Gulf and dates from July 1972, with 13 founder members, and the Clean Caribbean Cooperative, which covers most of the Caribbean (but not Cuba) and had nine founder members.

7–08 The role of industry is not limited to these sort of arrange-

ments, either. Since the foundation of TOVALOP,[4] the organisation which administers it—the International Tanker Owners' Pollution Federation (ITOPF), based in London—has developed
probably more expertise in oil spill control and response than any
other organisation. It is, therefore, now called upon to assist
governments and private interests alike in a very large number of
cases, and of recent years has had some involvement in almost
every major oil spill in the world.

2. THE INTERNATIONAL LEGAL FRAMEWORK

A. Customary International Law

7–09 Customary international law is certainly capable of providing a basis for the development of international rules relating to oil
spill response. It was seen in paragraphs 2–05 *et seq.*, above, that
the *Trail Smelter* and *Corfu Channel* cases in particular can found a
duty in international law at least to warn other states which may be
affected by a spill of its existence, and possibly also to contain its
spread. It is important to note that the *Corfu Channel* case turned
very much on the fact that the Albanian Government knew about
the minefield, and it was this knowledge which gave rise to the duty
to warn. Principle 21 of the Declaration of the United Nations Conference on the Human Environment, 1972 is relatively clear on the
duty to contain, although it is not clear whether this extends to the
classic oil spill from a ship:

> "States have, in accordance with the Charter of the United Nations
> and the principles of international law . . . the responsibility to
> ensure that activities within their jurisdiction and control do not
> cause damage to the environment of other States or of areas beyond
> the limits of national jurisdiction."

The uncertainties associated with formulating any such duty[5]
would be more important if it were not the case that, as with substantive duties relating to the protection of the marine environment
generally, the process of implementation envisaged by international law is the formulation of detailed rules by agreement.

---

[4] See paras. 12–04 *et seq.*, below.
[5] See para. 2–10, above. A. L. Springer, *The International Law of Pollution*, (Westport, Conn.
& London, 1983), after an extensive review of cases and treaties states at p.144: "While a
strong case can be made that a general duty exists to warn all potentially affected parties
of any serious environmental threat, it is far less clear that the state has any further obligations unless it or persons under its jurisdiction have contributed to the problem in some
way."

7–10   This is expressed both in the High Seas Convention 1958, Article 25(2) of which provides that all states shall co-operate with the competent international organisations in taking measures for the prevention of pollution of the seas by radioactive materials or other harmful agents, and by the practice of states since 1967 in developing the bilateral and regional instruments on co-operation discussed below. However, it was not until the Law of the Sea Convention 1982 that there was an instrument which set out a more specific framework.

B. The Law of the Sea Convention 1982

7–11   Article 194(1) provides that states shall take measures necessary to prevent, reduce and control pollution of the marine environment from any source using the best practicable means at their disposal, and shall endeavour to harmonise their policies in this connection. Article 194(2) picks up the theme of Principle 21 of the 1972 Declaration and provides that states shall take all measures necessary to ensure that activities under their jurisdiction or control are so conducted as not to cause damage by pollution to other states and their environment; but as with the 1972 Conference text of Principle 21, it does not say specifically that such activities are envisaged as including oil spill response activities. However, from these two Articles may be derived the main elements of the regime on response and co-operation envisaged in the Convention: not only a general duty to prevent and control, but a duty to harmonise policies and not to allow the spread of pollution already in existence.

7–12   These important, but general duties, are supplemented and made more specific by Section 2 of Part XII on Global and Regional Co-operation. Article 197 elaborates the principle that detailed standards are to be evolved by international agreement:

> "States shall co-operate on a global basis and, as appropriate, on a regional basis, directly or through competent international organisations, in formulating and elaborating international rules, standards and recommended practices and procedures consistent with this Convention, for the protection and preservation of the marine environment, taking into account characteristic regional features."

7–13   Article 198 is more specific by laying down a particular duty of notification in addition to the general duty in Article 194(2) noted above:

> "When a State becomes aware of cases in which the marine environment is in imminent danger of being damaged or has been damaged

by pollution, it shall immediately notify other States it deems likely to be affected by such damage, as well as the competent international organisations."

7–14   And Article 199 elaborates the general duty in Article 194(1) concerning co-operation and harmonisation of policies by providing that:

"In the cases referred to in article 198, States in the area affected, in accordance with their capabilities, and the competent international organizations shall co-operate, to the extent possible, in eliminating the effects of pollution or minimising the damage. To this end, States shall jointly develop and promote contingency plans for responding to pollution incidents in the marine environment."

7–15   Although each of these provisions contains language which mitigates its rigour, there is a collective message conveyed which, so far as it goes, is clear and unambiguous: states must notify others of impending danger, and they must at least co-operate in oil pollution response and harmonise their policies, using international agreement to elaborate detailed standards. It is no coincidence that, during the long period of negotiation of this Convention, this last is precisely what states had been doing.

3. SPECIFIC CONVENTIONS

7–16   A large number of bilateral[6] and multilateral agreements have been concluded since the *Torrey Canyon* went on the Seven Stones Reef in 1967; in addition, a certain amount of work has been done by the European Economic Community.[7] The following multilateral regional instruments have been adopted to date.[8]

---

[6] See, *e.g.* the Agreement between the United States of America and Canada on Great Lakes Water Quality of April 15, 1972 (23 *UST* 301, *TIAS* 7312); the Agreement between the same two states on the establishment of joint pollution contingency plans for spills of oil and other noxious substances of 19 June 1974 (*TIAS* 7861) and August 30, 1977 (*TIAS* 8957); and the Agreement of Co-operation between the United States of America and the United Mexican States Regarding Pollution of the Marine Environment by Discharges of Hyrocarbons and Other Hazardous Substances of April 24, 1980 (20 *ILM* 696) which entered into force on March 31, 1981.

[7] The European Economic Community has done a certain amount of work in the field of regional co-operation on enforcement of standards and pollution response. See, *e.g.* EEC Council Directive No.79/116 of December 21, 1978 and Directive No.79/1034 of December 3, 1979; Decision of December 3, 1981 establishing a Community information system for the control and reduction of pollution caused by hydrocarbons discharged at sea, and Decision of September 27, 1983 on drawing up of contingency plans to combat accidental oil spills at sea. See also Cremona, "The Role of the EEC in the Control of Oil Pollution," (1980) 17 *Common Market Law Review*, 171.

[8] And see generally for comment J. A. de Yturriaga, "Regional Conventions on the Protection of the Marine Environment," 162 *Recueil des Cours*, 1979–I, 319.

1969 The Agreement for Co-operation in dealing with Pollution of the North Sea by Oil (the 1969 Bonn Agreement); this Agreement will be superseded by the 1983 Bonn Agreement mentioned below when the latter enters into force.

1971 The Copenhagen Agreement on Co-operation in Taking Measures against Pollution of the Sea by Oil.

1973 The International Convention for the Prevention of Pollution by Ships.

1974 The Convention on the Protection of the Marine Environment of the Baltic Sea.

1976 The Convention for the Protection of the Mediterranean Sea against Pollution, and the Protocol thereto concerning Co-operation in Combating Pollution by Oil and Other Harmful Substances in Cases of Emergency.

1978 The Kuwait Regional Convention for Co-operation on the Protection of the Marine Environment from Pollution, and the Protocol thereto concerning Regional Co-operation in Combating Pollution by Oil and Other Harmful Substances in Cases of Emergency.

1981 The Convention for Co-operation in the Protection and Development of the Marine and Coastal Environment of the West and Central African Region, and the Protocol thereto concerning Co-operation in Combating Pollution in Cases of Emergency.

1981 The Convention for the Protection of the Marine Environment and Coastal Area of the South-East Pacific, and the Agreement on Regional Co-operation in Combating Pollution by Hydrocarbons or Other Harmful Substances in Cases of Emergency.

1982 The Regional Convention for the Conservation of the Red Sea and Gulf of Aden Environment, and the Protocol thereto concerning Regional Co-operation in Combating Pollution by Oil and Other Harmful Substances in Cases of Emergency.

1983 The Agreement for Co-operation in dealing with Pollution of the North Sea by Oil and Other Harmful Substances (the 1983 Bonn Agreement). The Convention for the Protection and Development of the Marine Environment of the Wider Caribbean Region, and the Protocol thereto concerning Co-operation in Combating Oil Spills.

7–17 All the instruments adopted after 1974 were developed as part of UNEP's Regional Seas Programme, and so not surprisingly they bear certain family likenesses. It is not, therefore, necessary to elaborate the provisions of every one, but instead certain of them will be chosen to enable the relevant themes to be observed.

A. The 1969 and 1983 Bonn Agreements

7–18 The parties to the Agreement for Co-operation in dealing with Pollution of the North Sea by Oil 1969 (the 1969 Bonn Agreement), which came into force on August 9, 1969 are all the North Sea coastline states, and the area of sea covered is the whole of the North Sea and the English Channel. By Article I the Agreement applies "whenever the presence or the prospective presence of oil polluting the sea . . . presents a grave and imminent danger to the coast or related interests of one or more Contracting Parties." Thus it does not apply to small slicks or spills or the prospect thereof, for these would not present a "grave" danger to the coasts or other interests; but it seems that fishing grounds and offshore installations would be within "related interests," and so serious danger to them is covered.

7–19 Each party undertakes to inform any other contracting party whenever it becomes aware of a casualty or the presence of oil slicks to which the Agreement applies (Article 5). By Article 5(2) there is a duty on parties to request[9] masters of their flag ships and registered aircraft to report to them all casualties causing or likely to cause marine oil pollution, and the presence, nature and extent of all slicks to which the Agreement applies, but there is no commitment by parties to undertake any systematic patrols.

7–20 Once a pollution situation, or a potential one, is discovered, Articles 6(1) to (4) apply to regulate the initial responsibility for assessment and observation. The North Sea and the Channel are divided into zones of responsibility, thus avoiding duplication of effort and providing the initial framework for a single command structure to be set up. In addition, once the situation is discovered, the state in whose zone the danger lies comes under a duty to keep it under observation. By Article 7, a party requiring assistance in disposal of oil floating on the surface of the sea or polluting its coast may call on the other contracting parties for help, starting with those which also seem likely to be affected, and parties so requested must use their best endeavours to bring such assistance as is within their power.

7–21 On September 13, 1983, in Bonn, parties to the 1969 Bonn Agreement and the European Economic Community (EEC) con-

---

[9] *Cf.* IMO Resolution A.147(ES.IV): " . . . Recommends to governments that they (a) *require* masters of all ships to report . . . "

cluded the Agreement for Co-operation in dealing with Pollution of the North Sea by Oil and Other Harmful Substances, which, when it enters into force, will supersede the 1969 Bonn Agreement. As the name of the 1983 Agreement indicates, it is no longer limited to oil but extends to other harmful substances also. The other main improvement on the 1969 Agreement is that the new Article 9 deals with who pays for assistance, whereas the original Agreement is silent on this. As would be expected, it is the requesting state which must bear the cost unless the assisting state took the action on its own initiative. Costs shall be calculated according to the law and current practice of the assisting state. As with other regional agreements discussed below, the parties agree to meet together regularly to supervise implementation of the agreement, review the effectiveness of measures taken and carry out such other functions as may be necessary. The 1983 Agreement does not, however enter into force until two months after all signatory states and the EEC have become parties. As at May 13, 1985, this had not occurred.

B. The 1971 Copenhagen Agreement

7–22 The Copenhagen Agreement on Co-operation in Taking Measures against Pollution of the Sea by Oil 1971, replacing an earlier Agreement on the same subject of December 8, 1967, is made between Norway, Denmark, Finland and Sweden, and is clearly inspired by the 1969 Bonn Agreement; indeed, its Preamble recites that it takes that Agreement into account. It came into force on October 16, 1971.

7–23 Like the 1969 Bonn Agreement, the parties undertake to inform each other of the sighting of any significant oil slick which may drift towards a party's territory (Article 1), and a threatened state may request help from another, who must do what is possible to render such assistance (Article 3). But no area of sea is divided into zones as in the 1969 Bonn Agreement, probably because the parties do not enclose an area of sea. This omission is effectively superseded in respect of the Baltic by the 1974 Convention on the Baltic Sea Environment, which divides the Baltic into surveillance zones.

7–24 In addition to providing for the exchange of information on various topics (Article 8) and for planning co-operation (Article 9), the Agreement places a duty on parties to maintain stocks of slick-fighting equipment (Article 4). This is an addition to the obligations contained in the 1969 Bonn Agreement, which contains no such provision. Another important difference is that Article 7 of this Agreement provides for co-operation in enforcing international regulations.

> "The Contracting States shall render assistance to each other in the investigation of offences against the regulations concerning pollution by oil which are presumed to have been committed within the territorial or adjacent waters of the Contracting States. Such assistance may include inspection of the oil record book, the ship's official log-book and the engine-room log, the taking of oil samples and so on."

This power to take oil samples (presumably limited to ships registered in a contracting state), is particularly interesting because it provided an important element lacking in the enforcement of preventive standards at the time.[10]

C. The 1973 Convention

7–25 The 1973 International Convention for the Prevention of Pollution by Ships was amended by the Protocol of 1978 relating thereto, and is known as MARPOL 73/78. It entered into force on October 2, 1983. Article 6(1) contains a general duty to co-operate in the detection of violations and in the enforcement of the provisions of the Convention. Article 8 and Protocol 1 lay down a more detailed set of duties relating to notification of a pollution incident. Notable differences from the 1969 Bonn Agreement are that the parties undertake to issue "instructions" to their maritime inspection vessels and aircraft and to other appropriate services to report pollution incidents (Article 8(4)), and masters or other persons having charge of ships[11] must report the particulars of such incidents in accordance with Protocol 1. While elaborate details of the nature of the report to be made are given in Protocol 1, the duty on states to co-operate with each other is limited to notifying the ship's flag state, and "any other State which may be affected" (Article 8(3)).

D. The 1974 Baltic Convention[12]

7–26 Annex VI of the Convention on the Protection of the Marine Environment of the Baltic Sea Area 1974 contains pro-

---

[10] See para. 3–110, above, for the current position.

[11] However, it is not made absolutely clear to which contracting state a master must make his report. Article 1(2) of Protocol 1 does not specify which of the named persons—owner, charterer, manager, operator or their agents—shall execute the master's duty in the event of an abandonment: are all of them to be charged with the duty?

[12] For comment see Boczek, "International Protection of the Baltic Sea Environment Against Pollution: A Study in Marine Regionalism," (1978) 72 *American Journal of International Law*, 782.

visions binding as between the parties (all the Baltic Sea coastal states) which are almost completely identical to the combined provisions of the 1969 Bonn Agreement and the above mentioned parts of the 1973 Convention. However, one important difference must be noted: in the Baltic Sea Convention the parties have undertaken to conduct regular surveillance patrols—see Regulation 3— because of the very special danger that the Baltic Sea is in.

E. The 1976 Convention for the Protection of the Mediterranean Sea against Pollution and the Protocol thereto Concerning Co-operation in Combating Pollution

7–27   This Convention, with its Protocols, was adopted at a Conference in Barcelona on February 16, 1976, and is the first to have been adopted under the auspices of the UNEP Regional Seas Programme. The Convention and Protocols entered into force on February 12, 1978 and the parties comprise nearly all the states bordering the Mediterranean Sea.

7–28   The Convention itself is a framework document in which the parties undertake general obligations to each other. Article 3 envisages that they will enter into bilateral, regional and subregional agreements for the protection of the marine environment of the Mediterranean Sea, in conformity with the Convention and international law. Article 4 provides for joint or individual combating of pollution, and Article 6 deals specifically with pollution from ships, providing for the effective implementation of rules which are generally recognised at the international level. Article 8 provides specifically for co-operation in taking measures to deal with pollution emergencies. But it is the Protocols which go into detail, the relevant one here being the Protocol concerning Co-operation in Combating Pollution of the Mediterranean Sea by Oil and Other Harmful Substances in Cases of Emergency.

7–29   The Convention and Protocol deal with the whole Mediterranean Sea, the largest area yet to be covered by an agreement of this type. So, in addition to making general undertakings to maintain and promote combat facilities and to develop and apply monitoring activities, the parties agree to establish regional centres, through which all information is to be disseminated (Article 6 of the Protocol). Calls for assistance may be channelled through these centres, and in certain circumstances a centre may co-ordinate an operation (Article 10 of the Protocol). The Regional Oil Combating Centre has now been established on Manoel Island, Valetta, Malta since December 1976. Its role is to facilitate co-operation between the parties in emergencies and to assist them in developing their own anti-pollution capability. This somewhat limited remit could usefully be extended to inspection and monitoring of compliance with generally accepted international pollution standards, incident review and an expanded training role.

7–30   Like the 1969 Bonn Agreement, Article 1 of the Protocol places upon parties a duty to co-operate with each other in cases of grave and imminent danger to a party's coast or related interests, but in addition to supplying a definition of related interests, it is expressed to apply if the danger is to "the marine environment" of a party. The danger must, however, be presented by massive quantities of oil or other harmful substances, so the Protocol is more restrictive in scope than the 1969 Bonn Agreement.

7–31   Article 9 of the Protocol provides that a party faced with a grave and imminent danger described in Article 1 shall, *inter alia*, "take every practicable measure to avoid or reduce the effects of pollution," and shall notify all the other parties either directly or through the Regional Centre of the assessment made of the situation and the action it has taken and intends to take. It shall continue to monitor the situation and report on it. Hence, there are here duties not only to warn but also to monitor and contain.

7–32   By Article 10 of the Protocol, a party may call for assistance if its coast is threatened in other circumstances, and Parties shall use their best endeavours to render the assistance so requested. The last point worthy of note is that the parties undertake to issue instructions to their ships and aircraft to report polluting accidents and the presence of spills.

7–33   It can therefore be seen that the Convention and Protocol represent a development of ideas contained in previous Conventions, such development generally consisting of a widening of the ambit of the undertakings given, and the setting up of a communications centre. These features have been further developed in certain subsequent UNEP Conventions.

F. The 1978 Kuwait Regional Convention for Co-operation on the Protection of the Marine Environment from Pollution and the Protocol thereto Concerning Co-operation in Combating Pollution

7–34   This Convention and the Protocol on Co-operation were adopted at a Conference in Kuwait on April 23, 1978 and entered into force on July 1, 1979, upon ratification by all parties. The waters covered can be broadly described as those of the Arabian Gulf, and the parties are the coastal states thereof. Like the 1976 Convention on the Mediterranean, this Convention and its Protocols were adopted under the auspices of the UNEP Regional Seas Programme, and so it is not surprising to find that the form and substance of the legal instruments is similar thereto.

7–35 While many of the essential provisions are therefore familiar, this text is a development of the 1976 text in a number of significant respects. Hence, there is the obligation to co-operate in preventing and combating pollution of the marine environment and to endeavour to harmonise policy for this purpose; but Article IX(a) of the Convention, in dealing with the taking of necessary measures to deal with emergencies, specifies that states should ensure that adequate equipment and qualified personnel are readily available to deal with pollution emergencies. In giving this obligation further elaboration, the Protocol requires parties to co-operate in taking necessary and effective measures in the case of a marine emergency, and defines marine emergency less restrictively than in the 1976 Convention, the essence being an incident resulting in substantial pollution or imminent threat of substantial pollution to the marine environment. Further, the obligation to harmonise policy is given more detailed elaboration in Article XII of the Protocol, to include distribution and allocation of stocks, training of personnel, surveillance and monitoring and other matters.

7–36 Article III of the Protocol provides for the establishment of the Marine Emergency Mutual Aid Centre (known as MEMAC), which now has its base in Bahrain. MEMAC's role is a little more widely defined than that of the Regional Oil Combating Centre set up under the 1976 Convention, and apart from liaison and assistance in communication, it includes the co-ordination of training programmes, the preparation of inventories of available personnel and equipment for marine emergency response and the preparation of reports on incidents.

7–37 Article X of the Protocol contains duties to warn other parties, and to monitor and combat spills, in terms very close to those of the 1976 Protocol. Article VII of the Protocol and Article IX(b) of the Convention contain provisions in familiar form on notification of spills, both for states and their ships and aircraft. In these respects, this Convention does not make any significant improvement on the 1976 text.

G. The 1983 Convention for the Protection and Development of the Marine Environment of the Wider Caribbean Region, and the Protocol thereto Concerning Co-operation in Combating Oil Spills

7–38   This Convention and Protocol were adopted at a Conference in Cartagena de Indias, Colombia, from March 21–24, 1983, and they represent the latest state of the art of the UNEP Regional Conventions. As with the 1976 Convention, the provisions relevant to oil spill response are very close to those of the Kuwait Convention, including the establishment of a Regional Centre and the now familiar duties to assess the need for oil spill response, to take action and to notify other contracting states. In one or two places, however, the emphasis is different.

7–39   Article 10 of the Convention contains the concept of specially protected areas with rare or fragile ecosystems, or the habitats of depleted, threatened or endangered species, although the establishment of any such area shall not affect the rights of other contracting states or third party states. More importantly, perhaps, Article 11 of the Convention contains a specific obligation of the parties individually and jointly to develop and promote contingency plans for pollution response. Article 3 of the Protocol provides that the preparation of such plans shall be part of the parties means of maintaining the means of responding to oil spill incidents.

7–40   Although the Protocol only deals with oil spill incidents, this phrase is defined in the least restrictive terms so far, so that the duty to respond and to co-operate in response applies to a discharge (or significant threat of a discharge) of oil, however caused, of a magnitude that requires emergency action or other immediate response. Hence the test here is not whether a grave and imminent danger exists, nor is there any restriction to cases of massive pollution: instead, the test focuses on the need for the response.

H. Conclusion

7–41   The process of elaborating regional agreements on response to pollution incidents has developed since 1969 at a pace as rapid as that of other branches of the international law relating to pollution from ships. The development of the infrastructural framework necessary to ensure an efficient response is, of course, a matter of implementing all the words in these Conventions. Fortunately, the opportunity to test what infrastructural developments there have been in real emergencies comes rarely. If, however, there is a criticism to be made of the instruments adopted so far, it must be that the obligations are sometimes phrased in a manner which could make subsequent non-performance a matter of choice, and that there has been until recently an insufficient emphasis on training and contingency plan development in the texts. The establishment of regional centres for spill response is encouraging, but these centres must be given a solid role to play for the majority of the time when there are no emergencies to deal with.

8. International Aspects of Salvage and General Average

SUMMARY

8–01 Salvage—the principle of rewarding those who voluntarily come to the aid of distressed vessels and their cargoes—has been a central feature of maritime laws since time immemorial. Today, as a result of commercial and environmental pressures, the concept of salvage has been extended to rewarding those who prevent or minimize damage to the environment from vessel pollution. Known as "liability salvage," this new form of salvage is a step towards ensuring the existence of a more viable salvage industry. Whether it will provide a complete solution to the problem posed by vessels stricken while laden with oil, however, remains to be seen.

1. INTRODUCTION

A. Salvage in Oil Pollution Cases is of International Concern

8–02 The decision to present salvage in the International Law section of this book was made in light of the growing importance which has been attached to the subject by the international community in the last ten to twenty years. Today it is obvious that whenever a vessel loaded with oil or some other type of hazardous cargo is stricken, numerous interests beyond the traditional ones of ship and cargo are threatened. There is, therefore, both a practical and legal interface between salvage law (which is still heavily influenced and, outside the United States, dominated by English law) and international conventions. In addition, the Legal Committee of IMO is now considering a new draft Salvage Convention, substantial provisions of which deal with the special case of damage to the environment.

B. Origin of Salvage and No Cure-No Pay

8–03 Salvage in the marine context has been known since at least 3 B.C., when the Rhodians compiled a comprehensive maritime code which included salvage.[1] Then, as now, salvage consisted of three basic elements: (1) property which was in danger or in imminent risk of danger; (2) voluntary action on the part of a person or persons who were under no pre-existing duty to aid the property[2]; and (3) successful salving of at least a portion of the property in danger. If a salvor could demonstrate that his efforts contained all three elements, he became entitled to a salvage award based on the concept of implied contract. In theory the contract arose between the salvor and the shipowner, and thus the salvor was said to take his award against the shipowner. In practice, however, cargo contributed to the salvor's award through general average adjustments,[3] and the salvor became free to pursue anyone into whose hands the property came.

8–04 What kinds of services constituted salvage was never precisely identified, although salvage services were recognised in numerous settings. Some of the more familiar circumstances leading to an award of salvage included: (1) towing, piloting or navigating a ship in danger to safety; (2) standing by a ship in danger; (3) getting a stranded ship afloat; (4) holding a stranded vessel in position; (5) beaching a vessel in danger of sinking; (6) raising a sunken ship; (7) bringing a derelict or wreck into safety; (8) supplying crew to a ship short of crew; (9) extinguishing a fire aboard a ship; and (10) saving a ship from an impending collision.[4] As can be seen from the foregoing list, a salvor could not claim an award for merely having come to the aid of a distressed vessel. What had to be demonstrated was that the would-be salvor's efforts had led to the actual salving of property. This requirement gave rise to the principle of "no cure-no pay." Under the rule of no cure-no pay, a salvor could not recover any award unless he actually salved some property out of which an award could be granted. This intensely practical principle achieved three goals. First, it tended to dissuade unsavoury or incompetent salvors. Secondly, it set for both courts and arbitrators an upper limit on the size of the award, since a salvor could not claim an award greater than the value of the prop-

---

[1] For a discussion of the Rhodian origins of salvage law, see 3A. M. Norris, *Benedict on Admiralty* §5, pp. 1–6 (1980). It should be pointed out, however, that some commentators have criticized the idea that the Rhodians created such a code. See R. Benedict, "The Historical Position of the Rhodian Law", 18 Yale L.J. 223 (1909).

[2] Persons who are under a pre-existing duty to save property are barred from claiming a salvage award. See *The Gregerso* [1973] 1 Q.B. 274; *The Resolute*, 168 U.S. 437 (1897).

[3] See paras. 8–55 to 8–63, below.

[4] These examples are drawn from K. McGuffie, *Kennedy's Civil Salvage* (4th ed. 1958), pp. 6–7.

erty he had salved.[5] In practice, however, awards were much lower. They tended to reflect a percentage of the salved property (known as the "arrived salved value"), and were usually set below twenty per cent.[6] The third advantage of the system was that it guaranteed salvors enforceable awards. Since salvage awards could be enforced either against the shipowner (by means of an *in personam* action), or against the salved property (through an *in rem* action), no successful salvor needed to worry about receiving an unenforceable award. For these reasons, a central feature of salvage became and remains today the doctrine of no cure-no pay.[7]

C. Effect of No Cure-No Pay in Pollution Cases where no Property is Salved

8–05 The requirement of salving property which had an arrived salved value could have particularly harsh results when a salvor undertook to act in a situation where vessel pollution was present. A particularly glaring example of this came in an incident involving two ships: the *Aegean Captain* and the *Atlantic Empress*.[8]

8–06 The incident involved the collision of two supertankers off the coast of Tobago in the Caribbean in July 1979. Although the *Aegean Captain* was towed to safety with minimal cargo loss, the *Atlantic Empress*, which was leaking badly and engulfed in flames, had to be towed 300 miles out to sea by the salvors so as to avert danger to the local coastal areas. Eleven days after the collision, the *Atlantic Empress* suffered explosions which caused the death of one salvor and the loss of valuable salvage and fire-fighting equipment. Because the salvors were working under a no cure-no pay agree-

---

[5] Although no salvage award has ever reached 100% of the value of the property salved, some judicial support exists for such an award. In the *Petition of the United States (Vessel Invincible)*, 229 F. Supp. 241, 245 (D. Or. 1963), the Court stated in dictum "that under unusual circumstances, the total value of the vessel might be awarded to the salvors."

[6] As explained in G. Gilmore & C. Black, *The Law of Admiralty* (2nd ed. 1975), §8–10, pp. 563, at one time the standard figure of recovery was set at 50 per cent. As steam replaced sail and the value of ships increased, the figure dropped to its present levels. Norris, above, note 1, contains in Appendix D a listing of American salvage awards which provides both the value of the salved property and the size of the salvor's award.

[7] Regardless of the political system in which they originate, salvage forms throughout the world are based on the principle of no cure-no pay. It is pointed out in A. Miller, "Lloyd's Standard Form of Salvage Agreement—LOF 1980: A Commentary," (1981) 12 J. Mar. L. & Com. 243 at p. 244 that the German, French, Chinese, Russian and English salvage forms all contain no cure-no pay provisions.

[8] The discussion of the *Aegean Captain-Atlantic Empress* incident which appears in paras. 8–06 to 8–07 is based on the account contained in P. Coulthard, "New Cure for Salvors?—A Comparative Analysis of the LOF 1980 and the C.M.I. Draft Salvage Convention" (1983) 14 J. Mar. L. & Com. 45 at p. 50. This excellent article was also relied on for other portions of this chapter.

ment, they received no remuneration for two weeks of extensive effort and expense.

8–07 The reason for this inequitable result was, of course, the equating of success with the salving of property having an arrived salved value. With the loss of the *Atlantic Empress* due to the explosions, no property was salved. Yet the salvors had clearly bestowed a benefit on the owners of the *Atlantic Empress*. Their actions had reduced damage to the local coast and had protected the shipowners from liability for ensuing pollution claims. To that extent there had been both benefit and success. Their voluntary efforts had also benefited the public interests ashore.

8–08 Before 1980, this position was reflected in Lloyd's Standard Form of Salvage Agreement ("Lloyd's Open Form", or "Lloyd's Form") which since 1890 was the most commonly used form of contract, and much favoured by salvors. The 1972 version of the Lloyd's Form provided in Clause 1 that the salvor was to use his best efforts to salve the ship and/or her cargo on a no cure-no pay basis. No distinction was made between types of salvage and no provision was made for the salvor to receive special compensation if he succeeded in preventing or reducing pollution damage.

8–09 Clause 15 provided that if the operations were "only partially successful," without any negligence or lack of ordinary skill and care on the part of the salvor, the salvor was entitled to receive a "reasonable remuneration." Clause 1 provided that in the event of success the salvor's remuneration would be set by arbitration in London. The amount of the salvor's remuneration if he was only partially successful was also to be determined by London arbitration.

8–10 Although the form did not attempt to detail how the arbitrators were to arrive at a remuneration amount that was reasonable, it was understood that the ordinary principles of salvage law were to apply. Thus, if the salvor undertook to salve both the ship and her cargo, but was only able to salve the cargo, his remuneration would be assessed on the merit of his services as a whole. His award would come out of the salved value of the cargo, however, since it was the only salved value against which an award could be made.

8–11 Accordingly, in a case where the salvor incurred substantial expenditure in using his best efforts to salve the ship and cargo but both were lost, the salvor could not receive any remuneration. This was true even if, as in the case of the *Atlantic Empress*, the salvor had conferred a benefit upon the underwriters responsible for pollution

damage claims, and the coastal state and its landowners by preventing or reducing pollution damage.[9]

D. Oil Pollution creates Other Problems for Salvors

8–12 Besides having to contend with the obstacles posed by a no
cure-no pay clause, a salvor acting in a situation where vessel pollution either might or had occurred had to contend with three other
factors which lessened his chances of receiving adequate compensation for his efforts.[10]

8–13 First, as the damage to the vessel increased, the size of a
potential spill increased. This in turn increased the threat of oil pollution, and reduced the potential salvable cargo (and hence the
potential award). Secondly, the existence of an oil spill rendered a
salvage operation physically and politically more difficult. The
operation became physically more difficult because of the attendant
dirt and smell. It became politically harder because an oil spill
which threatened a coastline tended to attract the attention of the
government whose coast was threatened. That government sometimes felt constrained to intervene to prevent or minimize the pollution.[11] This often involved giving directives to the salvor which
interfered with the successful salvage of the vessel.

8–14 Just how vexing this interference could become was illustrated by the so-called "Flying Dutchman" problem. This phenomenon occurred whenever a stricken tanker under tow was unable
to find a port willing to admit her for fear of the risk of pollution.
Even if admitted, the terms and conditions of admittance were
often so onerous as seriously to reduce the potential profit that the
salvor was likely to realize. In an extreme case, the salvor could be
totally deprived of his award by a government decree. In the *Christos Bitas* incident, off South Wales and Ireland in October 1978 for
example, the tanker was towed out to sea and sunk on the orders of
the British Government.

8–15 The third problem associated with the salving of a tanker
laden with oil stemmed from the fact that the salvor rarely received
more than half the value of the property saved as a salvage reward.
This limit could lead to a reluctance on the part of a professional
salvor to undertake the salvage of a tanker, which, together with

---

[9] *Ibid.* at p. 51.
[10] The discussion in paras. 8–12 to 8–14 of the problems which oil pollution creates for salvors is based on A. Bessemer-Clark, "The Role of Lloyd's Open Form" [1980]
L.M.C.L.Q. 297 at p. 298.
[11] See generally Chap. 6 for a discussion of the powers of intervention enjoyed by States.

the freight at risk and the cargo, was worth less money than might be needed in order to salve her successfully. The fact that oil cargoes are now very often worth more than the ship (a likelihood which increases with the deadweight capacity of the ship), due to the high price of oil and the currently low market values of oil tankers, does not in fact relieve this situation. If the cargo is lost or partly lost in the salvage process, the salvor will be unable to benefit to the extent of the lost cargo because the size of the reward is based upon the value of property actually saved. This problem will, however, be more apparent than real in many cases because salvors are used to taking risks. But where there is also a risk of oil pollution damage being caused, these considerations may tip the balance in favour of refusing the opportunity to salve.

E. The Concept of Liability Salvage and the Problems of Awarding It

8–16 A number of responses in the private law of salvage to the problems discussed above have been pursued by salvors in the past. The most ambitious one—least liked by shipowners and their insurers—can be called "liability salvage." Under this concept, the owner's potential oil pollution liabilities become the object of salvage in themselves, just like the ship and cargo, so that if they are saved, a salvage award is made in respect of them. In line with the insurance offered by hull underwriters for ship's contribution to salvage and with the insurance offered by cargo underwriters for cargo's contribution to salvage, according to this theory the shipowner's liability underwriter—normally his P. & I. Club—would cover this risk. A classic example of this full form of liability salvage proposal can be illustrated as follows.

8–17 A full laden tanker suffers a steering failure such as the one which was suffered by the *Amoco Cadiz*. She is very close to the rocky shore, and is drifting with the wind and tide towards the shore. A salvage tug takes her in tow and succeeds in towing her to a place of safety. No oil is spilled, and no expenses are incurred apart from the salvor's. Under the normal law of salvage, the salvor gets an award for salving the ship and her even more valuable cargo. Under the full liability salvage proposal, the salvor also gets an award because no oil pollution liability has resulted; and the same is true if oil pollution liability had been merely reduced, and not eradicated. The proposal requires that arbitrators have some means to identify what potential oil pollution liabilities are involved, and how they have been affected by the salvor's actions. However, if no oil pollution liability is saved, despite perhaps considerable effort by the salvor, he is left solely to his traditional award—there is no cure-no pay in pure liability salvage too! The authors are not aware of any real case where a salvor has tried, in an arbitration, to advance this concept with legal argument. It is a proposal made purely *de lege ferenda*.

8–18 A different proposal, which has been very much advanced in arbitrations, is that a salvage award made under traditional legal principles should be increased—"enhanced" is the word used—as a result of actions taken by the salvor. There are two forms of this enhancement argument. One says that regard should only be had to whether or not liabilities have been abated, and in this form the argument is clearly close to, but not identical with, the pure liability salvage concept mentioned above. Another form says that regard should be had to the efforts the salvor has made to abate the owner's liabilities. Here, the reason for taking the action becomes the subject of investigation (with all its attendant problems). In both forms, however, if no property with an arrived salved value is saved, there is no scope for enhancement.

8–19 There are difficulties in establishing an enhancement argument on a legal basis. As will be seen,[12] the 1910 Salvage Convention does mention the idea of a salvage act having a "useful result" justifying "equitable remuneration," but the Convention fails to mention the owner's liabilities as being among the considerations to be taken into account when fixing the award. In the English case of *The Whippingham*,[13] potential damage to pleasure yachts caused by the salved ship was taken into account in the calculation of the award. But apart from that case there is little to help the salvor.[14] Despite this, prior to 1980, when as we shall see[15] Lloyd's Open Form was changed to reflect the second version of the enhancement argument, arbitrators in London under the old version of the Form, applying traditional salvage law concepts, were widely reported to be taking into account the salvor's efforts to prevent the escape of oil in assessing salvage awards.

8–20 To cure the problem of the enhancement proposal that there can be no enhancement when no property with an arrived salved value is saved, a third proposal has been advanced under which, in these circumstances, the salvor nonetheless gets his

---

[12] See para. 8–41 below.

[13] (1934) 48 Ll. L. Rep. 49.

[14] In B. Sheen, "Conventions on Salvage" (1983) 57 Tul. L. Rev. 1387, 1405, *The Buffalo*, (1937) 58 Lloyd's List L.R. 302 (Adm. Div.) and *The Gregerso*, above, note 2, are suggested as providing further support for the concept of liability salvage.

[15] See paras. 8–24 to 8–37, below.

expenses reimbursed by the owner. This, too, has now been incorporated in the 1980 Lloyd's Open Form, but prior thereto there was no basis in English law—whether through "unjust enrichment" or *quantum meruit* or otherwise—which could base such a claim under a no cure-no pay contract.

F. Governmental Moves following the Amoco Cadiz

8–21 The *Amoco Cadiz* incident off Brittany, France in March 1978 focussed international attention on both the need for a viable salvage industry and on the shortcomings of the existing salvage law. Some countries began calling for a vigorous application of the principle of self-protection.[16] The French Government took the lead in this matter. It expressed its position by saying that:

> " . . . it is impossible any longer to tolerate the continuation of the existing regime in which one of the parties having a direct interest in the outcome of the assistance operations is not involved in any way in negotiating, deciding on or executing such assistance operations."

8–22 Accordingly, France argued for an expansion of the powers of intervention beyond the territorial waters contained in the 1969 International Convention Relating to Intervention on the High Seas in Cases of Oil Pollution Casualties. Under that Convention, parties:

> " . . . may take such measures on the high seas as may be necessary to prevent, mitigate or eliminate grave and imminent danger to their coastal or related interests from oil pollution or threat of pollution of the sea by oil, following upon a maritime casualty or acts related to such a casualty, which may reasonably be expected to result in major harmful consequences."

Joined by Mexico and Uruguay, France also suggested to IMO that a new Convention should be adopted in which the phrase "grave and imminent" danger was omitted from the above wording, and in which the powers were not limited to prevention of pollution of the sea nor to prevention of pollution by oil.

8–23 Besides requesting greater powers of intervention, France argued in favour of allowing the coastal state to "give such directions as it deems necessary and appropriate to the ship(s) and salvor(s) concerned and those ship(s) and salvor(s) shall take all reasonable and practical steps to comply with these directions."

---

[16] The discussion in paras. 8–21 to 8–23 was originally included in D. Abecassis, "Some Topical Considerations in the Event of a Casualty to an Oil Tanker" [1979] L.M.C.L.Q. 449 at pp. 452–453.

Recognising that a salvor acting under such direction might lose his right to salvage award because his actions were not voluntary,[17] it was further suggested that in such cases the salvor would still be entitled to an award. Under the proposal the salvor could proceed against either the owner of the ship(s) involved or the state which gave the orders.

2. LLOYD'S OPEN FORM

A. Lloyd's Open Form is Amended following the Amoco Cadiz

8–24   Against this background, the commercial interests in the oil and salvage industries decided to try to head off governmental action by putting their own house more in order. The first body to call for a revision of the Lloyd's Form was the Oil Companies International Marine Forum, which began to issue proposals substantially to alter or amend the Lloyd's Open Form shortly after the *Amoco Cadiz* incident. Responding to the pressure of the Forum and others, the Committee of Lloyd's set up a working group in early 1979 to study the question of revising the Lloyd's Form. The Committee was originally formed under the chairmanship of Mr. Barry Sheen, Q.C. Later Mr. Gerald Darling, Q.C., took over the chairmanship of the Committee.

8–25   Numerous changes were proposed to the Committee. One commentator has suggested that when all of the proposals made during this period are considered, they can be grouped into three broad areas of concern.[18] The areas were:

1. The general falling off in salvor's remuneration, despite the increasing values of ship and cargo. This phenomenon was caused by arbitrators reducing their percentage awards. Salvors contended that this meant that salvage was generating insufficient capital to provide for the new vessels and equipment necessary to salve tankers of the size of the *Amoco Cadiz*;

2. The belief that the lack of financial incentive to salve the stricken tanker had led, or at least could lead in some cases, to reluctance by salvors to undertake such salvage, or at best to delay the start of such salvage while an appropriate salvage contract was agreed upon; and

---

[17] Recall above, note 2.
[18] The three areas of concern (and the discussion which appears in paras. 8–26 to 8–30) were initially formulated by Bessemer-Clark, above, note 10, at pp. 298–301.

149

3. The feeling that in the case of the successful salvage of a stricken laden tanker one of the interests which benefited most from the salvage—the oil pollution liability underwriter—was not contributing.

8–26 Most of the proposals introduced at the Committee of Lloyd's sought to deal with all three of these concerns by forcing a new interest to contribute to salvage—the owner's oil pollution liability underwriter, which is usually his P. & I. Club. In order to get the P. & I. Club to contribute what was felt to be its "fair" share to the problem, numerous proposals were suggested. The most authoritative of such proposals was the one proposed by Gerald Darling's working party. It was composed of two separate and distinct amendments.

8–27 First, in respect of all vessels, there was to be an obligation on the salvor to prevent or reduce pollution. In return, such services would be regarded as salvage services for the purpose of an award. Secondly, in the case of stricken laden tankers, there was to be an additional fund (the "Pollution Fund") out of which the salvor was to be rewarded for his services in preventing, minimizing or controlling pollution even (and this would be the exception to the general rule of no cure-no pay) if there was no property salved.

8–28 Shipowners and their clubs viewed such proposals with alarm. In particular, they considered it inappropriate for the pollution liability underwriter to be required to contribute simply to make up the apparent shortfall in salvors' awards. Nevertheless, shipowners and clubs recognized the political difficulties faced by salvors of a stricken tanker. Although they contended that on no occasion had salvors hung back or delayed from a salvage operation by reason of the supposed shortcomings of the Lloyd's Open Form, they were prepared to offer an additional reasonable inducement if agreement on what was an acceptable inducement could be reached. However, they were strongly opposed to the idea of an oil pollution fund out of which would be paid an award fixed by reference to the potential pollution that had been avoided.

8–29 Shipowners and clubs took the position that while an award payable out of a known fund (the value of the property) might be an appropriate way to reward the traditional salvage services to ship and cargo, it was wholly unsuitable when applied to pollution avoidance. Besides the difficult problem of how anti-pollution services should be valued, they argued that no rational system could be devised to calculate such a fund. A recurring question throughout the discussion was whether it was the maximum liability which the owner might have faced had the salvage failed that was to be used in calculating the fund, or only a reasonable estimate of the exposure. Both presented severe difficulties of assessment in practice.

8–30   Acordingly, shipowners and their clubs felt that the proper way to approach the problem was to follow the practice used for all other questions of liability avoidance. In other words, they suggested that salvors be reimbursed the reasonable costs and expenses of the services rendered, plus an acceptable margin for profit. Based on the foregoing, the clubs proposed that where the property to be salved was a laden tanker, there should be an exception to the principle of no cure-no pay. Under the exception, if a salvor through no fault of his own failed to earn any award, or earned an award which failed to cover his expenses, he would receive a form of "safety-net." This safety-net would consist of his reasonable expenses as well as a mark-up which would not exceed 15 per cent. The safety-net would be reduced, however, by any award recovered on any part of the property salved. The cost of the safety-net would be borne by the shipowner, and would be paid by the shipowner's club.

8–31   Although shipowners and their insurers had balked at the idea of paying salvors anything for pollution control, their eventual willingness to do so was caused by the generally shared concern that failure to do so would lead to further intrusions like the previously described *Christos Bitas* incident. The end product was very much like the alternative proposed by the P. & I. Clubs. Entitled LOF 1980, or Lloyd's Open Form 1980, it came into effect on June 23, 1980.

B. The Provisions of Lloyd's Form 1980

8–32   Clause 1(a) of the new Form provides as follows:

"1.(a) The Contractor agrees to use his best endeavours to salve the   and/or her cargo bunkers and stores and take them to   or other place to be hereafter agreed or if no place is named or agreed to a place of safety. The Contractor further agrees to use his best endeavours to prevent the escape of oil from the vessel while performing the services of salving the subject vessel and/or her cargo bunkers and stores. The services shall be rendered and accepted as salvage services upon the principle of 'no cure-no pay' except that where the property being salved is a tanker laden or partly laden with a cargo of oil and without negligence on the part of the Contractor and/or his Servants and/or Agents (1) the services are not successful or (2) are only partially successful or (3) the Contractor is prevented

151

from completing the services the Contractor shall nevertheless be awarded solely against the Owners of such tanker his reasonably incurred expenses and an increment not exceeding 15 per cent. of such expenses but only if and to the extent that such expenses together with the increment are greater than any amount otherwise recoverable under this Agreement. Within the meaning of the said exception to the principle of 'no cure-no pay' expenses shall in addition to actual out of pocket expenses include a fair rate for all tugs craft personnel and other equipment used by the Contractor in the services and oil shall mean crude oil, fuel oil, heavy diesel oil and lubricating oil.''

The new Form is now expressed (in Clause 1(d)) to be governed by English law, and so the principles of that law relating to interpretation of contracts apply to it. A number of points raised by the new drafting call for comment.

THE SALVOR'S DUTY

8–33 From the oil pollution viewpoint the most significant aspect is that the salvor must now use his best endeavours to prevent the escape of oil from the ship while performing the salvage services. This adds a new dimension to the services to be rendered (although in practice most, if not all, salvors used to do this anyway), and it applies to all types of ships, not just to tankers, whether they are laden or in ballast. The duty does not, of course, extend to taking measures to prevent pollution by oil which has already escaped, although in fact this might be done pursuant to the carrying out of the specified duty—for instance, where oil floating around the ship must be dispersed before hot work could be undertaken on the hull plating. The focus, then, of this duty is the oil remaining in the ship. This duty is complemented by the new inclusion of a duty to use best endeavours to salve the ship's bunkers.

ENHANCEMENT OF AWARD

8–34 In return, the salvor receives new rights. The most important of these is that, whatever the type of ship and whatever her laden condition, the exercise of those best endeavours is agreed to be rendered and accepted "as salvage services," and therefore these endeavours become part of the criteria by which the size of any award which may be made is calculated. The new Form therefore adopts the concept of enhancement discussed above,[19] but does not adopt the full concept of liability salvage; there is no separate "pollution fund," nor is the potential pollution liability taken into

[19] See para. 8–18, above.

account. Indeed, there is no suggestion in the wording that the success or lack of success in actually preventing pollution is relevant to the enhancement to be made: what matters is the extent to which the best endeavours were exercised. There appears to be no requirement for enhancement that the best endeavours succeed in preventing the escape of oil from the ship.

The Safety-Net

8–35 Enhancement is of no use if no property with an arrived salved value is saved, and so the new Form provides, under its so-called "safety-net" provisions, for an exception to the no cure-no pay principle in certain fairly restricted circumstances. These are as follows:

(a) The property being salved must be a *tanker laden or partly laden with a cargo of oil*. Hence, the safety net does not apply to tankers in ballast or to ships laden with any cargo other than *oil*, which is restricted in definition but is very close to the concept of persistent hydrocarbon mineral oil used in the 1984 Liability Convention,[20] and to that used in the 1969 Intervention Convention.[21] A tanker in ballast which carries oil as slops would probably not be regarded as "partly laden."

(b) There must have been no negligence on the part of the contractor and/or his servants and/or agents which contributes to any of the following occurring:
(1) The services rendered are not successful;
(2) The services rendered are only partially successful;
(3) The contractor is prevented from completing his services.

The third possibility is clearly aimed at dealing with the "Flying Dutchman" type of case, such as the *Andros Patria* off Spain in December 1978 and the *Christos Bitas* case discussed above.[22] However, it could be better phrased. While it is clear that it would cover all cases where the salvage services are frustrated by government action, whether in exercise of the powers discussed in Chapter 6 or otherwise, it is not clear to what extent it would cover other circumstances (suppose the nearest port refused entry, and while haggling goes on, bad weather sinks the ship—has the contractor "been prevented," within the meaning of the Clause, from completing his services?). The other two possibilities are fairly clear, and must refer to the unsuccesful or partially successful salving of property. In these circumstances, subject to the proviso discussed below, the

[20] See paras. 10–13 and 10–120 *et seq.*, below.
[21] See para. 6–20, above.
[22] See para. 8–14, above.

contractor *shall* nevertheless be awarded solely against the tanker's owners:

   (i) his reasonably incurred expenses—and in this connection "expenses" include out-of-pocket expenses and a fair rate for all tugs, craft, personnel and oter equipment used by the contractor *in the services, i.e.* the entire services rendered under the agreement, and not just the service of exercising best endeavours to prevent the escape of oil from the ship; plus

   (ii) an increment which is not to exceed 15 per cent. of the reasonably incurred expenses, and which by impliction should not be zero (otherwise it would not be "an increment").

However, the proviso is that these amounts can only be awarded if and to the extent that such expenses together with the increment are greater than any amount otherwise recoverable under the agreement. Effectively, then, if some property if salved (situations (c)(2) and (3) above) the award made in respect thereof is deducted from the safety-net award.

THE FLYING DUTCHMAN

8–36 Another way in which this aspect of the problem is addressed in the new Form is that now, unless the salvor agrees to take the ship to a named place, he may take her to *a place of safety*. Hence, in the classic Flying Dutchman case, once the ship reaches sheltered waters, the salvor has completed his contract and becomes entitled to an award, notwithstanding that subsequently the ship may be towed out to sea and sunk.

CONCLUSION

8–37 The proof of the new Form will be how it stands up to usage. The salvage industry is known to be disappointed that it failed to adopt the full concept of liability salvage. The shipowner's liability insurers are pleased that they have escaped an obligation to contribute to salvage awards by the compromise device of the safety-net. The drafting of the Form—which also includes some other important new provisions discussed below[23]—is not as clear as it could be in places, but is probably clear enough to work well in nine out of ten cases. There is no doubt that the Form does address the oil pollution problems discussed in the Intoduction to this Chapter, but it does not provide the basis the salvage industry was seeking for long-term capital investment in the type of tug capable

---

[23] See, for example, the discussion in para. 8–77, below, regarding the new rights of salvors to limit their liability.

of turning round a fully laden VLCC in a Force 8 gale, nor does it provide a basis for solving the industry's long-term viability problems. But then, should it be expected to by bringing in the liability insurer as a contributor to salvage awards? That is a policy question beyond the scope of further discussion here.

3. THE LEGAL RESPONSE: THE 1981 DRAFT SALVAGE CONVENTION

8–38 The Lloyd's Revised Form came into effect in June 1980. It was the product of industry reaction to the fear that if industry did not regulate itself first, far more onerous legislation would emerge from coastal states.[24] Yet the speedy revision of the Lloyd's Form did not prevent steps from being taken on the international level also to address the issue of liability salvage. The result of these steps was the preparation in 1981 of a new Draft Salvage Convention.

A. The 1910 Brussels Convention on Salvage

8–39 In order to tell the full story of the preparation of the 1981 Draft Salvage Convention, one must begin with the 1910 Brussels Convention on Salvage,[25] for it became the starting point for the discussions concerning a new salvage convention, and the final product which emerged retains much of the old convention's language and spirit.

8–40 The heart of the 1910 Convention was Articles 2 and 8, which stated that:

"2. Every act or assistance or salvage which has had a useful result gives a right to equitable remuneration.

No remuneration is due if the services rendered have no beneficial result.

In no case shall the sum to be paid exceed the value of the property salved.

8. The remuneration is fixed by the court according to the circumstances of each case, on the basis of the following considerations:
(a) Firstly, the measure of success obtained, the efforts and deserts of the salvors, the danger run by the salved vessel, by her passengers, crew and cargo, by the salvors, and by the salving vessel; the time expended, the expenses incurred and

---

[24] See paras. 8–21 to 8–24, above.
[25] For a detailed discussion of the 1910 Convention, see I. Wildeboer, *The Brussels Salvage Convention* (1965).

losses suffered, the risks of liability and other risks run by the salvors, and also the value of the property exposed to such risks, due regard being had to the special appropriation (if any) of the salvor's vessel for salvage purposes;

(b) secondly, the value of the property salved . . . ."

8–41 As can be seen, Article 2 embodies the basic principle of no cure-no pay. As to the remuneration, Article 8 mentions the danger run by the salved vessel, by her passengers, crew and cargo and by salvors and the salving vessel. It does not, however, mention the danger run by the owners of any other property as a factor to be taken into account. Again, the Convention expressly mentions the risks of liability and other risks run by the salvors but is silent as to the risks of liability and other risks to which the owners of the salved property might be subject. The Convention refers to the value of the property salved but is silent as to the manner in which the salvage remuneration shall be borne by the different salved interests. Beyond the reference in Article 2 to "useful result" and "beneficial result" and the right to "equitable remuneration" there is no further guidance as to how that equitable remuneration should be assessed and borne, and none of the other provisions of the Convention provides any firm basis either for enhancement of awards for oil pollution liabilities abated, or for liability salvage in its fullest sense.

B. Preparation of the 1981 Draft Salvage Convention

8–42 Following the *Amoco Cadiz* incident in March 1978, discussions had begun within the Legal Committee of IMO on the problems raised by incidents involving stricken tankers laden with oil or other hazardous substances. As a result of these discussions, the CMI was asked to review the principles of private salvage law without dealing with matters of coastal state intervention or the control of salvage operations by public authorities in the context of intervention. The CMI convened an International Sub-Committee under the chairmanship of Professor Erling Selvig of the Norwegian Maritime Law Association to study the subject and prepare a report. This Sub-Committee was able to do so having the concluded text of the 1980 revision of Lloyd's Open Form before it. From the work of this Sub-Committee emerged a Draft Convention, the final text of which received final approval in May 1981 when the CMI met as a whole in Montreal. This draft is now under discussion at IMO's Legal Committee.

8–43 The heart of the Draft Convention, for purposes of the present discussion, is contained in Articles 1–1(4), 2–2(1), 3–1(1) and (2), 3–2, and 3–3.

8–44   Article 2–2(1) sets out the basic duty of the salvor by stating that:

"The salvor shall use his best endeavours to salve the vessel and property and shall carry out the salvage operations with due care. In so doing the salvor shall also use his best endeavours to prevent or minimize damage to the environment."

8–45   The phrase "damage to the environment" is defined in Article 1–1(4) as meaning:

" . . . substantial physical damage to human health or to marine life or resources in coastal or inland waters or areas adjacent thereto, caused by pollution, explosion, contamination, fire or similar major incidents."

8–46   The relevant conditions for reward remain those of no cure-no pay, and are set out in Articles 3–1(1) and (2):

"(1) Salvage operations which have had a useful result give right to a reward.
    (2) Except as otherwise provided, no payment is due under this Convention if the salvage operations have no useful result."

8–47   Article 3–2 deals with the factors which shall influence the size of the award:

"(1) The reward shall be fixed with a view to encouraging salvage operations, taking into account the following considerations without regard to the order in which presented below:
    (a) the value of the property salved,
    (b) the skill and efforts of the salvors in preventing or minimizing damage to the environment,
    (c) the measure of success obtained by the salvor,
    (d) the nature and degree of the danger,
    (e) the efforts of the salvor, including the time used and expenses and losses incurred by the salvors,
    (f) the risk of liability and other risks run by the salvors or their equipment,
    (g) the promptness of the service rendered,
    (h) the availability and use of vessels or other equipment intended for salvage operations,
    (i) the state of readiness and efficiency of the salvor's equipment and the value thereof.
    (2) The reward under paragraph 1 of this Article shall not exceed the value of the property salved at the time of the completion of the salvage operation."

8–48   Finally, Article 3–3 provides for an exception to the no cure-no pay principle providing for special compensation to the salvor in certain circumstances:

"(1) If the salvor has carried out salvage operations in respect of a vessel which by itself or its cargo threatened damage to the environment and failed to earn a reward under Article 3–2 at least equivalent to the compensation assessable in accordance with Article 3–3, he shall be entitled to compensation from the owner of that vessel equivalent to his expenses as herein defined.

(2) If, in the circumstances set out in paragraph 1 of Article 3–3 hereof, the salvor by his salvage operations has prevented or minimized the damage to the environment, the compensation payable by the owner to the salvor thereunder may be increased, if and to the extent that the tribunal considers it fair and just to do so, bearing in mind the relevant criteria set out in paragraph 1 of Article 3–2 above, but in no event shall it be more than doubled.

(3) 'Salvor's expenses' for the purpose of paragraphs 1 and 2 of this Article means the out of pocket expenses reasonably incurred by the salvor in the salvage operation and a fair rate for equipment and personnel actually and reasonably used in the salvage operations, taking into consideration the criteria set out in paragraph 1(g), (h) and (i) of Article 3–2.

(4) Provided always that the total compensation under this Article shall be paid only if and to the extent that such compensation is greater than any reward recoverable by the salvor under Article 3–2.

(5) If the salvor has been negligent and has thereby failed to prevent or minimize damage to the environment, he may be deprived of the whole or part of any payment due under this Article.

(6) Nothing in this Article shall affect any rights or recourse on the part of the owner of the vessel."

8–49 The above provisions, whose tortuous drafting betrays both the speed of production and the compromises made in the process, provide for a basic regime closely modelled on the 1980 Lloyd's Open Form concepts, but wider in scope. Thus, it is explicitly stated that a reward shall be fixed with the view of encouraging salvage operations; moreover, reference is made to the skill and efforts of the salvors in avoiding or minimizing damage to the environment in calculating the size of the award. Therefore, the Draft Convention takes up the idea of enhancement contained in Lloyd's Open Form 1980, and generalises it from an increase due to exercising best endeavours to prevent the escape of oil from the vessel to exercising best endeavours to minimise damage to the environment.

8–50 Clearly, the duty of the salvor to use best endeavours to prevent or minimise damage to the environment is much wider than the Lloyd's Form 1980 duty, which is limited to preventing the escape of oil from the vessel. But, apart from covering all types of ship and cargo, it is probably not as alarmingly wide as it may seem on first sight, for the duty is limited to being in the course of salvage operations: "In so doing . . . ," it says. Hence, the salvor is probably not required to try to initiate clean-up, lay booms and so on unless such actions would form part of the process of salving the vessel and property. But the drafting needs clarification, which IMO's Legal Committee will, no doubt, achieve.

8–51 It is certainly important that the text of the draft be clarified; otherwise, obligations forced upon a salvor by this new provision could be interpreted in an extremely onerous manner. The use of the phrase "best endeavours" in a purely objective sense may result in salvors being forced to provide some broadly defined standard of service, requiring an extensive and costly upgrading of procedures and equipment. On the other hand, the use of a more subjective test, in the nature of "best endeavours in the circumstances" may entail a great deal of dispute, while leaving the salvor open to allegations of negligence.[26]

8–52 Like LOF 1980, the Draft Convention contains an exception to the no cure-no pay principle. Articles 3–1(2) and 3–2(2) make it clear that the principle still applies to the major part of the salvage operations; however Article 3–3 allows for an amount to be awarded as special compensation for salvage operations involving a ship which "by itself or its cargo threatens damage to the environment." The measure of compensation is to be equal to the salvor's expenses, as defined, plus (where the salvage operations have succeeded in preventing or minimising damage to the environment) an increment of up to the amount of those expenses, but in any case it is only available when that amount will exceed any amount awarded by the usual considerations outlined in Article 3–2(1).

8–53 The effect of this is that the Draft Convention provides a greater range of situations which entitle the salvor to claim his expenses as special compensation, but a lesser likelihood that in any particular situation such "special compensation" will be increased to provide a profit margin. Consequently, despite the apparent superiority of the Draft Convention in terms of its broader application of a no cure-no pay exception, the question arises as to whether this provision will offer inducement satisfactory to salvors.

8–54 The Draft Convention is, indeed, only a draft. It will undergo considerable alteration in the course of debate in the IMO

[26] Coulthard, above note 8, at pp. 54–56.

Legal Committee and any subsequent Diplomatic Conference which may be convened. Hence, more detailed analysis of the draft is inappropriate here.

4. GENERAL AVERAGE

8–55   The previous sections dealing with the Lloyd's Open Form 1980 and the 1981 Draft Salvage Convention have focussed in the main on the relationship between the shipowner and the salvor. The shipowner will usually be able to collect back a portion of the salvage expense from the cargo through a general average contribution. The question arises, however, as to what extent, if any, the shipowner can claim the cost of anti-pollution measures he has taken as general average, and how the enhanced award under LOF 1980 would be treated in general average.

A. The Nature of General Average in the Salvage Context

8–56   Salvage expenses constitute an important type of expenditure which, while not strictly general average, often become general average by reason of the circumstances in which they are incurred. Any doubt regarding the world-wide acceptance of the practice of treating salvage services as general average was set to rest for all practical purposes by the introduction of Rule VI in the 1974 version of The York/Antwerp Rules.[27] This reads as follows:

> "RULE VI. SALVAGE REMUNERATION. Expenditure incurred by the parties to the adventure on account of salvage, whether under contract or otherwise, shall be allowed in general average to the extent that the salvage operations were undertaken for the purpose of preserving from peril the property involved in the common maritime adventure."

8–57   Rule VI is supplemented by three additional rules relevant to the present discussion: A, C and X(b). Those rules read as follows:

> "Rule A.
>   There is a general average act when, and only when, any extraordinary sacrifice or expenditure is intentionally and reasonably made or

---

[27] L. Burglass, *Marine Insurance & General Average in the United States*, 315 (2nd ed. 1981). The York-Antwerp Rules, which first appeared in 1887, are incorporated into most charterparties and bills of lading by choice of the parties. The 1974 York-Antwerp Rules were adopted by the CMI at its Hamburg Conference in 1974.

incurred for the common safety for the purpose of preserving from peril the property involved in a common maritime adventure.
Rule C.

Only such losses, damages or expenses which are the direct consequence of the general average act shall be allowed as general average. Loss or damage sustained by the ship or cargo through delay, whether on a voyage or subsequently, such as demurrage, and any direct loss whatsoever, such as loss of market, shall not be admitted as general average.
Rule X(b).

The cost of handling on board or discharging cargo, fuel or stores whether at a port or place of loading, call or refuge shall be admitted as general average, when the handling or discharge was necessary for the common safety or to enable damage to the ship caused by sacrifice or accident to be repaired, if the repairs were necessary for the safe prosecution of the voyage, except in cases where the damage to the ship is discovered at a port or place of loading or call without any accident or other extraordinary circumstances connected with such damage having taken place during the voyage . . . ."

B. Application of the General Average Rules to Salvage in Oil Pollution Cases

8–58   Just how the foregoing rules can operate is best explained by quoting the following passage from L. Buglass, *Marine Insurance and General Average in the United States*[28]:

"Another example of the operation of Rule C is the occasional allowance in general average of oil pollution expenses when such pollution is the direct consequence of a general average act. Thus, when a vessel is obliged to enter a port of refuge for the common safety following an accident which has resulted in the vessel leaking oil, any expenditure incurred to avoid or minimize such pollution or any liabilities arising from such pollution would be allowable in general average; such expenditure or liability would be the direct consequence of the general average act of entering the port of refuge. Similarly, any pollution directly resulting from the jettison of oil (whether cargo or bunkers) for the common safety would also be allowable in general average. The usual rules of consequential damage would, of course, be applicable and if the pollution did not occur for some time following the general average act and/or was remote geographically, the pollution might be too remote to be considered a direct consequence of the general average act and therefore might not meet the requirements of Rule C.

If a vessel leaking oil was able to proceed directly to her next scheduled port, and the leakage was not the result of a general average act,

[28] *Ibid.* at p. 196.

any expenditure or liability resulting from the leakage would not concern general average. Such expenses would be the inevitable result of the accident giving rise to the leakage and not the result of a general average act or the direct consequence of a general average act; such leakage would in fact be of a purely accidental nature.

Rule X(b), on the other hand, comes into play in those cases where the damage to the ship is merely discovered at the port or place of loading or call without any 'accident or extraordinary circumstance connected with such damage having taken place during the voyage' (that is to say, subsequent to the commencement of loading of cargo). The act of handling or discharging cargo is not allowable in general average."

8–59 Although the rules might appear straightforward, their application can be difficult in some pollution cases because the rules address the purpose of expenditures and hence are concerned with motives—which can be very mixed. In particular, a salvor's reason for undertaking a particular task may not be to increase the overall value of the salved property. This is well illustrated by the following example, given in a recent paper[29]:

"[A] vessel, fully laden with a cargo of gypsum, had just set sail from a Canadian port. Experiencing difficulties at sea, the vessel put back and was about to re-enter Halifax when, under distress of weather, it stranded. During the ensuing eight hours, the responsibilities were laid down and accepted by the various parties concerned. A [pre-1980] LOF salvage agreement was entered into; the Canadian Ministry of Transport entered the picture almost immediately, followed by the Department of the Environment. The report was that the vessel had let go a substantial quantity of oil at the moment of stranding. It was immediately obvious from wing tank and double bottom tank soundings that the vessel's bottom shell plating was fractured throughout most of its length. . . .

The Ministry accordingly ordered the shipowner to remove all the oil from the vessel *prior to* the commencement of salvage operations. The owner accordingly hired a company to remove approximately 570 tons of heavy oil, including 90 tons of diesel oil.

The authorities initially refused to allow the salvage consortium to attempt to refloat the vessel until the oil was out. This order was subsequently relaxed when it became apparent that the oil was being properly removed and the threat of pollution was being eliminated. The two operations, which is to say the removal of the oil and the jettisoning of the ship's cargo of gypsum, then ran side by side with the salvors casting out the cargo and fitting the tanks for air as the con-

[29] G. Freehill, "The Qualification of Oil Pollution Avoidance Expenditures as General Average reprinted in Proceedings of the International Symposium on Marine Salvage" (1979), New York, 86, at pp. 89–90 (W.F. Searle, ed., 1980).

tractor hired to strip out the oil continued to work. Equally of significance, considering what subsequently happened, is that by the time
the contractor had removed about 450 of the 570 tons of oil, the authorities recognized that the remaining oil was pressed up in the fractured tanks to the extent that it could not easily escape. The
authorities therefore became more concerned with the possibility of
the ship breaking its back and therby becoming an even greater pollution and obstruction risk. The oil removal operation gave way to
the refloating effort. The salvors immediately took over the fractured
tanks, made them tight at the main deck level, and then blew them
down to provide the necessary floatation. The successful refloating
followed several days later. Unfortunately, despite all precautions by
the shipyard, when the vessel was lifted onto the dry dock the fencing
arrangement broke, causing further extensive oil pollution of the
water, the shoreline and other facilities.

Although all of the ship's cargo had to be jettisoned in order to escape
this predicament, the vessel was eventually repaired and set sail. The
general average adjusters dealing with the case were eventually asked
to admit the following expenditures in general average:

    —The cost of removal of the oil while the vessel was aground—
$102,000.00.

    —Pollution and/or control expenses while the vessel was afloat
prior to docking—$25,000.00

    —The cost of cleaning up the oil which polluted the harbor outside
the floating dock following the collapse of the dam—$208,000.00."

The question then posed was whether the removal of the oil was a
salvage expense which could be properly included in general average.

8–60   One could answer "no" to the above question[30] by recalling
that the Canadian Government's objective was not to "preserve
from peril the property involved in a common maritime adventure," but was solely to prevent pollution. Thus, while the removal
of the oil became *sine qua non* of the advancement of the salvage
operation, it was not a part of that operation. To clarify this point,
if a salvor had approached the vessel unencumbered by any
governmental authority or regulations about pollution or by the
duty of best endeavours to prevent the escape of oil under LOF
1980, he might simply have sealed the fractured bunker tanks and
blown them down. The oil would have been pressed out through
the bottom fractures and into the harbour, instead of being pumped
out from the top into a tanker. All that the salvor would have
required was empty tanks for the insertion of compressed air. Push-

---

[30] The question was answered no by the Canadian counsel who had been retained. Its position is set out at page 91 of the paper, and is reprinted here with minor variations.

ing the oil and water out through the fractures would not have added anything to the salvage cost. The salvor would not have incurred the additional $100,000 cost of removing the oil by pumping.

8–61   On the other hand, one could answer "yes" to the question[31] by arguing that the salvage operation could *not* have proceeded without the removal of the oil because of the government restrictions. Unless and until the oil was removed, no one would have been permitted to go near the vessel, either to save her very valuable hull or the cargo. Thus, it could easily be said that the owners were motivated to salve both the ship and cargo. If the vessel was to be salved, the salvors had no alternative but to incur the $100,000 cost of removing the oil by pumping before anyone laid a hand on the ship.

C. Enhanced Salvage Awards under LOF 1980—Treatment in General Average

8–62   Another question regarding the application of the general average rules to salvage in oil pollution cases regards the duty of the salvor under LOF 1980 to use his best endeavours to prevent the escape of oil from the ship. Since Rule VI allows in general average only those expenditures which were incurred to preserve the property from peril, it might be argued in some cases that the salvor's actions, or part of them, were for the purpose of fulfilling this duty of best endeavours rather than for the purpose of preserving the property from peril, so that the resulting salvage award should be wholly or partly disallowed in general average. For instance, suppose $100,000 were expended by the salvor in sealing a tank known to contain only ten tons of crude oil, valued at under $2,000: could one really say that the salvor did that to save the oil from peril? The answer is probably no, even though it may well have the result of saving the ten tons of oil (to the benefit of its owner or underwriter).

8–63   The plain fact is that York-Antwerp Rules are ill-drafted to cope with situations which can now easily arise under LOF 1980, for a number of reasons. First, salvors are likely to act with mixed motives—to save property from peril and to prevent the escape of oil—but the Rules do not deal adequately with how to cope with

---

[31] The question was answered yes by the P. & I. Club. Its views are set out at page 92, and are reprinted here with minor variations. The General Average Adjuster ruled that the removal of the oil had been "government mandated" and therefore could not be included in general average.

this in general average. Secondly, to the extent that salvors act purely to prevent the escape of oil rather than to save property from peril, the expense would appear not to qualify for general average; this could arise not only in the context of an award having been enhanced, but also where salvage has been partly successful and the safety-net is awarded. Where the award has been enhanced, the amount of the enhancement will not be separately specified, either, and so the adjuster inclined to disallow that portion of it has his task made all the more difficult.

5. SALVOR NEGLIGENCE, INSURANCE AND THE RIGHT OF LIMITATION

8–64 We turn now to three subjects which directly involve the manner in which the salvor carries out his operation. The first is the case where the salvor has performed the operation in a negligent manner. In such a case the usual result will be that the salved property arrives in a less valuable condition than it otherwise would have been in at the conclusion of the salvage operation. The second subject is the ability of the salvor adequately to protect himself through insurance. This is an especially important concern for the salvor who undertakes to salve vessels fully or partly laden with oil or other hazardous substances, since there is always the risk that the salvor's actions will unintentionally aggravate the pollution threat to the marine environment. The final subject is the issue of the right of the salvor to limit his liability once it has arisen in connection with a salvage operation. Although shipowners have had a right to limit their liability since at least the nineteenth century, salvors have not been so fortunate.

A. Salvor Negligence

8–65 Little needs to be said here about salvor negligence and the consequences of such negligence, for the matter is now fairly well settled.[32] Any damage negligently done by a salvor to a vessel while rendering salvage services is taken into account first by the salvage award being based on a salved value reduced by the damage, secondly by deducting part or all of the damage from the salvage award and thirdly by the fact that the safety-net award under LOF 1980 is not available where there has been contributory negligence

---

[32] The subject is ably handled in Norris, above, note 1, §§120 to 123.

165

by the salvor.[33] Furthermore, it has been held that vessel owners are entitled to counterclaim for damages (including loss of profit, if any) arising from negligence of salvors, although it is said that the law casts a benevolent eye towards the conduct of a salvor.[34]

B. Salvor Insurance

8–66 The salvor's ability to secure adequate insurance became a contentious issue as their potential liabilities for negligence rose with the size of tankers.[35] Up until the time of the *Pacific Glory* casualty in the English Channel in October 1970, salvors had generally been able to obtain insurance against liabilities which might be incurred in connection with the salvage of a laden oil tanker. After the *Pacific Glory* incident, however, this cover became unavailable.

8–67 The withdrawal of oil pollution insurance for salvors brought a response from the International Salvage Union. In a Notice to Shipowners published in Lloyd's List on January 16, 1973, it was announced that in certain cases involving a laden tanker no salvage would be undertaken on a contractual basis unless the shipowner provided indemnity for oil pollution. Such indemnity was to be given by means of a "PIOPIC" or similar clause, and was to be countersigned by the shipowner's third party liability underwriter. The PIOPIC clause (Protection and Indemnity Oil Pollution Indemnity Clause) was a clause which obligated the shipowner's P. & I. Club to indemnify the salvor for all claims which might be brought against the salvor for any oil pollution caused by the salvor in the course of his operation.

8–68 Shipowners and their clubs were understandably reluctant to agree to the insertion of a clause such as PIOPIC because of the unlimited liability which such clauses placed on them. As a result, representatives of both shipowners and salvors engaged in further meetings in an attempt to resolve the problem. Finally, on February 20, 1975 a compromise was reached between the International Salvage Union and the London Clubs. Under the accord, the clubs agreed to a reinstatement of the pollution insurance which had previously been available to salvors. In return for this concession, the

---

[33] See para. 8–48, above.

[34] *Serviss* v. *Ferguson*, 84 F. 202 (2d Cir. 1897); *The Cape Race*, 18 F.2d 79 at p. 81 (2d Cir. 1927). See also J. Rudolph "Negligent Salvage: Reduction of Award, Forfeiture of Award or Damages?" (1976) 7 J. Mar. L. & Com. 419.

[35] The incidents recounted in paras. 8–66 to 8–68 were originally detailed in Miller, above, note 7, at p. 245.

salvors agreed not to ask for PIOPIC or similar clauses unless truly exceptional circumstances were present.

8–69 The agreement reached between the clubs and the International Salvage Union to replace contractual formulas such as PIOPIC contained two provisions[36]:

(1) regarding the civil liability of salvors for pollution damage, cover per salvage craft of up to $20 million, with an additional limit of $40 million per salvage operation where more than three craft were involved in one operation; and

(2) in respect of the same risk, but where the salvors were operating as contractors, cover per salvage operation of up to $20 million, with a deductible of $50,000.[37]

These new guarantees are said to have proved satisfactory to all interests concerned.[38]

C. Salvor Limitation of Liability

8–70 Although the availability of adequate insurance has been a continuing goal of professional salvage associations, their most important objective with regard to the salvage of vessels carrying oil or other hazardous substances has been the creation and implementation of a right to limit salvor liability. In order to understand the importance of this right, one must remember that shipowners have had a right to limit their liability since the earliest times. Salvors, however, have historically not been so protected.

8–71 For instance, the 1957 Brussels Convention on Limitation of Liability[39] did not recognize a specific right of limitation for salvors. This fact was made painfully clear in 1971, when the English House of Lords delivered its decision in a case known as the *Tojo Maru*.[40] In that case a crew member of the salvage tug had left the tug to perform work on the vessel being salved. In the course of his work the crew member negligently caused damage to that vessel. The court decided that in such circumstances the salvor could not limit his liability. The reason for this result was that the negligent act was neither done in the "management" of the salvage tug nor

---

[36] The discussion in this paragraph was prepared after a review of B. Dubais, "The Liability of a Salvor Responsible for Oil Pollution Damage" (1977) 8 J. Mar. L. & Com. 375 at p. 385.

[37] The amount of insurance available to salvors has steadily increased since 1975, and is now on par with that available to tanker owners. See para. 10–101, below.

[38] Dubais, above, note 36.

[39] See paras. 9–31 *et seq.*, below.

[40] [1971] 1 Ll. Rep. 341.

was it an act or omission of a person "on board" the tug, as was required by both the 1957 Convention and English law.

8–72 The decision, which was clearly correct on the drafting of the instruments concerned, caused alarm among both industry and governments. It took five years before a suitable opportunity to improve the position arose. The solution decided on was specifically to provide for salvor limitation in a new Convention on Limitation of Liability for Maritime Claims which was adopted in 1976. The Convention specifically recognized for the first time the special position of salvors, and included explicit language allowing salvors to limit their liability, whether or not operating from a ship.

8–73 Article 1(1) of the Convention states that shipowners and salvors may limit their liability in accordance with the rules of the Convention. Article 1(3) defines a salvor as:

> "any person rendering services in direct connection with salvage operations. Salvage operations shall also include operations referred to in Article 2, paragraph 1(d), (e) and (f)."

8–74 Article 2, paragraph 1 defines a salvage operation to include:

> "(d) claims in respect of the raising, removal, destruction or the rendering harmless of a ship which is sunk, wrecked, stranded or abandoned, including anything that is or has been on board such ship;
>
> (e) claims in respect of the removal, destruction or the rendering harmless of the cargo of the ship;
>
> (f) claims of a person other than the person liable in respect of measures taken in order to avert or minimize loss for which the person liable may limit his liability in accordance with this Convention, and further loss caused by such measures.
>
> Article 2, paragraph 2, provides that claims set out in paragraph 1 shall be subject to limitation of liability even if brought by way of recourse or for indemnity under a contract or otherwise. However, claims set out under paragraph 1(d), (e) and (f) shall not be subject to limitation of liability to the extent that they relate to remuneration under a contract with the person liable."

8–75 An important exception to the rights of shipowners and others to limit their liabilities in relation to salvage is contained in Article 3 of the Convention. Article 3 provides that the rules of the Convention do not apply to:

> "(a) claims for salvage or contribution in general average;
>
> (b) claims for oil pollution damage within the meaning of the

International Convention on Civil Liability for Oil Pollution Damage, dated November 29, 1969 or of any amendment or Protocol thereto which is in force;

(c) claims subject to any international convention or national legislation governing or prohibiting limitation of liability for nuclear damage;

(d) claims against the shipowner of a nuclear ship for nuclear damage;

(e) claims by servants of the shipowner or salvor whose duties are connected with the ship or the salvage operations, including claims of their heirs, dependents or other persons entitled to make such claims, if under the law governing the contract of service between the shipowner or salvor and such servants the shipowner or salvor is not entitled to limit his liability in respect of such caims, or if he is by such law only permitted to limit his liability to an amount greater than that provided for in Article 6."

8–76 The Convention has not yet come into effect because it has not been ratified by the requisite number of states.[41] A matter of concern to salvors in these otherwise helpful provisions is the exception in Article 3 of claims for oil pollution damage within the meaning of the International Convention on Civil Liability for Oil Pollution Damage.[42] The result is that states becoming party to the 1976 Convention are free to regulate such claims against salvors as they please. By default, they may make no provision at all, and so they may be unlimited.

8–77 The principles embodied in the Convention have been adopted into both the Lloyd's Open Form 1980 and the 1981 Draft Salvage Convention. The adoption occurs in the LOF in Clause 21, which states that:

"The Contractor shall be entitled to limit any liability to the Owners of the subject vessel and/or her cargo bunkers and stores which he and/or his Servants and/or Agents may incur in and about the services in the manner and to the extent provided by English law and as if the provisions of the Convention on Limitation of Liability for Maritime Claims 1976 were part of the law of England."

[41] Article 17(1) of the Convention provides that the "Convention shall enter into force on the first day of the month following one year after the date on which twelve States have either signed it without reservation as to ratification, acceptance or approval or have deposited the requisite instruments of ratification, acceptance, approval or accession." As of February 21, 1985 eleven states had become parties to the Convention: the Bahamas, Denmark, Finland, France, Japan, Liberia, Norway, Spain, Sweden, the United Kingdom and the Yemen Arab Republic.

[42] This matter is further analyzed in paras. 9–36 to 9–38, below.

The adoption occurs in the Draft Salvage Convention in Article 5–1, which states that:

> "1. A contracting State may give salvors a right of limitation equivalent in manner and extent to the right provided for by the 1976 Convention on the Limitation for Maritime Claims."

8–78 The incorporation of the 1976 Limitation Convention into the LOF 1980 and 1981 Draft Salvage Convention leaves room for problems in the future.[43] For example, under the LOF 1980, a conflict of laws problem may arise. Although arbitrations under the LOF 1980 are to take place in London, and English law is to apply, it does not follow that disputes arising from the contract will go to an English court. In the absence of a clear choice having been expressed by the parties, a court's jurisdiction over a contract depends on factors such as the place of business of the parties and the place where the contract was to be performed. Thus, it is possible for a court other than an English court to find that it has jurisdiction over, for example, a suit for oil pollution damage arising from salvor negligence in the course of salvage rendered under LOF 1980. Of course, that court may still apply the law of the contract, as set out in Clause 21. However, where a court is not so bound to recognize the law of the contract, Clause 21 may be struck down as contrary to the law of the forum. Much more importantly, though, the contractual defence afforded by Clause 21 binds only the parties to the contract, even in an English court.[44]

8–79 The incorporation of the 1976 Limitation Convention into the 1981 Draft Salvage Convention presents problems similar to those present under LOF 1980. Article 4–5(1) provides a number of alternative jurisdictions from which the plaintiff may choose in bringing an action. The use of the term "plaintiff" clearly indicates that this applies to any of the parties concerned, including salvors, and the salvor's liability for negligence is clearly stated in Articles 3–3(6), 3–7, and 3–3(5). Since Article 5–1 provides that a contracting state may give salvors a right of limitation equal to the right provided for in the 1976 Convention, clearly the contracting state has an option of doing so. There is no guarantee, however, that the jurisdiction selected by a plaintiff will be one which has provided the salvor with such a right.

---

[43] The problems caused by the incorporation were first noted by Coulthard, above, note 8, at pp. 61–63.

[44] Suits against a salvor can arise in two distinct ways: (1) in an action brought against the salvor by the owner of the salved ship or cargo; and (2) by a lawsuit instituted against the salvor by a third party. Clause 21 would not apply to a party who was not bound by the salvage agreement.

6. SALVAGE AS A PREVENTIVE MEASURE

8–80 It will be seen in paragraph 10–61 below that a shipowner who has taken what are defined as "preventive measures"[45] may claim the cost of them against his own limitation fund set up under the 1969 Liability Convention; and he may also claim them against the IOPC Fund under the 1971 Fund Convention.[46] Preventive measures are so defined that they only apply to a situation where oil has already escaped from the ship, but this will change under the 1984 Protocols to those Conventions, which apply from the time there is a grave and imminent threat of pollution damage being caused.[47] The question therefore arises whether the cost of salvage operations could qualify as preventive measures, as defined in these instruments.

8–81 Preventive measures are defined in all four instruments by Article I(7) of the Liability Convention, as follows (the change of applicability to pure threat situations being effected by the 1984 instruments by a change in the definition of *incident*):

> "Preventive measures' means any reasonable measures taken by any person after an incident has occurred to prevent or minimize pollution damage."

Hence, the test here is one of purpose, just as the test in general average is one of purpose. As we saw there, motives for doing any particular act can be mixed, and the possible intervention of a state, or applicable regulations, or the new duties on salvors under Lloyd's Open Form 1980 can further complicate matters. The point that needs to be made here is that, although there does not seem to have been much difficult in the past, the difficulty is conceptually there and must be increased by the new salvor duties discussed above.

8–82 It would probably have been thought hitherto that anything which was salvage was not preventive measures. P. & I. Clubs in particular would have been keen to maintain that principle, because they do not contribute to salvage. But now that they will be paying up under the safety-net from time to time, they will no doubt be keen to claim such expenditure against the IOPC Fund in appropriate cases as a preventive measure; and if they do, they would have to show that, in the circumstances of the case, the salvor's actions were being done to prevent or minimize pollution

---

[45] See paras. 10–47 to 10–57, below.
[46] See para. 11–13, below.
[47] See paras. 10–131 to 10–133 and 11–87, below.

damage. The Convention does not state that the action need be done *solely* for that purpose, which may help them in such an argument.

8–83   An even stronger argument might be advanced by those who have contributed to an enhanced award made under Lloyd's Open Form 1980, for the enhanced portion of the award. The enhancement is only related to the endeavours the salvor has made in preventing the escape of oil from the ship, and so must *prima facie* be recoverable as a preventive measure. However, the problem will be that it will probably not be shown as such in the award, and so quantifying it will present a difficulty. But the IOPC Fund has shown itself to be a flexible instrument—unlike CRISTAL[48]—and so this kind of claim could be settled by negotiation.

8–84   Even aspects of the salvage award itself *not increased by enhancement* might qualify as preventive measures in a situation where action was taken by the salvors to prevent pollution damage. One example is the case cited above.[49] The more the interests of the coastal state interfere with what the commercial parties do in salvage, the more likely it is that a part of the salvage award will be attributable to things done pursuant to saving oil pollution damage.

---

[48] *Cf.* paras. 11–55 *et seq.*, and 12–21, below.
[49] See para. 8–59, above.

9. The Framework of Liability in International Law

SUMMARY

9–01   In this chapter we consider the liability of a state in international law for damage caused by oil pollution. This could arise under an international law obligation prescribing standards of behaviour before a spill occurs—in which case a distinction must be made between an incident involving "public" ships and one involving other ships—or after a spill occurs. While there may be liability arising in connection with duties to enforce pollution standards laid down in specific Conventions, there would not normally be any liability for oil pollution caused by private ships and there are considerable problems in finding a basis for liability in the case of public ships. After the spill, the affected state may have certain obligations to notify other states likely to be affected, or possibly to prevent it spreading, for breach of which liability may result, but the likelihood of a breach is low. Instead, the tendency has been for states to move towards concerted action to place a special, common liability regime on the commercial interests involved in the carriage of oil by sea, rather than to assume the role of becoming responsible for pollution damage themselves. In this regard the international community still have a considerable way to go before achieving the goal of a united, global regime.

1. INTRODUCTION

9–02   An important distinction must be made at the outset between liability which may arise under a system of municipal law and that arising under international law. A state may be liable under municipal law for oil pollution damage, if in the circumstances it is not immune from liability in that legal system on account of its status as a sovereign state. This liability may arise as a result of the provisions of an international Convention (such as the Liability Convention discussed in Chapter 10) being incorporated into municipal law, or as a result of provisions having an

173

indigenous origin (such as the provisions of United States law discussed in Part IV). A state may also be liable to another state under international law for the breach of an obligation owed to that other state. This liability is separate from, and irrespective of, any liability under a system of municipal law which may arise. It is this liability which is discussed in this chapter.

9–03 The other topic considered in this chapter is the obligation in international law which one state may owe to another to make provision in its municipal law for the liability of those responsible for oil pollution damage arising from an incident or of others chosen for the purpose. As we will see, the practice of states in dealing with the need for compensation for oil pollution damage has been to concentrate upon this type of obligation, rather than to assume direct liabilities themselves.

2. AN INTERNATIONAL LAW CLAIM ARISING FROM AN OIL POLLUTION INCIDENT

9–04 There is nothing in the concept of state responsibility in international law to prevent one state claiming reparation from another state in connection with an oil pollution incident. It is, however, a prerequisite that a breach of an international obligation can be shown. Once this is done, an obligation to make reparation arises automatically.[1] The most recent expression of this general principle of international law in the sphere of the marine environment is contained in Article 235(1) of the Law of the Sea Convention, 1982:

> "States are responsible for the fulfilment of their international obligations concerning the protection and preservation of the marine environment. They shall be liable in accordance with international law."

The question, then, in any particular case is whether the oil pollution incident constitutes a breach of a state's obligations concerning the protection and preservation of the marine environment, which

---

[1] E. Jimenez de Arichaga, "International Responsibility", in M. Sorensen (ed.), *Manual of Public International Law* (1968 London and New York), p. 533, citing the *Chorzow Factory* case, (1928) P.C.I.J. Ser. A, No 17 and the *Corfu Channel* case, (1949) I.C.J. Rep; L. Oppenheim, *International Law*, 8th ed., H. Lauterpacht, (1955 London), Vol. 1, pp. 336 *et seq.*; I. Brownlie, *Principles of Public International Law*, (3rd ed., 1979, Oxford, p. 433, citing the *Spanish Zone of Morocco Claims*, U.N.R.I.A.A. ii, 615, and the *Chorzow Factory* case; A.L. Springer, *The International Law of Pollution—Protecting the Global Environment in a World of Sovereign States*, Westport Connecticut and London, 1983, pp.125 *et seq.*, citing Article 1 of the International Law Commission's draft principles of state responsibility:"Every internationally wrongful act of a State entails the international responsibility of that State."

were discussed in Chapter 2. It is helpful to consider the principle of liability first in relation to behaviour before the spill, and then in relation to behaviour after.

A. Before a Spill

PRIVATE SHIPS

9–05 It might be thought that liability could arise in connection with a ship flying the flag of the state concerned. However, most incidents involve ships owned and operated by companies or other entities for the acts or omissions of which states are not generally liable under international law.[2] This might be otherwise if sailing the seas gave rise to a much greater risk of oil pollution, but customary law is tolerant of a degree of ordinary user which comprehends a certain level of pollution.[3] This must include occasional accidental oil pollution from ships, the risk of which is still not sufficiently high to engender responsibility for an occurrence, and the small amount of pollution resulting from operational (*i.e.* intentional) discharges occasioned by each individual ship (notwithstanding the cumulative, substantial effect of such discharges of the world fleet).

9–06 The obligations of states with respect to ships flying their flag are limited very largely to making provision in their municipal law for standards of pollution prevention and punishing offenders, and as regards their nationals or those present within their jurisdiction, to making provision for civil liability for oil pollution damage. It is most unlikely that in practice there would be a breach of such an obligation which could form the basis of an international claim. Therefore, in relation to what one might call "private" ships, something more would be required than the mere fact that the ship sails under the flag, or is owned or operated by a national, of the state

---

[2] As to the circumstances in which an act or omission can be imputed to a state, see generally E. Jimenez de Arichaga, "International Responsibility," in M. Sorensen (ed.), *Manual of Public International Law*, London and New York, 1968, pp. 544 *et seq.*; B. Cheng, *General Principles of Law as Applied by International Courts and Tribunals*, 1953 London, p. 180 *et seq.*; L. Oppenheim, *International Law*, 8th ed. (H. Lauterpacht, ed.), (1955, London), pp. 340 *et seq.*; D. P. O'Connell, *International Law* (2nd ed., 1970, London) Chap. 30; I. Brownlie, *Principles of Public International Law* (3rd ed., 1979, Oxford) pp. 445 *et seq.*

[3] I. Brownlie, "A Survey of International Customary Rules of Environmental Protection," in L .A. Teclaff and A. E. Utton (eds.), *International Environmental Law*, New York and London, 1974, p. 2; E. D. Brown, "The Role of Law in the Prevention of Oil Pollution", in J. Wardley-Smith (ed.), *The Prevention of Oil Pollution*, (1979, London), p. 268 comments that no guidance is offered about what would constitute an unreasonable discharge. And see paras. 2–11 *et seq.*, above.

concerned. While customary international law does not provide any such additional obligation, Conventional law might.

SPECIFIC UNDERTAKINGS—SURVEY AND CERTIFICATION

9–07 For instance, under MARPOL 73/78, all ships (except small ones) are required to be surveyed and inspected as part of the enforcement mechanism of that Convention.[4] Ships complying with the Convention are issued with an International Oil Pollution Prevention Certificate. Even if the survey or issue of the certificate is actually carried out by persons who are not officers of the state (such as a classification society to whom the task has been delegated), the state fully guarantees the completeness of the survey and inspection, and assumes full responsibility for the certificate.[5] These provisions therefore place upon states party to the Convention an obligation in international law, owed to the other states parties, to ensure that such Certificates are only issued to ships which comply with the requirements laid down in the Convention. If, in contravention thereof, a ship is issued with a certificate and is allowed to proceed to sea as a result, and if its non-compliance subsequently causes oil pollution damage to a state party, then the flag state may become liable to make reparation to the state which has suffered the damage. Similar specific obligations are undertaken by states in other international Conventions—for instance, SOLAS 74/78 [6]—and there is a general obligation in Article 219 of the Law of the Sea Convention 1982 upon port states which do inspect a ship and find that it threatens damage to the marine environment to prevent the ship from sailing.

9–08 It is not only flag states who may find that provisions such as these form the basis of an international law claim against them. A recent feature of the international regime of enforcement of pollution standards has been a desire to move away from exclusive flag state jurisdiction towards port or coastal state jurisdiction.[7] Under MARPOL 73/78, port states are given certain powers of inspection,[8] for instance, to see if the ship possesses the necessary certificate; if it does not, "the Party carrying out the inspection shall take such steps as will ensure that the ship shall not sail until it can proceed to sea without presenting an unreasonable threat of harm to

---

[4] Regulation 4 of Annex I. These and other enforcement provisions of MARPOL and other relevant conventions are analysed in paras. 3–97 *et seq.*, 4–38 *et seq.* and 4–58 *et seq.* above.

[5] Regulations 4(3)(e) and 5(2).

[6] Regulations 6, 10, 12 and 13 of Chapter I, and Article I (*b*).

[7] See generally Chapter 5.

[8] Article 5.

the marine environment." If a contravention of a duty such as this were to cause a subsequent oil pollution incident, the basis for an international law claim against the port state would exist.

9–09 Obligations such as these are specific enough to form the basis of a claim, but the claim would arise in respect of the ship not because it was owned or operated by nationals of the state concerned, or because it flew the flag of that state, or even because the ship was at the time in the port of the state concerned, but because officials of the state, or those to whom the state had delegated authority, had failed to keep a specific obligation undertaken by the state in the Convention. Even then, causation would have to be shown. This could present something of a problem in a particular case, since many incidents involve the negligence of those on board and this may be a *novus actus interveniens.*

PUBLIC SHIPS

9–10 The position is different, however, where the ship is what might be loosely called a "public ship." Here, the problem of attribution does not arise because those who are responsible for the condition of the ship and for its navigation are public servants, agents of the state for whose actions, except in very unusual circumstances,[9] the state is responsible in international law. Examples of such ships are warships, naval auxiliaries and such less important ships (from the oil pollution viewpoint) as fisheries protection vessels and customs patrol ships. If a ship on commercial service were owned or operated directly by the state (as opposed to a state-owned entity such as a national shipping company) then that ship, too, would in this sense be a "public ship."

9–11 Is there, then, any obligation upon states in international law which a public ship could breach if it caused oil pollution to another state? It is necessary again to examine the principles discussed in paragraphs 2–05 to 2–13 to see if any extend to providing a sufficiently firm ground for a claim. It can be readily seen that doctrines suitable for creating a duty to protect the environment by making regulations are much less suited to grounding liability claims. The problems are substantial: for instance, there is the difficulty of establishing with sufficient precision any applicable general principles of law. The customary doctrines present problems of

---

[9] Principle 7 is also of interest: "States shall take all possible steps to prevent pollution of the seas by substances that are liable to create hazards to human health, to harm living resources and marine life, to damage amenities or to interfere with other legitimate uses of the sea."

definition of scope, and although gaps have been filled by the almost universal acceptance of certain Conventions, there must remain grave doubt about whether any such principles are accepted by states as applicable to public ships. There have always been exclusions of public ships, except those in commercial service, from the relevant Conventions dealing with prevention and liability, suggesting strongly that different standards apply to them.[10]

9–12 Until recently, the tendency has been to exempt certain classes of public ship altogether, but the Law of the Sea Convention 1982 has seen the emergence of a new approach to liability for public ships. Two provisions are relevant. Article 236 uses almost identical language to Article 3(3) of MARPOL 73/78 in providing as follows:

> "The provisions of this Convention regarding the protection and preservation of the marine environment do not apply to any warship, naval auxiliary, other vessels or aircraft owned or operated by a State and used, for the time being, only on government non-commercial service. However, each State shall ensure, by the adoption of appropriate measures not impairing operations or operational capabilities of such vessels or aircraft owned or operated by it, that such vessels or aircraft act in a manner consistent, so far as is reasonable and practicable, with this Convention."[11]

This exemption covers all public ships except those on commercial service, and of course the obligation is redolent with defences (*e.g* a state could say that since the obligation is only to adopt "appropriate measures," the measures adopted were "appropriate"; other measures would have "impaired operations or operational capabilities"; the negligence of those involved was not inconsistent with the Convention and so it was consistent; those involved did all that was reasonable and practicable in the circumstances).

9–13 A much fiercer-looking provision is found in Article 31, which states:

> "The flag state shall bear international responsibility for any loss or damage to the coastal State resulting from the non-compliance by a warship or other government ship operated for non-commercial purposes with the laws and regulations of the coastal State concerning

---

[10] Prevention: Article II(1)(*d*) of the 1954 Oil Pollution Convention as amended in 1962 and 1969; Regulation 3(*a*)(i) of Chapter I of SOLAS 74/78; Liability: Article XI(1) of the 1969 Civil Liability Convention; Article 4(2)(*a*) of the 1971 Fund Convention.

[11] The principle can in fact be traced to Article VII(4) of the 1972 London Dumping Convention: " . . . each party shall ensure by the adoption of appropriate measures that such vessels and aircraft owned or operated by it act in a manner consistent with the object and purpose of this Convention . . . ." Similar wording is to be found in many of the regional Conventions discussed in Chapter 7.

passage through the territorial sea or with the provisions of this Convention or other rules of international law."

However, for responsibility to follow under this provision, there must first have been a breach of a coastal state's law or of international law, and as we have seen both have traditionally not applied to ships on government non-commercial service.[12]

9–14 As to public ships on commercial service, it is just possible that liability might be established for operational discharges which breach the provisions of the 1954 Convention, but this would not apply to most cases of accidental discharges, since the 1954 Convention exempts many (but not all) cases of accidental discharge from its provisions.[13]

9–15 The conclusion must be that there are many difficulties associated with establishing the existence of a sufficiently definite obligation upon a state in respect of its own ships, even where they are engaged on commercial service, that the chances of such a claim succeeding when liability is disputed are small. However, this is not to say that a state would not wish, for reasons of good relations and comity, to pay compensation for oil pollution damage, or to make other reparation therefor, if a public ship were the cause.

B. After a Spill

9–16 The question which arises after a spill is, are there any obligations on either the flag state or the nearest coastal state to do anything about it—for instance, to minimise the spillage or clean it up at sea so that oil does not threaten the coasts of other states, or the marine environment generally?

FLAG STATES

9–17 In so far as the flag state is concerned, the analysis appears to be little different after the fact of the spill than before it. The only difference relates to the fact that Article 2 of the High Seas Convention 1958 requires that the freedom of navigation on the high seas be exercised "with reasonable regard to the interests of other States," and Article 87(2) of the Law of the Sea Convention 1982 is to similar effect, requiring "due regard." If the flag state knows about the spill, it is possible to construct from this a duty at least to notify other states, whose interests may be affected, of the fact of the

---

[12] See also Article 42(5) to similar effect.
[13] Article X.

spillage, but it would be difficult to take it further. This is now rein-forced by Article 198 of the Law of the Sea Convention, 1982:

> "When a State becomes aware of cases in which the marine environ-ment is in imminent danger of being damaged or has been damaged by pollution, it shall immediately notify other States it deems likely to be affected by such damage, as well as the competent international organisations."

COASTAL STATES

9–18 The principles discussed in paragraphs 2–05 *et seq.* and 7–09 *et seq.*, above, offer a similar basis for saying that the coastal state has at least a duty to notify others likely to be affected, but whether they stretch further to impose upon a coastal state the obli-gation to contain or minimise the spill is quite another matter. Of much greater significance here are the provisions of the various regional and multilateral Conventions discussed in Chapter 7. All of these impose obligations of notification—although in differing circumstances—some of which are sufficiently specific to be capable of founding an international claim.[14] There are also vary-ing degrees of obligation to combat the pollution (which would have the effect of reducing any possible threat to neighbouring states) and to assist others in so doing, but none are really specific or objective enough to be of much use in this context. Even if they were, as with breach of the duty to notify, there would be the almost insoluble problem of causation (to what extent did the few days' delay in notification cause the damage to the other state?) and quantification of damage. It is also important to realise that these rather minimal obligations are extremely unlikely to be violated in practice, so that discussion of potential state responsi-bility in connection therewith is somewhat academic.

C. Conclusion on International Law Claims

9–19 It can be seen from the above that international law has not received any general doctrine capable of clearly defining an obli-gation which rests upon states to compensate any other state in respect of oil pollution damage, and that states have very rarely undertaken treaty obligations which are capable of supporting an international law claim for such damage. Consequently, it is only in the rare case that a state could find itself liable in international law.

---

[14] *e.g.* Article IX(*b*) of the Kuwait Regional Convention 1978 and Article VII(2) of the Pro-tocol thereto.

The reason for this is that, in responding to the emergence of the problem of oil pollution from ships, states have not seen themselves in the role of the one who must compensate. Equally importantly, it has not been the practice of states to raise such claims.

9–20 Rather, states have adopted the role of evolving and policing preventive regimes (as discussed in Chapters 2 to 5), of protecting themselves after the occurrence of an emergency (as discussed in Chapter 6), and of co-operating with other states on a regional basis in effecting these roles (as discussed in Chapter 7). The role of compensation has been left to the commercial parties involved in the carriage of oil by sea, and the means chosen to effect this has been municipal law. To this end states have evolved various mutual obligations to put into their municipal law rules which will ensure that, in appropriate circumstances, compensation is available to both states and private claimants for the oil pollution damage they may have suffered. It is to these obligations that we now turn.

3. OBLIGATIONS IN INTERNATIONAL LAW TO MAKE PROVISION FOR LIABILITY

9–21 It will by now come as no surprise to the reader to discover that there can be found no principle of customary international law which would bind a state to enact any particular liability legislation. Such obligations are to be found in international Conventions for the unification of private law and, more recently, in regional co-operation Conventions. The evolution of the international response to the problem of compensation has been recent, swift and dramatic.

9–22 As explained in Chapter 10, it was the *Torrey Canyon* incident in 1967 which provided the political pressure for states to respond with a new and global regime for the compensation of oil pollution damage caused by ships. Prior to that, compensation was governed by the various general rules of tort law in each state, and only a minor element of unification was provided by the Conventions on limitation of liability which we are about to discuss. There was no instrument obliging parties to create a special liability regime for oil pollution from ships. The impetus at IMO following the *Torrey Canyon* incident resulted in the adoption of the 1969 Liability Convention, only the second of its kind in the field of environmental law specifically to create a compensation regime for damage caused by ship.[15] As explained in Chapter 11, the 1971 Fund Convention followed soon as a result of a compromise between opposing groups at the 1969 Conference. At about the same time growing public awareness of the problems posed by other forms of pollution lead to the elaboration of new multilateral treaties on these subjects too,[16] so that liability for oil pollution from ships came to be seen as a specific part of a developing whole.[17] This movement to create a new international regime for the management of the world's resources was most importantly and articulately expressed in the recommendations of the Stockholm Conference on the Human Environment in 1972.[18] Perhaps partly in response to that Conference, the emphasis then shifted, so far as liability for oil pollution from ships was concerned, to the development of regional Conventions.

9–23 The result is that in 1984 there are several international instruments under which parties accept obligations to make liability rules in their municipal law. The 1969 and 1971 Conventions, with their Protocols of 1984, are discussed at length in Chapters 10 and 11. These obligations are specific, and actually lay down the rules to be implemented, leaving relatively restricted scope for parties to depart from or interpret the detailed provisions. The same is true of the limitation Conventions discussed below. By comparison, the other instruments express general obligations to co-operate in the development of such rules.

9–24 The earliest of these other provisions is contained in Article 17 of the Convention on the Protection of the Marine Environment of the Baltic Sea Area 1974:

[15] The 1962 Convention on the Liability of Operators of Nuclear Ships was the first to create a special liability regime, although there had been conventions in 1924 and 1957 to provide for unification of rules limiting the liability of ships, as discussed in paras. 9–29 *et seq.*, below.

[16] Notably the 1971 Convention Relating to Civil Liability in the Field of Maritime Carriage of Nuclear Material, the 1972 Convention on the Prevention of Marine Pollution by Dumping of Wastes and Other Matter (the London Dumping Convention), the 1974 Convention on the Protection of the Environment between Denmark, Finland, Norway and Sweden, and the 1974 Convention for the Prevention of Marine Pollution from Land-based Sources (the Paris Convention). Regional Conventions discussed in Chapter 7 are also part of the development of international environmental law which took place at this time.

[17] For an elaboration of this theme see K.Hakapaa, *Marine Pollution in International Law— Material Obligations and Jurisdiction*, Helsinki, 1981, Chaps. 1 and 4; A. L. Springer, *The International Law of Pollution—Protecting the Global Environment in a World of Sovereign States*, Westport Connecticut and London, 1983, esp. Chap. 5 on State Responsibility.

[18] Principle 22 is of particular interest in this connection: "States shall co-operate to develop further the international law regarding liability and compensation for the victims of pollution and other environmental damage caused by activities within the jurisdiction or control of such States to areas beyond their jurisdiction."

"The Contracting Parties undertake, as soon as possible, jointly to develop and accept rules concerning responsibility for damage resulting from acts or omissions in contravention of the present Convention, including, *inter alia*, limits of responsibility, criteria and procedures for the determination of liability and available remedies."

This theme was taken up two years later in Article 12 of the Convention for the Protection of the Mediterranean Sea Against Pollution, 1976, which is in almost identical terms, and again in 1978 in Article XIII of the Kuwait Regional Convention for Co-operation on the Protection of the Marine Environment from Pollution, which is, however, drafted in wider terms in that it also extends to cover liability for acts not prescribed by the Convention itself:

"The Contracting States undertake to co-operate in the formulation and adoption of appropriate rules and procedures for the determination of:
  (a) civil liability and compensation for damage resulting from pollution of the marine environment, bearing in mind applicable international rules and procedures relating to those matters; and
  (b) liability and compensation for damage resulting from violation of obligations under the present Convention and its protocols."

9–25 A significant feature of this rather general wording is the mention of "applicable international rules and procedures." It can only refer, in the case of oil pollution from ships, to the 1969 and 1971 Conventions, and in so doing exhorts the parties to take note of them. In fact, it seems that these and the earlier provisions may have borne some little fruit in the attraction of recent signatures to the 1969 and 1971 Conventions.[19] If such signatures can indeed be so attributed, it is no mean feat. These provisions may at least be taken as evidence of a growing feeling amongst the international community of states that pollution liabilities are well tackled on a regional basis, even if that is achieved through the wider means of ratifying global treaties.

9–26 However, the relevant provision of the Law of the Sea Convention 1982 does not make any specific mention of a regional approach to this matter. Instead, Article 235 makes the most speci-

---

[19] As at December 31, 1983, a total of five instruments of ratification or acceptance to the 1969 Liability Convention had been deposited by signatories to any of the three regional conventions after their respective entries into force, and the corresponding figure for the 1971 Fund Convention is seven. That left relatively few states still to become party to the 1969 Convention (11, from the Gulf and Mediterranean), but more for the 1971 Convention (17, from all three regions).

fic mention of the aims of liability regimes yet adopted in a general treaty:

> "2. States shall ensure that recourse is available in accordance with their legal systems for prompt and adequate compensation or other relief in respect of damage caused by pollution of the marine environment by natural or juridical persons under their jurisdiction.
>
> 3. With the objective of assuring prompt and adequate compensation in respect of all damage caused by pollution of the marine environment, States shall co-operate in the implementation of existing international law and the further development of international law relating to responsibility and liability for the assessment of and compensation for damage and the settlement of related disputes, as well as, where appropriate, development of criteria and procedures for payment of adequate compensation, such as compulsory insurance or compensation funds."

9–27 A significant feature here is that compensation or other relief should be "prompt and adequate." Perhaps it is no coincidence that the primary motivation for the 1984 Conference to revise the 1969 and 1971 Conventions is to improve the adequacy of the compensation available. The obligation is put in strong terms, too: states "shall ensure" that "recourse is available," in contrast to the regional Conventions quoted above, where the obligation is merely to co-operate in the formulation of rules. The Article goes on to speak of implementing existing international law, which must refer to the wider ratification of existing multilateral or bilateral treaties—a theme emphasised in Resolution A.500 by the Assembly of the International Maritime Organisation in January 1982. Particularly interesting is the specific role mentioned for compulsory insurance and compensation funds, for these are both major features of the 1969 and 1971 regimes.

9–28 All these provisions give respectability to the view that the community of states now expects to see the development of a special liability regime for oil pollution from ships in the municipal law of all states, even if this expectation has not achieved the status of a binding obligation of international law. Furthermore, the provisions of all the instruments mentioned above, culminating in the Law of the Sea Convention 1982, are evidence of trends in international environmental law towards the provision of prompt and adequate compensation by means of international Conventions for the unification of private law. The most important of these are the 1969 and 1971 Conventions (as amended), which are examined in detail in Chapters 10 and 11. It is necessary, however, first to look briefly at the Conventions on limitation of liability, in so far as they relate to oil pollution claims.

184

4. CONVENTIONS ON LIMITATION OF SHIPOWNERS' LIABILITY

A. Introduction

9–29   The origins of the shipowner's right to limit his liability, and the early policies on which the right was based, are not wholly clear.[20] This has not, however, prevented the international community from developing the concept in three multilateral Conventions. The first was in Brussels in 1924, and gave rise to the International Convention for the Unification of Certain Rules Relating to the Limitation of the Liability of Owners of Sea-going Vessels 1924. The second was in 1957, again in Brussels, and gave rise to the International Convention Relating to the Limitation of the Liability of Owners of Sea-going Ships 1957. This was amended as to the units of account to be used by a Protocol adopted in 1979. Lastly, a Conference held in London in 1976 adopted the Convention on Limitation of Liability for Maritime Claims 1976.

9–30   These Conventions, which deal with the limitation of third party liabilities generally, are for a number of reasons an integral part of the international framework of liability for oil pollution, of which the 1969 and 1971 Conventions are the core. First, the concept of limitation in the 1969 Convention can only be properly understood as a special development of the concept embodied in the 1924 and 1957 Conventions. Second, there are states which are party to these instruments which are not party to the 1969 or 1971 Conventions; relationships inter se, and between them and the states party both to the latter Conventions and these instruments, will be governed by these instruments.[21] Thirdly, for states which are party both to the 1969 and 1971 Conventions and to these instruments, liabilities for oil pollution damage arising outside the provisions of the 1969 and 1971 Conventions may be limitable according to these instruments.[22]

9–31   The co-existence of the 1969 and 1971 regimes with these

---

[20] Marsden's *The Law of Collisions at Sea* (11th ed.) (ed., K. C. McGuffie) (1961, London) paras. 168–174, mentions a number of possibilities: the contract of commande (joint venture of shipowners and merchants), protection, *noxae deditio* (*deodand*), division of loss and arrest.

[21] For instance, in 1980 the United Kingdom was a party to the 1969 Convention and to the 1957 Convention. Iran was a party only to the 1957 Convention. Limitation of oil pollution liabilities for Iranian ships in the United Kingdom would therefore be in accordance with the 1957 Convention.

[22] For instance, oil pollution damage caused by escape of fuel oil from a ship not carrying oil in bulk as cargo: see paras. 10–07 *et seq.*; but *cf.* paras. 10–118 *et seq.*

three Conventions means that for any given oil pollution incident, not only will the liability of the various possible defendants (*e.g.* shipowner, shipmanager, salvor, bareboat charterer, shipbuilder, etc.) be governed by different principles, but the question of whether they can limit their liability, and if so to how much, will be governed by different national laws enacted in accordance with the various limitation Conventions. This creates a hideous complexity. It is, however, important to appreciate that none of these three Conventions makes any provision for the creation of liability; they deal only with the limitation of liability that arises according to principles of national law.

9–32   The problem facing the delegates to the 1924 Conference was primarily one of lack of uniformity, and so the instrument adopted was one which sought a compromise between the interests of the shipper, the passenger, the shipowner and his insurer, within the context that in many countries the shipowner could limit his liability to the value of his ship, its freight and accessories.[23] In 1957, however, the problem was rather different, the main task being to revise the limits upwards. Since the value of the ship had proved a difficult figure to ascertain, limits were set by reference to the tonnage of the ship. This introduced a principle which is now fundamental to this type of limitation, but which has been extensively criticised in recent times[24] because the capacity of a ship to cause damage—especially oil pollution damage—is only very loosely connected with its size. This, and the fact that at neither Conference was oil pollution regarded as a problem requiring special treatment, underline the relative inappropriateness of these Conventions for their current role in the international regime of oil pollution liability.

9–33   By 1976, so many problems had developed within the existing system that a complete overhaul was called for. The two previous Conferences had proceeded on the basis that a shipowner's interest in the maritime adventure should be limited to the actual or presumed value of his ship, however that be ascertained. A new policy formed the basis of the discussion in 1976, namely that limitation should be at a figure as high as that against which the shipowner can insure at reasonable cost. This principle gave rise to a dramatic increase in the real value of the limits, whereas the 1957

---

[23] C. J. Colombos, *The International Law of the Sea* (6th ed., 1967, London) pp. 352 *et seq.* In fact, this is still the case today in some countries, including the United States under the Limitation Act 1851.

[24] Notably in the deliberations of the Legal Committee of IMO in the long run-up to the 1984 Conference.

186

limits had done little more than restore real values and change the method of evaluation. At the same time, the system of relating the limitation amount to the tonnage of the ship was retained.

A detailed analysis of these three Conventions is not required here; it will suffice to mention just three topics.

B. The Right to Limit

9–34 The 1924 and 1957 Conventions clearly grant the right to limit in respect of claims for oil pollution damage[25] so that the idea of limiting such claims was already well established in international law by the time of the 1969 Convention. However, when the 1976 Conference met, the 1969 Convention had already entered into force, the 1971 Convention had been adopted and was slowly attracting ratifications, and oil pollution was well established as a special case. Article 3(*b*) therefore provides that the rules of the Convention shall not apply to claims for oil pollution damage within the meaning of the 1969 Convention or of any amendment or Protocol thereto which is in force.[26]

9–35 The inclusion of this provision and its exact wording were matters of considerable concern at the Conference. Some states were worried about being committed to become a party to the 1969 Convention if they became party to the 1976 one, but the wording of Article 3(*b*), while not free from doubt, is at the same time suf-ficiently unclear to leave open to states the opportunity not to ratify the 1969 Convention. As to the scope of the provision,[27] it would seem that a claim for loss or damage described in the 1969 Conven-tion as oil pollution damage[28] is one to which the 1976 Convention does not apply, whether or not the 1969 Convention is in force in the territory where the loss or damage was suffered. The test is,

---

[25] Although the wording of the 1957 Convention gave rise to certain problems for salvors: see paras. 8–70, *et seq.*, above.

[26] Two types of vessel in connection with which oil pollution may arise are treated differ-ently. By Article 15(5)(*b*), floating platforms constructed for the purpose of exploring or exploiting the natural resources of the sea-bed or subsoil thereof are excluded from the Convention. By Article 15(4), "The Courts of a State Party shall not apply this Conven-tion to ships constructed for, or adapted to, and engaged in, drilling: (*a*) when that State has established under its national legislation a higher limit of liability than that otherwise provided for in Article 6; or (*b*) when that State has become party to an international Con-vention regulating the system of liability in respect of such ships."

[27] See generally E. Selvig, "The 1976 Limitation Convention and Oil Pollution Damage," [1979] L.M.C.L.Q. 21; Selvig rightly points out the gaps created by Article 3(*b*) and argues for states which become party to the 1976 Convention to legislate for yet another limitation fund to cover these gaps.

[28] "Oil pollution damage" is not, in fact, defined in the 1969 Convention, but "oil" and "pollution damage" are. See paras. 10–10 and 10–47 *et seq.*, below.

what type of damage is the claim for? States which become party to the 1976 Convention may therefore regulate the limitation of such claims as they please; if they are also party to the 1969 Convention, they must regulate them also in accordance with that Convention.

9–36 The result is that some claims for oil pollution damage which are limitable under the 1957 Convention are not limitable under the 1976 Convention and are therefore open to regulation *sui generis* (and by default, this may mean they are not limitable at all). States becoming party to the 1976 Convention must therefore consider how they will regulate the limitation of such claims. An example would be a claim for oil pollution damage caused by a laden tanker and suffered in a state party to only the 1976 Convention.[29] Another would be a claim for oil pollution damage caused by a laden tanker, suffered in any territory, but made against the ship's bareboat charterer rather than its owner or his servants or agents.[30] By contrast, if the same oil had escaped from a dry cargo ship's bunker tanks, it would be limitable under the 1976 Convention because (a) it is not a claim for oil pollution damage "within the meaning of" the 1969 Convention and (b) it is therefore not excluded from the application of the 1976 Convention. This gives a foretaste of the extreme complexity which the international law on oil pollution liability has now reached, due to the patchwork way in which the regulation of that liability has developed and to the birth of many of the provisions in compromises between opposing groups of states.

C. Conduct Barring the Right to Limit

9–37 There has been a continual development in the provisions denying the right to limit, although a trait common to all three Conventions and to the 1969 Convention is that the phrases used have, to varying degree, been left open to interpretation by courts. In the 1924 Convention, under Article 2(1) the right to limit was not available in respect of "obligations arising out of acts or faults of the owner of the vessel." This idea was developed and re-expressed in Article 1(1) of the 1957 Convention, under which limi-

---

[29] Because damage of this type caused by laden tankers is pollution damage within the meaning of the 1969 Convention. Had the state also been party to the 1969 Convention, then this claim would be limitable under the 1969 Convention in that state.

[30] Because again the type of damage is pollution damage within the meaning of the 1969 Convention and because the 1969 Convention does not regulate limitation of claims for such damage against bareboat charterers—see paras. 10–21 *et seq.*—and so this type of claim is excluded from the rules of the 1976 Convention even if the damage was suffered in a state party to both Conventions.

tation was available "unless the occurrence giving rise to the claim resulted from the actual fault or privity of the owner." The result was that by the time the 1976 Conference came along, courts had so expanded the concept of fault of the owner that in some jurisdictions limitation had either become uncommon or unpredictable. The 1976 Conference therefore borrowed a concept from two earlier maritime Conventions[31] and, by Article 4, provided that:

> "A person liable shall not be entitled to limit his liability if it is proved that the loss resulted from his personal act or omission, committed with the intent to cause such loss, or recklessly and with knowledge that such loss would probably result."

9–38 This wording is very close in intent and effect to the English law concept of "wilful misconduct," which governs the question of when conduct of the assured invalidates insurance cover.[32] Since most policies of insurance covering liability to third parties are governed by English law, in most cases there would be a right to limit under the 1976 Convention while the insurance policy subsists. For most practical purposes, this right is unbreakable. As we shall see, this thinking was repeated when the 1984 Conference amended the 1969 Convention,[33] so that, even though this wording does not currently have that much impact on the question of oil pollution liabilities (due to Article 3(b) of the Convention, discussed above), it will do so in future when the 1984 Protocol enters into force. The pattern that emerges, therefore, is of an increasingly unimpeachable right to limit; but this has been one side of the coin, the other being steadily rising limits of liability.

D. The Limits of Liability

9–39 The problems associated with expressing limits of liability which give a fair result when translated into the different currencies of the parties and of ensuring that those limits hold their value over time are dealt with in paragraphs 10–65 *et seq.* The point that should be made here is that the real current values of these limits has risen over time, as illustrated broadly by Table 9.1. Apart from the caveats expressed in the footnotes to the table, it should be borne in mind that, from the oil pollution viewpoint, the majority of

---

[31] See Article 7 of the International Convention for the Unification of Certain Rules Relating to the Carriage of Passengers by Sea, 1961, and Article 7 of the International Convention for the Unification of Certain Rules Relating to Carriage of Passenger Luggage by Sea, 1967.

[32] See Marine Insurance Act, 1906, s.55(2)(a). The difference between this concept and wilful misconduct is further discussed in paras. 10–148 *et seq.*, below.

[33] See para. 10–147.

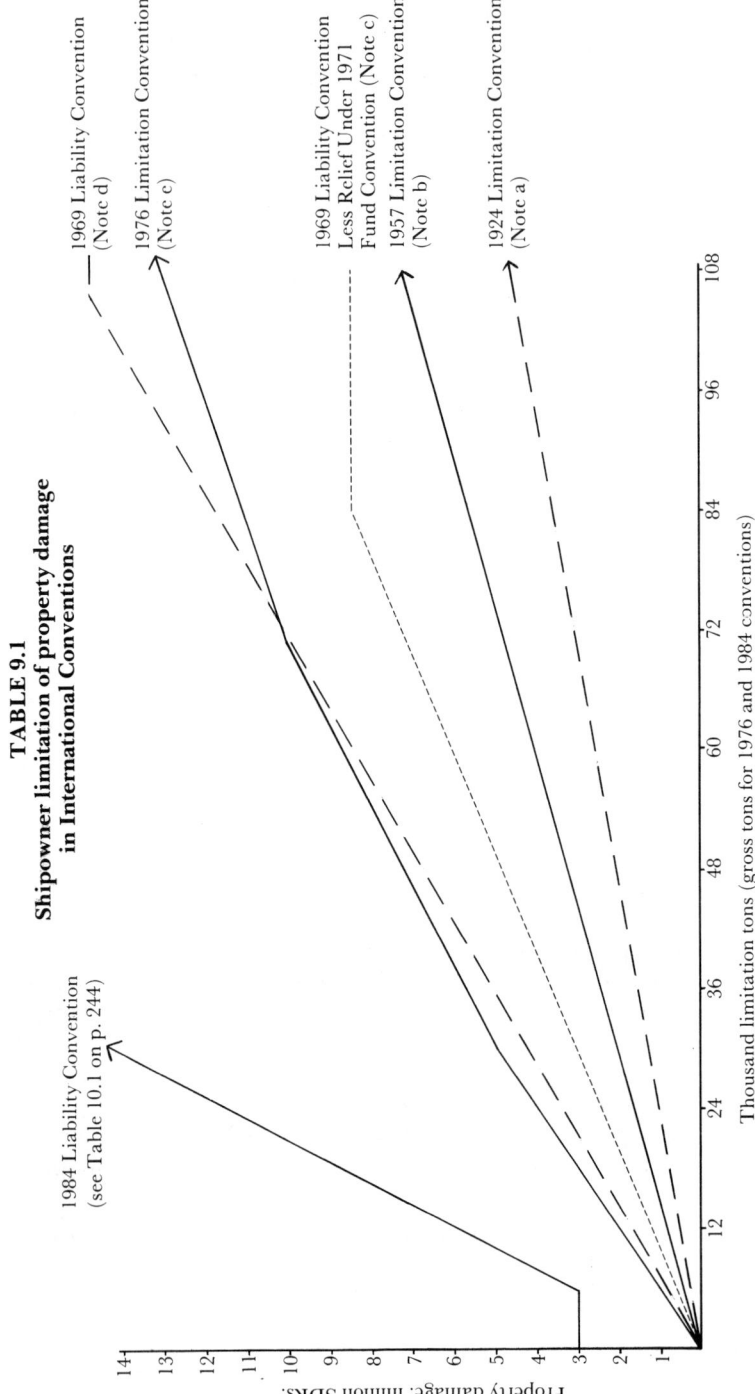

**TABLE 9.1
Shipowner limitation of property damage
in International Conventions**

Notes to Table 9.1

(a) The gold value of the pound sterling in the 1924 Convention has for purposes of this Table been taken at SDR 43.

(b) The limitation amounts in the 1957 Convention have been taken in SDRs in accordance with the 1979 Protocol amending that Convention.

(c) Relief under the 1971 Fund Convention has been calculated in SDRs in accordance with the 1976 Protocol to that Convention.

(d) The limitation amounts for the 1969 Convention have been expressed in SDRs in accordance with the 1976 Protocol to that Convention.

(e) The limitation for the 1976 Limitation Convention and the 1984 Liability Convention is slightly different to that used for the others. This has been ignored.

---

incidents will not be limitable under the 1976 Convention and that the question of limitation will either be governed by the 1969 Convention (as amended) for states party thereto, or by the *lex fori* for other states.

5. CONCLUSION

9–40 The development of international environmental law has been gathering impetus steadily since the early 1960s, and liability for oil pollution from ships has been in the forefront of this development. This has taken place in the elaboration of two types of international Convention: general instruments of global or regional application, dealing with a wide range of environmental problems and including oil pollution from ships, and specific instruments creating very detailed rules which states undertake to give effect to in their municipal law. There has been a growing awareness that liability is an important issue—as important as prevention, which was discussed in Chapters 2 and 5—although states have generally not seen it as part of their role to assume liability themselves, except, perhaps, where they undertake a specific duty in relation to policing a system of prevention. Rather, they have agreed to harmonise their national law on liability for oil pollution from ships, and to place such liability upon the commercial interests involved, by means of multilateral Conventions. The idea of limiting liabilities for oil pollution damage considerably preceded the idea of creating a special regime of liability for oil pollution damage, although this was really a by-product of the idea of generally limiting liabilities of those involved in shipping. The pre-existence of these Conventions on limitation, and the adoption of the 1976 Convention, have added to the complexity of the international law of liabilities.

9–41 It was said in paragraph 9–28, that there was a trend in international law towards the provision of a liability regime which achieved prompt and adequate compensation of claims for environmental damage, including oil pollution damage. As at December 31, 1984, there were 55 out of a possible 150 parties to the 1969 Convention, and 30 parties to the 1971 Convention. This means that there is still a long way to go before this trend becomes a universal practice—and hence, takes its place in the body of customary international law.

10. International Conventions on Civil Liability for Oil Pollution Damage

SUMMARY

10–01 This chapter analyses in detail the scope and impact of the 1969 Liability Convention and all its Protocols, including that of 1984 which creates the 1984 Liability Convention. The strict liability for oil pollution damage which the 1969 Liability Convention creates and places upon the owner of the ship is so defined that it does not in fact apply in a number of important areas, for instance:

1. Oil escaping from river and lake vessels, offshore installations and pipelines.

2. Oil escaping from dry cargo ships and tankers not carrying oil in bulk as cargo.

3. Damage caused by non-persistent oils.

4. Damage suffered by installations outside the territory or territorial sea of a contracting state and all damage suffered on the territory or territorial sea of a non-contracting state.

5. Claims against persons other than the registered owner, his servants or agents.

Contracting states are, therefore, free to regulate these areas under national law. The 1984 Protocol, on the other hand, extends into areas 2, 4 and 5.

10–02 Other features of the 1969 Liability Convention are that the limits of liability prescribed are higher than those existing under Conventions discussed in Chapter 9, and the limitation fund which may be established is a global one applying to damage suffered in all contracting states. The liability is backed up with compulsory insurance. Within its scope, therefore, the 1969 Liability Convention is a revolutionary instrument in international law.

10–03 However, it had not been in operation long before states began to feel that in some respects, notably regarding the limits of liability, it required amendment. This was achieved in 1984 by a Protocol adopted as part of an overall review of the compensation of oil pollution damage available under this Convention and the 1971 Fund Convention discussed in Chapter 11. The 1984 Protocols to both Conventions strengthen the links between the regimes of shipowner and oil industry liability, and make consideration of only one in isolation increasingly difficult.

1. INTRODUCTION

10–04 It was the *Torrey Canyon* disaster in March 1967 which highlighted the need for a new international regime on two major subjects: the rights of a coastal state to intervene in case of an oil pollution threat and civil liability for oil pollution damage. Following deliberations in IMO's Legal Committee, a Diplomatic Conference[1] met in November 1969 to adopt two new instruments. The first of these dealt with the right to intervene, and has already been discussed.[2] The second instrument, the International Convention on Civil Liability for Oil Pollution Damage 1969 ("the Liability Convention") is the subject of this chapter, together with its amending Protocols of 1976 and 1984. In 1976 a Protocol was adopted amending the unit of account in which limits of liability are expressed, it is dealt with together with the main Liability Convention in paragraphs 10–65 to 10–71. In 1984 a further Protocol was adopted; the provisions of this Protocol are, however, of much wider and more complex effect, and so these are treated separately in paragraphs 10–112 *et seq.*

10–05 The Liability Convention contains many features which at the time were regarded as fairly radical in maritime law. The fundamental principle of the Convention is that the shipowner shall not be responsible according to his fault or culpability, but shall instead be strictly liable for oil pollution damage, subject only to certain very limited exceptions. He may limit his liability, but to an amount which exceeds that to which he would have been entitled to limit under the 1957 Brussels Convention discussed in paragraph 9–39 above. The Convention attempts to attract all the litigation in a particular case to the jurisdiction in which the limitation fund is established and to ensure that litigation is only instituted in a state where pollution damage has been suffered. Most of the ships covered by the Convention are required to take out insurance against their limit of liability under the Convention, and a plaintiff may sue the insurer directly, without need to sue the ship or its owner. The Liability Convention entered into force on June 19,

---

[1] The Official Records of the Conference are published by IMO. Documents reproduced in the Official Records and cited herein are followed by the abbreviation *OR* and the page number at which they appear.

[2] Paras. 6–14 *et seq.*, above.

1975, and the 1976 Protocol on April 8, 1981. The 1984 Protocol is not expected to enter into force until the late 1980s.

2. SPHERE OF APPLICATION

A. The Rule of Liability: Article III(1)

(1) THE WORDS USED TO EXPRESS THE RULE OF LIABILITY

10–06 Article III(1) provides that:

"Except as provided in paragraphs 2 and 3 of this Article, the owner of a ship at the time of an incident, or where the incident consists of a series of occurrences at the time of the first such occurrence, shall be liable for any pollution damage caused by oil which has escaped or been discharged from the ship as a result of the incident."

It can be seen at once that the rule is tortious in nature, and is one of strict liability within its sphere of operation. That sphere is delineated not only by the exceptions contained in paragraphs 2 and 3 (discussed below) but also by the words used in the body of Article III(1).

Ship

10–07 Ship is defined in Article I(1) as "any sea-going vessel and any seaborne craft of any type whatsoever, actually carrying oil in bulk as cargo." This includes tankers and combination carriers laden with bulk oil cargo, probably from commencement of loading to completion of final discharge, but they are excluded when they are carrying oil only as slops. Thus, where an OBO has changed from an oil cargo to a dry cargo and has retained oil residues on board in a slop tank, any subsequent escape of oil from the slop tank or bunker tanks would not be covered. A tanker on a ballast voyage is also not covered—hence, the damage caused by bunker oil escaping from the *Olympic Bravery*, which went aground off Ushant, France in March 1976, would not have been covered by the Convention because she carried no oil as cargo at the time. By contrast, if the vessel carries oil in bulk as cargo, the Convention applies to oil pollution caused by any oil emanating from it (whether cargo, bunker or, probably, slop oil) and not just to oil cargo.

10–08 A significant proportion of the world's shipping is excluded from the ambit of the Convention by this definition. Dry cargo ships (except those very few which carry oil cargo in their deep tanks), river and lake vessels (because they are not sea-going),

195

and seaborne offshore installations such as drilling barges (because they do not carry oil as cargo) are all outside the scope of the Convention. However, the converted tankers which store oil at certain offshore installations are a more difficult case; while they are clearly sea-borne craft, can they be said to be "actually carrying oil in bulk as cargo"? There is no definition of what constitutes a cargo, but most would probably take it to be something which is on board for the purpose of being moved by that ship (even though at times during loading, discharge and intermediate port calls the ship is stationary), rather than something which is on board for the purpose of being stored prior to movement by another ship; and on this basis such vessels would be excluded.

10–09 The scope of the Convention was expanded from this limited position by the 1984 Protocol; see paragraphs 10–118 *et seq.*, below.

Oil

10–10 The scope of Article III(1) is further limited by the definition of oil in Article I(5): " . . . any persistent oil such as crude oil, fuel oil, heavy diesel oil, lubricating oil and whale oil, whether carried on board a ship as cargo or in the bunkers of such a ship." The listed oils are merely examples, and the definition could be clearer.

10–11 Whether slop and bilge oils are included depends on whether the clause at the end of the definition is interpreted restrictively or illustratively. No assistance is afforded by the *travaux preparatoires*[3]; the corresponding definition in the Intervention Convention[4] omits the clause altogether and so includes these oils. To interpret the definition here to exclude them would be unnecessarily arbitrary, and the better interpretation is that they are included. The question is important only in the unusual case of an escape or discharge of slops or bilges from a ship laden with oil cargo, but some states such as the United Kingdom felt the need for specific inclusion in their enabling legislation.[5]

Persistent

10–12 A much more important question is what is meant by persistent, which is also not defined. It might be thought that an oil is persistent for this purpose if it has actually caused damage, but this

---

[3] The phrase began life as "whether carried as cargo or in bunkers" in draft Article I (LEG/CONF/4, *OR* 437 at p. 444) and was amended only by the Drafting Committee.

[4] Discussed in para. 6–21, above.

[5] Merchant Shipping (Oil Pollution) Act 1971, s.1.

test is applicable only *ex post facto* and leaves open the question in cases where only preventive measures are claimed. Shipowners and insurers need to know before the oil is loaded if it is persistent. Clearly, in relation to any substance the questions of whether any proportion remains on or in the water column after evaporation has ceased, and how long complete evaporation takes, are relevant to its persistence. The concept of persistence is thus inextricably linked with that of volatility. In practice, most, if not all substances regarded by industry as being persistent contain or consist of the fractions which do not distil below a given temperature.

10–13 The International Oil Pollution Compensation Fund[6] has produced a practical study[7] which adopts the approach of seeking to define what is non-persistent, so that all other oils are persistent:

> " 'Non-persistent oil' is oil which, at the time of shipment, consists predominantly of non-residual fractions and of which more than 50 per cent. by volume distils at a temperature of 340 degrees Centigrade when tested by the ASTM Method D 86/78 or any subsequent revision thereof."

Application of this definition to that in Article I(5), leads to the following broad categorisation (which guides the practice of the IOPC Fund):

PERSISTENT	NON-PERSISTENT
Crude Oils	Gases (*e.g* LNG and LPGs)
Marine Diesel Oil	Gasolenes
No. 4 Fuel Oil	White Spirit
Most Lubricating Oils	Kerosenes (*e.g.* Aviation Fuels)
Residual Fuels (*e.g.* Marine Fuel Oil, Bunker 'C', Nos. 5 & 6 Fuel Oils)	Distillate (*e.g.* Gas Oil, Automotive Diesel, No. 2 Fuel Oil)
Bitumen	Gasolene Blending Components
Aromatic Tar	
Whale Oil	

There are some surprises in this list. Some substances which one might expect to see in the persistent list, such as creosote, are not there because they are not oils at all. By contrast, whale oil is included because it is specifically listed in the definition. It is the only non-hydrocarbon oil.[8] This anomaly is removed by the 1984 Protocol: see paragraph 10–120, below.

---

[6] The IOPC Fund is discussed in detail in Chapter 11.

[7] *A Non-technical Guide to the Nature and Definition of Persistent Oil*, attached to FUND/A.4/11.

[8] It was included at the Conference at the request of the Japanese because it had the same viscosity and persistence as heavy oil and was carried in bulk: LEG/CONF/C.2/SR.14, *OR* 701 at p. 713. But the issue was hardly debated.

Escape or discharge

10–14 For the Convention to apply, the oil which causes the pollution damage or occasions the preventive measures must have "escaped or been discharged" from the ship, so that both accidental and intentional discharges are covered; but the discharge must have been "from the ship." Consequently, where the oil has escaped due to a burst in a pipe connecting a tanker to a terminal or other installation (*e.g* a single buoy mooring or even another ship) the Convention does not apply.

10–15 The requirement that there has been an escape or discharge which causes the pollution damage or preventive measures means that the Convention does not apply to what are known as "pure threat" situations. A stranded tanker may present the threat of an escape of oil, but at the moment no oil has escaped. Coastal states and owners faced with such a situation may incur substantial expenditure to prevent the escape of oil, but if they do the Convention gives them no remedy. This is being challenged by litigation in the United Kingdom in the unusual and curious case of the *Tarpenbeck*.[9]

10–16 The tanker *Tarpenbeck*, loaded with about 1,600 tonnes of lubricating oil, collided on June 21, 1979 with the Royal Fleet Auxiliary *Sir Geraint*, off Selsey Bill, England. The cargo tanks of the *Tarpenbeck* remained undamaged and no cargo was spilled, but some non-persistent light diesel oil from the bunker tanks escaped to sea. The *Tarpenbeck* was towed to a sheltered bay and the cargo was successfully pumped out. There is a dispute over whether any cargo escaped during the pumping operation. The United Kingdom Government and some local authorities incurred costs in taking certain preventive measures, the total of which claims, when added to the owner's costs of towing away the ship and pumping her out, greatly exceeded the ship's limitation amount under the Liability Convention, to which the United Kingdom is a party. The International Oil Pollution Compensation Fund[10] has contested these claims on the grounds that there was no spill of persistent oil, so that neither the Liability Convention nor the 1971 Fund Convention applies.

10–17 While the text of the Convention is clear that preventive measures are not recoverable where there is no spill of persistent oil

---

[9] The account of the facts of this case, and of the dispute over those facts, is taken from the following IOPC Fund Documents: Annual Reports 1982 & 1983; FUND/EXC.4/3; FUND/EXC.7/2.

[10] See note 6, above.

at all, there is still the question of whether the cost of preventive measures taken before such a spill does occur are recoverable. This is not quite so clear, as the relevant definitions (dealt with in detail below) are circular. Preventive measures are defined to be those taken after an incident has occurred, incident means any occurrence which causes pollution damage and pollution damage includes the costs of preventive measures! Perhaps the better view is that here, too, the pre-spill measures are irrecoverable because the actual words which impose liability in Article III(1) say that the pollution damage (including preventive measures) for which the owner is liable is that which is caused by an escape or discharge of oil.

10–18 The exclusion of pure threats has been remedied by the 1984 Protocol to the Liability Convention: see below, paragraphs 10–124 *et seq*. Pure threats are also covered by TOVALOP: see below, paragraph 12–08.

Incident

10–19 The Convention will further apply only where the oil has escaped "as a result of the incident." Incident is defined widely in Article I(8) to be " . . . any occurrence, or series of occurrences having the same origin, which causes pollution damage." This means that the Convention is capable of covering pollution damage caused by the escape of oil from the sunken wreck of a tanker, since such an escape would clearly have its origin in the casualty causing the wreck; but the question of when the elapse of time interrupts the series is not elaborated in the text, and so would be one for the courts in each case.[11] *A fortiori*, the Convention covers preventive measures such as pumping out the oil from the sunken wreck of a tanker, or sealing such a wreck. This kind of action is not uncommon.[12] However, it would seem that there is no intention that a collision between two ships, both of which spill oil, is one incident. The spill of oil from each ship would be an incident in each case.

10–20 The definition of incident was altered by the 1984 Protocol to cover "pure threat" situations; see paragraphs 10–121 *et seq*., below.

---

[11] Hence, conceptually the Convention would cover pollution damage caused by any oil which may escape from the sunken parts of the *Castillo del Bellver*, a Spanish flag 220,000 tonne V.L.C.C. In that case, the ship exploded off Cape Town, South Africa on August 7, 1983, and subsequently broke in two. The bow section containing a very substantial quantity of oil was towed out to deep water and sunk. The stern section sank in situ. Both sections could leak oil after a period on the bottom.

[12] For instance, the *Arrow* case off Nova Scotia in February 1970; the *Tanio* case off Brittany, France, in 1980/81.

(2) PERSONS LIABLE AND EXEMPT FROM LIABILITY

Owners and others

10–21 By Article III(1), it is the owner of the ship who is liable. *Owner* is defined in Article I(3) as:

> " . . . the person or persons[13] registered as the owner of the ship or, in the absence of registration, the person or persons owning the ship. However, in the case of a ship owned by a State and operated by a company which in that State is registered as the ship's operator, 'owner' shall mean such company."

It is important to appreciate that the Convention places no liability whatsoever upon any other person. The owner, on the other hand, is liable irrespective of his residence or domicile, or of the state in which his ship is registered. Hence, liability (as opposed to the right to limit dealt with in paragraphs 10–59 *et seq.*) is placed upon an owner whether or not the state of his ship's flag is a contracting state.

Channelling of liability

10–22 In an attempt to channel the liability to the owner, Article III(4) provides that:

> "No claim for compensation for pollution damage shall be made against the owner otherwise than in accordance with this Convention. No claim for pollution damage under this Convention or otherwise may be made against the servants or agents of the owner."

In so far as the incident is covered by the Convention, the first limb of this provision completely eradicates claims against the owner under general principles of law—such as common law or civil code claims. Thus a person who has suffered pollution damage within the meaning of the Convention[14] must rely only on the Convention for his remedy against the owner for that damage. If, therefore, the damage suffered is pollution damage but does not attract the rule of liability because one of the specific exemptions discussed below applies, he has no other remedy against the owner (although he may have remedies under general principles against the ship's managers, bareboat charterers, builders or other person).

10–23 Claims against the owner must therefore be divided at the outset: those relating to pollution damage and those relating to

---

[13] By Article I(2): " 'Person' means any individual or partnership or any public or private body, whether corporate or not, including a State or any of its constituent subdivisions."

[14] See below, paras. 10–47 *et seq.*

other damage. The Convention limitation fund[15] is available for the former only; with respect to the latter, other principles of liability will apply and there may be a right to limit under another Convention, such as those discussed in paragraphs 9–29 *et seq.*, above.

10–24 This system breaks down, however, where the plaintiff who has suffered pollution damage in a contracting state decides to sue the owner, his servants or agents in a non-contracting state whose law is not, of course, in conformity with the Convention. This occurred in the *Amoco Cadiz* case.[16]

10–25 The second limb of Article III(4) eradicates all claims against the servants or agents of the owner if such claims are for pollution damage as defined in the Convention. This means that states whose existing legislation gives a statutory right of recovery against the ship's agent resident in that state must change that legislation when they become party to the convention to avoid being in breach of their treaty obligations to other parties. This does not mean that a ship's agent could not be required to give a contractual indemnity in respect of any pollution damage which may be caused by the ship to the port authority as a condition of a licence to trade in that port, nor does it mean that a registered owner cannot take such an indemnity from a charterer or manager. It simply denies a tortious or statutory right of recovery by those suffering pollution damage. The question of exactly who is a servant or agent of the owner is not defined in the convention and so is left to municipal law.

10–26 The scope of Article III(4) is therefore somewhat limited: it does not offer any protection to salvors, charterers and others. Where such a person is responsible for the spill, claimants will have claims against both the owner, under this Convention, and such other person, under general principles of municipal law. Hence in the *Tanio* case, the French Government and others who suffered pollution damage sued a wide range of defendants under French general principles of law, including the two bareboat charterers, the technical managers, the classification society and the ship repairer,[17] as well as suing the registered owner under the Convention.

---

[15] See below, paras. 10–76 *et seq.*
[16] See below, para. 10–88.
[17] See IOPC Fund Document FUND/EXC.9/3. The *Tanio*, a Malagasy flag tanker, broke her back off Brittany, France, on March 7, 1980, causing widespread pollution to the coastline of France and minor pollution to the coasts of the Channel Islands. See IOPC Fund Annual Reports 1981 to 1984.

10–27 It is, of course, open to states to make provision in their municipal law for the liability or protection of those who are not the servants or agents of the owner. Several states have done this, to varying degrees protecting them.[18]

10–28 Article II(4) is effectively subject to Article III(5), which preserves the owner's rights of recourse against third parties. Hence, the owner may, for instance, still pursue the owner of a ship which collided with his, causing it to spill oil.

10–29 The partial nature of the scope of the protection afforded by Article III(4) lead to it being considerably widened by the 1984 Protocol as part of a package deal, the centrepiece of which was a substantial increase in the limits: see paragraphs 10–125 *et seq.*, below.

State-owned ships

10–30 Article XI(1) provides that:

> "The provisions of this Convention shall not apply to warships or other ships owned or operated by a State and used, for the time being, only on Government non-commercial service."

This exclusion is in keeping with the role which states have seen for themselves in the regulation of oil pollution[19]; it was felt to be necessary to enable the Convention to attract ratifications. One delegate at the Conference pointed out that the attempt to extend the Convention on the Liability of Operators of Nuclear Ships 1962 to warships and state-owned ships had led to the complete failure of that Convention, which not a single state had ratified.[20]

10–31 Article XI(2) provides that:

> "With respect to ships owned by a Contracting State and used for commercial purposes, each State shall be subject to suit in the jurisdictions set forth in Article IX and shall waive all defences based on its status as a sovereign State."[21]

This provision had a stormy history at the Conference because the U.S.S.R. objected to it all along, on the grounds that in her view it contradicted the general public international law doctrine of sover-

---

[18] *e.g.* the United Kingdom: see para. 15–97; the Bahamas have a provision identical to the U.K.'s in section 26 of the Bahamas Merchant Shipping (Oil Pollution) Act 1976; section 269 of the Norwegian Merchant Shipping Act and section 269 of the Danish Merchant Shipping Act completely exempts any charterer or salvor of the ship from which the discharge is made.

[19] See paras. 9–40 *et seq.*, above.

[20] Mr. Zhudro of the U.S.S.R. LEG/CONF/C.2/SR.13, *OR* 694 at p. 696.

[21] Article IX refers to contracting states where pollution damage has been suffered: see below, paras. 10–84 *et seq.*

eign immunity.[22] The U.S.S.R. objection has continued to the point of making a reservation when she ratified the Convention,[23] which by December 31, 1984 had drawn objections from eleven states; a similar reservation by the German Democratic Republic had drawn seven such objections.

10–32 It is possible that the U.S.S.R. and the G.D.R. are not intending to seek to avoid their ships being liable under the Convention,[24] but rather to avoid being dragged into court against their will, so that in practice a limitation fund would be established and, if the right to limit was denied, claims would be paid in full. In a recent case, the *Jose Marti*, which ran aground off Sweden on January 7, 1981, there is an opportunity to test the U.S.S.R. attitude. There, the Swedish Government have sued the owner, a U.S.S.R. state shipping company, for the resultant pollution damage, but no limitation fund had been established by December 31, 1983 because the owner claims to be exempt from liability altogether on the grounds that Article III(2) applies (government negligence in the maintenance of navigational aids: see below paragraphs 10–38 *et seq*).[25] In the earlier case of the *Antonio Gramsci* (February 27, 1979), the Latvian Shipping Company did establish a limitation fund, but the case is not really in point since the fund was established in the People's Court at Riga in the U.S.S.R.[26]

10–33 The question whether a state would, but for the provisions of Article XI, succeed in claiming sovereign immunity is one for municipal law in each case. As to United Kingdom law, see below Paragraph 16–39.

---

[22] The U.S.A. proposed the provision, LEG/CONF/4, *OR* 437 at p. 500 and LEG/CONF/ C.2/WP.14, *OR* 569. It was discussed in Committee; LEG/CONF/C.2/SR.13, *OR* 694 *et seq.*, and a Russian objection defeated. Article I(3) was amended in an attempt to draw the Convention into line with the systems in socialist states; LEG/CONF/C.2/WP.20, *OR* 575, which was adopted LEG/CONF/C.2/SR.15, *OR* 714 at p. 715, but this did not prevent the U.S.S.R. issuing a declaration at the end of the Conference that it did not agree with Article XI(2)—LEG/CONF/WP.19, *OR* 76.

[23] "The Union of Soviet Socialist Republics does not consider itself bound by the provisions of Article XI, paragraph 2 of the Convention, as they contradict the principle of judicial immunity of a foreign state." IMO Circular CLC/Circ. 13 of June 30, 1975.

[24] In LEG/CONF/WP.19, *OR* 76, the U.S.S.R. stated that, *inter alia*, Article XI(2) was "of no practical value. Any owner of a ship, and under paragraph 2 and 3 of Article I, States or State Companies registered as operators fall under this definition of an owner, in order to limit his liability under the Convention has to constitute a fund, as required by article V, in the Court . . . of one of the States mentioned in Article [IX] of the Convention. In such a case, a Court of that State will be fully competent to consider all the aspects of liability, calculation of damages, as well as division and distribution of the fund." But this ignores the situation where the right to limit is denied.

[25] IOPC Fund Annual Report 1983.

[26] IOPC Fund Annual Report 1978/9.

B. The Geographical Scope of the Rule of Liability: Article II

10–34 Article II limits the geographical scope of the Convention:

"This Convention shall apply exclusively to pollution damage caused on the territory including the territorial sea of a Contracting State and to preventive measures taken to prevent or minimise such damage."

The sole criterion is, therefore, territorial: the nationality, domicile or residence of the defendant is irrelevant. Under this provision, if a ship goes aground on the high seas and oil pollutes the shores of State A (a contracting party) and State B (a non-contracting party), the Convention will apply only to the former damage, but this would include any pollution damage suffered in State A by nationals of any state. Additionally, reasonable preventive measures taken anywhere which are designed to prevent or minimise pollution damage in State A are covered, irrespective of who took them. On the other hand, action to protect interests outside the territorial sea of a contracting state is not covered and is outside the scope of the Convention. Offshore installations, such as rigs and single buoy moorings, fishing grounds for sedentary and free-swimming species and artificial islands might well lie outside the territorial sea, which Article 3 of the Law of the Sea Convention 1982 limits to 12 miles.

10–35 Although states are free, within the limits of their jurisdiction in international law, to make provision in their national law for the recovery of pollution damage to such interests outside the territorial sea and preventive measures taken to protect them, the limited scope of Article II led to its amendment at the 1984 Conference: see paragraphs 10–131 *et seq.*, below.

C. Specific Exemptions from Liability: Articles III(2) & (3)

10–36 Article III(2) contains certain exceptions which in 1969 represented uninsurable risks. Since the Convention makes insurance against a certain amount of liability under it compulsory,[27] as a practical matter those risks were excluded. Hence, by Article III(2):

"No liability for pollution damage shall attach to the owner if he proves that the damage:
(a) resulted from an act of war, hostilities, civil war, insurrection

---

[27] See paras. 10–96 *et seq.*, below.

or a natural phenomenon of an exceptional, inevitable and irresistible character, or"

(b) . . .

(c) . . .

10–37 The first part of this exception does not give rise to any real conceptual difficulty. In relation to the natural phenomenon referred to, the key word is "irresistible," and the exception is more limited than the more familiar concept of act of God. It seems clear that the phrase does not cover hurricanes, for these are negotiable by some ships, but it would cover tidal waves. It has been suggested that for the exception to apply, the defendant must show that in no circumstances could anyone have avoided the accident.[28] In any event, it would appear that it would not be sufficient to prove that the accident could not have been avoided by the exercise of ordinary care and maritime skill.[29]

10–38 The exceptions now contained in sub-paragraphs (*b*) and (*c*) were known at the Conference as "the British exceptions," as the United Kingdom delegation made it clear that insurance was available on the London market if, but only if, those exceptions to liability were adopted.[30] Hence the owner will escape liability if he proves that the damage:

"(b) was wholly caused by an act or omission done with intent to cause damage by a third party, or

(c) was wholly caused by the negligence or other wrongful act of any Government or other authority responsible for the maintenance of lights or other navigational aids in the exercise of that function."

10–39 The inclusion of the word *wholly* keeps outside the scope of the exclusion all cases where there are contributory causes, for instance where a government deliberately damages a stricken ship, as happened in the *Torrey Canyon* case in 1967 and in the *Amoco Cadiz* case in 1979, for there the discharge will be at least partly caused by the initial casualty. The wording of sub-paragraph (*c*) has given rise to some lawsuits in Sweden concerning its interpretation. In 1977 the *Tsesis*, a U.S.S.R. flag ship, went aground in Swedish waters. The owner claimed to be exonerated under Article III(2)(*c*) because the Swedish government had failed properly to maintain maritime charts being used aboard, and these charts were "other navigational aids." The Supreme Court of Sweden delivered

---

[28] M. Forster, *Civil Liability of Shipowners for Oil Pollution*, [1973] J.B.L. 23 at p. 26.

[29] Which is the common law defence of inevitable accident: see *Marsden, The Law of Collisions at Sea*, 11th edition (Ed. K. C. McGuffie), London, 1961, para. 9–15.

[30] LEG/CONF/C.2/WP.35, *OR* 596.

judgement for the owner on these grounds. In an almost identical case, the *Jose Marti*, a U.S.S.R. flag ship owned by the same company as owned the *Tsesis*, went aground on an unmarked rock in a narrow channel off Dalaro, Sweden on January 7, 1981. The owner claimed to be exonerated, as in the *Tsesis* case, under Article III(2)(c) but the Swedish Government and local authority plaintiffs claim that the pilot was at least in part at fault. This case is still continuing.[31]

10–40 Suppose a government fails to mark a wreck at all, as opposed to failing to maintain the light which it has placed there. Article III(2)(c) would not absolve the owner, but then Article III(3) would probably achieve the same result, for this provides that:

> "If the owner proves that the pollution damage resulted wholly or partially either from an act or omission done with intent to cause damage by the person who suffered the damage or from the negligence of that person, the owner may be exonerated wholly or partially from his liability to such person."

10–41 This provision was introduced[32] on grounds of fairness, despite anxieties on the part of some that it would weaken the concept of strict liability.[33] The use of the words "may be" rather than "shall be" leaves some discretion for the court.

3. DAMAGES RECOVERABLE

A. Types of Damage Suffered

10–42 There are many types of damage which might be suffered or claimed following an oil pollution incident, which can be broadly categorised as follows:

(a) Costs of preventing or minimising the effects of the spill (whether undertaken before or after the spill, if any). This may include such items as slick surveillance, laying protective booms, clean-up at sea, loading the oil remaining on board into another ship, refloating the ship and towing it to a place of repair or lightening.

(b) Costs of restoring sea or fresh water (although this may also

---

[31] The accounts of these two cases are taken from IOPC Fund documents FUND/EXC.4/2, FUND/EXC.7/2 & FUND/EXC.9/2.
[32] LEG/CONF/C.2/WP.41, *OR* 601.
[33] LEG/CONF/C.2/SR.18, *OR* 738.

be categorised under category (a)), land, beaches and personal property to their condition before the spill.

(c) Loss of use of the resources in category (b) before their restoration is complete (including loss of livelihood or loss of net income), or, where restoration is impossible, permanent loss of value and loss of use.

(d) Loss of amenity.

(e) Personal injury caused by contact with spilled oil.

(f) Other damage, sometimes called "environmental damage," which does not fit into any of categories (a) to (e).

(g) Interest.

10–43 In relation to categories (c) and (d) there often arises a severe problem of quantification; for instance, where a resource like the seashore cannot be valued (either for rent or sale) because there is no market for seashore as such, how can the loss of use of it be evaluated? Even where property is in use as a revenue earner, such as a fishing boat or camp site, evaluating three weeks' loss of use can still be difficult. This is, however, less of a problem in relation to categories (a) and (b), where the costs can be broadly subdivided into two:

1. Additional Costs, such as work done by contractors, salaries of persons specially employed to deal with the spill, overtime and travel for personnel permanently employed, costs of materials used (*e.g.* dispersant, fuel and clothing), depreciation of equipment used, oil disposal costs and other costs which would not have been incurred if the spill had not occurred.

2. Fixed Costs, such as salaries of persons permanently employed (*e.g.* military personnel), capital costs of equipment used, and other overheads.[34]

10–44 In relation to all items, there is also the question of what causal relationship has to be shown between the incident and the loss. This is the traditional problem of remoteness of damage which has caused so much difficulty to courts around the world.[35] The problem may be illustrated as follows: suppose the spill causes widespread pollution to five miles of prime holiday resort beaches in a small country heavily dependent on foreign tourism, at the period when foreign tourists are booking holidays. As a result very few tourists come that year, and bookings for the next three years

---

[34] The analysis of pollution damage by fixed and additional costs has been developed particularly by the OECD; see e.g. *Combating Oil Spills*, (OECD, 1982, Paris), and *Compensation for Pollution Damage* (OECD, 1981, Paris).

[35] For the position in the United Kingdom, see para. 15–110, and in the United States, paras. 21–47 *et seq.*

turn out to be lower than hoped. The sea-front hotels lose business. So do those in the inland town fifteen miles away. So do the whole-salers supplying food to these hotels, the souvenir sellers, local fishermen and the national airline. Which of these have suffered loss which is too remote (even if quantifiable), and in respect of which years?

10–45 In many legal systems, there is also the question of whether the person claiming the loss must have any particular type of interest (such as ownership, possession or control) in order to claim the loss. This may be characterised as the *locus standi* question.

10–46 It is necessary to bear all these distinctions and problems in mind in analysing the drafting of the Convention on the subject.

B. The Rule in the Convention

10–47 Article III(1) imposes liability on the shipowner only for pollution damage. This is defined in Article I(6) and (7) as follows:

> "6. 'Pollution damage' means loss or damage caused outside the ship carrying oil by contamination resulting from the escape or discharge of oil from the ship, wherever such escape or discharge may occur, and includes the costs of preventive measures and further loss or damage caused by preventive measures.
>
> 7. 'Preventive measures' means any reasonable measures taken by any person after an incident has occurred to prevent or minimise pollution damage."

10–48 There are, therefore, three separate elements to the definition:
  1. Loss or damage by contamination; and this includes
  2. Costs of preventive measures, and
  3. Further loss or damage caused by preventive measures.
Each will be examined in turn.

(1) Loss or Damage by Contamination

10–49 It is clear that fire damage caused by the initial or subsequent ignition of the oil[36] is not caused by contamination, and so is irrecoverable under the Convention. But damage by oil contamination following a fire or explosion aboard ship is; for instance, in

---

[36] Oil floating on the water has subsequently ignited and caused damage in at least three reported cases: *The Wagon Mound* [1961] A.C. 388 and (No. 2) [1967] 617; *The Kazimah* [1967] 2 Lloyd's Rep. 163; and *Eastern Asia Navigation Co. Ltd.* v. *Fremantle Harbour Trust Commissioners* (1951) 83 C.L.R. 353. Also in some unreported ones, *e.g. Betelgeuse* in Bantry Bay, Ireland, January 8, 1979.

the *Betelgeuse* case in Bantry Bay, Ireland in January 1979, an explosion and fire led to the ship breaking its back on the jetty and oil escaped to cause widespread pollution in the Bay. Among many similar cases was the *Castillo del Bellver*, off Cape Town, South Africa in 1983. In both these cases (and others) there was also reported to have been damage caused onshore by rain contaminated by soot and/or oil, and smoke damage. Damage in such cases caused by oil droplets, as opposed to soot or smoke, would be covered by the Convention.

10–50 It may be inferred that personal injury caused by unignited oil is recoverable where this is caused by contamination, but precisely how far other types of loss or damage are recoverable is not spelled out in the definition. The use of the phrase "loss or damage" indicates that more than just physical damage caused by contamination is allowable. But that is where the Convention stops. It does not attempt to solve the issue of causation and remoteness noted above, nor does it deal with *locus standi* or the problems of quantification. All this is left to the interpretation of national laws.

10–51 This very wide area left open by the text of the Convention means that owners are vulnerable to claims of a speculative nature in the "environmental damage" class, or those where quantification is claimed to be allowable by means of abstract models. Examples of this are afforded by two cases from widely divergent jurisdictions. In the *Antonio Gramsci* case,[37] a U.S.S.R. flag tanker went aground off Ventspils, U.S.S.R. on February 27, 1979, causing pollution of the U.S.S.R. waters. About five weeks later, oil from the ship reached the Swedish Archipelago off Stockholm, causing considerable pollution. The Finnish coast was also polluted. The Soviet Ministry for Conservation and Control and Utilisation of Water claimed against the limitation fund set up in the People's Court at Riga, U.S.S.R., an amount for damage to resources and for costs and expenses in restoring the polluted water to a clean condition. The claim, based on the relevant Soviet legislation, was calculated as follows:
   (a) Estimate the quantity of oil spilled.
   (b) Assume it disperses into the marine environment to 50 parts per million.
   (c) Estimate the quantity of water affected by dividing (a) by (b).
   (d) Multiply (c) by an amount per cubic metre read from tables giving expenses per cubic metre for restoration.The result is

---

[37] The account of this case is taken from IOPC Fund Document FUND/A/ES.1/9 and IOPC Fund Annual Report 1979. The case is further discussed below in para. 11–57.

the amount of damage inflicted on the state by the pollution! The Court at Riga allowed the Soviet Ministry's claim.

10–52 The other case is the *Zoe Colocotroni*.[38] That case, which was not under the Convention but under local Puerto Rican legislation, saw claims being made for damage to wild flora and fauna (most of the latter being minute organisms). The United States Court of Appeals, First Circuit, found diminution of value inappropriate to quantify the damage to the marine environment (there was no market for the mangrove swamp concerned even before the oil spill). Instead, it remitted the case back to the court below to determine damages on the basis of "the cost reasonably to be incurred by the sovereign . . . to restore or rehabilitate the environment in the affected area to its pre-existing condition, or as close thereto as is feasible without grossly disproportionate expenditures." The problem was, that such restoration was impossible. The case was settled out of court.

10–53 The Convention's definition of pollution damage is so vague it is really not a definition at all. This lack of clarity, and the desire of states firmly to establish the scope of recoverable damage led the 1984 Conference to adopt an amendment to the definition: see paragraphs 10–137 *et seq.* But this does not mean the definition has no meaning at all. The intention in 1969 was not to create an international basis for purely spurious claims. Such claims are not claims for pollution damage under the Convention.

(2) COSTS OF PREVENTIVE MEASURES

10–54 The requirement of reasonableness in Article I(7) is an important safeguard. Governments and local authorities faced with an oil pollution incident are often tempted by political pressures to do things which are quite unreasonable simply in order to be seen to be doing something. An example is the *Eleni V* spill, off the east coast of England in 1978, when detergent was sprayed at sea on an oil/water emulsion (called "mousse") which could not be affected by such spraying. But the question of what is reasonable, and how far the cost of a measure is to be considered in deciding whether it was reasonable, is nowhere defined in the Convention and so is again a matter for national law.

10–55 Preventive measures undertaken by the owner may be included in the owner's limitation fund (see below, paragraph 10–61) in which connection the distinction between preventive measures and salvage of ship and/or cargo arises.

---

[38] *Commonwealth of Puerto Rico et al.* v. *S.S. Zoe Colocotroni* (1980) 628 Fed.R.2d 652.

(3) FURTHER LOSS OR DAMAGE CAUSED BY PREVENTIVE MEASURES

10–56 In 1969 the detergents used to disperse oil slicks were often highly toxic to marine life, and in some cases were capable of causing more environmental damage than the oil! Although modern dispersants are of much reduced toxicity, it is still possible that their use could cause physical damage, and loss consequent thereon could be suffered. To be recoverable, this loss or damage does not itself need to be caused by contamination but merely by the preventive measures.

(4) CONCLUSION ON DAMAGES RECOVERABLE

10–57 A comparison of the wording of the Convention and the damage which may be suffered shows that the Convention is capable of wide interpretation, and that it leaves most of the issues to national law for solution. In particular, it does not seek to address issues of causation, *locus standi*, quantification of damage and interest, and only to a limited extent does it address the issue of type of damage. Where such issues are addressed, such as the specification of preventive measures, the Convention gives no guidance on the question of the recoverability of fixed costs, nor on what measures are reasonable. As we shall see below in paragraphs 11–53 *et seq.*, this fact, coupled with the equal vagueness of national legislation in many countries, means that in cases where the International Oil Pollution Compensation Fund is involved, the policy of the IOPC Fund in allowing claims prior to adjudication by a court is of fundamental importance, for it is that which in practice determines how much claimants receive (unless they want the trouble and expense of fighting it out in court). In cases where the IOPC Fund is not involved, the policy of the Protection and Indemnity Club involved is equally important.

10–58 The unsatisfactory nature of the definition of pollution damage led to its amendment by the 1984 Protocol: see paragraphs 10–137 *et seq.*, below.

4. LIMITATION OF LIABILITY

A. The Right to Limit

10–59 By Article V(1), "The owner of a ship shall be entitled to limit his liability under this Convention in respect of any one incident . . . " It is only the owner who may limit, and it is only his liability under the Convention which he may limit under Article V. However, in contrast to the position under the 1957 Convention, he may limit his liability in all contracting states to the single amount discussed below, and does not have to limit liability separately in each state where suit is brought.[39-40] Just as the Convention imposes liability on owners not resident or domiciled in a contracting state, so it accords them the right to limit.

10–60 It is, of course, only upon the owner that the Convention imposes liability,[41] but the absence of the right on the part of others to limit in this limitation fund means that oil pollution plaintiffs can recover more than the Article V limitation amount. This will occur when some person other than the registered owner has become legally liable to the plaintiffs under general principles of law. Such a person may, for instance, be the manager, charterer, builder or classification society of the ship. There have been two notable cases where such others have been sued——the *Amoco Cadiz*,[42] where the managers and the builder were sued in the United States in respect of a spill off France in 1979, and the *Tanio*,[43] where a large number of defendants, including the bareboat charterers, managers and ship repairers, were sued in France in respect of a spill off France in 1980.

10–61 However, this situation is likely to arise less often once the 1984 Protocol enters into force, since under the amendments adopted by it far more persons are protected from liability for pollution damage—see paragraphs 10–125 *et seq.*, below.

10–62 In marked contrast to the position under other Conventions,[44] Article V(8) of this Convention provides that:

> "Claims in respect of expenses reasonably incurred or sacrifices reasonably made by the owner voluntarily to prevent or minimise pollution damage shall rank equally with other claims against the fund."

This provision is designed to provide the owner with an incentive to act. However, the restriction to the prevention or minimisation of

---

[39-40] Articles V(1), VI(1)(a) and IX(3).

[41] See para. 10–21.

[42] See para. 21–61 *et seq.*, below. The question of how the suit was brought in the United States when France was at the time a party to the Convention is discussed in paras. 10–87 *et seq.*

[43] Not yet reported, and the case is still *sub judice*. For information see IOPC Fund Document FUND/EXC.9/3.

[44] *e.g.* those discussed above in paras. 9–29 *et seq.*

pollution damage restricts the application of Article V(8) to preventive costs incurred by the shipowner after the oil has escaped.[45]

10–63   Costs incurred in salvage attempts, having as their object the saving of ship or cargo, are also excluded, because in theory at least, they are not undertaken to prevent or minimise pollution damage. The problem is that in many cases the motives for trying to get the ship off the rocks, or for lightening her, are mixed. As was seen in paragraphs 8–80 *et seq.*, if a salvage award was made before 1980 pursuant to a Lloyd's Open Form or otherwise, then a court would be unlikely to find that that cost was within Article V(8) or was a preventive measure within the meaning of Article I(7). But now, under the 1980 Lloyd's Open Form, if the award is enhanced as a result of pollution prevention and the amount of the enhancement is declared, there would seem to be at least a *prima facie* case that the amount of the enhancement would qualify as preventive measures. In such a case, it might be arguable that the owner's share of the award was covered by Article V(8).

10–64   It is possible that a problem will arise with the interpretation of the word "voluntarily." If, for instance, under the law of the flag or of the coastal state the owner is under a duty to take some action in the event of an oil spill,[46] it might be said that such measures were not taken voluntarily.

B. The Limits of Liability: Units of Account

10–65   By Article V(1), the limits to which a shipowner may limit his liability under the Convention in respect of any one incident are "an aggregate amount of 2000 francs for each ton of the ship's tonnage. However, this aggregate amount shall not in any event exceed 210 million francs." The franc referred to is the gold (or Poincare) franc, which "shall be converted into the national currency of the state in which the fund is being constituted on the basis of the official value of that currency" on the date of constitution of the fund.[47] Hence, the basis for the conversion is left substantially to national law. Where an incident causes pollution damage in more than one contracting state, the court charged with distribut-

---

[45] See above, paras. 10–14 *et seq.*
[46] For instance, under Article 39 of the Japanese Marine Pollution Prevention Law 1970 (Law No. 136), the owner of a ship is in certain circumstances under a duty to remove spilled oil. If he fails in his duty, it seems he may be denied the right to limit liability.
[47] Article V(9), which also defines the franc.

ing the limitation fund[48] would be faced with solving the problem of translating the limit into two currencies at once.[49]

10–66 The gold franc was adopted at the 1969 Conference in the hope that it would ensure a measure of uniformity in the real value of the limits, wherever the owner sought to limit his liability. The world currency crisis which led to the flotation of the major currencies on international markets destroyed the efficacy not only of the gold franc but of its conversion at an official rate, and so at a Conference held in London from November 17–19, 1976, there was adopted a Protocol to the Convention which alters the unit of account from the gold franc to the Special Drawing Right (SDR) as defined by the International Monetary Fund (IMF).[50]

10–67 Hence, Article II of the 1976 Protocol (which entered into force on April 8, 1981 but does not bind states party only to the Liability Convention) replaces the amounts of 2000 francs and 210 million francs by 133 SDRs and 14 million SDRs respectively. These amounts must be converted into national currency of the state in which the fund is being constituted on the date of constitution of the limitation fund. If the state concerned is a member of the IMF then the method of valuation is that applied by the IMF; if it is not, the state is free to determine the method itself, or, if its national law does not permit the use of SDRs in this way, then it can retain the use of gold francs, converting them "in such a manner as to express in the national currency of the Contracting State as far as possible the same real value" for the amounts as is expressed in the Protocol in SDRs.

10–68 The whole idea of using such a unit of account in the Protocol is that for any contracting state the amounts, expressed in the currency of that state, vary from time to time with the international strength and weakness of the currency as measured by its relationship with the basket of currencies on the basis of which the IMF values the SDR. It is, therefore, expedient for a state to declare from time to time what is the value of the SDR (or, if they still use

---

[48] See paras. 10–76 et seq., below.

[49] This occurred in the *Tanio* case in 1980, where the coasts of France and the United Kingdom Channel Islands were polluted. The limitation fund was established in France, the Article V(1) limit being converted into French Francs. The problem of the Channel Islands claimants' losses being in pounds sterling was solved by all claimants agreeing that their loss be expressed in French Francs at a particular date.

[50] On the problems of units of account in international transport Conventions, see T. M. C. Asser, *Golden Limitations of Liability in International Transport Conventions and the Currency Crisis* (1973/4) 5 J. Mar. L. & Comm. 646; P. P. Heller, *The Value of the Gold Franc—a Different Point of View* (1974/5) 6 J. Mar. L. & Comm. 73; A. Tobolewski, *The Special Drawing Right in Liability Conventions: An Acceptable Solution?* [1979] L.M.C.L.Q. 169.

the gold franc, the gold franc) in its national currency. This practice is adopted by the United Kingdom, for instance.[51]

10–69   The structure of the limitation amounts is both traditional, in that it is related to the limitation tonnage of the ship,[52] and new, in that it introduces a maximum limit once 105,000 limitation tons is reached. 105,000 limitation tons is approximately the size of a 220,000 deadweight tons Very Large Crude Carrier, and in 1969 this was the maximum insurable liability (then, about U.S.$14 million). In 1969, this represented a substantial increase in the limits over those of the 1957 Convention. Not only is this limitation fund reserved exclusively for oil pollution damage claims under this Convention (so that oil pollution claimants do not compete with personal injury, cargo and other claimants) but it roughly doubles the amounts; this is represented graphically in Table 10.1 below. In fact, the owner may in most cases claim back a portion of his liability under the Convention either from the IOPC Fund or CRISTAL.[53]

10–70   There is, however, an important interaction between the Liability Convention and others, notably the 1957 Convention. Article XII of the Liability Convention provides that it shall supersede earlier Conventions, but only to the extent that such Conventions would be in conflict with it; nothing affects the obligations of contracting states to non-contracting states arising under such Conventions. Hence, a state party to both the 1957 and 1969 Conventions must still honour its obligations to states party only to the 1957 Convention. This means it should allow the owners of ships registered in a state party only to the 1957 Convention to limit their liability arising under Article III(1) of the 1969 Liability Convention at the levels and in accordance with the 1957 Convention.[54]

10–71   The capacity of the insurance markets has expanded dramatically since 1969, and this enabled the 1984 Conference to adopt radically increased limits in the 1984 Protocol; see below, paragraphs 10–141 *et seq.* Discussion of the adequacy of the various limits is postponed until then.

---

[51] This is done by the regular issue of Sterling Equivalents orders.

[52] By Article V(10), the ton referred to in Article V(1) is the net tonnage plus that amount deducted from the gross tonnage on account of engine room space for the purpose of ascertaining net tonnage.

[53] See paras. 11–17 *et seq.* (IOPC Fund) and para. 12–29 (CRISTAL).

[54] Hence the adoption by the United Kingdom and the Bahamas, for instance, of special provisions to this effect: section 8A of the U.K. Merchant Shipping (Oil Pollution) Act, 1971 inserted by section 9 of the U.K. Merchant Shipping Act 1974 (see also paras. 15–136 *et seq.*), and section 28 of the Bahamian Merchant Shipping (Oil Pollution) Act 1976.

215

C. Conduct Barring the Right to Limit

10–72 By Article V(2):

> "If the incident occurred as a result of the actual fault or privity of the owner, he shall not be entitled to avail himself of the limitation provided in paragraph 1 of this Article."

This phraseology is identical to that contained in the 1957 Convention,[55] and so the right to limit hereunder is likely to be denied according to the same principles and in the same situations as under that Convention. Article V(2) must therefore be regarded as just as breakable as Article I(1) of the 1957 Convention, particularly in view of the pace at which courts have developed the scope and level of duties to which shipowners are subject. It is also as likely to give rise to wide variations in interpretation as the 1957 Convention has done.[56]

10–73 There is no definition of what the phrase means, so that interpretation is left to national courts and legislatures. This vagueness is enhanced by the difference in the English and French texts (both of which, under Article XXI, are equally authentic), for the French text refers to "une faute personnelle" in place of "actual fault or privity." It therefore lacks the distinct idea of the privity of the owner in the fault of another, although the simple phrase "faute personnelle" gives ample scope to courts to include this.

10–74 Another item left to national law is the question of burden of proof of the fault and of causation, and this too enhances the width of interpretation. For instance, in the United Kingdom the owner must show, on a balance of probabilities, that the incident did not occur as a result of his actual fault or privity[57] if he wishes to limit his liability; in France, he may do so unless the victim proves that the incident was caused by a "faute personnelle" of the owner.[58]

10–75 The basis for denial of the right to limit is changed fundamentally by the 1984 Protocol: see below, paragraph, 10–147 *et seq*.

---

[55] See para. 9–37.

[56] This width of interpretation can be observed even within the confines of one jurisdiction—see for instance the development of the United Kingdom law on the subject, paras. 15–139 *et seq*.

[57] *The Norman* [1960] Lloyd's Rep. 1 (H.L.), affirmed in *The Marion* [1984] 2 Lloyd's Rep. 1, [1984] 2 W.L.R. 942 (H.L.). These cases are on the 1957 Convention legislation; the 1969 Liability Convention legislation has not yet been tested on the point.

[58] Article 58, Loi No. 67–5 du 3 janvier 1967, the 1957 Convention legislation.

216

D. Establishment of the Limitation Fund

10–76 Article V(3) requires that the owner must constitute a fund for the total sum representing the limit of his liability with the court or other competent authority of any one of the contracting states in which action has been brought under the Convention, either by depositing the sum or by producing a guarantee acceptable to the court. This is the only way to gain the benefit of limitation of liability. In most jurisdictions, the question of denial of the right to limit under Article V(2) will only arise after the establishment of the limitation fund. Strictly speaking, Article V(3) requires that the establishment of the limitation fund should only take place once action under the Convention has been brought, but in practice it seems courts sometimes allow an owner to establish a fund before this.[59] This rule is changed by the 1984 Protocol; see paragraphs, 10–151 *et seq.*, below.

10–77 If the right to limit is granted, under Article V(4) the fund shall be distributed among the claimants in proportion to the amounts of their established claims, and there are provisions for subrogation in certain cases (Articles V(5) to (7)). Unfortunately, in many cases where claims are likely to exceed the limitation amount, the inexpert presentation of claims means that it is a long time before it can be established whether the limitation amount will be exceeded and if so, what is the amount of each claim, so that distribution of the limitation fund is delayed. Since the Convention is silent on the question of who shall get the benefit of any interest accruing on the limitation fund between the date of its establishment and of its distribution, this is left to national law; and so in states where it is the victims who get the interest, there is some small consolation for delay. The result is that, on the whole, victims are reimbursed quicker in cases where there is no question of the limitation amount being exceeded.

10–78 Establishing a limitation fund has the effect that, unless the right to limit is subsequently denied, it becomes the sole source of satisfaction of claims under the Convention, and any arrested property must be released.[60] Claims for damage other than pollution damage suffered in contracting states, and all claims for damage suffered in non-contracting states, are unaffected by the establishment of a limitation fund under this Convention. Consequently, where an incident pollutes the shores of one contracting

---

[59] An example was the *Tanio* case off Brittany, France in 1979.
[60] Articles VI and IX(3).

state and one non-contracting state, the owner's liability can exceed the Convention amount because he does not have the privilege of establishing only one fund, whereas had only contracting states been affected, a fund established in any such affected state would limit claims in all of them. This illustrates an advantage to shipowners if as many states as possible become party to the Convention.

5. JOINT AND SEVERAL LIABILITY

10–79 By Article IV:

> "When oil has escaped or been discharged from two or more ships, and pollution damage results therefrom, the owners of all the ships concerned, unless exonerated under Article III, shall be jointly and severally liable for all such damage which is not reasonably separable."

The preconditions for the application of Article IV are, therefore, (1) there must have been an escape or discharge of oil, as defined in the Convention,[61] from two or more ships, as defined in the Convention,[62] and (2) the pollution damage resulting must be not reasonably separable. This makes the application of Article IV much more restrictive than at first sight. It is therefore insufficient, for instance, for there to have been a collision between a laden tanker and a dry cargo ship (as in the well publicised *Eleni V/Roseline* collision off the Norfolk, England coast in May 1978), or between a laden tanker and an unladen one (as in the less well-known *Gino/Team Castor* collision off Ushant, France in April 1979), or between two laden tankers from only one of which oil escapes. If the discharge is of a non-persistent oil, as in the *Vera Berlingieri/Emmanuel Delmas* collision off Fiumicino, Italy in June 1979 (where gasolene and gasoil escaped from the former), again Article IV will not apply.

10–80 Perhaps the most significant case not covered by Article IV is where only one ship is involved, but several defendants connected therewith are sued. In this, as in all the above cases, the question of whether the defendants are jointly, severally or jointly and severally liable to the victims of oil pollution are covered by principles of national law.

10–81 The situation which is clearly envisaged by Article IV is

---

[61] See above, paras. 10–14 *et seq.*
[62] See above, paras. 10–07 *et seq.*

where two tankers (more is exceptionally unlikely) carrying oil in bulk as cargo collide, and oil from both of them spills into one slick. Although this would constitute two *incidents* and each ship has a separate limitation fund, if it could not be reasonably established what pollution had been caused by which oil, the victims would not have to establish the fact because Article IV gives them a right to proceed against either or both for all of their loss.

10–82 Even where Article IV does apply, there are matters arising in connection with the incident which are not dealt with by the Convention, and therefore are left to national law. Such matters include the principles on which separation of damage will be made, the question of contribution between two or more defendants and the effect of one or both of the ships limiting liability.[63]

10–83 The scope of Article IV has been considerably widened by the 1984 Protocol to cover pure threats: see below, paragraph 10–121.

6. JURISDICTIONAL PROVISIONS

A. Forum for the Plaintiff's Action

10–84 By Article IX:

"1. Where an incident has caused pollution damage in the territory, including the territorial sea of one or more Contracting States, or preventive measures have been taken to prevent or minimise pollution damage in such territory including the territorial sea, actions for compensation may only be brought in the Courts of any such Contracting State or States. Reasonable notice of any such action shall be given to the defendant.

2. Each Contracting State shall ensure that its Courts possess the necessary jurisdiction to entertain such actions for compensation."

The policy of Article IX(1) is to attempt to solve for plaintiffs the problems associated with bringing an action in the jurisdiction where the defendant happens to be. This problem arises, for instance, where claims are brought under the Collision Convention, 1952, as was seen in paragraphs 5–37 *et seq.*, above. For most oil pollution claimants, Article IX(1) means suit may be brought at home. Article IX(2) ensures that in respect of claims under the

---

[63] As to United Kingdom law on these items, see paras. 15–131 *et seq.*

Convention the plaintiff can always serve his writ out of the jurisdiction.[64]

10–85 The corollary to giving jurisdiction to the place where the damage was suffered is to deny it to other fora—hence the use of the word "only" in Article IX(1). But there are two problems with this wording.

10–86 The first is that, taken literally, Article IX(1) goes too far, by denying the right to bring an action for compensation outside the place where damage has been suffered in all cases where "pollution damage" has been suffered; Article IX(1) would therefore prohibit, on its face, the bringing of an action for pollution damage against a demise charterer, manager or other person responsible for the incident in a contracting state other than the one in which the damage has been suffered. This cannot have been the intention of the Conference, since the scheme of the Convention is to place liability upon the shipowner and to deal with the way in which that liability is to be established. Thus it appears reasonable to interpret the phrase "actions for compensation" as meaning "actions for compensation under this Convention." Unfortunately, while the 1984 Conference changed the scope of application of this provision—see paragraph 10–131, below—it did not take the opportunity to improve the drafting in this respect.

10–87 But even this interpretation cannot affect the loophole which forms the second problem. If the shipowner is domiciled in a non-contracting state there is unlikely to be any provision in the law of that state to prevent the plaintiff suing him there—indeed, the law of that state would in the normal course of events grant jurisdiction over the defendant.[65] The Conference was aware of this problem, and of the fact that it was powerless to do anything about it.[66]

10–88 This has been dramatically illustrated in the *Amoco Cadiz* case off France (a contracting state) in 1978, where French plaintiffs, including the French Government, instituted actions in the United States (which was not a contracting state) where the managers of the ship were domiciled. The other defendants

---

[64] See paras. 16–26 *et seq.*, for an analysis of jurisdictional issues under U.K. law.

[65] It is noticeable that the fundamental rule adopted by the EEC Convention on Jurisdiction and the Enforcement of Judgements in Civil and Commercial Matters 1968 ("the Brussels Convention"), is that persons domiciled in a contracting state shall, whatever their nationality, be sued in the courts of that state; everything else is a special exception to that rule (see Article 2). In tort claims, the place where the harmful event occurred is a concurrent, not exclusive, jurisdiction (Article 5(3)).

[66] LEG/CONF/C.2/SR.19, *OR* 745 at p. 755.

included the owner, registered in Liberia (a contracting state) and the ship builder, registered in Spain (also a contracting state). The United States Court of Appeals rejected the plea of *forum non conveniens* made by the ship builder,[67] and the District Court for the Northern District of Illinois proceeded to enter judgement against the owners, managers and their parent company.[68]

Unless non-contracting states are prepared to grant a stay on the ground of *forum non conveniens* there is no way Article IX(1) can apply outside the states party to the Convention.

10–89 Article IX(1) is widened in scope by the 1984 Protocol, but only to reflect the increase in the scope of Article II: see paragraph 10–131, below.

B. Time Limit for the Plaintiff's Action

10–90 Article VIII states that:

> "Rights of compensation under this Convention shall be extinguished unless an action is brought thereunder within three years from the date when the damage occurred. However, in no case shall an action be brought after six years from the date of the incident which caused the damage. Where this incident consists of a series of occurrences, the six years' period shall run from the date of the first such occurrence."

10–91 In major cases plaintiffs are not uncommonly dilatory in adhering to this rule, which focuses attention on when damage can be said to have occurred. The Convention does not define this, leaving it for national law. Since the Convention deals only with "pollution damage," it is to this that "the damage" must refer, but that definition, as we have seen in paragraphs 10–47 *et seq.*, deals with different types of damage and loss, each of which is capable of being suffered at a different moment.[69] The vagueness of the Convention on this point is compounded in states which have a normally very long period of prescription for tort claims, for they seldom have any developed case-law on the date when damage is taken to have occurred. Such states, when becoming party to the Convention, would do well to add a definition in their enabling legislation.

---

[67] Original hearing (N. D. Ill, 1979) *In Re Oil Spill by the Amoco Cadiz off the Coast of France on March 16, 1978*, 491 F.Supp. 170.

[68] For a full account of this case see below, paras. 21–61 *et seq.*

[69] For instance, is the cost of clean-up incurred (a) when the oil hit the coast, (b) when the contract for beach cleaning was signed, (c) when the invoice was rendered, (d) or paid? When are the costs of employed personnel incurred? These and similar questions arise when a plaintiff fails to institute an action within three years of the incident.

10–92 It is, however, clear that the thing which is required for the time bar to be met is the bringing of an action under the Convention. The only actions possible under the Convention are (1) an action for pollution damage against the owner,[70] and (2) an action for the same against the owner's insurer.[71] The Convention implies, therefore, that it is not enough simply to inform the owner of the claim, or to register it with the liquidator of any limitation fund which may have been set up in response to an action brought by some other plaintiff, for neither of these are actions under the Convention. Again, the action must be brought "thereunder," and so must be one brought in accordance with Article IX; but whether a class action brought by a plaintiff on behalf of himself and all those similarly situated would satisfy the requirements of Article VIII in respect of the latter persons is left to national law.[72]

C. Recognition and Enforcement of Judgments

10–93 Article X provides that:

> "1. Any judgement given by a Court with jurisdiction in accordance with Article IX which is enforceable in the State of origin where it is no longer subject to ordinary forms of review, shall be recognized in any Contracting State, except:
> (a) where the judgement was obtained by fraud; or
> (b) where the defendant was not given reasonable notice and a fair opportunity to present his case.
> 2. A judgement recognized under paragraph 1 of this Article shall be enforceable in each Contracting State as soon as the formalities required in that State have been complied with. The formalities shall not permit the merits of the case to be re-opened."

10–94 This important provision attempts to solve the plaintiff's problems of recognition and enforcement of judgements abroad. The Convention could hardly have forced the plaintiff to sue in a state where damage had been suffered and then left him to enforce his judgment in the state where the defendant has his assets as best he may. However, here too there is a problem with the non-contracting state. Suppose the defendant has his assets in such a state: there will be no such provision in the law of that state corresponding to Article X. In such a case, the plaintiff will have to take

---

[70] Article III(1); above, para. 10–21.
[71] Article VII(8); below, para. 10–108.
[72] Class actions were brought in the United States following the *Amoco Cadiz* incident in 1978. See above, para. 10–88.

his defendant where he finds him, and hope that the law of that state does not place too many difficulties in his way to enforcing his judgment.

10–95 It is important to appreciate that the Convention creates its own internal regime of recognition and enforcement: it does not apply to judgements obtained in non-contracting states against the owner or other persons responsible for the spill.[73] These, therefore, are subject to the normal rules applying in contracting states to claims outside the Liability Convention, so that, for instance, they would be governed by purely national law (or, in the EEC, by the Brussels Convention of 1968 to the extent that that Convention applies).

7. COMPULSORY INSURANCE AND DIRECT RECOURSE

A. The Duty to Insure

10–96 Compulsory insurance is not a new concept in international law,[74] but it is unusual. In the legal committee which prepared the draft Convention, there was a division of opinion as to whether or not to adopt the idea,[75] and this continued until nearly the end of the Conference.[76] The drafting of these provisions suffers from the haste with which they were eventually debated.

10–97 The formula worked out is contained in Article VII(1):

> "The owner of a ship registered in a Contracting State and carrying more than 2000 tons of oil in bulk as cargo shall be required to maintain insurance or other financial security, such as the guarantee of a bank or a certificate delivered by an international compensation fund, in the sums fixed by applying the limits of liability prescribed in Article V, paragraph 1 to cover his liability for pollution damage under this Convention.

10–98 The advantages of compulsory insurance become obvious when it is appreciated that there are many one-ship or small fleet

---

[73] Hence, Article X would not require the recognition and enforcement in Spain of the judgement delivered in the United States against the Spanish ship builder in the *Amoco Cadiz* case. See above, para. 10–88.

[74] See, *e.g.* the International Convention for the Unification of Certain Rules relating to Damage Caused by Aircraft to Third Parties on the Surface 1933 (the Rome Convention).

[75] LEG/CONF/4, *OR* 437 at p. 465; LEG/CONF/C.2/SR.14, *OR* 701 at p. 702.

[76] Adopted by 30 votes to 3, LEG/CONF/C.2/SR.14, *OR* 701 at p. 705.

companies in existence whose total assets are insufficient to cover the oil pollution damage which could be caused. But the duty to insure does not arise in all cases where, if the ship caused pollution damage, there would be liability under the Convention: only ships carrying more than 2,000 tons of oil in bulk as cargo must insure. The figure of 2,000 tons was controversial at the Conference, but seems to have been chosen so as to exclude the bulk of the coastal trade, and those dry cargo ships which occasionally carry up to about 2,000 tons of oil in their deep tanks.[77]

10–99 In practice, almost all tankers are insured for liabilities under the Convention either with a Protection and Indemnity Club which is a member of the International Group of Protection and Indemnity Clubs, or with the International Tanker Indemnity Association of Hamilton Bermuda. These insurance companies, each of which functions on the mutual principle, share amongst themselves their oil pollution liabilities to an extent which varies from time to time, and then re-insure such liabilities on the open market. At present, they do not provide cover beyond the limit of the market re-insurance, so that the amount of cover available (which has risen steadily over the years in real terms) is limited to the availability of market re-insurance. For the insurance year beginning February 20, 1984 the available limit was U.S.$300 million.

10–100 The Convention does not require an owner to be insured beyond the amount to which he could limit liability under Article V(1).[78] In theory, therefore, victims who broke the owner's right to limit could find that his oil pollution cover is limited to the limitation figure. In practice, though, most owners take out the full availability of cover.

B. Enforcement: Certificates and Recognition

10–101 Compulsory insurance has its difficulties, and these were all vociferously mentioned at the Conference. The most serious one is the problem of enforcement. The solution adopted was that every ship to which the Convention applies must, under Article VII(2), be issued with "a certificate attesting that insurance or other financial security is in force in accordance with the provisions of this Convention," and this must be done by the state of the ship's

---

[77] LEG/CONF/C.2/SR.14, *OR* 701 at p. 708.
[78] See paras. 10–54 *et seq.*

registry.[79] The form of certificate is laid down,[80] and it must be carried on board.[81]

10–102   Article VII(6) provides that the state of registry shall determine the conditions of issue and validity of the certificate, and that those issued by a contracting state shall be accepted by other contracting states as having the same force as their own. There was felt to be a problem for the authorities of a flag state to estimate the financial security of an insurer who is resident abroad, acting under foreign law and insurance conditions,[82] so Article VII(7) provides that "a Contracting State may at any time request consultation with the State of a ship's registry should it believe that the insurer or guarantor named in the certificate is not financially capable of meeting the obligations imposed by this Convention."

10–103   Articles VII(10) and (11) provide the teeth of the enforcement system. By Article VII(10), "a Contracting State shall not permit a ship under its flag to which this Article applies to trade unless a certificate has been issued under . . . this Article." This is intended to give states the power to stop a ship putting to sea unless she has on board a valid certificate, although the wording is not so clear.

10–104   There was considerable anxiety at the Conference that restrictions placed on ships registered in contracting states would make them competitively at a disadvantage.[83] In fact these fears are groundless, for under Article VII(11) contracting states must ensure that all ships carrying more than 2,000 tons of oil in bulk as cargo, wherever registered, which enter or leave their ports or offshore terminals within the territorial sea have the required insurance.

10–105   The effect of Article VII(11) is that if any tanker, even one registered in a non-contracting state, wishes to trade to a contracting state, it must have a certificate. Under Article VII(2) the certificate must be issued by the flag state, but under Article VII(7) only certificates issued by or certified under the authority of

---

[79] This phrase is defined in Article I(4).

[80] Article VII(2) and (3) and the Annex.

[81] Article VII(4). The certificate is not to be confused with the so-called "blue card" which members of the International Group of Protection and Indemnity Clubs issue, and which ought not to be carried on board. The blue card is addressed to the appropriate authority of the state which will issue the certificate, and it certifies that there is in force in respect of the ship named therein, whilst in the specified ownership, a policy of insurance satisfying the requirements of Article VII.

[82] LEG/CONF/4/Add.1, *OR* 503 at p. 505.

[83] LEG/CONF/4/Add.1, *OR* 503 at p. 506; LEG/CONF/C.2/SR.14, *OR* 701 at p. 702.

another contracting state must be regarded as having the same force as those issued by the recognising state. It follows that certificates issued by non-contracting states to their own ships need not be recognised, and this led to difficulties in the early days following the entry into force of the Convention.[84] These provisions were amended by the 1984 Protocol: see paragraph 10–155, below.

10–106 Certain minor problems are dealt with by special provisions. Article VII(12) deals with government-owned ships—if they are not insured, they must carry a certificate stating that the ship is owned by the state and that its liability is covered up to the Article V(1) limit. Article VII(5), which suffers from tortuous drafting, deals with the problem of change of ownership by apparently requiring that the insurance cover the new owner for at least three months.

C. Proceeds of Compulsory Insurance

10–107 In jurisdictions where the shipowner may limit his liability to the value of the ship and freight following the casualty, it has happened that claimants are faced with next to nothing, while the owner collects the hull insurance. To avoid any possibility of claimants under the Convention failing to get the benefit of the compulsory insurance, Article VII(9) provides that:

> "Any sums provided by insurance or by other financial security maintained in accordance with paragraph 1 of this Article shall be available exclusively for the satisfaction of claims under this Convention."

However, this does not, of course, cover hull insurance!

D. Direct Proceedings against Insurers

10–108 Making direct action against insurers a possibility, against the will of the insurer and by operation of law, is exceptionally rare in insurance law and practice in common law countries,[85] but not so in civil law systems. But at the Conference, the French and others regarded it as vital: "Shorn of the right of direct action in all cases, insurance would lose its prime benefit."[86] So Article VII(8) provides that proceedings may be made directly against the insurer

---

[84] For a history of this, see a 1984 Conference paper, LEG/CONF.6/48.

[85] For instance, in United Kingdom law it is only possible in the case of the bankruptcy or winding-up of the assured: see the Third Parties (Rights Against Insurers) Act 1930.

[86] France: LEG/CONF/4, *OR* 437 at p. 467; Sweden, Greece, Liberia and F.G.R.: LEG/CONF/C.2/SR.14, *OR* 701 at p. 705.

or other persons providing financial security for the owner's liability for pollution damage.

10–109   But the insurer is not left defenceless by any means. Article VII(8) gives him the right to limit his liability to the Article V(1) amount, and this right is absolute—it applies even in the case of the actual fault or privity of the assured owner. Article V(11) provides the insurer with the right to constitute a limitation fund in any event.

10–110   Article VII(8) also affords the insurer all the defences which the owner could have invoked,[87] except the bankruptcy or winding-up of the owner, and the additional defence that the pollution damage resulted from the wilful misconduct of the owner, but no other defences are allowed. This position can only be explained by reference to what the London insurance market would accept: Lord Devlin of the United Kingdom explained that "insurers would insist on that minimum defence in order to allay their fear that a shipowner might decide deliberately to destroy, wreck or strand his ship in order to collect his insurance money."[88] That this fear is not groundless is illustrated by the recent case of the *Salem*,[89] which was deliberately scuttled to try and cover an elaborate cargo fraud, but also doubtless with this in mind. However, the small amount of oil which escaped in the process did not give rise to a claim for pollution damage.

10–111   By not allowing other defences, however, the Convention does remove from the insurer the important protections which he would otherwise have had.[90]

8. THE 1984 PROTOCOL: INTRODUCTION

10–112   Just as the 1969 Liability Convention began with the *Torrey Canyon* incident, so the 1984 Protocol to it began with the *Amoco Cadiz* incident. Both cases caused public outcry, in response to which states at IMO undertook fundamental reviews of international maritime law relating to oil pollution. The fact that this time it took six years, instead of two, reflects the fact that there was

---

[87] *i.e.* those under Article III(2) and (3): see above, paras. 10–36 *et seq.*

[88] Most Protection and Indemnity policies are governed by English law. Under section 55(2)(*a*) of the Marine Insurance Act which governs all time policies, the insurer is not liable for any loss attributable to the wilful misconduct of the assured.

[89] [1983] 1 Lloyd's Rep. 342 (H.L.).

[90] *e.g.* sending the ship to sea in an unseaworthy state with the privity of the assured: section 39(5) of the United Kingdom Marine Insurance Act 1906.

227

widespread disagreement about the changes which ought to be made. Another difference is that the task of revision was undertaken with two conventions in view which had been adopted relatively recently (and which had entered into force even more recently) instead of the *tabla rasa* of 1967.

10–113 By January 1982, the Legal Committee of IMO had narrowed the discussion of possible amendments to about a dozen items which were refined and elaborated over the following two years for the Diplomatic Conference which met at IMO in London for four weeks from April 30, 1984.[91] The most important of those items was the limits of liability. In this, as in nearly all other matters, the proposed items of amendment were seen as a "package deal" incorporating both Conventions. Although in form the instruments adopted are Protocols to the existing 1969 Liability Convention and the 1971 Fund Convention, in substance they are new Conventions which together deal comprehensively with liability and compensation for oil pollution from ships.[92] Before study of the 1984 Protocol to the 1969 Convention, one therefore does well to study the 1971 Convention.[93]

10–114 The other subjects which the 1984 Conference had on its agenda were extension of the Conventions to cover unladen tankers, "pure threat" situations, non-persistent oils and damage suffered beyond the territorial sea; making the shipowner's right to limit less breakable and improving the provisions channelling liability to the owner; clarifying the definition of pollution damage and reducing the grounds on which the owner could be exonerated from liability. In addition, there were the questions of when the owner should be allowed to establish his limitation fund, how to keep the limits of liability up to date from time to time, and, perhaps conceptually the most difficult of all the subjects, how to effect the transition between the old order and the new regime (which became known as the "treaty law" problem). Of all these items, only the extension to non-persistent oils and the narrowing of the grounds for exoneration failed to attract sufficient support at the Confer-

---

[91] The deliberations of the Legal Committee over the two years from January 1982 are conveniently set out in a 1984 Conference Document, LEG/CONF.6/7. For an assessment of the pre-conference position, see R. H. Ganten, "Oil Pollution Liability: Assessment of Possible Revisions," [1984] *Oil & Petrochemical Pollution* 13; and for an account of the Conference and its decisions see R. H. Ganten, "Oil Pollution Liability: Amendments Adopted to Civil Liability and Fund Conventions," [1984] *Oil & Petrochemical Pollution* 41.

[92] Both the legal committee and the conference declined to adopt the method of having an entirely new Convention. This compounded the "treaty law" problems discussed in paras. 10–156 *et seq.*, 11–81 *et seq.* and 11–98 *et seq.*

[93] See Chap. 11.

ence.[94] The others are all discussed below, following the order in which the provisions they amend were discussed above.

10–115   From the above list, it can be seen that only the issue of the limits of liability were directly raised by the facts of the *Amoco Cadiz* incident,[95] the rest having grown out of the discussions in the Legal Committee as time went on. Other issues which that incident does raise are either on the Legal Committee's agenda for future attention (like the reform of salvage law and practice), or have been quietly forgotten (like the question of how to prevent plaintiffs in a contracting state from suing an owner in any non-contracting state where he can be found).

10–116   The 1984 Protocol will enter into force for those who have ratified or acceded to it twelve months following the date on which ten states including six each with not less than one million units of gross tanker tonnage have become parties to it.[96] This is not expected to occur until the late 1980s. The transitional provisions and the deliberate choices given to states[97] mean that the unamended 1969 Liability Convention is capable of continuing to exist in force indefinitely alongside the 1984 Liability Convention (as Article 11 of the 1984 Protocol bids us call the 1969 Convention as amended by the 1984 Protocol). The prospect therefore exists of some states being party only to the 1969 Liability Convention; others, to the 1984 Liability Convention; yet others, to both. The co-existence, for a while, of the 1971 Fund Convention further adds to the complexity. What is already a complicated and ill-understood international regime of liability and compensation for oil pollution damage is therefore destined to become much more so.

9. THE 1984 PROTOCOL: SPHERE OF APPLICATION

10–117   The 1984 Protocol only amends certain of the Articles of the 1969 Convention, so that the amended text—called by Article 11 of the 1984 Protocol the International Convention on Civil Liability for Oil Pollution Damage, 1984—has the original scope of the 1969 Convention except where a specific amendment has been made.

---

[94] Extension to non-persistent oils was rejected in Committee of the Whole, LEG/CONF.6/C.2/SR.3/PROV. Article III(2)(c) was deleted in Committee of the Whole, LEG/CONF.6/C.2/SR.10/PROV., but this deletion failed to attract the necessary majority in Plenary: LEG/CONF.6/SR.5/PROV.

[95] See paras. 21–61 *et seq.*, below.

[96] Article 13 of the 1984 Protocol.

[97] See paras. 10–158 *et seq.*, below.

Ship

10–118 By Article 2(1) of the 1984 Protocol, the definition of *ship* in Article I(1) of the 1969 Convention[98] is replaced by the following:

> "1. 'Ship' means any sea-going vessel and sea-borne craft of any type whatsoever constructed or adapted for the carriage of oil in bulk as cargo, provided that a ship capable of carrying oil and other cargoes shall be regarded as a ship only when it is actually carrying oil in bulk as cargo and during any voyage following such carriage unless it is proved that it has no residues of such carriage of oil in bulk on board."

Whereas the 1969 definition only covered ships carrying oil in bulk as cargo, this definition extends the Convention to unladen tankers carrying slops or those carrying only bunkers. Combination carriers which can carry oil in bulk as cargo are now covered not only when they are doing so, but on any[99] voyage following such carriage until the slops are fully discharged. The burden of proof that the slops are no longer aboard will, on this wording, lie with the party in whose interests it is to prove it: the owner or the 1984 IOPC Fund.[1]

10–119 Dry cargo ships which can carry oil in bulk in their deep tanks are now to be treated like combination carriers; the rest are still excluded, as are river and lake vessels. The new definition appears more capable of covering the converted tankers which store oil at certain offshore installations or elsewhere, for the test now is not whether they are carrying oil in bulk as cargo, but whether they are constructed or adapted to do so, irrespective of their actual use at the time of the incident.

Oil

10–120 By Article 2(2) of the 1984 Protocol the definition of "oil" in Article I(5) of the 1969 Convention[2] is replaced by the following:

> "5. 'Oil' means any persistent hyrdocarbon mineral oil such as crude oil, fuel oil, heavy diesel oil and lubricating oil, whether carried on board a ship as cargo or in the bunkers of such a ship."

---

[98] See paras. 10–07 *et seq.*, above.

[99] "Any" was preferred to "the": LEG/CONF.6/C.2/SR.18/PROV.

[1] The 1984 IOPC Fund might want to prove it because if the residues were not on board there can be no question of compensation being due from it: see generally Chap. 11. The conference intended the burden to be on the owner as a general rule: LEG/CONF.6/C.2/SR.18/PROV.

[2] See paras. 10–10 *et seq.*, above.

The anomaly of whale oil has been removed, and the definition has been marginally clarified by the complimentary inclusion of "hyrocarbon mineral," but there is still no definition of what is persistent and what is not. The existing practice will therefore continue to be important. However, after years of debate in the Legal Committee of IMO, the Conference speedily decided against including non-persistent oils.[3]

Incident: pure threats

10–121   By Article 2(4) of the 1984 Protocol the definition of *incident* in Article I(8) of the 1969 Convention is replaced by the following:

> "8. 'Incident' means any occurrence, or series of occurrences having the same origin, which causes pollution damage or creates a grave and imminent threat of causing such damage."

Articles III(1) and IV of the 1969 Convention underwent consequential amendment.

10–122   This amendment has the important effect of extending the ambit of the Convention to the time before an escape or discharge of oil actually occurs. In this respect the Convention is brought into line both with TOVALOP and CRISTAL[4] and with the 1969 Intervention Convention.[5] The wording does not require the subsequent discharge or escape of oil for preventive measures taken immediately the threat appears to be recoverable under the Convention: indeed, it will always be the hope that such an escape will not occur and very often the preventive measures will have the prevention of any escape as their object.

10–123   However, the preventive measures taken before an escape will not be recoverable unless the threat of causing pollution damage is grave and imminent. These words were preferred by the Conference over the wider word "serious."[6] The only way in which pollution damage could occur (apart from the preventive measures which it includes) is if there is an escape or discharge of oil. So the Convention cannot apply at least until there is a grave and imminent threat of an escape or discharge of oil. But more than that is necessary. If there are no grounds for believing that the escape of oil would cause any pollution damage, the necessary threat does

---

[3] It became clear that there would not be the required majority in Plenary (two-thirds of states present and voting, including one half of contracting states present and voting): LEG/CONF.6/C.2/SR.3/PROV.

[4] See paras. 12–08 and 12–23, below.

[5] See paras. 6–14 *et seq.*, above.

[6] LEG/CONF.6/C.2/SR.18/PROV.

231

not exist; and while "grave" does not really add much to the concept of "threat," "imminent" certainly does. As the CMI pointed out,[7] it means "near at hand . . . on the point of happening." So if the prospect of a discharge is too remote in time the Convention does not apply.

10–124 Some states were worried about the kind of situation which arose in the *Tanio* case off Britanny, France in 1980 or in the *Castillo del Bellver* case off Cape Town, South Africa in 1983, where part of the wrecked ship sinks with oil aboard. There is no "imminent" threat of an escape of oil from the sunken part, but it is rightly feared that in time the wreck will leak. Why should not the costs of pumping out or sealing the wreck be recoverable then?[8] In the majority of such cases, there will almost undoubtedly have been an escape of oil prior to the sinking, pollution damage will have been caused, and so the question of whether the costs of pumping out the wreck are recoverable will probably revolve around whether this is a reasonable preventive measure to take, and not whether an incident has occurred.

Channelling of liability

10–125 By Article 4(2) of the 1984 Protocol, the provisions of Article III(4)[9] which attempt a measure of channelling of liability to the owner are replaced by the following:

> "4. No claim for compensation for pollution damage may be made against the owner otherwise than in accordance with this Convention. Subject to paragraph 5 of this Article,[10] no claim for compensation for pollution damage under this Convention or otherwise may be made against:
>
> (a) the servants or agents of the owner or the members of the crew;
> (b) the pilot or any other person who, without being a member of the crew, performs services for the ship;
> (c) any charterer (howsoever described, including a bareboat charterer), manager or operator of the ship;
> (d) any person performing salvage operations with the consent of the owner or on the instructions of a competent public authority;
> (e) any person taking preventive measures;
> (f) all servants or agents of persons mentioned in subparagraphs (c), (d) and (e);

---

[7] LEG/CONF.6/39.
[8] See, *e.g.* the intervention of the F.G.R. at LEG/CONF.6/C.2/SR.17/PROV.
[9] See paras. 10–22 *et seq.*, above.
[10] Which preserves the owner's rights of recourse.

unless the damage resulted from their personal act or omission, committed with the intent to cause such damage, or recklessly and with knowledge that such damage would probably result."

10–126 This is a departure from the old text of considerable significance; its impact on the total recovery plaintiffs might expect was recognised by the many states who declared that their attitude to how widely it was drawn depended on the outcome of discussions on the limits.[11] The provision is designed to obviate the need for those mentioned to take out insurance against oil pollution liability.[12] The list does not cover quite a number of those who may become liable under general principles of law, such as the builder, repairer and classification society of the ship, those who recruit personnel for the ship (unless they are managers or operators as well), the owner of the radar and other hired equipment aboard ship, the shipper of goods aboard the ship (unless he is also a charterer) and, last but not least, those concerned in any vessel which collides with the ship; but the Conference rejected a proposal which would have protected everyone except the owner.[13]

10–127 In giving protection to a wider range of persons than those mentioned in the 1969 text, the Protocol has built upon a trend which a number of states had initiated in their national law[14]; for states are free to regulate the liabilities of those not mentioned in Article III(4).

10–128 A comparison of the new and old[15] texts shows that while the position of the owner is unchanged, the servants and agents of the owner are marginally less well protected than they are under the 1969 Convention because now the proviso at the end applies to them. However, as we shall see,[16] the chances of this proviso applying are in practice very slight indeed. The others mentioned are put in a much more favourable position than before.

10–129 The effect of including salvors and those taking preventive measures is to encourage them to take action without fear of subsequently being on the wrong end of a lawsuit. The wording relating to salvors is curious in referring to the undertaking of operations "with the consent of the owner"; it is unclear whether the consent of the master is sufficient (since, even in these days of

---

[11] See for instance LEG/CONF.6/C.2/SR.11/PROV.

[12] LEG/CONF.6/7, p. 22.

[13] LEG/CONF.6/C.2/WP.24, an OCIMF proposal, gained no support and the new text was adopted: LEG/CONF.6/C.2/SR.11/PROV.

[14] *e.g.* the United Kingdom, Bahamas, Denmark, Norway.

[15] See paras. 10–22 *et seq.*, above.

[16] See paras. 10–147 *et seq.*, below.

sophisticated telecommunications, the salvage may be undertaken without initial reference to the owner). There is an argument that if the master can, under his general authority,[17] bind the owner to the salvage contract when his ship is in danger, the salvage operations are undertaken with the owner's consent.

10–130 The somewhat redundant, specific mention of the bareboat charterer originated at a time when some states were advocating joint and several liability for the bareboat charterer with the owner, and it thus provided a clear alternative philosophy, which won the day at the Conference.[18]

Geographical scope

10–131 By Article 3 of the 1984 Protocol, Article II of the 1969 Convention is replaced by the following:

> "This Convention shall apply exclusively:
>> (a) to pollution damage caused:
>>> (i) in the territory, including the territorial sea, of a Contracting State, and
>>> (ii) in the exclusive economic zone of a Contracting State, established in accordance with international law, or, if a Contracting State has not established such a zone, in an area beyond and adjacent to the territorial sea of that State determined by that State in accordance with international law and extending not more than 200 nautical miles from the baselines from which the breadth of its territorial sea is measured;
>> (b) to preventive measures, wherever taken, to prevent or minimise such damage."

Article IX(1) of the 1969 Convention[19] is consequentially amended in scope.

10–132 This provision was one of the most contentious at the Conference, and its eventual adoption in Plenary on a roll-call vote[20] just met the required majorities. It is possible that some of the relatively high number of abstentions on the adoption of the whole text of the Protocol (sixteen) were due to the inclusion of this provision.

10–133 Those in favour of extending the geographical scope of

---

[17] See, *e.g.* G. Brice, *Maritime Law of Salvage* (1983, London) para. 434; *The Unique Mariner* (No. 1) [1978] 1 Lloyd's Rep. 438.
[18] LEG/CONF.6/C.2/SR.11/PROV.
[19] See paras. 10–84 *et seq.*, above.
[20] LEG/CONF.6/SR.5/PROV.

the Convention[21] advanced largely legal reasons in justification of the idea. They felt that recent developments in international law gave states the jurisdiction to protect the environment within the exclusive economic zone. They clearly had in mind not only the provisions of the Law of the Sea Convention 1982[22] but the existing assumption by many states of jurisdiction over areas beyond the territorial sea. Clearly, if this Convention were to extend to the exclusive economic zone, it would help to establish the concept as a creature of international law, even if the Law of the Sea Convention 1982 did not enter into force. Although it was specifically stated at the Conference by only one delegate, and then in a somewhat veiled reference,[23] this is precisely what those against the extension feared. The reasons publically advanced against extension were pragmatic and no doubt also genuinely held: there was a danger that such an extension would lead to unnecessary preventive measures being taken on the high seas and thus to exorbitant claims; experience had shown that in relation to oil (as opposed to hazardous and noxious substances) the nature of the risks did not warrant it, particularly having regard to the fact that preventive measures taken on the high seas were already recoverable if taken for the purpose of preventing pollution damage in the territorial sea or territory of a contracting state.[24]

10–134 The provision finally adopted is less extensive than at one stage proposed by the advocates of extension, a proposal to cover the Continental Shelf of a Contracting State, including artificial islands and other installations under the coastal state's jurisdiction, being withdrawn in order to attract more support.[25] Its curious drafting is also the product of compromise. Some states objected to the use of the words "exclusive economic zone," since

---

[21] The states of North, Central and South America and Australasia were the main protagonists.

[22] *e.g.* Article 56(1): "In the exclusive economic zone, the coastal state has . . . (*b*) jurisdiction as provided in the relevant provisions of this Convention with regard to: . . . (iii) the protection and preservation of the marine environment . . . ."

[23] *Per* Mr. Carly of Belgium at LEG/CONF.6/C.2/SR.5/PROV.: " . . . Extension of the Convention's scope to areas beyond the territorial sea would incur the risk that, in subsequent years, some countries might claim that those waters were their property, since they could obtain compensation for damage occurring therein; they might indeed claim that the 200-mile zone comprised territorial waters. Belgium attached great importance to the freedom of the high seas. Many States were now claiming rights over all kinds of areas which had formerly been high seas."

[24] See especially LEG/CONF.6/C.1/SR.4 and 5/PROV; LEG/CONF.6/C.2/SR.5, 19, 24, 25, 26 and 27/PROV for the public debate on the issue and LEG/CONF.6/SR.5/PROV for the voting in Plenary.

[25] Paragraph (c) of LEG/CONF.6/C.2/WP.27 was withdrawn: LEG/CONF.6/C.2/SR.24/PROV.

many states had not declared one, and hence the addition of the alternative formulation. The reference to the zone being "established in accordance with international law" leaves open the question of whether the exclusive economic zone has yet reached the status of incorporation into international law.

10–135 The effect of this new provision is to allow certain claims for loss or damage which are not valid under the 1969 Convention, namely (i) loss or damage caused by contamination outside the territorial sea but inside the 200 mile zone; in this connection it is important to realise that the state concerned must either have established an exclusive economic zone or have made a determination of an area within which the new provisions are to apply; and (ii) preventive measures taken anywhere to prevent or minimise such damage. Whether in practice this will lead to any significant increase in the cost of claims remains to be seen. Since the vast majority of oil pollution incidents occur in ports or other restricted waters (most of which will be within the territorial sea), it may be thought unlikely.[26] But nonetheless, the distinct possibility remains that in individual cases states may be tempted to respond to pressure of public opinion by pursuing slicks blowing out to sea with unnecessary (and possibly ineffective) detergent sprayers, or to bring claims for damage to fisheries which might not otherwise have been brought. There may also be claims for preventive measures to protect offshore installations.

10–136 One interesting question which does not require the occurrence of an incident to test it and which the text leaves open, is whether in international law a state currently has jurisdiction to allow such claims beyond the territorial sea. This was discussed in paragraphs 5–42 to 5–44. While it would not arise in connection with a ship flying the flag of another party to the 1984 Protocol (for such a state would, in becoming party, have acquiesced in the assumption of such jurisdiction by other parties in respect of its ships) the new text could form the basis of international disputes by non-parties to the 1984 Protocol—just as, at present, both non-parties to the 1969 Convention and parties thereto may dispute the

---

[26] In the debate on this provision in committee, the observer from ITOPF Limited, whose major function is to attend serious oil spills, stated that it had analysed the 70 spills it had attended from 1978 to 1983, and found that "More than 90% of oil spills of any size from tankers occurred during routine cargo handling operations, normally in port; tanker accidents such as collisions and groundings comprised more than 80% of all spills in excess of 5,000 barrels; and 75% of those larger spills occurred in ports, port approaches and restricted waters, virtually all constituted a threat to the territorial sea . . . ": LEG/CONF.6/C.2/SR.5/PROV.

right of a state to declare a territorial sea in excess of that permitted in international law and then apply the 1969 Convention to it. For reasons such as this, it is possible that many states will either be cautious in ratifying the 1984 Protocol, or they may ratify it but only declare an area of application considerably less wide than the 200 mile maximum of the amended Article II, until the status in international law of the exclusive economic zone becomes clearer.

Pollution damage

10–137 By Article 2(3) of the 1984 Protocol, the definition of "pollution damage" in Article I(6) of the 1969 Convention is replaced by the following:

> "6. 'Pollution damage' means:
> (a) loss or damage caused outside the ship by contamination resulting from the escape or discharge of oil from the ship, wherever such escape or discharge may occur, provided that compensation for impairment of the environment other than loss of profit from such impairment shall be limited to costs of reasonable measures of reinstatement actually undertaken or to be undertaken;
> (b) the costs of preventive measures and further loss or damage caused by preventive measures."

10–138 This was another subject which gave the Conference great difficulty. The legal committee had adopted a definition which had been elaborated by the International Maritime Committee (the CMI), but the Conference was unable to agree on it and so set up a working group to elaborate a new definition. In this group, there was a broad measure of agreement on what the definition should try to achieve[27]: it should cover property damage and personal injury, loss of use or exploitation by people who normally use or exploit the polluted environment (such as, in particular, hoteliers), costs of reinstatement of the environment or alternatively compensation for impairment thereof, and costs of preventive measures. There was also a desire to disallow speculative or theoretical damages such as those which were the subject of the *Zoe Colocotroni* and *Antonio Gramsci* cases.[28] However, questions of causation caused particular difficulty, and there was no agreement about how to express these general aims. The working group therefore elaborated two texts[29] of which the Committee eventually

---

[27] Author's personal notes.
[28] See paras. 10–51—10–52 above.
[29] LEG/CONF.6/C.2/2.

chose this one, predominently, it seems, because it bore a closer resemblance to the existing text than the other alternative based on the CMI draft.[30] It is, inevitably, a compromise between those looking for a text which was readily suited to national interpretation and those seeking the opposite.

10–139 Sub-paragraph (*b*) is identical to the existing 1969 text and so needs no further comment.[31] The first part of sub-paragraph (*a*) is also identical to the 1969 text, so that all that the new definition does is add the proviso thereto. This proviso makes it very clear that compensation for impairment of the environment is included within "loss or damage," but is intended to eliminate the speculative and theoretical claims referred to above. Again, the inclusion of the phrase "other than loss of profit" makes it clear that loss of profit is allowable as a type of "loss or damage" and confirms the present practice of the IOPC Fund which was elaborated to the Working Group.[32] However, it does not solve the questions of causation discussed above in connection with the 1969 text.[33] Loss of profit had to be excluded from the proviso because it was felt that hoteliers, fishermen and others could be said to be claiming for "impairment" of the environment, and it was not desired to limit claims by them. The final phrase—"or to be undertaken"—may puzzle some. This was designed to meet the problem, which had been encountered by the IOPC Fund in the *Tanio* case in Brittany, France in 1980, that some claimant authorities faced with having to reinstate the environment may not have the funds to do so until paid by the shipowner. Although the drafting does not require that the claimant repay any sums received if he does not go ahead and make the reinstatement after being paid, it was suggested by an Observer in the Plenary that the claimant ought to have taken all necessary steps for remedial work to be undertaken as soon as money became available.[34] The fact that this drew no criticism is indicative of the feeling at the conference and in the working group that "to be undertaken" implies a very high degree of likelihood and intention that the work really will be done.

---

[30] For the committee debate, see LEG/CONF.6/C.2/SR.3, 4, 15, 16, 17 and 24/PROV. Another reason for the choice of this text was that the alternative text, which referred to damage or loss actually sustained as a *direct result* of contamination, may in some states have meant that many hoteliers and others involved in tourism might not have been able to recover.

[31] See paras. 10–47, 10–54 *et seq.*, above.

[32] The IOPC Fund's practice on pollution damage claims is discussed below in paras. 11–60 *et seq.*

[33] See para. 10–44, above.

[34] LEG/CONF.6/SR.4/PROV, by the International Group of P. & I. Clubs.

10–140 The new definition is still very vague, leaving a great deal to national courts and legislatures. While some progress is made towards solving the *locus standi* and speculative claim issues, the issue of causation is not solved, nor is there any mention of interest or quantification. It is therefore of somewhat marginal utility, and states ratifying the 1984 Protocol are well advised to draw up their own detailed rules.

10. THE 1984 PROTOCOL: LIMITATION OF LIABILITY

A. Amended Limits of Liability

10–141 The single most important purpose of the 1984 Conference was to amend the limits of liability for both the 1969 and 1971 Conventions. While all states were agreed that the limits needed raising—although it must be said that uncompensated claims under the present system were estimated at relatively low amounts[35]—that was where agreement ended. There was a particularly wide diversity of view. At one end of the scale, the Japanese (who were particularly important because to date they had consistently been the largest single contributor to the IOPC Fund, contributing some 39 per cent. of its costs) led a group favouring figures like U.S.$ 30 million for the Liability Convention and U.S.$ 100 million for the Fund Convention. Proposals ranged from there through various computations to the high range, favoured by the United States, Canada and others, of around U.S.$ 70 million for the Liability Convention and over U.S.$ 225 million for the Fund Convention.[36] Another issue divided states too: although all were agreed that the "small ship" presented a special problem because it was relatively easy for its limit to be exceeded, and that the solution was to provide a minimum limit for all ships below a specified size, there was no agreement either on what this limit should be or on the maximum size of ship it should apply to.

---

[35] The most comprehensive information on spill costs was supplied by the International Group of P & I Clubs in LEG/CONF.6/14, updated by LEG/CONF.6/C.2/WP.34. The P. & I. Clubs analysed some 17,000 oil pollution incidents which occurred between 1970 and 1982. Spill costs were estimated in constant 1983 dollars. The 1969 and 1971 Convention maximum limits were then applied as if both Conventions had applied to all spills and as if in all cases exceeding the owner's limit he had been entitled to limit. Of the total estimated costs of U.S.$ 1,259 million, 90% was covered by the two Conventions and just 10% would have been uncompensated, arising from just two incidents.

[36] For the substantive debate on the limits see LEG/CONF.6/C.2/SR.5, 6, 8, 15, 22, 28 and 29/PROV.

239

Proposals ranged from 2,000 to 30,000 tons for the ship size, and from U.S.$ 3 million to 17.5 million for the amount.[37]

10–142 The differences were discussed largely in informal groups throughout the conference, so that the real story does not appear from the Summary Record. In the end, it looked as if Committee 2 might get stuck on the issue and endanger the whole conference, since Committee 1, dealing with the Hazardous and Noxious Substances side of the conference agenda, had already faced up to the fact that agreement could not be reached on an HNS Convention at that conference. So the Chairman of Committee 2, Mr. Mans Jacobsson of Sweden, was asked to produce a proposal, which the committee rapidly adopted on the penultimate day as the best compromise which could be achieved.[38] But that was not the end of the story. The Japanese in particular were very unhappy about the compromise,[39] and there were sixteen abstentions when it came to adopting the whole Protocol, some of which undoubtedly will have been caused by fears that the limits are too high.

10–143 Hence, Article 6 of the 1984 Protocol replaces Article V(1) of the 1969 Liability Convention[40] with the following:

> "1. The owner of a ship shall be entitled to limit his liability under this Convention in respect of any one incident to an aggregate amount calculated as follows:
> (a) 3 million units of account for a ship not exceeding 5,000 units of tonnage;
> (b) for a ship with a tonnage in excess thereof, for each additional unit of tonnage, 420 units of account in addition to the amount mentioned in subparagraph (a);
> provided, however, that this aggregate amount shall not in any event exceed 59.7 units of account."

The unit of account provisions are, by Article 6(4), the same as those in the 1976 Protocol to the Liability Convention,[41] and replace those of Article V(9) of the 1969 Convention. The unit of tonnage is also changed, in line with the 1976 Convention on Limitation of Liability for Maritime Claims, to be the gross tonnage calculated in accordance with the International Convention on

---

[37] In fact, the Oil Companies International Marine Forum had suggested 50,000 tons and U.S.$ 35 million minimum, in LEG/CONF.6/INF.3, but this never had support among states. The Canadians provided the highest figures amongst the states.

[38] The Chairman's proposal, LEG/CONF.6/C.2/WP.44, was adopted at LEG/CONF.6/C.2/SR.28 and 29/PROV.

[39] LEG/CONF.6/C.2/SR.28/PROV. and LEG/CONF.6/SR.5/PROV.

[40] See paras. 10–65 *et seq.*, above.

[41] See paras. 10–67 *et seq.*, above.

Tonnage Measurement of Ships, 1969.[42] This makes accurate direct comparison between the 1969 and 1984 limits difficult.

10–144 The upper limit of 59.7 million SDRs is diplomatically chosen below 60 million, no doubt to please some important group of states for whom 60 million was too much. The small ships limit is at the lower end of the range debated. The limit goes between the two points on a straight line basis. The maximum is reached at 140,000 gross tons, which is a larger V.L.C.C. It is unlikely that there will be all that many such ships in service come the late 1980s when the 1984 Protocol may come into force. A more interesting figure to look at would be that for a ship of some 83,000 tons. This is approximately the tonnage at which the maximum 1969 Convention liability is reached after taking into account the shipowner relief under the Fund Convention.[43] It is also more the size of ship which is expected to be predominent in the long-haul tanker trade in the 1990s. The limit of liability for this ship is around 36 million SDRs. A comparative representation of the limits is shown in Table 10.1.

10–145 The P. & I. Clubs have analysed some 17,000 oil spills covering 1970 to 1982 as if the new limits had applied to them, estimating the cost of each spill in 1983 money. On this basis, the share of total oil spill costs which would have been borne by the shipowner rose from some 47 per cent. (under the 1969 Liability Convention limits) to at least 68 per cent., and the cover provided by the Fund Convention would have been required for an average of only 1.5 incidents per year.[44] There can be little doubt, then, that the 1984 Liability Convention will provide such an attractive level of cover that many states will not feel the need to become party to the 1984 Fund Convention, and will rely on the revision provisions of the 1984 Liability Convention to keep them covered as time progresses.[45]

10–146 For the application of the new limits in cases where a state is party to both the 1984 and 1969 Conventions, see paragraphs 10–158 *et seq.*, below. The relationship between the new and old limits is further discussed in connection with the 1984 Protocol to the Fund Convention: see below, paragraphs 11–87 *et seq.*

---

[42] Article 4(5) of the 1984 Protocol replacing Article V(10) of the 1969 Liability Convention.
[43] See para. 11–17 *et seq.*, below.
[44] Private communication from The Britannia Steam Ship Insurance Association Limited. Based on past experience of oil spills, the new Liability Convention limit would have covered some 68% of the damage, and the new limit for the Fund Convention, 30%.
[45] See para. 10–154, below.

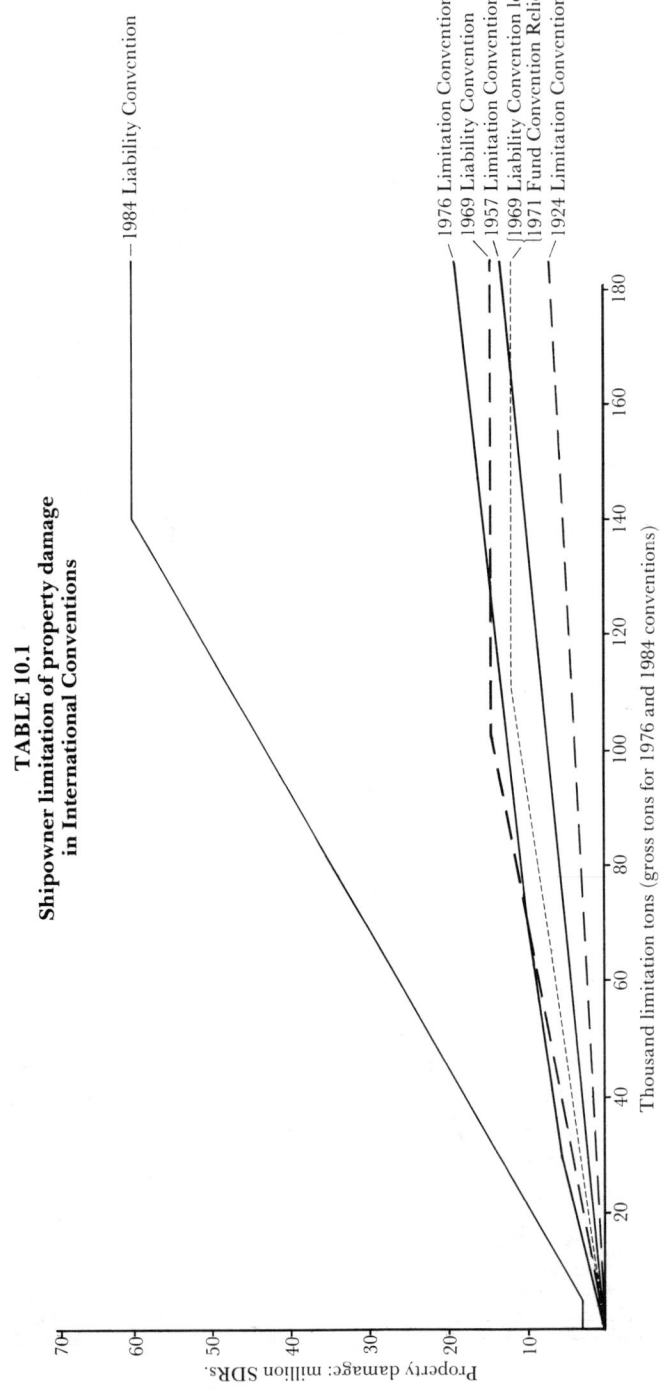

TABLE 10.1
Shipowner limitation of property damage in International Conventions

Notes to Table 10.1

(a) The gold value of the pound sterling in the 1924 Convention has for purposes of this Table been taken at SDR 43.

(b) The limitation amounts in the 1957 Convention have been taken in SDRs in accordance with the 1979 Protocol amending that Convention.

(c) Relief under the 1971 Fund Convention has been calculated in SDRs in accordance with the 1976 Protocol to that Convention.

(d) The limitation amounts for the 1969 Convention have been expressed in SDRs in accordance with the 1976 Protocol to that Convention.

(e) The limitation ton for the 1976 Limitation Convention and the 1984 Liability Convention is slightly different from that used in the others. This has been ignored.

B. Conduct Barring Limitation

10–147   By Article 6(2) of the 1984 Protocol, Article V(2) of the 1969 Convention[46] is replaced by the following:

> "2. The owner shall not be entitled to limit his liability under this Convention if it is proved that the pollution damage resulted from his personal act or omission, committed with the intent to cause such damage, or recklessly and with knowledge that such damage would probably result."

There is also a consequential amendment of Articles V(11) and VII(8) of the 1969 Convention.

10–148   This provision is borrowed from the Convention on the Limitation of Liability for Maritime Claims 1976[47] and was an integral part of the package deal under which the owner's limits were so substantially raised. The concept is close to, but not identical with, the English law concept of wilful misconduct, which governs the question of when the assured's conduct invalidates the insurance contract.[48] Since most Protection and Indemnity policies insuring oil pollution damage under the Convention are governed by English law, in most cases the cover will be intact when there is a right to limit, and the cover will fail if the right to limit is broken: this alone may discourage victims from attempting to challenge in court the owner's right to limit under these new provisions.

10–149   The new test is not identical to wilful misconduct, for it focuses attention specifically on oil pollution. The facts of the *Salem*

---

[46] See paras. 10–72, above.

[47] See paras. 9–37 *et seq.*, above.

[48] See Marine Insurance Act 1906, s.55(2)(a). For cases on wilful misconduct, see M. J. Mustill and J. C. B. Gilman (eds.), *Arnould's Law of Marine Insurance* (16th ed., 1981, London), para. 786; and E. R. Hardy Ivamy, *Marine Insurance* (3rd ed., 1979, London), p. 262.

case[49] illustrate this well. There, the owner and crew conspired to scuttle the ship, clearly wilful misconduct, and some oil escaped in the process (which did not, it seems, cause any pollution damage). But it is debateable whether it could be shown in that case that there was either intent to cause pollution damage, or both reckless-ness and knowledge that such damage would probably result, par-ticularly as the ship was sunk well out to sea. There may therefore be cases where, despite wilful misconduct, an owner retains the right to limit; and it is difficult to think of a case where the owner could not limit, yet would not be guilty of wilful misconduct under the P. & I. policy.

10–150 The burden of proof, as with the 1969 Convention, is left to national law: the Conference specifically rejected a proposal which would clearly have put it on the owner.[50] Also left to national law is the question of when an act will be deemed a personal act of the owner in the normal case of the owner being a company, an issue which arose also under the 1969 text in connection with actual fault or privity. The case which is least unlikely to occur is where a ship makes a deliberate discharge of oil residues, for instance in order to prepare for a dry-docking, under standing or express instructions from the shore-based establishment of the owner; but even there, the likelihood of the damage exceeding the new limits of liability is fairly remote. It may be concluded that for all practical purposes, the new right to limit may be regarded as unbreakable.

C. Establishment of the Limitation Fund

10–151 Although the Conference rejected a proposal that the establishment of a limitation fund be optional,[51] it did adopt Article 6(3) which amends Article V(3) of the 1969 Convention[52] to the effect that the fund may be established before any action under the Convention is brought. This gives the owner an initiative he did not have before, which may be valuable in a case where he wishes to take a view on currency movements, for the date of constitution of the fund is the date of conversion of the SDR into local currency.[53] The possibility of establishing the compulsory limitation fund before action is brought has the added advantage, in cases where the IOPC Fund is also involved, that claims may all be

---

[49] *The Salem* [1983] 1 Lloyd's Rep. 342 (H.L.).
[50] LEG/CONF.6/C.2/SR.13/PROV.
[51] LEG/CONF.6/C.2/SR.13/PROV.
[52] See para. 10–76, above.
[53] Article V(9) of the 1969 Convention, in both its original and amended forms.

settled without actions actually being brought in court; for the IOPC Fund cannot accurately calculate its own liability until that of the owner is known, and for that the establishment of the limitation fund is needed.[54]

10–152   For the application of the amended Article V(3) where a state is party to both the 1984 and the 1969 Convention, see below, paragraphs 10–158 and 10–160.

D. Revision of Limits in Future

10–153   The only way in which the limits of the 1969 Convention could be amended was to have a full-scale Diplomatic Conference in accordance with Article XVIII. Such a Conference is costly and more difficult to arrange than a meeting of a Committee of IMO. Both the MARPOL 1973 Convention and the SOLAS 1974 Convention contain provisions under which specified parts of the text may be amended by a procedure simpler than the convening of a Diplomatic Conference.[55] Accordingly, Article 15 of the 1984 Protocol provides for the limits to be revised by a simplified procedure under which an amendment adopted by the Legal Committee of IMO is deemed to be accepted eighteen months after being notified to all parties to the 1984 Protocol unless at least a quarter of them object.

10–154   The detailed provisions of Article 15 do not require attention here,[56] except to note that (i) no amendment may be considered before December 1, 1989 or before the 1984 Protocol enters into force, whichever is the later, nor less than five years from the date of entry into force of a previous amendment under Article 15; (ii) no limit may be increased beyond 6 per cent. per year from December 1, 1984 to the date of adoption of the amendment, or three times the 1984 limits, whichever the less,[57] and (iii) the Legal Committee must, in considering the adoption of an amendment, take into account the experience of incidents, changes in monetary values, the effect of the proposed amendment on the cost of insurance and the relationship between the limits in this 1984 Protocol and in the 1984 Protocol to the Fund Convention. This last provision will not really stop states in the Legal Committee doing anything they would otherwise like to do to the limits, but the former

---

[54] See paras. 11–50 *et seq.*, below.
[55] Article 16 of MARPOL 1973 and Article VIII of SOLAS 1974.
[56] For the debate on Article 15 see LEG/CONF.6/C.2/SR.6, 7, 8, 12, 22 and 23/PROV.
[57] Since 1.06 to the power 19 = 3 , the 6% limit will cease to operate as a restriction 19 years after December 1, 1984—in 2003.

one is certainly capable of so doing, and this, taken with certain restrictions on the timing and periods for entry into force of amendments, mean that the provision is relatively conservative in its approach.

11. THE 1984 PROTOCOL: COMPULSORY INSURANCE AND DIRECT RECOURSE

10–155 The 1969 Convention's system of compulsory insurance and certification is retained under the 1984 Protocol, and the opportunity is taken in Article 7 to amend the provisions of Article VII of the 1969 Convention in so far as they relate to ships flying the flag of non-contracting states; these ships may now obtain a certificate from any contracting state to the 1984 Protocol, and it must be recognised and accepted by other contracting states as having the same force as their own. There is a new model certificate annexed which is capable of use under both the 1969 and 1984 regimes, and the Conference adopted a Resolution calling upon states party to the 1969 and 1984 regimes respectively to recognise each other's certificates.[58]

12. THE 1984 PROTOCOL: TREATY RELATIONS, TRANSITIONAL PROVISIONS AND FINAL CLAUSES

A. Introduction

10–156 Both the Diplomatic Conference and the Legal Committee before it agonised over a problem of quite exceptional difficulty: how to integrate the new regime with the old. This problem would have been considerably easier if there had been only the 1969 Convention to amend. Instead, the Conference had to amend the 1971 Fund Convention as well and the amendments were seen right from the start as a "package deal" because of the close relationship between the two. This close relationship was characterised not only by the 1971 Convention being supplementary to the 1969 Convention, but by the fact that only parties to the 1969 Convention can be or remain parties to the 1971 Fund Convention.[59] The Legal Committee had decided early on that, rather than simply adopt an entirely new instrument designed to replace the two Conventions—as, for instance, the MARPOL 1973 replaces the Inter-

---

[58] For the debate on these provisions see LEG/CONF.6/C.2/SR.14, 20, 23, 24 and 26/PROV.

[59] Articles 37(4) and 41(4) of the 1971 Fund Convention.

national Convention for the Prevention of Pollution of the Sea by Oil 1954—the existing legal structure should be retained, with states remaining free to become party either to the Liability Convention regime only, or to both regimes. The problem was how to devise a way of enabling the new regime for both Conventions to enter into force as soon as possible, retain the maximum flexibility for states to choose the regime to which they became a party, and to enable parties to the 1971 Fund Convention to become parties to the revised Liability Convention before the revised Fund Convention came into force. If by now readers are confused, they will have appreciated why the subject was left at the Diplomatic Conference to the committee on Final Clauses!

10–157 The Conference adopted a number of provisions to solve these problems, only a few of which require a mention here: the others may be postponed until the Fund Convention has been discussed.[60] However, it is important to appreciate now that the scheme of these provisions is that, for an indefinite period of time, which may be very considerable indeed, the old 1969 Liability Convention regime will continue to exist alongside the new 1984 Liability Convention regime, in that states may be party to either or both. This is at the heart of what became known as the "phased-in" solution to the problems described above. An essential feature of this "phased-in" solution was intended to be that there was a transitional period during which the new regime "topped-up" the old one.

B. Transitional Provisions

10–158 Article 9 of the 1984 Protocol inserts a new Article into the Convention, called XII bis, as follows:

"Article XII bis
Transitional Provisions

The following transitional provisions shall apply in the case of a State which at the time of an incident is a Party both to this Convention and to the 1969 Liability Convention:
  (a) where an incident has caused pollution damage within the scope of this Convention,[61] liability under this Convention shall be deemed to be discharged if, and to the extent that, it also arises under the 1969 Liability Convention;

---

[60] See paras. 11–100 *et seq.*, below.
[61] Article 11 of the 1984 Protocol expressly provides that the 1969 Convention as amended by the 1984 Protocol shall be known as the International Convention on Civil Liability for Oil Pollution Damage, 1984.

(b) . . . .[62]

(c) in the application of paragraph 4 of Article III of this Convention the expression 'this Convention' shall be interpreted as referring to this Convention or the 1969 Liability Convention, as appropriate;

(d) in the application of paragraph 3 of Article V of this Convention the total sum of the fund to be constituted shall be reduced by the amount by which liability has been deemed to be discharged in accordance with sub-paragraph (a) of this Article."

10–159 A typical case where this will apply is where a state which is already party to the 1969 Convention wants to apply the new regime itself as soon as possible but at the same time does not want to break off treaty relations with other parties to the 1969 Convention. It therefore becomes party to the 1984 Convention by ratifying the 1984 Protocol but does not denounce the 1969 Convention. Both Conventions will apply in that state until the state denounces the 1969 Convention either voluntarily or compulsorily.[63] The effect of Article XII bis(*a*) is that a shipowner does not have to pay twice in that state—once, under the 1969 Convention and once under the 1984 Convention—but that instead the 1984 Convention acts to increase the liability of certain owners, both in terms of scope and amount, to that indicated by the 1984 Convention. "Certain owners" is used here advisedly, because the owner of a ship registered in a state which is party "only" to the 1969 Convention still has the right to the application of the 1969 Convention in other states party thereto, and by Article 12(5) of the 1984 Protocol to the Liability Convention a state party only to that Protocol (*i.e.* only to the 1984 Liability Convention) is not bound by the 1969 Liability Convention. The following cases illustrate the complexity of the co-existence of the regimes: in each case, state A is party only to the 1969 Convention, state B only to the 1984 Convention and state C is party to both Conventions. The ship has a limitation amount under Article V of the 1959 Convention of 1.3 million SDRs, and under Article V of the 1984 Convention of 5.1 million SDRs.

10–160 **Example 1.** The ship is registered in state A and pollutes state C. The owner is liable under the 1969 Convention in state C up to 1.3 million SDRs, and by Article XII bis(*c*) he cannot be otherwise liable in state C.

Example 2. Now the ship is registered in state B and pollutes

---

[62] See paras. 10–162, below.
[63] Under Article 31 of the 1984 Protocol to the Fund Convention: see below, para. 11–100.

state C. The owner is liable under the 1984 Convention in state C up to 5.1 million SDRs, and by Article XII bis(c) he cannot be otherwise liable in state C.

Example 3. The ship is still registered in state B but now it pollutes states A and C. In state A, the ship will incur a liability under the 1969 Convention of up to 1.3 million SDRs, and the owner may establish a limitation fund of that amount. In state C, the ship is liable under the 1984 Convention, but under Article XII bis(a) this liability is reduced by the 1.3 million SDRs of liability incurred under the 1969 Convention in state A. In state C the owner may therefore establish a limitation fund which, under Article XII bis(d) is the 1984 Convention amount of 5.1 million SDRs less 1.3 million SDRs, equals 3.8 million SDRs.

Example 4. This time the ship is registered in state C, and it pollutes the shores of states A and B. In state A it is liable under the 1969 Convention up to 1.3 million SDRs. In state B it is liable under the 1984 Convention up to 5.1 million SDRs, but Article XII bis(a) does not reduce the owner's liability in state B because state B is not party to both Conventions. Hence, in this case the owner may not aggregate his liability, may not establish one limitation fund, but is liable separately in each state to a total of 6.4 million SDRs.

Example 5. This time the ship is registered in state A, and it pollutes the shores of states B and C. In state B the owner is liable under the 1984 Convention (5.1 million SDRs), but in state C he is liable under the 1969 Convention because both states A and C are parties thereto and so they have agreed that the 1969 Convention governs their relations. Therefore in state C the owner must establish a separate fund of 1.3 million SDRs. He is therefore liable to a higher aggregate amount (6.4 million SDRs) than if state A had not been party to either Convention, for in that case he would have been liable in both states B and C under the 1984 Convention with its single 5.1 million SDR limit.

10–161 From this it can be seen that the co-existence of the two regimes leads to a hideous complexity, the enabling legislation for which will require the most careful drafting in the case of any state which wishes to be party to both Conventions. This complexity not only gives rise to such anomalous surprises as Example 5, above, but it is made worse by the fact that some states party to either or both of these Conventions may continue to be party to the 1957 or 1976 limitation Conventions,[64] to which there will be contracting states not party to either or both of the 1969 or 1984 Conventions.

[64] See paras. 9–29 et seq., above.

10–162 As if this was not bad enough, the possibility exists that a state will be party, not only to the 1969 and 1984 Liability Conventions, but also to the 1971 and 1984 Fund Conventions. Typically, this might occur where a state which is party to both the 1969 Liability Convention and the 1971 Fund Convention wishes to apply the regime of the 1984 Liability Convention as soon as possible. Now, in order to encourage the entry into force of the two 1984 regimes, it is envisaged that this possibility shall exist only for a limited time. This is effected by Article 12(4) of the 1984 Protocol to the Liability Convention, which provides that a state party to the 1971 Fund Convention can only become party to the 1984 Protocol to the Liability Convention (*i.e.* to the 1984 Liability Convention) if it becomes a party to the 1984 Fund Protocol at the same time.[65] It then finds that, under Article 31 of the 1984 Protocol to the Fund Convention there comes a time when it must denounce the 1969 Liability Convention.[66] Until that time, a state may be party to all four Conventions: the 1969 and 1984 Liability Conventions and the 1971 and 1984 Fund Conventions. During this period, Article XII bis(*b*) of the 1984 Liability Convention provides that where an incident has caused pollution damage within the scope of the 1984 Liability Convention, liability remaining to be discharged after the application of Article XII bis(*a*)[67] arises under the 1984 Liability Convention only to the extent that the pollution damage remains uncompensated after application of the 1971 Fund Convention. This may be illustrated as follows.

10–163 **Example 6.** State A is a party only to the 1969 Liability Convention, State B is a party only to the 1984 Liability Convention and state D is party to both of those and to the 1971 Fund Convention. A ship which is registered in state B pollutes states A and D. The owner is liable in state A under the 1969 Convention, say for 1.3 million SDRs. He is liable in state D under the 1984 Liability Convention, but only after application of the 1971 Fund Convention, whose current limit is 45 million SDRs[68] including the 1.3 million SDRs, so that the IOPC Fund will pay a further 43.7 million SDRs to victims in state D. Article XII bis(*b*) says that in state D the owner only has liability under the 1984 Liability Convention to the extent that the pollution damage in state D remains unsatisfied.

---

[65] It has the alternative of denouncing the 1971 Fund Convention with effect from the entry into force for that state of the 1984 Liability Convention.

[66] See para. 11–100, below.

[67] See paras. 10–158 *et seq.*, above.

[68] As at December 31, 1984.

Example 7. Had the ship in Example 6 been registered in state A the result would have been that the owner would have been liable in both states A and D under the 1969 Convention to a total of 1.3 million SDRs, and the IOPC Fund would have come in in state D for a further 43.7 million SDRs. The owner would not be liable further in state D.

C. Final Clauses: Entry into Force

10–164   The final clauses of the 1984 Convention are the final clauses of the 1984 Protocol. Some have been mentioned above. The only other provision requiring mention here is Article 13(1) of the 1984 Protocol which says that the Protocol shall enter into force twelve months following the date on which ten states, including six each with not less than one million gross tanker tons, have deposited the necessary instrument with the Secretary-General of IMO. It took nearly six years for the 1969 Convention, with a slightly less rigorous requirement, to enter into force, so this Protocol is not expected to do so before the late 1980s.

11. The 1971 Fund Convention and the International Oil Pollution Compensation Fund

SUMMARY

11–01 This chapter analyses in detail the scope and impact of the 1971 Fund Convention and its two Protocols, the main one being that of 1984 which creates the 1984 Fund Convention. The 1971 Fund Convention is designed to provide compensation, on the basis of strict liability with very few exceptions, supplementary to that provided by the 1969 Liability Convention. The 1984 Fund Convention has the same objective with respect to the 1984 Liability Convention.

11–02 The 1971 Fund Convention creates a new international organisation—the IOPC Fund—to effect the twin purpose of compensating victims, as described above, and of indemnifying the shipowner against a part of his liability under the 1969 Liability Convention. This organisation, which is a revolutionary feature of the international law on liability for oil pollution, has functioned extremely well and very efficiently in the six years it has been in existence so far, and has built up an unrivalled expertise in claims handling.

11–03 The 1984 Protocol to the 1971 Fund Convention, which creates the 1984 Fund Convention, envisages a new legal entity— the IOPC Fund 1984—to effect its purpose of compensating victims (the purpose of indemnifying the shipowner for part of his loss having been abolished in connection with a substantial rise in limits). It is envisaged that the IOPC Fund 1984 will be administered by the same Secretariat as that which administers the IOPC Fund, and there are other transitional provisions, of considerable complexity, whose purpose is to ensure a smooth transition between the existing 1971 regime and the 1984 regime.

1. INTRODUCTION

11–04 At the Conference in 1969 which adopted the 1969 Liability Convention there had been a protracted debate on the main

252

issues of strict or fault liability and of who was to bear that liability—the shipowner, the cargo owner or both.[1] At one stage it had begun to look as if the negotiations might completely break down in deadlock, but a series of votes on the important principles[2] evinced a majority view in favour of strict liability on the ship, combined with liability on the cargo interests in the form of a fund. In order to facilitate this compromise formula, a Working Group was set up to examine the question of liability based on an international fund. It was apparent by the time the group produced its report[3] that there was no hope of formulating an instrument to set up such a fund, and so the Conference adopted a Resolution[4] that IMO put the matter in hand immediately, and call a Diplomatic Conference not later than 1971 to consider and adopt a suitable Convention.

11–05 The origins of the 1971 Fund Convention are thus to be found in the conflicts at the 1969 Conference: without the promise of a fund, the 1969 Conference would very probably have failed to adopt an instrument at all. The Resolution, "recognising the view having emerged during the Conference that some form of supplementary scheme in the nature of an international fund is necessary to ensure that adequate compensation will be available for victims of large-scale oil pollution incidents," directed that the future work on the fund should be conducted in accordance with two guiding principles: (1) that victims should be fully and adequately compensated under a system based upon the principle of strict liability; and (2) that the fund should in principle relieve the shipowner of the additional financial burden imposed by the Liability Convention. These two principles are reflected in the International Convention on the Establishment of an International Fund for Compensation for Oil Pollution Damage 1971 ("the 1971 Fund Convention"), which was adopted on December 18, 1971 by a Diplomatic Conference held in Brussels.[5] The device adopted to fulfil these guiding principles was to establish an international fund for the compensation of pollution damage, to be named "The International Oil Pollution Compensation Fund" ("the IOPC Fund").

11–06 The provisions of the Fund Convention are directly

---

[1] LEG/CONF/C.2/SR.2 to 13, *OR* 623 *et seq.*
[2] LEG/CONF/C.2/SR.9, *OR* 662.
[3] LEG/CONF/C.2/WP.45, *OR* 604. Presented to Committee of the Whole II, LEG/CONF/ C.2/SR.20, *OR* 762, and to Plenary LEG/CONF/SR.5, *OR* 100.
[4] *OR* 185.
[5] The Official Records of the 1971 Conference are published by IMO. Documents reproduced in the Official Records and cited herein are followed by the abbreviation *OR* and the page number at which they appear. All the 1971 Conference documents have the prefix LEG/CONF.2/.

tailored to supplement those of the 1969 Liability Convention, so that in most cases the same definitions are adopted. The basic principle is that, where liability under the Liability Convention ends, the IOPC Fund's liability begins, although there are a few cases where the IOPC Fund is liable but the owner is not liable under the Liability Convention. The IOPC Fund's liability is strict, subject to very limited defences. In fulfilment of its twin functions, both the victims of oil pollution damage and the owner of the ship causing it may be claimants against the IOPC Fund. The IOPC Fund acquires its funds by levying contributions from those who have received crude oil and fuel oil in the territory of contracting states. The IOPC Fund is governed by an Assembly of all contracting states to the Fund Convention. Hence, while it is states who govern the IOPC Fund, it is largely oil companies who contribute to it. Only states party to the 1969 Liability Convention may become party to the 1971 Fund Convention.

11–07 The 1971 Fund Convention entered into force on October 16, 1978. In 1976 a Protocol was adopted amending the unit of account used in the Convention, and this is dealt with below in the body of the discussion of the Convention at paragraph 11–42. In 1984 the Diplomatic Conference already discussed at paragraphs 10–112 *et seq.* adopted another Protocol to the Convention which is of such significance that it is dealt with separately below, at paragraphs 11–72 *et seq.* The 1984 Protocol is not expected to enter into force until the late 1980s.

2. COMPENSATION FOR POLLUTION DAMAGE: ARTICLE 4

A. Compensation for Pollution Damage

11–08 Article 4(1) provides as follows:

"For the purpose of fulfilling its function under Article 2, paragraph 1(*a*),[6] the Fund shall pay compensation to any person suffering pollution damage if such person has been unable to obtain full and adequate compensation for the damage under the terms of the Liability Convention,
  (a) because no liability for the damage arises under the Liability Convention;
  (b) because the owner[7] liable for the damage under the Liability

---

[6] " . . . to provide compensation for pollution damage to the extent that protection afforded by the Liability Convention is inadequate."

[7] Defined in Article 1(2) as having the same meaning as in the Liability Convention: see para. 10–21.

Convention is financially incapable of meeting his obligations in full and any financial security that may be provided under Article VII of that Convention[8] does not cover or is insufficient to satisfy the claims for compensation for the damage; an owner being treated as financially incapable of meeting his obligations and a financial security being treated as insufficient if the person suffering the damage has been unable to obtain full satisfaction of the amount of compensation due under the Liability Convention after having taken all reasonable steps to pursue the legal remedies available to him;

(c) because the damage exceeds the owner's liability under the Liability Convention as limited pursuant to Article V, paragraph 1, of that Convention[9] or under the terms of any other international Convention in force or open for signature, ratification or accession at the date of this Convention.[10]

Expenses reasonably incurred or sacrifices reasonably made by the owner voluntarily to prevent or minimise pollution damage shall be treated as pollution damage for the purposes of this Article."

11–09 The major requirement is that the claimant has suffered *pollution damage*, which is defined in Article 1(2) as having the same meaning as in the Liability Convention and which has already been discussed.[11] This ties Article 4 to the scope of the Liability Convention to the extent that it is limited by the definition therein of pollution damage, which itself is limited by the definitions of *ship* and *oil* and by the requirement of an escape or discharge.[12] However, as regards the definition of *oil*, Article 1(2) limits it for purposes of the Fund Convention to "persistent hydrocarbon mineral oils," the effect of which is to exclude whale oil and any other non-mineral oils from its ambit, although the question of what is persistent remains. The practice of the IOPC Fund as to what constitutes pollution damage is further discussed below.[13]

If this pollution damage has been suffered within the geographical scope of Article 4,[14] and the claimant has been unable to obtain full and adequate compensation for it for one of the stated reasons, the IOPC Fund will, subject to the exceptions discussed below,[15] be strictly liable to him.

---

[8] See para. 10–96, above.
[9] See paras. 10–65 *et seq.*, above.
[10] Such as the 1957 Brussels Convention discussed in paras. 9–31 *et seq.*, above, but not the 1976 Limitation Convention since that was opened for signature after December 18, 1971.
[11] See paras. 10–47 *et seq.*, above.
[12] See paras. 10–06 *et seq.*, above.
[13] See paras. 11–53 *et seq.*, below.
[14] See para. 11–14, below.
[15] See paras. 11–15 *et seq.*, below.

11–10   The most important of the three reasons is that contained in *sub-paragraph (c)*, and to date it is under this heading that all Article 4 claims against the IOPC Fund have arisen.[16] The application of the first limb is clear and it means that usually the IOPC Fund comes in at the point where the damage exceeds the Liability Convention limit of 2,000 gold francs (or 133 SDRs) per limitation ton. The position where the owner is not entitled to limit his liability, due to his actual fault or privity, is discussed in paragraph 11–52 below, but of course in that case the IOPC Fund is unlikely to have to pay anything because the claimants should have received *full and adequate* compensation for their pollution damage from the owner. Since only states party to the Liability Convention may become party to this Convention, the cases in which the second limb of sub-paragraph (c) applies may, however, not be so obvious. This may be illustrated by the following example.

Example

State A is a party to the International Convention Relating to the Limitation of the Liability of Owners of Sea-going Ships 1957 ("the 1957 Convention"). State B is a party to that Convention, and to the 1971 Fund Convention (and hence to the 1969 Liability Convention also). A ship registered in State A pollutes state B. The owner is liable in state B under the Liability Convention, but is entitled to limit his liability in state B under the 1957 Convention to 1,000 gold francs per ton.[17] The IOPC Fund will compensate victims above this limit under Article 4(1)(c).

11–11   *Sub-paragraph (b)* would cover the situation where the owner is insolvent and either he is uninsured against his liability under the Liability Convention or that insurance has failed. The insolvency of the owner is not uncommon, especially in the case of one-ship companies where the ship is lost in the polluting incident, as occurred in the *Globe Asimi* case in November 1981.[18] But the other condition is much less likely to apply. This could occur because the insurer has become insolvent—unlikely but possible[19]—or because the insurer is not obliged to pay on the policy, due, for instance, to the wilful misconduct of the assured.[20] The

---

[16] Brief details of the claims handled by the IOPC Fund are given each year in the IOPC Fund's annual report.

[17] See para. 10–70, above.

[18] For the facts of this case see the IOPC Fund's Annual Reports, 1983 and 1984.

[19] In 1984 a Protection and Indemnity Club was put into a creditor's liquidation, and members were forced to seek pollution liability cover elsewhere, but no claims against the IOPC Fund resulted from this.

[20] Due to the application of s.55(2)(a) of the United Kingdom Marine Insurance Act 1906, which governs most pollution liability policies. See further paras. 10–148 *et seq.*

result of sub-paragraph (b) is that if an owner who is liable under the Liability Convention does not have assets sufficient to meet that liability, the IOPC Fund will pay the claimant—and where nothing is recovered from the owner it pays for all the damage.

11–12 *Sub-paragraph (a)* would, without more, apply in all the cases in which the owner is exempted from liability under Articles III(2) and XI(1) of the Liability Convention.[21] However, Article 4(2) of the Fund Convention exempts the IOPC Fund from liability in the following terms:

> "(2) The Fund shall incur no obligation under the preceding paragraph if:
> (a) it proves that the pollution damage resulted from an act of war, hostilities, civil war or insurrection or was caused by oil which has escaped or been discharged from a warship or other ship owned or operated by a State and used, at the time of the incident, only on Government non-commercial service; or
> (b) the claimant cannot prove that the damage resulted from an incident involving one or more ships."

Hence, the IOPC Fund is exonerated not only where the claimant is unable to identify the ship causing the damage, but in two cases covered by Articles III(2) and IX(1) of the Liability Convention, namely, (i) where the pollution damage resulted from an act of war or similar act, and (ii) where the pollution was caused by a warship or other ship on Government non-commercial service. In consequence, sub-paragraph (*a*) will apply to make the IOPC Fund liable:

(1) where the damage resulted from a natural phenomenon of an exceptional, inevitable and irresistible character;
(2) where the damage was wholly caused by an act or omission done with intent to cause damage by a third party; and
(3) where the damage was wholly caused by the negligence or other wrongful act of any government or other authority responsible for the maintenance of lights or other navigational aids in the exercise of that function.[22]

The least unlikely of these three cases is the last one, but, as has already been pointed out in connection with the Liability Conven-

---

[21] For Article III(2), see paras. 10–36 *et seq.*, above; for Article IX(1), see para. 10–30, above.

[22] This somewhat motley collection of cases results from the fact that those at the Conference who wanted the IOPC Fund to be exempted only in the most restricted of cases ultimately held sway over those who reasoned for wider exemption or complete identity with the exemptions of the shipowner. For the debate see LEG/CONF.2/C.1/SR.5 to 8, *OR* 341–361; SR.10 and 11, *OR* 374–384.

tion,[23] the inclusion of the word "wholly" considerably reduces the scope even of this case.

B. Preventive Measures by the Shipowner

11–13   Perhaps the most significant provision for shipowners and plaintiffs alike is the inclusion of the final sentence of Article 4(1). This was absent from the draft Convention, and was proposed by Norway.[24] It was argued that, just as such measures ranked as claims for limitation purposes under Article V(8) of the Liability Convention,[25] and thereby operated as an incentive to the owner to act, so the owner should be able to claim for them against the IOPC Fund. The intention is that the shipowner shall have a claim against the IOPC Fund for the amount he has not notionally recovered in his own limitation fund, rather than have a claim for the full cost against the IOPC Fund.[26] In practice, it is very often the owner's insurer, his Protection and Indemnity Club, which takes such measures, and the insurer would not in fact need this provision in order to claim against the IOPC Fund because he is therefore a "person" who has suffered "pollution damage" in his own name; but sometimes the insurer would use the owner's name.

C. Geographical Scope

11–14   Article 3(1) provides that compensation under Article 4 applies exclusively to pollution damage caused on the territory including the territorial sea of a contracting state and to preventive measures taken to prevent or minimise such damage. Therefore, the benefits of the convention are conferred upon those in contracting states irrespective of the flag of the ship causing the damage or the domicile or residence of the owner. This is a major incentive to states to become party to this Convention. The only limitation is, therefore, territorial in nature and the scope of Article 4 is identical to that of the Liability Convention.[27] The Conference rejected a proposal that the scope be extended to cover damage suffered on the high seas by a contracting state or person resident therein,[28]

---

[23] See para. 10–39, above.

[24] LEG/CONF.2/3, *OR* 55. The proposal was not contentious at the Conference: LEG/CONF.2/C.1/SR.5, *OR* 339.

[25] See para. 10–62, above.

[26] As the United Kingdom puts it in s.4(6) of the Merchant Shipping Act 1974: " . . . he shall be in the same position with respect to claims against the Fund . . . as if he had a claim in respect of liability under . . . " the Liability Convention.

[27] See paras. 10–34 *et seq.*, above.

[28] Note 2 of LEG/CONF.2/3, *OR* 46, discussed and, with other similar proposals, rejected: LEG/CONF.2/C.1/SR.18 and 19, *OR* 449–463.

although thirteen years later at the 1984 Conference a similar amendment succeeded; see paragraph 11–78, below.

D. Exemption of the IOPC Fund

11–15 Apart from the specific exemptions contained in Article 4(2) mentioned in paragraph 11–12, Article 4(3) provides that:

> "If the Fund proves that the pollution damage resulted wholly or partially either from an act or omission done with intent to cause damage by the person who suffered the damage or from the negligence of that person, the Fund may be exonerated wholly or partially from its obligation to pay compensation to such person provided, however, that there shall be no such exoneration with regard to such preventive measures which are compensated under paragraph 1. The Fund shall in any event be exonerated to the extent that the shipowner may have been exonerated under Article III, paragraph 3, of the Liability Convention."[29]

The drafting of this provision has given rise to problems. The original version did not contain the proviso. Norway proposed a proviso protecting preventive measures "by the owner"[30] to encourage rapid clean-up operations by him even where the incident had been caused by his negligence. But the words "by the owner" were withdrawn in Plenary because it was reasoned that there should be no discrimination against victims who were negligent: they should be equally encouraged to take preventive measures.[31] The resulting drafting is a nonsense: the proviso to the first sentence says preventive measures are always recoverable, the second sentence effectively says that "in any event" they are not always recoverable! It is clear from the summary record that the intent is that the proviso has sway over the second sentence. This was put beyond doubt for the future by a change made in the 1984 Protocol.[32]

11–16 However, the 1984 Protocol does not make a more serious problem which has already presented itself to the IOPC Fund any clearer. In practice, it may be very difficult to tell whether any given loss is a "preventive measure"[33] or some other type of pollution damage. A government which unnecessarily tries to clean a beach which ought to have been left to natural renewal could argue that that was expenditure on "preventive measures," and so

---

[29] As to Article III(3) of the Liability Convention, see paras. 10–40 *et seq.*, above.
[30] LEG/CONF.2/WP.7, *OR* 622.
[31] LEG/CONF.2/SR.4, *OR* 677.
[32] See para. 11–80, below.
[33] Defined in Article 1(2) as having the same meaning as in the Liability Convention: see paras. 10–47 and 10–54, above.

recoverable, because the cleaning was undertaken to "minimise pollution damage." The IOPC Fund, on the other hand, might argue that it was simply an item of clean-up cost, not a "preventive measure" as such. Each case can only be resolved on its own special facts.

3. SHIPOWNER RELIEF: ARTICLE 5

A. Indemnity against Liability

11–17   By Article 5(1) and (7):

"1. For the purpose of fulfilling its function under Article 2, paragraph 1(b),[34] the Fund shall indemnify the owner and his guarantor,[35] for that portion of the aggregate amount of liability under the Liability Convention which:
  (a) is in excess of an amount equivalent to 1,500 francs[36] for each ton of the ship's tonnage[37] or of an amount of 125 million francs, whichever is the less, and
  (b) is not in excess of an amount equivalent to 2,000 francs for each ton of the said tonnage or an amount of 210 million francs whichever is the less,
provided, however, that the Fund shall incur no obligations under this paragraph where the pollution damage resulted from the wilful misconduct of the owner himself.

7. Expenses reasonably incurred and sacrifices reasonably made by the owner voluntarily to prevent or minimise pollution damage shall be treated as included in the owner's liability for the purposes of this Article."

11–18   Since sub-paragraph (b) is the maximum limit of liability

---

[34] " . . . to give relief to shipowners in respect of the additional financial burden imposed on them by the Liability Convention . . . ."

[35] Owner *and* his guarantor is original drafting. The question of who is to receive the indemnification could become contentious if the owner is insolvent, as happened in the *Globe Asimi* case in November 1981: see IOPC Fund Annual Report 1983 and FUND/EXC.10/5. Since Article 5 provides for an indemnity, there is an argument that if the guarantor actually establishes the limitation fund in its own name under Article V(11) of the Liability Convention it is he who should receive the indemnity, and if the limitation fund is established in the owner's name under Article V(3), then the owner gets it. It cannot have been the intention that the IOPC Fund must pay twice under Article 5, once to the owner's liquidator, once to the guarantor.

[36] By Article 1(4), "franc" is defined as having the same meaning as in the Liability Convention: see paras. 10–65 *et seq.*, above. The franc is replaced as a unit of account by the 1976 Protocol: see below, paras. 11–34 *et seq.*

[37] By Article 1(5) ship's tonnage is defined to have the same meaning as in the Liability Convention: see para. 10–69.

under Article V(1) of the Liability Convention,[38] subject to the maximum liability of the IOPC Fund being reached[39] a shipowner who is entitled to avail himself of the limitation provisions of the Liability Convention, of compensation under Article 4 of this Convention and of the indemnity under Article 5 has an effective oil pollution liability limited to 1,500 francs per ton or 125 million francs, whichever the less, for any one incident. This may be seen as follows: the Liability Convention limit is 2,000 francs per ton or 210 million francs, whichever is the less; Article 5 reimburses the shipowner down from that limit to 1,500 francs per ton or 125 million francs, whichever is the less; and Article 4(1) compensates the shipowner for clean-up costs he has incurred which are not notionally compensated in the Liability Convention limitation fund.

11–19 It will be recalled from paragraph 11–05 that the intention of the 1969 Conference had been that the shipowner should be relieved of the "additional financial burden" imposed by the Liability Convention, which taken strictly would mean that he should be put into the same position as if he could have limited his oil pollution liability under the 1957 or 1924 Limitation Conventions, and in the same "property damage" limitation fund. The draft Convention presented to the Conference could not easily achieve the "same limitation fund" point, but it did refer to the 1,000 franc limit of the 1957 Convention.[40] Both the principle of shipowner relief and the level to which the shipowner should be relieved were extremely contentious at the Conference, and were only resolved at the end by a compromise between those supporting shipping interests and those supporting oil industry interests. The result adopted is extremely close to the level of liability which owners had at that time voluntarily undertaken in TOVALOP,[41] which formed the basis of a number of submissions.[42]

B. Exemption of the IOPC Fund

11–20 There are limitations on the right of the shipowner to relief under Article 5, designed to encourage compliance with the major oil pollution prevention and ship safety conventions. Under Article 5(4), the list of these conventions may be kept up to date by the

---

[38] See para. 10–65, above.
[39] See paras. 11–24 *et seq.*, below.
[40] LEG/CONF.2/3, *OR* 64.
[41] See Chapter 12, below.
[42] For the debate on Article 5, see LEG/CONF.2/C.1/SR.12, *OR* 390, SR.22 and 23 *OR* 483–501; and LEG/CONF.2/SR.4, *OR* 678. See also LEG/CONF.2/5 at *OR* 191, and LEG/CONF.2/C.1/WP.31 at *OR* 272.

261

Assembly, which had by December 31, 1984 done this in respect of a number of them, so that at that date Article 5(3) may be taken to have the following effect (even where the flag state is not a party to any relevant convention):

> The Fund may be exonerated wholly or partially from its obligations under paragraph 1 towards the owner and his guarantor if the Fund proves that as a result of the actual fault or privity of the owner:
>
> (a) the ship from which the oil causing the pollution damage escaped did not comply with the requirements laid down in:
>
>   (i) the International Convention for the Prevention of Pollution from ships, 1973 as amended by the 1978 Protocol relating thereto[43];
>
>   (ii) the International Convention for the Safety of Life at Sea, 1974 as modified by the Protocol of 1978 relating thereto[44];
>
>   (iii) the International Convention on Load Lines, 1966;
>
>   (iv) the Convention on the International Regulations for Preventing Collisions at Sea, 1972[45];
>
>   (v) any amendments to the above-mentioned Conventions which have been determined as being of an important nature in accordance with their provisions and which have been in force for at least twelve months at the time of the incident[46]; and
>
> (b) the incident or damage was caused wholly or partially by such non-compliance.

11–21 The IOPC Fund must therefore prove three things in order to avoid paying an indemnity which would otherwise be payable: first, that an event occurred (or failed to occur) which constituted the actual fault or privity of the owner; secondly, that as a result of that event or failure the breach of a provision of one of the named Conventions occurred; and thirdly, that that breach wholly or partially caused the incident or damage. Hence, what seems on the face to be a fairly wide opportunity for the IOPC Fund to escape liability is on closer examination rather narrow. There had been no such cases as at December 31, 1983 out of the 10 cases of Article 5 indemnity claimed or paid by the IOPC Fund by then.

---

[43] See Chapter 3 above.

[44] See Chapter 4 above. Algeria has declared under Article 5(4) that she does not accept this updating in respect of her flag ships, so SOLAS 1960 still applies to such ships until this objection is withdrawn.

[45] See paras. 4–25 *et seq.*, above.

[46] The Assembly determined that the amendments adopted by IMO's Maritime Safety Committee in Resolution MSC.1(XLV) on November 20, 1981 were not covered by this provision: FUND/A.4/16.

C. Geographical Scope of Article 5

11–22   Article 3(2) provides that Article 5 indemnification shall only be available in respect of pollution damage caused on the territory, including the territorial sea, of a state party to the Liability Convention by a ship registered in or flying the flag of a contracting state to this Convention, and of preventive measures taken to prevent or minimise such damage. This limitation to the ships of contracting states is another incentive to states to become party to the Convention.

D. Abolition of Shipowner Relief

11–23   Article 5 was deleted by the 1984 Protocol as part of an overall increase in shipowner and Fund Convention limits: see below paragraph 11–76.

4. THE LIMIT OF ARTICLE 4 COMPENSATION AND UNITS OF ACCOUNT

A. Calculation and Effect of the Amount of the Limit

11–24   Article 4(4)(a), which is tortuously drafted, provides for a maximum amount of compensation in respect of any one incident[47] of 675 million gold francs.[48] This amount includes "the amount of compensation actually paid under the Liability Convention for pollution damage," part of which may have been the subject of a claim by the owner under Article 5. It is therefore important to appreciate that the amount victims can receive from the IOPC Fund *is not affected by whether the IOPC Fund has paid a claim to the owner under Article 5*. In order to work out how much victims can recover from the IOPC Fund, simply subtract from 675 million francs the amount of compensation actually recovered by victims under the Liability Convention.

11–25   The deduction from 675 million francs of the amount actually paid under the Liability Convention means that calculation of the exact amount payable must wait for the ascertainment of

[47] "Incident" is by Article 1(2) defined to have the same meaning as in the Liability Convention: see para. 10–19, above.
[48] The original figure was 450 million gold francs, but this was increased by the Assembly of the IOPC Fund under Article 4(6) to 675 million gold francs with effect from April 20, 1979. See para. 11–31, below.

the amount actually paid under the Liability Convention. Thus where, as in the *Tanio* case in France in 1980, the limitation fund established by the shipowner under the Liability Convention accrues interest for the account of the victims prior to distribution, that interest does not lead to any increase in the amount recovered by the victims; in such a case, every franc of interest reduces the amount to be paid by the IOPC Fund by one franc, and only when the limitation fund is fully paid out can the IOPC Fund calculate its limit. This inevitably delays a final payment by the IOPC Fund until after the disbursement of the limitation fund.

11–26 Another consequence of the use of the phrase *actually paid* would appear to be that the cost of preventive measures taken by the shipowner, which rank as claims under the limitation fund by virtue of Article V(8) of the Liability Convention,[49] are not to be taken into account in calculating the Article 4(4)(a) limit: an owner cannot really pay himself under the Liability Convention. However, the costs of action taken by his P and I Club are to be taken into account, because in law the shipowner pays them like any other claim. This is really an unwarranted distinction, which must be regarded as a drafting mishap, and in practice it is overcome by interpreting "actually paid" to mean "actually paid or deducted."

11–27 Article 4(4)(a) refers to the amount actually paid under the Liability Convention for pollution damage caused in the *territory* of the contracting states. This must, in the light of Article 3, be regarded as a drafting error which slipped through the Conference without notice, since both sub-paragraphs of Article 3 refer to *territory including the territorial sea*. It cannot have been the intention that costs and expenses incurred in the territorial sea, and preventive measures taken anywhere but on the territory of states, shall not be taken into account when computing the Article 4(4)(a) limit, for this would require an onerous and in many cases unreal division of the pollution damage costs.

11–28 There is a special limit of 675 million francs provided for in Article 4(4)(b) for the case where the pollution damage has been caused by a natural phenomenon of an exceptional, inevitable and irresistible character.[50] Without such a provision, if the phenomenon had caused more than one tanker to cause pollution damage, the Article 4(4)(a) limit would have applied to each tanker, for each tanker's discharge would constitute a different "incident." Hence,

---

[49] See para. 10–62, above.
[50] As to the meaning of this phrase, see para. 10–37, above.

Article $4(4)(b)$ creates a single, special limit in such a case, irrespective of how many ships spill oil.

11–29   The effect of the limit to Article 4 compensation is represented graphically in Table 11.1.

11–30   The only case prior to June 30, 1984 in which the IOPC Fund's limit has actually been exceeded is the *Tanio* case, off Brittany, France in March 1980.[51] Of course, the Fund Convention does not apply in all states where oil is spilled, and so the adequacy of the limit can only be estimated, by applying the limit theoretically to all known spills. This has been done by the International Group of P and I Clubs, who concluded that in relation to some 17,000 spills between 1970 and 1982 only two incidents would not have been fully compensated, amounting to an estimated U.S. $120 million uncompensated.[52] The limit was substantially raised at the 1984 Conference: see paragraphs 11–87 *et seq.*, below.

B. Revision of the Article 4 Limit

11–31   The Article 4 limit is set by the Convention at a figure which must be between 450 million and 900 million gold francs, and which began at 450 million. Under Article 4(6), the Assembly of the IOPC Fund may set the limit prospectively, and not retrospectively, having regard to the experience of incidents and in particular to the amount of damage resulting therefrom and to changes in monetary values. Such a decision requires, under Articles $32(c)$ and $33(1)(a)$, a three-fourths majority of those *present* (as opposed to those present *and voting*).

11–32   Following the *Amoco Cadiz* incident off Brittany, France, in March 1978 the French Government proposed to the Assembly of the IOPC Fund, as soon as it convened following the entry into force of the Fund Convention in October of that year, that the limit be raised to 900 million gold francs.[53] At its second session in April 1979, this proposal was very narrowly defeated and instead as a compromise the limit was raised with effect from April 20, 1979 to 675 million gold francs.[54] The French again proposed to increase the limit to the maximum possible in October 1981,[55] but the proposal was again defeated due to failure to attract the required majority, under the influence of the argument that it would be

---

[51] See IOPC Fund Annual Reports 1980–1984.
[52] See note 1, para. 10–141, above.
[53] See IOPC Fund Document OPCF/A.I/14/1.
[54] See IOPC Fund Document FUND/A.2/16/1.
[55] See IOPC Fund Document FUND/A.4/15.

TABLE 11.1

Illustration of the division of pollution damage between the shipowner (under the 1969 Liability Convention) and the 1971 Fund Convention

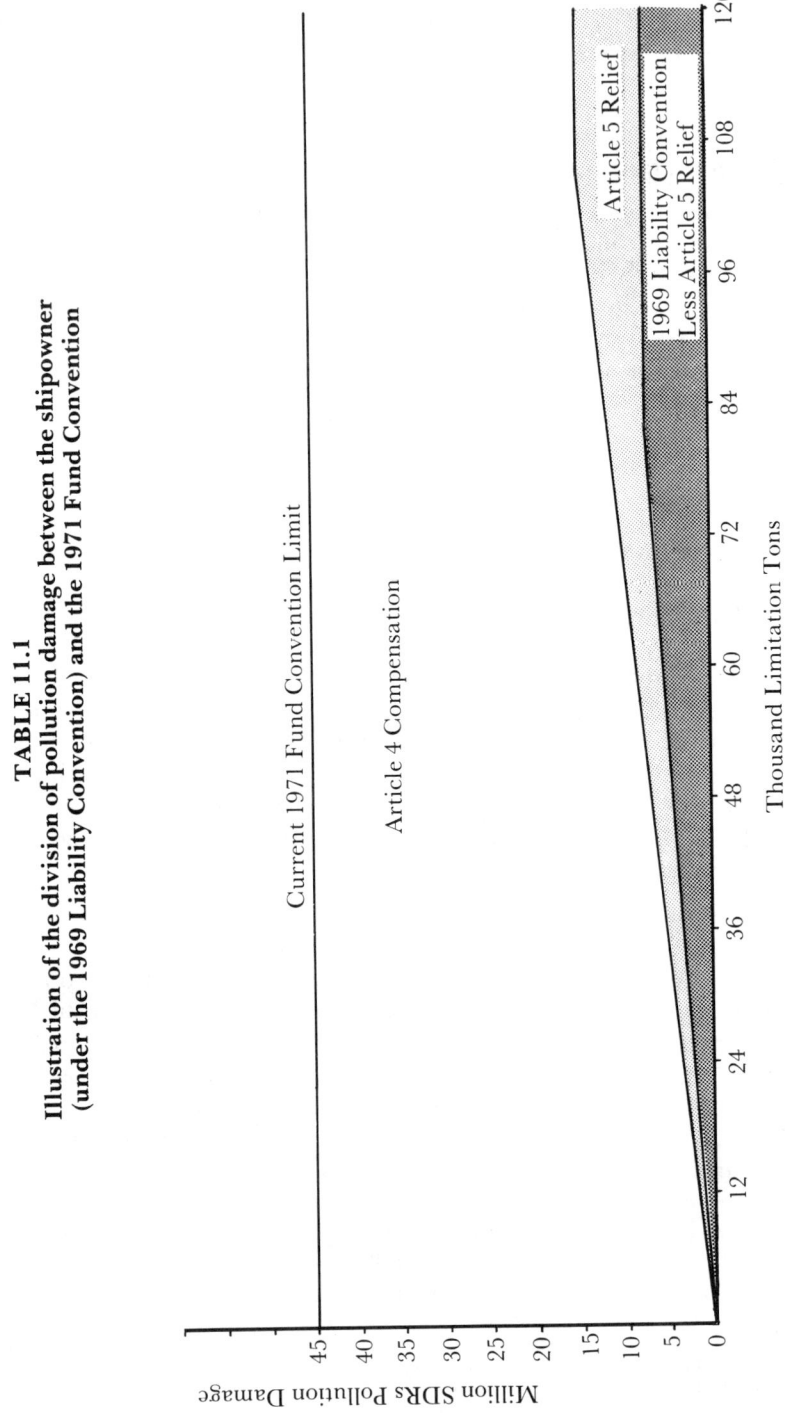

Current 1971 Fund Convention Limit

Article 4 Compensation

Article 5 Relief

1969 Liability Convention
Less Article 5 Relief

Thousand Limitation Tons

Million SDRs Pollution Damage

Notes to Table 11.1

(a) The current limit and the amount of relief under Article 5 of the 1971 Fund Convention has been calculated in SDRs in accordance with the 1976 Protocol to that Convention.

(b) The limitation amounts for the 1969 Liability Convention have been expressed in SDRs in accordance with the 1976 Protocol to that Convention.

(c) The graph assumes that the shipowner's rights to limit under the Liability Convention and to claim relief under Article 5 of the 1971 Fund Convention have not been denied.

---

unduly burdensome on the oil industry to increase the limit without a corresponding increase of the limit under the 1969 Liability Convention.[56]

11–33 The system of revising the Article 4 limit was completely changed by the 1984 Protocol: see paragraph 11–91, below.

C. Units of Account

THE UNITS

11–34 The unit of account in which the Article 4 limit is expressed is the familiar gold franc.[57] In 1976 a Protocol to the Convention had been adopted which changed the expression of the various amounts from gold francs to SDRs, as defined by the International Monetary Fund. This Protocol had not entered into force by the time the IOPC Fund became a reality upon entry into force of the 1971 Convention, and indeed as at December 31, 1984 it still had not done so. This leads to the somewhat anomalous situation that the Liability Convention limit, which of course affects the amount actually paid out by the IOPC Fund, may be expressed and calculated by reference to the SDR now that the 1976 Protocol to that Convention has entered into force, but that the Article 4 limit under this Convention is still calculated by reference to the gold franc.

11–35 The 1976 Protocol follows the 1976 Protocol to the Liability Convention, discussed above,[58] in expressing the gold franc amounts in Articles 4 and 5 of the Convention in SDRs on the basis of 15 gold francs equals 1 SDR. Hence, for instance, 450 and 900 million francs become respectively 30 and 60 million SDRs, and 675 million francs would on the same basis be 45 million SDRs.

---

[56] See IOPC Fund Document FUND/A.4/16.

[57] As to the gold franc, see para. 10–65 and references thereunder.

[58] See paras. 10–65 *et seq.*

The Assembly of the IOPC Fund has adopted two Resolutions recommending states to ratify the 1976 Protocol,[59] but its entry into force has been held up by the fact that it requires as parties states in which at least 750 million tons of contributing oil[60] have been received, and the realities are currently such that in practice this means that if a small number of major oil importing states do not ratify, it cannot enter into force.

CONVERSION OF THE UNITS INTO NATIONAL CURRENCIES

11–36   The Convention is curiously silent on how the gold franc is to be converted into national currencies, as is the 1976 Protocol on how the SDR is to be converted, and these omissions must be seen as drafting errors. The definition of "franc" in Article 1(4) refers to Article V(9) of the Liability Convention,[61] which contains a provision requiring the conversion of the shipowner's limit into national currency of the state in which the limitation fund is being constituted "on the basis of the official value of that currency . . . on the date of the constitution of the fund." Unfortunately, this provision is not entirely helpful in the case of the Fund Convention because the shipowner's limit is not the Article 4 limit, nor in every case will the shipowner have constituted a limitation fund, and even if he has the Assembly of the IOPC Fund might rightly feel that that date is less relevant in the case of the Fund Convention than others such as the date when the Assembly votes the contributions to pay the claims, or the date of the incident. While the date for conversion is therefore open on the relevant texts, there must be at least a presumption that the intent of the Fund Convention is that the basis for conversion should be an "official" rate.

11–37   There has been only one case to date where the IOPC Fund's limit has had to be determined, and that is the *Tanio* case off Brittany, France in March 1980. In that case the shipowner established a limitation fund and all claimants against the IOPC Fund agreed with the IOPC Fund that the Article 4 limit should be converted as at that date.[62]

11–38   States have adopted different practice on conversion.[63] Several states (*e.g* the Netherlands, the Scandinavian states and the Federal Republic of Germany) have adopted national legislation

---

[59] See IOPC Fund Documents OPCF/A.1/Res.1 and FUND/A/ES.1/13, Annex II.
[60] As to "contributing oil," see paras. 11–60 *et seq.*, below.
[61] See para. 10–65, above.
[62] See IOPC Fund Document FUND/EXC.10/WP.1.
[63] See IOPC Fund Document FUND/A.4/13.

following closely the method contained in the 1976 Protocols,[64] while others fix the rate by order from time to time (*e.g.* Japan and the United Kingdom). Yet others have no specific national legislation. Aware of the problems of converting the gold franc into national currencies, the Assembly of the IOPC Fund passed Resolution 1 as early as November 17, 1978, adopting the following method of interpreting the franc provisions in the Convention: first an amount in gold francs shall be converted to SDRs at the rate of 15 gold francs equals one SDR, and then the SDR amount is converted into the relevant national currency in accordance with the method of evaluation applied by the International Monetary Fund in effect at the applicable date. This interpretation binds the Director of the IOPC Fund but does not bind either states or claimants; however, as a guide to the recommended practice of contracting states it is of great importance, particularly as it was adopted unanimously.

11–39 The IOPC Fund itself has the problem of converting amounts expressed in gold francs into the currency in which it draws up its accounts and levies contributions, which is pounds sterling, the currency of its headquarters state. This problem was solved by the adoption of Regulation 2 of the IOPC Fund's Internal Regulations, which provides for the same method as in Resolution A.1(I) to be used. The SDR to pounds sterling rate is published daily by the IMF (as, indeed, is the SDR rate for other currencies).

11–40 The problems of choosing a date for conversion of the limit were solved prospectively in the 1984 Protocol by a special provision, which provides for conversion on the date of the decision of the Assembly as to the first date of payment of compensation: see paragraph 11–89, below.

D. Distribution of the IOPC Fund's Limit

11–41 By Article 4(5):

> "Where the amount of established claims against the Fund exceeds the aggregate amount of compensation payable under paragraph 4, the amount available shall be distributed in such a manner that the proportion between any established claim and the amount of compensation actually recovered by the claimant under the Liability Convention and this Convention shall be the same for all claimants."

The drafting here in the English text is seriously defective, although

---

[64] See para. 10–67, above.

the French text (which by Article 48 is equally authentic) is clearer.[65] Not only is the use of the phrase "proportion between" unhappy, but the inclusion of the Liability Convention in the formula gives rise to serious problems of interpretation in a case where an individual's claim for pollution damage established against the IOPC Fund is different to that established against the shipowner's limitation fund. This could happen, for instance, where a claimant's claim against the limitation fund was valid and in time, but was partly time barred against the IOPC Fund due to the appropriate action or notification under Article 6 being given too late. In such a case, for the wording of Article 4(5) to be adhered to strictly, a claimant may have to pay back to the IOPC Fund some of his recovery under the Liability Convention! There is also the question of whether, in the case of a shipowner claimant, sums offset against the limitation fund pursuant to Article V(8) of the Liability Convention[66] are sums "actually recovered" thereunder.[67]

11–42 These difficulties were solved in the 1984 Protocol by omission of the reference to the Liability Convention (see paragraph 11–92, below); but they can only be solved for the present by claimants and courts alike adopting a pragmatic attitude to the interpretation of Article 4(5), even though it does some violence to the words; this would effectively treat the claims against the fund as a separate group, and distribution of the available amount would be made *pro rata* without trying to take into account their actual recovery from the limitation fund. Following the equally authentic French text assists such an approach.

5. CLAIMS AGAINST THE IOPC FUND

A. Jurisdictional Provisions

11–43 The provisions of Article 7, dealing with the place of action against the IOPC Fund, are complicated. One reason for such complexity is the possibility that an action for Article 5 indemnity can be brought against the IOPC Fund in respect of damage suffered in a state party to the Liability Convention but not to the Fund Convention; another is that a single incident can give rise to claims against the owner (under the Liability Convention) and

---

[65] As became clear in the *Tanio* case referred to above. There, the problems were resolved by mutual agreement amongst the IOPC Fund and claimants, on the basis of the French text.

[66] See para. 10–62, above.

[67] See paras. 10–86 *et seq.*, above.

against the IOPC Fund (under this Convention) in respect of damage suffered in several states, some of which are party to both Conventions and some of which are party only to the Liability Convention.

11–44 The object of jurisdictional provisions limiting the place of action should be to facilitate the litigation and to draw it all before one court. Article IX of the Liability Convention[68] did this reasonably well for that Convention but the same purpose could not be achieved in all cases here, for the above reasons. The rules elaborated in Articles 7(1) and (3) accordingly provide that, if possible, a claim against the IOPC Fund in respect of any damage should be made in the same jurisdiction as the claim in respect of the same damage is made against the owner under the Liability Convention; but that if that jurisdiction is one where the Fund Convention is not in force, then the claim against the IOPC Fund must be made either before a court of the state where the IOPC Fund has its headquarters—the United Kingdom—or before the courts of a contracting state in which pollution damage has also been suffered as a result of the incident.

11–45 The Convention does as much as it can to keep the litigation in the court where the Liability Convention limitation fund has been established, but could not, of course, make any very simple provision for the case where this court is in a jurisdiction in respect of which the Fund Convention is not in force, nor for the case where there was no limitation fund under the Liability Convention (for instance because it was a clear case of the owner being exempted from liability under Article III(2))[69].

11–46 The detailed provisions of Article 7 do not require further discussion. Article 8 makes provision for recognition and enforcement of judgements against the IOPC Fund as in Article X of the Liability Convention,[70] so that again the system created is self-contained, and is not dependent upon the operation of other treaties relating to recognition and enforcement of judgements such as the 1968 Brussels Convention on Jurisdiction and the Enforcement of Judgements in Civil and Commercial Matters.

B. Time Limit for the Plaintiff's Action

11–47 By Article 6(1), plaintiffs are given a period of three years from the date when the pollution damage occurred, but not more

---

[68] See para. 11–26, above.
[69] See paras. 10–36 *et seq.*, above.
[70] See paras. 10–93 *et seq.*, above.

than six years after the date of the incident causing the damage, before their claim against the IOPC Fund is extinguished—the same as in the Liability Convention.[71] Hence, the clock starts ticking when the damage occurred, which, as with the Liability Convention, is not elaborated and so is left to national law. Plaintiffs are given a novel choice of two ways in which to ensure that the three year period which starts with the occurrence of the damage stops running: either they may bring an action under Article 4 (or, as the case may be, Article 5), or they may give a notification to the IOPC Fund pursuant to Article 7(6). This latter alternative can save the plaintiff having to bring an action against the IOPC Fund if he has already brought an action against the owner or his guarantor under the Liability Convention.[71] In the case of an owner claiming indemnification under Article 5, Article 6(2) gives the claimant a further six months from the date when he acquired knowledge of the bringing of an action against him under the Liability Convention.

C. IOPC Fund Settlement Procedures

11–48   A detailed examination of the policy and practice of claims handled by the IOPC Fund is outside the scope of this work. It is, however, dealt with by other writers and in other documents.[72] The IOPC Fund has produced a Claims Manual which is sent to all those who contact it with a claim which sets out the simple procedure in clear language. Claims are examined carefully, not with a view to trying to reduce them, but to see if they are justified under the Convention as enshrined in the relevant national law. As at June 30, 1984 it had always been possible to reach agreement with claimants on the size of their claim.

11–49   Co-operation with the owner's P and I Club is normally very close, and is usually based on an ad hoc agreement for each case, based upon a standard Memorandum of Understanding signed with the International Group of P & I Clubs on November 5, 1980.[73] One feature of this co-operation is that claims are met with the mutual consent of the IOPC Fund and the Club, as soon as possible. Whenever practicable the same lawyers and surveyors

---

[71] See paras. 10–90 *et seq.*, above.

[72] See E.D.Brown, "The International Oil Pollution Compensation Fund: An Analytical Report on Fund Practice," (1983) *Oil and Petrochemical Pollution*, 269; R.H.Ganten, "The International Oil Pollution Compensation Fund," in S. Mankabady, *The International Maritime Organisation*, (1984, Beckenham (England)) 318. See also IOPC Fund Documents FUND/WGR.5/1 & 2, FUND/A.4/10 and 16.

[73] The text is at Annex II of the IOPC Fund Annual Report 1980 and at FUND/A/ES.1/3.

are used. The result is that normally it is possible to ensure that claimants present their claims only once, and find that it is satisfied without them having to prove it separately against the ship in the limitation proceedings and against the IOPC Fund. They negotiate the claim once, receive compensation, and leave it to the Club and the IOPC Fund to work out the distribution amongst themselves.

NEED FOR ESTABLISHMENT OF THE LIABILITY CONVENTION LIMITATION FUND—PREPAYMENT POSSIBILITIES

11–50 Because the IOPC Fund will in nearly all cases be paying out on top of a sum paid out by the shipowner, it cannot finally calculate how much it owes a claimant until the Liability Convention limitation fund has been established and, in the case of a limitation fund which accrues interest for the benefit of the claimants, paid out. It is, therefore, impossible for the IOPC Fund to make a final payment to claimants until the limitation fund has at least been established. It does, however, have two options for ensuring that in appropriate cases claimants receive money as soon as possible. One is to make a provisional payment before all the information necessary to calculate the entitlement of a claimant is available,[74] but this is only done where the Director is satisfied that the owner is entitled to limit and only in order to mitigate undue financial hardship. This may be done even before the establishment of the owner's limitation fund, but the payment is restricted to a maximum of 60 per cent. of the claimant's likely eventual recovery. The other option is to use the power contained in Article 4(8) to provide credit facilities, on conditions to be laid down in the Internal Regulations, to enable a claimant to take preventive measures. However, the situations in which the Assembly has decided to make such facilities available are also rather restrictive: under Internal Regulation 12, application must be made by a contracting state which must be "in imminent danger of substantial pollution damage." Only the state may be extended such credit, and not a private claimant such as a local authority.

11–51 Hence, where a claimant other than a contracting state has suffered pollution damage but does not have the money to clean it up or otherwise mitigate its effects, he may be faced with the problem not only of getting the job done at all, or in reasonable time, but also in quantifying his loss so as to bring a claim and share in the distribution of the shipowner's limitation fund. Some

---

[74] For the detailed regulations on partial payment, see the IOPC Fund's Internal Regulations 8.6 and 8.7.

local authorities in the *Tanio* case found themselves in this situation, and the IOPC Fund tried to help by agreeing estimates and paying up on the basis of them. An attempt to alleviate this position was made by the 1984 Protocol in its re-definition of "pollution damage": see paragraphs 10–139 and 11–78.

AFTER ESTABLISHMENT OF THE LIABILITY CONVENTION LIMITATION FUND

11–52 Where the limitation fund is established, the IOPC Fund will proceed with payment of claimants as expeditiously as it can, and to date it has a commendable record as can be seen from perusal of its Annual Reports. The usual procedure, pursuant to agreement with the P and I Club concerned, is for the limitation proceedings to be adjourned after the limitation fund has been established to allow the IOPC Fund to intervene if considered necessary to break the right to limit. Compensation will then be assessed and paid as if the right to limit had been established. Where the shipowner's right to limit is broken after the IOPC Fund has made partial or final payment, the IOPC Fund will seek to recover all its expenditure from the owner pursuant to its agreement with the Club and using its powers of subrogation.[75] Where the right to limit is broken before payment, it would not expect to be liable to claimants at all, unless the owner is incapable of paying claims in full—which may easily occur if for one reason or another[76] his insurance fails—in which case Article 4(1)(*b*) would apply. The result of these arrangements is considerably to accelerate the process of paying claimants.

ADMISSIBILITY OF CLAIMS: PRACTICE ON POLLUTION DAMAGE

11–53 On November 5, 1980 the IOPC Fund concluded a Memorandum of Understanding with the International Group of P and I Clubs which records some of the general aspects of the co-operation noted above.[77] Of particular importance is Clause 6, which states that:

> " . . . the Clubs and the IOPC Fund will exchange views and will consult with one another when an incident occurs so that the term 'pollution damage,' which has the same definition in the Civil Liab-

---

[75] For subrogation, see para. 11–59, below.

[76] *e.g.* wilful misconduct of the assured, or sending the ship to sea in an unseaworthy state with the privity of the assured.

[77] The text is available as Annex II to the IOPC Fund's Annual Report 1980 and at FUND/A/ES.1/3.

ility Convention and the Fund Convention, receives the same interpretation by the Clubs and by the IOPC Fund."

The types of pollution damage which may be suffered were analysed in paragraphs 10–42 *et seq.*, above. Because of the vagueness
of the definition in the Convention noted above in paragraphs
10–47 *et seq.*, and because of the IOPC Fund's policy of reaching
agreement with claimants without recourse to the courts wherever
possible, the practice of the IOPC Fund on what items are admissible is crucial. Clause 6 of this Memorandum of Understanding
makes this practice even more important, for it affects the practice
of the insurers belonging to the International Group of P and I
Clubs.

(i) *Preventive measures and restoration of the environment: additional and
fixed costs*

11–54 In relation to the costs of preventive measures and restoration of the environment the IOPC Fund has had experience of
claims both for additional costs (those incurred solely as a result of
the incident) and fixed costs (those relating to the incident but
which would also have arisen had the incident not taken place).
Reasonable claims for the former are usually paid and may include
the following[78]:
  (a) work carried out by contractors with regard to the ship or the
      polluted area;
  (b) salaries and allowances for personnel especially employed for
      the purpose of dealing with the spill;
  (c) additional salaries and allowances (*e.g.* overtime and travel
      costs) paid to personnel under permanent contract;
  (d) costs of material bought specifically for the incident and used
      only for it (*e.g.* dispersant);
  (e) allowance for more rapid depreciation of materials used;
  (f) additional costs of fuel used in the operation;
  (g) costs of cleaning, repairing and replacing equipment;
  (h) destruction of oil collected;
  (i) taxes, duty and fees on all items;
  (j) additional insurance costs for duration of the operation.

11–55 Fixed costs may include the following[78]:
  (a) salaries and allowances for personnel permanently employed
      by the authorities and engaged in the operation, including
      management for operations and control units;
  (b) social costs and 'overheads' for the above personnel;

---

[78] See FUND/WGR.5/2.

(c) capital costs for equipment used and standby ships;

(d) costs for storing and maintaining equipment;

(e) costs for testing the equipment;

(f) insurance costs (covering damage to personnel or equipment);

(g) taxes duties and fees for these items.

A Working Group of the IOPC Fund, whose work was generally endorsed by the Assembly,[79] acknowledged that a state or local authority could, by making certain choices about how it responded to the incident, bring its claim more within one or other type of cost: for instance it could hire in outside labour (additional costs) or use its own (fixed costs). This no doubt influenced the majority who felt that a reasonable proportion of fixed costs should be recoverable, in the interests of encouraging states to maintain a response force, but that in the calculation of the figure only those expenses which correspond closely to the clean-up period in question and which do not include remote overhead charges should be included.[80] Ultimately the matter is governed by whether these items are allowable under the national law applicable. The IOPC Fund has paid claims for fixed costs to the Swedish government in the *Antonio Gramsci* case in 1979[81] and the *Furenas* case in 1980, to the French government in the *Tanio* case in 1980 and to the Japanese authorities in a number of cases.

(ii) *Loss of livelihood and loss of income: "economic loss"*

11–56 Another item of pollution damage mentioned in paragraph 10–42, above is loss of use, including loss of livelihood and loss of net income. This is sometimes referred to as economic loss, and most often arises in connection with claims by fishermen, hoteliers and others in the tourist trade. In principle, the IOPC Fund accepts such claims, although they present certain practical problems. Like fixed costs, they are often difficult to quantify. The IOPC Fund adopts a pragmatic approach and will pay up on reasonable estimates agreed with the claimant in the settlement negotiation. For instance, there have been many claims by fishermen in Japan for lost income, and the IOPC Fund requires evidence of previous years' catches and sales to compare with the performance in the year of the spill; then other factors are taken into account, such as the area of fishing ground affected and differences in market prices before arriving at an estimate of what has

---

[79] See FUND/A.4/16.

[80] See FUND/A.4/10.

[81] See para. 10–51, above for a summary of the facts.

been lost as a result of the incident.[82] Claims of hoteliers and others in the tourist industry are negotiated in a similar way, but in all cases the claimant must show a direct relationship between the pollution incident and his loss. Hence, the issue of remoteness of damage discussed above in paragraph 10–44, is dealt with in a similarly ad hoc, ragmatic way.

(iii) *Environmental damage*

11–57 The problem of so-called environmental damage illustrated in paragraphs 10–51 *et seq.*, above arose in the *Antonio Gramsci* case in 1979. The U.S.S.R. was not a member of the IOPC Fund but it was a party to the Liability Convention. Its claim, described in paragraph 10–51, was allowed by the U.S.S.R. Court where the owner had established his limitation fund, and so it consumed some 75 per cent. of that fund.[83] This reduced the amount available for the Swedish claimants, and since Sweden was a member of the IOPC Fund at that time their claim on the IOPC Fund was therefore increased. Although these claims were settled, on October 10, 1981 the Assembly passed Resolution 3 confirming its intention that the assessment of pollution damage "is not to be made on the basis of an abstract quantification of damage calculated in accordance with theoretical models." This means that, if a claim like the U.S.S.R. claim is raised directly against the IOPC Fund it will resist it. As the Working Group set up by the Assembly subsequently concluded, the possible effect of the oil on the environment after cleaning and restoration could not give rise to a claim unless it affected economic interests, and "it would be impossible to identify the person having a legal right for such a claim if no such interests were affected." It was not the intention in 1971 to set up an international system for the compensation of spurious claims. They are not "pollution damage" within the definition in the Convention.[84]

(iv) *Interest*

11–58 If under national law the IOPC Fund is liable to pay interest to claimants, it will, and has done so in the *Antonio Gramsci, Furenas,* and *Hosei Maru* cases in 1979, 1980 and 1980 respectively. However, where the Article 4 compensation limit is reached, claims for interest made by all claimants would not significantly alter the

---

[82] See further E.D.Brown, "The International Oil Pollution Compensation Fund, An Analytical Report on Fund Practice," *Oil & Petrochemical Pollution*, (1983) 269 at p. 271.
[83] See FUND/EXC.2/5.
[84] See paras. 10–47 *et seq.*, above.

amount each claimant actually recovers, and so in the *Tanio* case in 1980 interest, though claimed, was by agreement with all claimants not paid.

(v) *Release: subrogation and recourse*

11–59   Under Internal Regulation 8.4.3, it is a condition for the Director making a final settlement with a claimant to obtain a full and final release from the claimant in respect of claims arising from the incident. Where the IOPC Fund may wish to avail itself of its rights of subrogation, it will also require the claimant to execute any document or render any assistance necessary to enable it so to do under the applicable national law. Article 9 of the Convention deals with these rights, giving the IOPC Fund the claimant's rights against the owner under the Liability Convention and in respect of rights against any other person placing the IOPC Fund in a position not less favourable than that of an insurer. The IOPC Fund has used these rights in the *Tanio* case in 1980 to pursue the registered owner (seeking to break his right to limit), the charterers, the ship's last repairer, its classification society and those who supervised the repairs.[85]

6. CONTRIBUTIONS TO THE IOPC FUND

A. The Duty to Contribute

INTRODUCTION

11–60   Broadly speaking, the IOPC Fund is financed by contributions which, by Article 10, must be paid by any person who receives in a contracting state contributing oil which has been transported by sea at some point in its journey to the receiver. "Contributing oil" is carefully defined in Article 1(3) and is limited to crude oils (including topped and spiked crudes) and fuel oils. Thus, although the IOPC Fund pays out for oil pollution damage caused by the escape of all persistent hydrocarbon mineral oils,[86] it is only the receivers of these two types who contribute.[87] It is the receivers who must pay the contribution and not the contracting state (except in exceptional cases discussed below in connection

---

[85] See FUND/EXC.9/3.

[86] See para. 11–09, above.

[87] Hence lubricating oils, bitumen and aromatic tar would not be contributing oil. Crude and fuel oils account for the vast majority of persistent oils carried by sea. To have extended the contribution system further would have created additional administrative expense and had a minimal impact on the contributions paid by crudes and fuel oils.

with the role of states). An important and valuable minimum, introduced on the grounds of minimising administrative costs,[88] is provided: only those who have received in the relevant calendar year more than 150,000 metric tonnes of contributing oil need contribute, although the receipts of associated persons are to be aggregated to prevent evasion.[89] This minimum means that certain states with very small imports of crude and fuel oil can become party to the Convention and get its full protection without any person in that state having to contribute to the IOPC Fund. The combined effect of the definition of contributing oil and the 150,000 tonnes per year minimum is that it is very largely oil companies and the owners of power plants who contribute.

WHO MUST CONTRIBUTE AND IN RESPECT OF WHICH OIL

11–61 A difficulty has arisen in the interpretation of the tortuous drafting of Article 10(1) dealing with the oil in respect of which contributions must be made. Article 10(1) states that contributions shall be made by any person who "has received" contributing oil "in the ports or terminal installations" of a contracting state, and also any person who "has received" contributing oil "in any installations situated in the territory of that Contracting State" where it "has been carried by sea and discharged in a port or terminal installation of a non-Contracting State. . . . " The Convention is silent on what constitutes a receipt, who is the receiver and what is to be considered a "port or terminal installation." The problems arise particularly in relation to oil which is discharged from a ship into another ship for re-export or onward carriage inland, and where oil is discharged into the facilities of one person for the account of another.

11–62 Accordingly in 1980 the Assembly set up a Working Group to study these problems,[90] which found that the practice of states on these points varied. It adopted an agreed interpretation which was approved by the Assembly in October 1980,[91] and although it does not bind states[92] it is very much designed to guide them in interpreting Article 10. This interpretation, which is

---

[88] LEG/CONF.2/C.1/SR.13, *OR* 404.

[89] Article 10(2)(*b*) states that: "Associated person" means any subsidiary or commonly controlled entity. The question whether a person comes within this definition shall be determined by the national law of the state concerned.

[90] The Working Group's Report is FUND/A/ES.1/8.

[91] FUND/A/ES.1/13.

[92] Indeed, in the United Kingdom, for instance, ship to ship transfer within a port is considered a receipt, which is against the Assembly's approved interpretation.

printed in full on the IOPC Fund's reporting documents, states that (a) discharge into a floating tank (which does *not* include a ship unless it is not ready to sail) within the territorial waters of a contracting state is a receipt, whether or not the tank is connected with on-shore installations by pipeline; (b) in-port traffic is not carriage by sea; (c) ship to ship transfer is not a receipt wherever it happens and even if overland pipelines are used, unless the oil passes through a storage tank, in which case it is a receipt. This does not solve the problem of who is the receiver, which the Assembly tried to avoid by concluding that "within the scope of Article 10 . . . Contracting State should have a certain flexibility to adopt a practical reporting system . . . and that, failing payment by persons reported other than the physical receivers, the physical receivers should ultimately be liable for contributions irrespective of whether the persons reported have their place of business or residence in a Contracting State or not."[93]

INITIAL AND ANNUAL CONTRIBUTIONS

11–63 The IOPC Fund levies two types of contribution. Initial contributions, fixed by the Assembly at its first session in November 1978, are 0.04718 gold francs per tonne of contributing oil. By Article 11 this contribution is payable only once, within three months of the entry into force of the Convention for the new contracting state, and is levied on the contributing oil received in the calendar year preceding such entry into force.[94] Initial contributions were prospectively abolished by the 1984 Protocol: see paragraph 11–96, below.

11–64 Annual contributions are levied to enable the IOPC Fund to meet its administrative and claims expenditure each year. For this purpose, Article 12 provides for an annual budget to be produced which is divided into two: (i) major claims, that is those where the aggregate expenditure for any individual incident exceeds 15 million gold francs; and (ii) the general fund, which is all other expenditure including other claims. The reason for this division is that different groups of contributors contribute to each of the two types of expenditure. A contributor only contributes to a

---

[93] FUND/A/ES.1/13.

[94] Ironically, this has meant that some receivers effectively paid initial contributions twice. Receivers in the Federal Republic of Germany paid them when F.R.G. became a party in 1978; then when Italy became a party in 1979, Italian oil terminal operators passed on a part of their initial contributions as pipeline charges to their customers in the F.R.G. But the Assembly declined to adopt an interpretation of Article 11 which would avoid this: FUND/A/ES.1/13.

major claims fund in respect of contributing oil received in his contracting state in the year preceding the incident *if that state was a party to the Convention at the date of the incident*.[95] Hence, when a state becomes a party its contributors do not have to pay for major claims which occurred before then. All other expenditure (the general fund) is contributed to on the basis of contributing oil received in contracting states in the preceding calendar year.

11–65 Annual contributions are calculated in pounds sterling by dividing the budgets by the amount of contributing oil received in the relevant year,[96] and are due on January 15, following the Assembly's decision to levy them.[97] Hence, for most states the contributions to the IOPC Fund must be found in foreign currency. The amount may be deductible from corporation or similar taxes.

B. The Role of States

11–66 In so far as contributions are concerned, contracting states have three main roles. The first is to provide the mechanism whereby the IOPC Fund can raise its contributions by reporting to it those persons liable to contribute and the amount of contributing oil they have received.[98] The second is purely optional: under Article 14 a state may itself assume the obligation to contribute in respect of any person.

11–67 The third is contained in Article 13(2):

> "Each Contracting State shall ensure that any obligation to contribute to the Fund arising under this Convention in respect of oil received within the territory of that State is fulfilled and shall take any appropriate measures under its law, including the imposing of such sanctions as it may deem necessary, with a view to the effective execution of any such obligation; provided, however, that such measures shall only be directed against those persons who are under an obligation to contribute to the Fund."

This provision comes in an Article dealing generally with unpaid or late paid contributions, and could be taken to amount to a guarantee of last resort by each contracting state that contributions due will be paid, depending on whether the obligation to take "any appropriate measures" is taken as additional to the obligation to

---

[95] Article 12(2)(*b*).
[96] Article 12(3).
[97] Internal Regulation 3.8. Interest is due on late payments: Article 13(1) and Internal Regulation 3.10.
[98] Article 15 and Internal Regulation 5. The 1984 Protocol provides a sanction for non-fulfillment of this duty: see para. 11–97, below.

ensure that contributions are paid, or as the means by which that obligation is to be implemented. The discussion at the Conference is not helpful on the point,[99] and merely evinces an intention that states should be free to determine what appropriate measures should be taken, but Article 13(3) clearly envisages that a contributor might become insolvent, at which point it could be said that the state has failed to "ensure" that his obligation to contribute has been discharged. Whatever interpretation is favoured, the primary obligation is clear, and onerous. The state, if not a guarantor, is certainly envisaged as policeman. Internal Regulation 3.9 places upon the Director the duty to notify a state of a non-payment and to "request advice on the action to be taken to ensure that the obligations of that contributor are fulfilled."

C. The Costs So Far

11–68 The annual contributions levied by the IOPC Fund to date are given in Table 11.2. From this is can be seen that the largest call so far in any one year has been in 1983, when the call for the *Tanio* disaster was made, and even then the total amounted to only 2.6 pence per tonne of contributing oil. This compares with FOB crude oil prices during 1983 of the order of £130 to £150 per tonne (with fuel oil costing more), so that the pollution damage cover provided by the IOPC Fund cost the industry a maximum of 0.02 per cent. of FOB values in its most expensive year so far. By comparison, the aggregate of all risks and war cargo insurance, which includes cargo's contribution to salvage and general average, would not often dip below 0.1 per cent. of FOB values.

11–69 A situation could arise under the 1971 Fund Convention in its last days that all the states with contributing oil receipts have left and only those without such receipts would remain. If an incident occurred in a remaining member state, the IOPC Fund would have a liability but no means of meeting it. Very careful co-ordination of the denunciations of the 1971 Fund Convention will be required to avoid this absurdity, probably utilising the accelerated denunciation provisions of Article 42.

7. BECOMING A PARTY TO THE FUND CONVENTION

11–70 By Article 37(4), only parties to the Liability Convention may become party to this Convention. States therefore have a

[99] LEG/CONF.2/C.1/SR.15.

Table 11.2—Annual Contributions Levied by the IOPC Fund

Year	Contribution levied (£)		Contribution per tonne (£)
1979	750,000	(a)	0.0008455
1980	800,000	(a)	0.0008441
	9,200,000	(b)	0.0117659
1981	500,000	(a)	0.0005690
1982	600,000	(a)	0.0007197
	260,000	(c)	0.0003160
1983	1,000,000	(a)	0.0011931
	20,000,000	(d)	0.0211079
	3,106,000	(e)	0.0037776

Notes to Table 11.2
(a) General fund.
(b) *Antonio Gramsci* major claims fund—incident 27.2.79.
(c) *Fukutoku Maru No.8* major claims fund—incident 3.4.82.
(d) *Tanio* major claims fund—incident 7.3.80.
(e) *Fukutoku Maru No.8* and *Ondina* major claims funds—incidents 3.4.82 and 3.3.82.

---

number of choices open to them. A state can become a party to the Liability Convention but not to this Convention; as at December 31, 1984, some 25 states had taken this course, presumably because they preferred to rely on CRISTAL[1] and their own national law instead. A state can become party to both Conventions, as some 29 states had done, or it may become party to neither, and rely on their own national law and, while they last, TOVALOP and CRISTAL.[2] This last category contained at that date some states heavily exposed to oil pollution, including the United States and Canada (although both stated at the 1984 Conference their intention to ratify). Now that the 1984 Protocols are in place, states must work out not only what protection they need at present, and how best to provide it, but what they need in future. Discussion of the full range of options is therefore deferred until after the 1984 Protocol to this Convention has been examined: see paragraph 11–98, below. For the present, however, since the 1984 Protocols are not expected to enter into force until the late 1980s, each state must ask what are

[1] See Chapter 12.
[2] See Chapter 12.

the costs and benefits of joining the Fund Convention. In this connection the following considerations, among the many which will apply in the case of individual states, are relevant to such a cost/benefit analysis for any state.

11–71   (i) The likely benefits cannot be properly assessed until an attempt has been made to assess the exposure of the state to the risk of oil pollution damage. In this connection, the effect which a serious spill could have on the local or national economy is relevant. Risk is probability of occurrence times consequence of occurrence.

(ii) The cost of acquiring the benefits are likely to be borne largely by private interests (shipowners and oil receivers), who may be able to deduct these from tax and so share them with the state. In the case of the Liability Convention, most ships in international trade and many in domestic trade will already have entered a P and I Club for pollution risks including liabilities under the Liability Convention, so the incremental cost will relate only to those which before a ratification do not have such cover. Contributions to the IOPC Fund are in pounds sterling; to CRISTAL they are in U.S. Dollars. There is some scope for ensuring that exporters pay the CRISTAL contributions, but not in the case of the IOPC Fund.

(iii) The international legal regime of the two Conventions covers all ships, irrespective of their flag, and all persistent hydrocarbon mineral oil, irrespective of its ownership, whereas PLATO, TOVALOP and CRISTAL[3] only apply if the private interests involved have chosen this. This is part of an overall difference between the legal regime and the voluntary one: states are involved in the former. In particular, states, who are the largest claimants for oil pollution damage, participate in the claims settlement procedures of the IOPC Fund through the Executive Committee and Assembly. What the IOPC Fund does is open to international public scrutiny.

(iv) The limits under the international legal regime, when expressed in national currency, vary with the relationship that currency has with a basket of major currencies (if the 1976 Protocols are adopted) rather than with the U.S. Dollar, as in the case of the voluntary schemes.

(v) There have always been some differences in the scope of the international legal regimes and the voluntary ones: see further Chapter 12.

(vi) The IOPC Fund has built up a unique reputation for speed

[3] See Chapter 12.

of response. Its slowest case was the *Tanio*, a really major pollution disaster whose claims documentation ran to over 20,000 documents, and still it managed to pay up most of its available money four years after the incident; this would have been much sooner if the claims had been presented in a form which had been easier to evaluate and if the Article 4 limit had not been reached. All other cases have been settled within one or two years. The IOPC Fund has acquired an expertise in pollution claims handling possessed by no other organisation.

(vii) Most states will have to prepare special legislation to ratify the Liability Convention and the Fund Convention, and even if constitutionally this is not a requirement, the Conventions are so full of ambiguities that they would be well advised to do so. In addition, there is a certain administrative burden associated with the issue of insurance certificates under the Liability Convention and with the reporting obligations under the Fund Convention. Although states are not obliged to participate in the organs of the IOPC Fund, which meet in London, participation represents both an opportunity to extend international relations and an administrative burden.

8. THE 1984 PROTOCOL: INTRODUCTION TO THE IOPC FUND 1984

THE 1984 FUND CONVENTION

11–72 The background to the 1984 Conference which adopted the 1984 Protocol to the Liability and Fund Conventions has already been explained in paragraphs 10–112 *et seq.*, above. The 1984 Protocol to the Fund Convention takes the form of amendments thereto, and Article 27 of the 1984 Protocol provides that the 1971 Convention as amended by the Protocol shall be known as the International Convention on the Establishment of an International Fund for Compensation for Oil Pollution Damage, 1984—the 1984 Fund Convention. The 1984 Protocol only amends certain of the Articles of the 1971 Convention, so that the 1984 Fund Convention has the original scope of the 1971 Convention except where a specific amendment has been made. Many of the amendments are drafting amendments consequential upon the main changes effected by the 1984 Protocol, and do not require a mention here.

11–73 The 1984 Fund Convention, to which states may only become party if they are parties to the 1984 Liability Convention, is not expected to enter into force until the late 1980s.[4] The transitional provisions and the deliberate choices given to states mean that it is capable of existing indefinitely alongside the three other regimes—the 1969 and 1984 Liability Conventions and the 1971 Fund Convention—although it is expected that the 1971 Fund Convention will have a relatively short life after the 1984 Fund Convention enters into force (possibly as little as eighteen months).[5]

ESTABLISHMENT AND ORGANISATION OF THE IOPC FUND 1984

11–74 Article 3 of the 1984 Protocol establishes an entirely new legal entity, called "The International Oil Pollution Compensation Fund 1984" (herein the IOPC Fund 1984), whose purpose is to provide compensation for pollution damage to the extent that the protection afforded by the 1984 Liability Convention is inadequate. Although it is envisaged that the Secretariat of the IOPC Fund 1984 will be none other than that of the existing IOPC Fund,[6] the new organisation is a completely new legal entity. The Conference considered the alternative of having the new regime administered by the existing legal entity of the IOPC Fund, but concluded that it would not be legally possible.[7] One reason for this is that the new Convention will inevitably enter into force before all parties to the 1971 Fund Convention have become parties to it, and this was inconsistent with having a single Assembly for all and with the general principle that an amendment of a treaty can only bind parties to that amendment.[8]

11–75 The IOPC Fund 1984 will have a virtually identical internal structure and organisation to the existing IOPC Fund, except in one important particular: it will not have an Executive Committee.[9] Claims approval and other functions fulfilled by that organ of the existing IOPC Fund will be fulfilled in the IOPC Fund 1984

---

[4] See para. 11–99, below. The Plenary of the Conference adopted the Protocol by 44 votes to 0, with 21 abstentions. The 21 abstentions give grounds to fear that a number of important states may hold back from ratifying the Protocol, at any rate, until the 1971 IOPC Fund becomes too small in membership to make it worthwhile.

[5] See further para. 11–104, below.

[6] See further para. 11–93, below.

[7] See LEG/CONF.6/C.2/SR.12/PROV.

[8] The 1971 Fund Convention does not have a "rapid amendment" clause whereby states are bound by an amendment which has entered into force unless they make a specific declaration to that effect or denounce the Convention.

[9] Article 17 of the 1984 Protocol. The reasons for the abolition of the Executive Committee are set out at LEG/CONF.6/C.2/SR.12/PROV, where it is debated. The two main objects were to give the administration of the new Convention greater flexibility, and to strengthen the authority of the Assembly and make it more flexible.

by the Assembly and by ad hoc subsidiary bodies set up by the Assembly.[10]

THE PURPOSES OF THE IOPC FUND 1984—ABOLITION OF SHIPOWNER RELIEF

11–76 As part of the "package deal" on limits which was eventually agreed at the Conference, there is to be no Article 5 shipowner relief[11] under the 1984 Fund Convention. Accordingly, Article 7 of the 1984 Protocol deletes Article 5 altogether and all references to the indemnification function of the IOPC Fund are also deleted. This leaves the IOPC Fund 1984 with just one substantive function, to provide compensation for pollution damage to the extent that the protection afforded by the 1984 Liability Convention is inadequate.

9. THE 1984 PROTOCOL: COMPENSATION FOR POLLUTION DAMAGE—ARTICLE 4

SCOPE OF ARTICLE 4

11–77 Article 4(1) is left unamended by the 1984 Protocol, although some of the most important words it uses are redefined, which has the effect of widening the scope of the new Fund Convention to complement that of the 1984 Liability Convention. The most important change is that references in Article 4(1) to the "Liability Convention" are replaced by references to the "1984 Liability Convention,"[12] which means that claimants must show that they have been unable to obtain full and adequate compensation for their damage under the terms of the 1984 Liability Convention, which has wider scope and much higher limits than the old Liability Convention.[13]

11–78 By Article 2(3) of the 1984 Protocol, several changed definitions bring the scope of the 1984 Fund Convention into line with that of the 1984 Liability Convention. Hence, the change in the definition of "incident" which extended the scope of the 1984 Liability Convention to "pure threat" situations[14] extends the scope of the 1984 Fund Convention also. Similarly, now that "ship" in the

---

[10] Article 18(2) of the 1984 Protocol.
[11] See paras. 11–17 *et seq.*, above.
[12] Article 6(1) of the 1984 Protocol.
[13] As to the scope of the 1984 Liability Convention, see paras. 10–126 *et seq.*, above, and as to the limits, see paras. 10–141 *et seq.*, above.
[14] See paras. 10–121 *et seq.*, above.

1984 Liability Convention covers tankers and certain combination carriers after they have carried oil in bulk as cargo,[15] so does the 1984 Fund Convention. The alteration in the definition of "pollution damage" wrought by the 1984 Liability Convention[16] applies here too, and by Article 4 of the 1984 Protocol the 1984 Fund Convention is given the same extended geographical scope as that of the 1984 Liability Convention.[17]

11–79 Apart from those changes, Article 4(1) continues to apply as it does under the 1971 Fund Convention.[18] There are, however two exceptions to this rule, one relatively minor, concerning the exemption of the IOPC Fund 1984, the other of considerable importance, concerning the "transitional period."

EXEMPTION OF THE IOPC FUND 1984

11–80 It will be remembered that the exemption of the IOPC Fund in cases where the claimant had himself been wholly or partly to blame for his loss was, in respect of preventive measures, somewhat unclear due to bad drafting.[19] Article 6(2) of the 1984 Protocol takes the opportunity to put this right by making it entirely clear that in Article 4(3) of the 1984 Convention the IOPC Fund 1984 will never be exonerated with regard to preventive measures which may have been undertaken, and this would appear to be so both where the cause of the incident is the negligence of the claimant and where the cost of preventive measures is unduly high as a result of the negligence of the claimant.[20] This certainly provides all and sundry with an incentive to take preventive measures, but it could still lead to problems in distinguishing preventive measures from other items of pollution damage.[21]

COMPENSATION DURING THE TRANSITIONAL PERIOD

11–81 It will be remembered that the 1984 Conference adopted a "phased-in" solution to the treaty law problems associated with the transition from the old regime to the new one.[22] An essential feature of this solution is that there is a transitional period after the 1984 Fund Convention comes into force during which the criteria

---

[15] See paras. 10–118 *et seq.*, above.
[16] See paras. 10–137 *et seq.*, above.
[17] See paras. 10–131 *et seq.*, above.
[18] See paras. 11–08—11–13, above.
[19] See para. 11–15, above.
[20] For the debate on this, see LEG/CONF.6/C.2/SR.17, 21 and 22/PROV.
[21] See para. 11–16, above.
[22] See paras. 10–157 *et seq.*, above.

for a claimant to have a valid claim upon the IOPC Fund 1984 are different to those which apply normally (*i.e.* after the end of the transitional period). By Article 26 of the 1984 Protocol, a new Article 36 bis is inserted into the Convention as follows:

> "The following transitional provisions shall apply in the period, hereinafter referred to as the transitional period, commencing with the date of entry into force of this Convention and ending with the date on which the denunciations provided for in Article 31 of the Protocol of 1984 to the 1971 Fund Convention[23] take effect:
>
> (a) . . .[24]
>
> (b) Where an incident has caused pollution damage within the scope of this Convention, the Fund shall pay compensation to any person suffering pollution damage only if, and to the extent that, such person has been unable to obtain full and adequate compensation for the damage under the terms of the 1969 Liability Convention, the 1971 Fund Convention, and the 1984 Liability Convention, provided that, in respect of pollution damage within the scope of this Convention in respect of a Party to this Convention but not a Party to the 1971 Fund Convention, the Fund shall pay compensation to any person suffering pollution damage only if, and to the extent that, such person would have been unable to obtain full and adequate compensation had that State been party to each of the above-mentioned Conventions.
>
> (c) . . .[25]
>
> (d) . . .[26]"

11–82 This rule has two limbs, the main rule and the proviso. The operation of the main rule is occasioned by the fact that during this transitional period it is possible for a state to be a party to *all four Conventions*, as was seen in paragraphs 10–162 *et seq.*, above, and the idea is simply that the 1984 Fund Convention comes in on top of the other three during this time.[27] There are thus four layers for a claimant in such a state during this transitional period: first, the 1969 Convention; then the 1971 Fund Convention; then the 1984 Liability Convention; and lastly the 1984 Fund Convention. This avoids multiple recovery by claimants in such states.

11–83 The proviso, however, is not quite so easy to understand. It says that, in effect, during the transitional period states not party

---

[23] See para. 11–100, below.

[24] Amendment to Article 2 consequent on paragraph (*b*).

[25] Discussed in para. 11–88, below.

[26] Amendment to Article 9(1) consequent on paragraph (*b*).

[27] It was seen in para. 10–162 that, by Article XII bis(*b*) of the 1984 Liability Convention, that Convention would itelf operate on top of the 1969 and 1971 Conventions in this situation.

to the 1971 Fund Convention will be deemed to be party thereto, so that in fact the protection that Convention would have provided is uncovered. The effect of this is illustrated by the shaded triangle in Table 11.3 (see paragraph 11–90, below). The only clue to the purpose of this proviso which the *travaux preparatoires* contain is in the summary of the drafts and discussion in the IMO Legal Committee.[28] From this it is clear that originally, the proposal was that, during the transitional period, *only parties to the 1971 Fund Convention* would be eligible to become parties to the 1984 Protocol. Some states felt that this was not only unorthodox, but it could completely block the entry into force of the 1984 Protocol. Nonetheless, it was also felt that unrestricted access to the 1984 Protocol during the transitional period could result in practical difficulties, particularly in respect of contributions.[29]

11–84 Accordingly, a proposal of the Federal Republic of Germany in substantially the form of this proviso[30] found its way into the draft Convention, and there it stayed, its substance undebated at the Conference.[31] The problem with deeming provisions is that they are very difficult to apply in practice. In order to work out what can be claimed against the IOPC Fund 1984 in a state *not* party to the 1971 Fund Convention, it is necessary to go through the following steps.

1. How much money would that person have recovered if the state had been party to the 1971 Fund Convention (which means, of course, that it would also have been a party to the 1969 Liability Convention)?

2. How much money has that person actually recovered under the 1984 Liability Convention?[32]

3. What is the extent of his pollution damage as defined in the 1984 Fund Convention?

---

[28] LEG/CONF.6/7, page 78 and Annex 5, page 8.

[29] The difficulties envisaged may have been related more to fairness than practicalities. If contributors in a state party to both the 1971 and 1984 Fund Conventions had to contribute to the IOPC Fund 1984 in respect of damage suffered in a state party to only the 1984 Fund Convention for amounts above the 1984 Liability Convention level, but contributors in the latter state contributed to the IOPC Fund 1984 in respect of damage suffered in the former state only for amounts above the 1971 Fund Convention level, it might have been thought unfair.

[30] Legal Committee document LEG 50/4/12.

[31] The clause was discussed at LEG/CONF.6/FC/SR.5/PROV but the substance of this part was not.

[32] Because the state is not actually a party to the 1971 Fund Convention, Article XII bis(*b*)—see para. 10–162, above—will not apply; but Article XII bis(*a*)—see paras. 10–158 *et seq.*, above—would apply if the state was in fact a party to the 1969 Liability Convention.

4. Amount 3 minus amount 2 minus amount 1, if a positive sum, may be claimed against the IOPC Fund 1984.

11–85 As if this were not difficult enough in itself, step 1 above is a purely hypothetical exercise. In order to do it, you must know the totality of all the other claims actually made against the shipowner under the 1969 Liability Convention and against the IOPC Fund under the 1971 Fund Convention; then, you must re-work the distribution of the funds supposedly available thereunder to work out what the claimant would have got if he had claimed under them (and that will be impossible where in fact no shipowner limitation fund has been established under the 1969 Liability Convention for the simple reason, for instance, that no pollution damage was suffered in a state party to that Convention in the incident). The result is nightmarish, but not bad enough to warrant a state waiting until the end of the transitional period before ratifying the 1984 Fund Convention, for that would be to delay its early participation in the increased limits of the IOPC Fund 1984. There could also be difficulties in applying this proviso in practice, precisely because the IOPC Fund which administers the 1971 Fund Convention will not be a party to any claims negotiation with the claimant, but will be deemed to have been; although the fact that it is envisaged that the same Secretariat will staff the IOPC Fund 1984 and the IOPC Fund will help.

10. THE 1984 PROTOCOL: THE LIMIT OF ARTICLE 4 COMPENSATION

11–86 As was seen in paragraphs 10–141 and 10–142. the single most important purpose of the 1984 Conference was to amend the limits of liability for both Conventions. The limits finally adopted were, as explained there, the result of a compromise. The compromise on the question of the Fund Convention limit which was finally adopted originated with a proposal of the United States (supported by others),[33] and this proposal was primarily designed to attract the United States into the new Fund Convention. The original proposal was to have a basic coverage for the IOPC Fund 1984 of 150 million SDRs, and an expanded coverage of 200 million if and when participation in the new Convention became broad enough to make this amount economically sound. The idea was that the expanded coverage would only apply when the total of con-

---

[33] LEG/CONF.6/C.2/WP.36.

tributing oil received by terminals in three member states reached 600 million tonnes. Thus, "States would participate in the expanded coverage only after the Fund had been joined by the United States and at least two other large oil-receiving countries— for example, Japan and France or Italy and the Netherlands." In the process of compromise over the limits both for this Convention and the Liability Convention, the basic coverage was reduced to 135 million SDRs, but the expanded coverage remained at 200 million SDRs.[34]

CALCULATION AND EFFECT OF THE AMOUNT OF THE LIMIT

11–87   Hence, Article 6(3) of the 1984 Protocol replaces Article 4(4) with an entirely new limit. Article 4(4)(a) provides that the aggregate amount of compensation payable under the new Article 4, including the amount of compensation actually paid under the 1984 Liability Convention shall not exceed 135 million units of account (SDRs),[35] and the same limit is set by Article 4(4)(b) for damage resulting from a natural phenomenon of an exceptional, inevitable and irresistible character.[36] But in each case, by Article 4(4)(c) that maximum increases to 200 million SDRs "with respect to any incident occurring during any period when there are three Parties to this Convention in respect of which the combined relevant quantity of contributing oil received by persons in the territories of such Parties, during the preceding calendar year, equalled or exceeded 600 million tons."

11–88   During the transitional period, the way in which the limit is calculated has to be altered to accommodate the operation of Article 36 bis(b),[37] and so Article 36 bis(c) provides that in the application of Article 4 the amount to be taken into account in determining the aggregate amount of compensation payable by the IOPC Fund 1984 shall also include "the amount of compensation actually paid under the 1969 Liability Convention, if any, and the amount of compensation actually paid or deemed to have been paid under the 1971 Fund Convention."

11–89   The Conference took the opportunity to put right two blemishes in Article 4 of the 1971 Fund Convention. The first is the problem of interest on the owner's limitation fund, discussed at

---

[34] LEG/CONF.6/C.2/WP.44, debated at LEG/CONF.6/C.2/SR.28 and 29/PROV.

[35] Article 2(4) of the 1984 Protocol, amending Article 1(4) of the Fund Convention, defines the unit of account as having the same meaning as in Article V(9) of the 1984 Liability Convention: see paras. 10–143 and 10–67 *et seq.*

[36] For an explanation of this separate limit, see para. 11–28, above.

[37] See para. 11–81, above.

paragraph 11–25, above. The new Article 4(4)(*d*) thus provides that no account of such interest shall be taken in computing the Article 4 limit, so that the IOPC Fund 1984 will not benefit from this interest as the IOPC Fund does. The second is the problem of the date to be taken for conversion of the unit of account into national currencies, discussed at paragraph 11–36, above. This, the new Article 4(4)(*e*) provides, shall be the date of the decision of the Assembly of the IOPC Fund 1984 as to the first date of payment of compensation.

11–90   There can be little doubt that the new figures will prove adequate to cover most spills until at least the earliest possible time for their revision.[38] Figures produced by the P & I Clubs tend to confirm this: analysing some 17,000 oil spills covering 1970 to 1982 as if the new basic limits had applied to them, and estimating the cost of each spill in 1983 money, only one case would not have been fully compensated, representing some 2 per cent. of the total, and even this would have been fully compensated if the expanded limit of SDR 200 million applied.[39] The IOPC Fund 1984 can expect to be involved in far fewer spills than the present IOPC Fund: a rough notional application of the new limits to the nineteen cases in which the IOPC Fund had actually been involved by the time of the Conference[40] shows that only one—the *Tanio* case off France in 1980—would definitely have exceeded the new 1984 Liability Convention limits and one other may just have done. The new limits are illustrated in Table 11.3.

REVISION OF THE ARTICLE 4 LIMIT

11–91   The system of revision of limits contained in Article 4(6) of the Fund Convention[41] is completely altered by the 1984 Protocol, Article 33 of which replaces it with a system almost identical to that of the 1984 Liability Convention.[42] This is deliberate, to facilitate the changing of limits in both Conventions at the same time. The only difference between the two sets of provisions concerns the considerations which the Legal Committee of IMO must take into account when acting on a proposal to amend the limits. For the Fund Convention, there is no requirement to take into account the possible effect on the cost of insurance, this being a fairly irrelevant consideration in the context of the Fund Convention. A proposal to

---

[38] See para. 11–91, below.
[39] Private communication from the Britannia Steam Ship Insurance Association Limited.
[40] See LEG/CONF.6/C.2/INF.4 for figures.
[41] See paras. 11–31 *et seq.*, above.
[42] See paras. 10–153 *et seq.*, above.

TABLE 11.3

Illustration of the Division of Pollution Damage Between the Shipowner (under the 1984 Liability Convention) and the 1984 Fund Convention

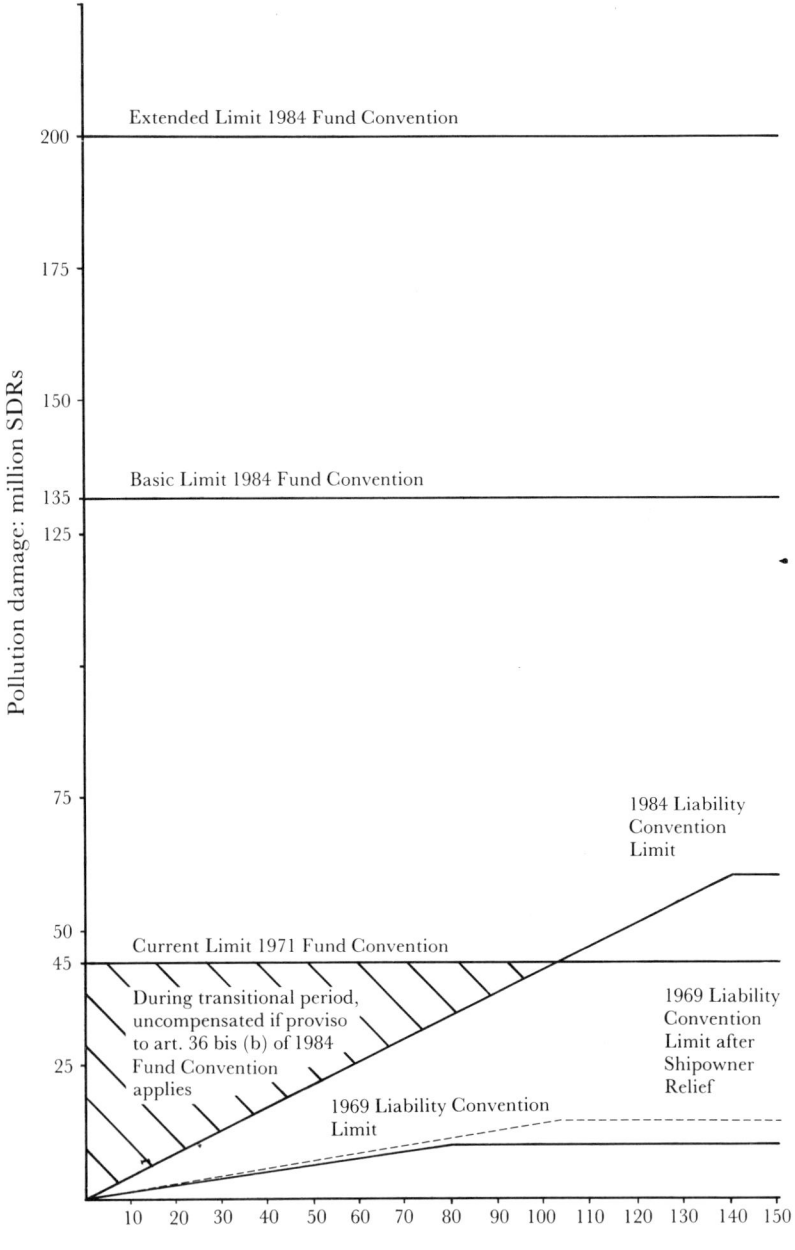

Thousand limitation tons

Notes to Table 11.3

(a) The current limit and relief under Article 5 of the 1971 Fund Convention has been calculated in SDRs in accordance with the 1976 Protocol to that Convention.

(b) The limitation amounts for the 1969 Liability Convention have been expressed in SDRs in accordance with the 1976 Protocol to that Convention.

(c) The graph assumes that the shipowner's rights to limit under the Liability Convention and to claim relief under Article 5 of the 1971 Fund Convention have not been denied.

---

replace it with the possible effect on the cost of oil was defeated in Committee,[43] so that only the experience of incidents, changes in monetary values and the relationship between the Fund Convention limits and the Liability Convention limits remain.

DISTRIBUTION OF THE ARTICLE 4 LIMIT

11–92 The Conference took the opportunity to put right the rather serious drafting problem contained in Article 4(5) of the 1971 Fund Convention[44] by providing that in a case where the limit is reached, "the amount available shall be distributed in such a manner that the proportion between any established claim and the amount of compensation actually recovered by the claimant under this Convention shall be the same for all claimants."[45] Hence, the reference to the Liability Convention which had created the problem in the old Article 4(5) has now gone, and it is clear that established claims against the IOPC Fund 1984 are to be met *pro rata*.

11. THE 1984 PROTOCOL: CLAIMS AGAINST THE IOPC FUND 1984

CLAIMS PROCEDURE

11–93 Since it is envisaged that the IOPC Fund 1984 will be staffed by the same persons as the IOPC Fund,[46] in order to achieve continuity in claims practice and experience, the only changes which at this stage can be predicted are those specifically

---

[43] LEG/CONF.6/C.2/SR.23/PROV.

[44] See paras. 11–41 *et seq.*, above.

[45] In Article 6(4) of the 1984 Protocol.

[46] The 1984 Protocol makes provision for this so far as the IOPC Fund 1984 is concerned by inserting a new Article 36 ter into the Convention, and in so far as the IOPC Fund is concerned, adopted a Resolution requesting the Assembly of that body to take the necessary steps to enable a smooth transition to be made.

provided for in the new Convention. There are two of these. The first is contained in the definition of "pollution damage" which we have seen[47] says that the costs of impairment of the environment (other than loss of profit therefrom) shall be limited to "costs of reasonable measures of reinstatement actually undertaken *or to be undertaken*." These last words were designed to assist a claimant against both shipowner and the IOPC Fund 1984 in cases where he did not have enough cash to do the work prior to reimbursement by them. This will enable the IOPC Fund 1984 to make payment in certain cases where previously it would have to wait until the claim could be quantified by the actual expenditure.

11–94 The other provision is contained in Article 6(5) of the 1984 Protocol, enacting a new Article 4(6) in these terms:

> "6. The Assembly of the Fund may decide that, in exceptional cases, compensation in accordance with this Convention can be paid even if the owner of the ship has not constituted a fund in accordance with Article V, paragraph 3, of the 1984 Liability Convention . . . .[48]"

This was adopted after a proposal that the establishment of the limitation fund be optional for the shipowner under the 1984 Liability Convention had been rejected,[49] because the need to wait for the establishment of this fund had, as noted above,[50] created problems for the IOPC Fund in the past in some cases.[51]

11–95 The claims procedure of the IOPC Fund 1984 should, therefore, be even more flexible and expeditious than that of the IOPC Fund. This was certainly the intention of the Conference, something which no doubt the Assembly of the IOPC Fund 1984 will bear in mind when interpreting what is an "exceptional" case.

12. THE 1984 PROTOCOL: CONTRIBUTIONS TO THE IOPC FUND 1984

11–96 Two changes are made to the substantive provisions of the 1971 Fund Convention. The first, and most important, is that the IOPC Fund 1984 will not require and may not levy initial contributions.[52] As the Director of the IOPC Fund 1984 remarked, "the

---

[47] See paras. 11–78 and 10–137 *et seq.*, above.
[48] See paras. 10–151 and 10–76 above.
[49] LEG/CONF.6/C.2/SR.13/PROV.
[50] para. 11–50, above.
[51] LEG/CONF.6/C.2/SR.18/PROV.
[52] The relevant provisions of the 1984 Protocol are Articles 12 to 14, amending Articles 10 to 12 of the Fund Convention.

very high known level of the initial contributions had been an obstacle to the ratification of the Fund Convention."[53] Their abolition was therefore seen as a method of facilitating the entry into force of the 1984 Convention, and the IOPC Fund 1984 will raise its early expenditure by annual contributions.

11–97 The other change is that the reporting role of states[54] is given a sanction which previously it lacked. Article 16 of the 1984 Protocol, adding a new paragraph 4 to Article 15 of the Fund Convention, provides that where a contracting state does not fulfil its obligations to submit the appropriate returns, and this results in a financial loss for the IOPC Fund 1984, that state, if the Assembly so decides, shall be liable to compensate the IOPC Fund 1984 for such loss.

13. THE 1984 PROTOCOL: FINAL CLAUSES AND CHOICES OPEN TO STATES

11–98 The final clauses of an oil pollution Convention are not normally of much concern. Those of the 1984 Protocol are different, because they affect not only its entry into force but also the choices which states have for effecting oil pollution liability cover for their coastlines. The adoption by the Conference of a "phased-in" solution to the treaty law problems[55] described in paragraphs 10–156 and 10–157, above, means that these choices are wide and complicated. Because of the indissoluble links between the Liability Convention and this one, it will be necessary in this connection to consider also some of the final clauses of the 1984 Liability Convention.

ENTRY INTO FORCE

11–99 The final clauses of the 1984 Fund Convention are those of the 1984 Protocol.[56] Because the 1984 Fund Convention is supplementary to the 1984 Liability Convention, by Article 28(4) only parties to the latter may become party to the 1984 Fund Convention. The entry into force requirement was seen by many states as

---

[53] LEG/CONF.6/C.2/SR.18/PROV.

[54] See para. 11–66, above.

[55] The debate at the Conference was short and not particularly informative:LEG/CONF.6/ C.2/SR.9/PROV, and LEG/CONF.6/FC/SR.1 to 7/PROV. The clearest explanation of the alternative solutions to the treaty law problems is contained in a United Kingdom paper, LEG/CONF.6/9. A summary of the previous discussions at IMO's Legal Committee is contained in LEG/CONF.6/7, pp. 68 *et seq.*

[56] Article 26 of the 1984 Protocol, inserting a new Article 36 quater into the Fund Convention.

being linked with the issue of limits,[57] and the requirements themselves were surprisingly contentious.[58] But they had to be chosen so
that the IOPC Fund 1984 would be workable economically, and
the compromise finally adopted was as follows: by Article 30, the
1984 Fund Convention will not enter into force until twelve months
after the date on which at least eight states, having received in the
preceding calendar year at least 600 million tonnes of contributing
oil, have become parties; and, of course, this must not be before the
1984 Liability Convention enters into force. The 1984 Fund Convention is not expected to enter into force until the late 1980s.

THE "PHASED-IN" SOLUTION

11–100 The heart of the "phased-in" solution is contained in
Article 31, which provides that if a state is party to either or both of
the 1969 or 1971 Conventions, it *must* denounce them within six
months of the fulfillment of the following conditions: when at least
eight states, having received in the preceding calendar year at least
750 million tonnes of contributing oil, have deposited instruments
of ratification, etc. The denunciation must take effect exactly twelve
months after the expiry of that six-month period (which we will call
the "denunciation date" for ease of reference). This provision
therefore sets a limit to the time which a state can be party to the
1984 Fund Convention *and* the original Conventions. The period
leading up to the denunciation date is called by Article 36 bis the
"transitional period," during which the 1984 Conventions, and in
particular the 1984 Fund Convention, operates somewhat differ-
ently.[59] For states party to both original Conventions, the phases of
operation using this "phased-in" solution may be illustrated as follows, assuming, as is most likely, that the 1984 Liability Convention will enter into force before the 1984 Fund Convention, rather
than at the same time.

```
1969 Convention       ----------------------------------|
1971 Convention       ----------------------------------|
1984 Liability Convention   | -------------------------------------
1984 Fund Convention              | ------------------------
```

Time   PHASE 1 | PHASE 2 | PHASE 3 | PHASE 4

---

[57] For instance, if a particular state found the limits adopted too high for its liking, then that
state would want to delay the entry into force of the 1984 Fund Convention, and so would
advocate stiff entry into force requirements; and vice versa for a state which liked the
limits adopted. See LEG/CONF.6/C.2/SR.24/PROV, where the debate was adjourned to
enable it to take place after the resolution of the issue on limits.

[58] LEG/CONF.6/C.2/SR.25/PROV.

[59] For the application of the 1984 Liability and Fund Conventions during the transitional
period, see paras. 10–158 *et seq.*, 11–81 *et seq.*, and 11–88, above.

The denunciation date is the end of Phase 3/beginning of Phase 4, and the transitional period defined by Article 36 bis is Phase 3.

THE DELAYED RATIFICATION/ACCESSION OPTION

11–101   But the "phased-in" solution is not the only one which is open to states already party to one or both of the original Conventions. Articles 30(4) to (6) give a state a remarkable option: when depositing its instrument a state may make a revocable declaration that the instrument shall not take effect until the date, after the entry into force of the Convention, when parties must denounce the 1969 and 1971 Conventions (if party thereto). Articles 13(2) and (3) of the 1984 Liability Convention are to the same effect, and a declaration (or withdrawal thereof) under the 1984 Liability Convention is deemed to be one also under the 1984 Fund Convention. Use of these provisions enables a state to achieve the effect, for itself, of what was called at the Conference the "delayed denunciation" solution to the treaty law issues. This is relevant for states already party to either or both of the 1969 and 1971 Conventions. In other words, such a state, by using these provisions, may remain party to the earlier Conventions without applying the new Conventions until the last moment possible, but at the same time their instrument contributes to the criteria for the trigger date for compulsory denunciation of the earlier Conventions. This may be illustrated for a state party to both original Conventions, as follows.

```
1969  Convention           ---------------------------------------- |
1971  Convention           ---------------------------------------- |
1984  Liability Convention                                          |-------------
1984  Fund Convention                                               |-------------

Time                       PHASE 1                                  | PHASE 4
```

The denunciation date is at the end of Phase 1/beginning of Phase 4. There is no transitional period *for this state* during which the old and new regime exist together, but the 1984 Conventions will of course have entered into force for *other states party thereto* prior to the denunciation date, which is when they enter into force for this state.

THE IMMEDIATE DENUNCIATION OPTION

11–102   Yet a third effect is possible for a state party to both the old regimes, which was known at the Conference as the "immediate denunciation" solution. This is simply achieved by the state ratifying the 1984 Protocols and at the same time denouncing the 1969 and 1971 Conventions with effect from the date when the 1984

Liability Convention enters into force. Since the 1984 Fund Convention is almost undoubtedly going to enter into force after the 1984 Liability Convention, this would in fact lead to a gap in the cover afforded to this state, for there would be a period when the 1984 Liability Convention applied but not the 1984 Fund Convention. This effect was the main reason why at (and indeed before) the Conference this approach was not favoured as a compulsory method of solving the treaty law problems. The effect for a state adopting this option may be illustrated as follows, assuming again that the 1984 Liability Convention enters into force before the 1984 Fund Convention.

```
1969  Convention       _____|
1971  Convention       _____|
1984  Liability Convention            |-------------------------------
1984  Fund Convention                                   |-----------

Time                   PHASE 1        |   PHASES 2 & 3   |   PHASE 4
```

Here, the denunciation date is irrelevant for this state (because it has denounced the 1969 and 1971 Conventions before the entry into force of the 1984 Fund Convention). During Phases 2 and 3 (which here are but one Phase), only the 1984 Liability Convention applies in that state.

THE LIABILITY CONVENTION ONLY OPTION

11–103   The last important option might have been better mentioned in connection with the 1984 Liability Convention, rather than here, for it is the option not to ratify the 1984 Fund Convention at all, but simply to rely on the extended cover provided by the 1984 Liability Convention and any additional or concurrent cover provided by the voluntary schemes.[60] For states party to the 1971 Fund Convention, however, this option cannot be enjoyed along with continued membership of that Convention, for Article 12(4) of the 1984 Liability Convention provides that such a state must either become a party to the 1984 Fund Convention at the same time as it becomes a party to the 1984 Liability Convention, or it must denounce the 1971 Fund Convention to take effect on the date when the 1984 Liability Convention enters into force for that state.

CONCLUSION

11–104   The first three options illustrate the need for every state to consider its exposure to oil pollution, as recommended in para-

---

[60] See Chapter 12 below.

graph 11–70 above, not only in the short term but also longer term. In this connection it must be recalled[61] that the vast majority of oil pollution incidents are, on past experience, likely to be covered by the 1984 Liability Convention limits, and that the IOPC Fund 1984 is likely to be involved in only a handful of cases each year; in this respect, it will be more of a "disaster fund" than its predecessor. The choice which a state makes will, therefore, be governed by its perceived need for cover and the timing of that need, balanced against the perceived costs of acquiring that cover. States considering the Liability Convention-only option in particular should assume that the 1971 Fund Convention will have a fairly short life after the 1984 Fund Convention enters into force, and certainly after the denunciation date, for it will then have relatively few parties with enough contributing oil to make it viable.

14. CONCLUSION ON THE LIABILITY AND FUND CONVENTIONS

11–105 From Chapters 10 and 11 it can be seen that the twin pillars of the liability and compensation regime have been a resounding success. They are truly revolutionary instruments of international law, creating a regime of strict liability with very few exceptions, and making provision for governments and citizens alike to have relatively quick and assured compensation for their oil pollution losses. The texts of the Conventions themselves are not free from ambiguities, although some are cleared up by the 1984 Protocols, and this means that particular care must be taken in ratifying legislation. This will be especially so in the case of the 1984 Protocols for the additional reason that the co-existence of regimes creates additional complexity. Further, states should consider the extent to which they wish to make provision in their national legislation for the special regulation of the problem of oil pollution from ships and installations not covered by the instruments discussed above, and the relationship which that has to ratification of the 1976 Limitation Convention.

11–106 The criticisms which can be made of these instruments is small, and so it is a matter of some regret that more states have not ratified them. It is to be hoped that the 1984 Protocols will induce even wider participation than that which the 1969 and 1971 Conventions have enjoyed to date, bearing in mind the trend, observed

---

[61] See paras. 10–145 and 11–90.

in Chapter 9[62] that the community of states now expects to see the development of a special liability regime, providing prompt and adequate compensation, by means of international Conventions for the unification of private law. Quite what the fate of the 1984 Protocols will be is further discussed in Chapter 12 in connection with the most recent developments to the voluntary schemes.

[62] See paras. 9–21—9–28.

12. Voluntary Compensation Schemes

SUMMARY

12–01   In this chapter, an overview is given of the voluntary compensation schemes set up by the oil and tanker industries which have continued to co-exist alongside the Liability and Fund Conventions discussed in Chapters 10 and 11. The earliest scheme, known as TOVALOP, follows the form and substance of the Liability Convention very closely, and indeed offers some improvement on it in terms of its width of scope. The second scheme, known as CRISTAL, follows the form and substance of the Fund Convention in a general way, but has quite a number of significant differences from it. The two schemes operate together as an integrated whole, but they do not apply to cases actually covered by their respective international legal counterparts: a claimant cannot recover under both the Fund Convention and CRISTAL, for instance, but he can recover under the Liability Convention and CRISTAL if the Fund Convention does not apply to the case. In 1985, a new scheme known as PLATO was devised to replace TOVALOP, and CRISTAL Revised '85 was devised as a very substantial amendment to CRISTAL. These two new schemes, if implemented in 1986, will apply in all states irrespective of whether they are party to the 1969 Liability and 1971 Fund Conventions or not. They would therefore represent a serious challenge to states considering implementation of the 1984 Conventions and in particular the 1984 Fund Convention.

1. INTRODUCTION

12–02   It was seen in Chapters 10 and 11 that the *Torrey Canyon* incident in 1967 produced a great improvement in the international law relating to oil pollution; it also played its part in changing the practice of the oil and tanker industries and in influencing their attitudes to the problem of oil pollution. Following the incident, it was felt by many in the industry that constructive action was needed to fill the gaps in the law, and that waiting for the entry into force of any international treaties which may be adopted was not good enough, both from the point of view of the plaintiff, who needed compensation, and of the industry, in which there were many who felt the need to respond to public opinion positively and to arrange insurance for any removal costs voluntarily incurred.

12–03   There were also some who felt that, if industry responded with a voluntary compensation scheme for tanker owners, treaties might either be regarded as unnecessary by states, or might be influenced in form and substance by the voluntary scheme adopted. In the former respect, this view has been proved to be manifestly mistaken, yet it has not prevented the oil industry from mounting a threat to the 1984 Protocol to the Fund Convention which, if successful, may ensure that it does not enter into force; in the latter respect, it proved fairly accurate. The influence of the concepts used in the original TOVALOP and CRISTAL can be clearly seen in the Civil Liability and Fund Conventions, and there are, as will shortly appear, many points of detail in common. But whereas industry began by leading governments, after a while it dropped behind, and until 1985 it showed a tendency to recognise the inevitable and to alter its voluntary schemes in response to governmental developments. The advent of PLATO and CRISTAL Revised '85 is evidence of a new resurgence of earlier attitudes to pollution compensation within, in particular, the oil industry.

2. TOVALOP

A. Introduction

12–04   On January 7, 1969 the Tanker Owners Voluntary Agreement Concerning Liability for Oil Pollution (TOVALOP) was signed, initially by the B.P. Tanker Company Ltd., Esso Transport Company Inc., Gulf Oil Corporation, Mobil Oil Corporation, Shell International Petroleum Company Ltd., Standard Oil Company of California and Texaco Inc. The agreement came into operation on October 6, 1969 when 50 per cent. of the tankers of the world (as measured by gross registered tonnage) became subject to it. By August 1972 this figure had risen to 99 per cent.; this figure subsequently fell back to around 92 per cent., mainly due to an increase in laid-up tonnage. The figure varies from time to time, but it can be said at any one time that almost every seagoing tanker, OBO, ore/oil carrier and oil barge in the world is entered in TOVALOP. For this, if for no other reason, the agreement is of importance.

12–05 The first form of TOVALOP[1] was a scheme whereby participating tanker owners agreed amongst themselves to reimburse national governments for expenses reasonably incurred by them to prevent or clean up pollution of coast lines as the result of the negligent discharge of oil from one of their tankers, which was presumed to be at fault unless this was disproved. There was a U.S.$100 per grt or $10 million limit, whichever the less. The first members were, of course, all oil companies, and the fact that they happened to own large fleets of tankers does not stop them behaving like oil companies. The early TOVALOP can therefore be viewed as an attempt to influence the international community of states to think in terms of oil pollution liability resting solely on the tanker owner, with liability being based on fault, with relatively modest limits, and with compensable damage being limited to clean-up and removal costs.

12–06 As was seen in Chapter 10, the 1969 Conference which adopted the Liability Convention did not go along with this view all the way. For a long time, TOVALOP remained unamended, but in 1972 (before the Liability Convention entered into force) it was expanded to cover pure threat situations, in addition to those where oil had actually been spilled. In 1975 the Liability Convention entered into force, at which point the two alternative schemes clearly showed the Convention ahead in its ability to compensate the victims of oil pollution. But in May 1978, in response to this situation, TOVALOP underwent a fundamental change with effect for incidents occurring on or after June 1, 1978 which dramatically improved the scope of its cover.

B. The Agreement

12–07 This account must be read in the context of developments discussed below in paragraphs 12–39 *et seq.* Hence, the current TOVALOP may not apply to incidents on or after June 1, 1986. Put in its most simple form, the basis on which the tanker owner voluntarily undertook to accept the cost of clean-up changed from one of negligence with the burden of proof reversed, to strict liability; further, the limits were raised from U.S. $100 per ton or $10 million (whichever the less) to $160 per ton or $16.8 million (whichever the less), and other changes were made so that the agreement mirrored the Liability Convention very closely. The

---

[1] For comment see G.L. Becker, "A Short Cruise on the Good Ships TOVALOP and CRISTAL" (1974) 5 J. Mar. L. & Comm., 609. The first version of TOVALOP is reproduced at (1969) 8 I.L.M. 497. Copies of the current version may be obtained from the International Tanker Owners' Pollution Federation Limited in London.

agreement now contained (and still contains) two main limbs. The first is Clause IV, by which each party[2] to the agreement undertakes that, subject to the terms and conditions of the agreement, he will assume liability for pollution damage[3] caused by oil[4] which has escaped or which has been discharged from any tanker[5] owned[6] by him and which has been involved in an incident, and that he will also assume liability for the cost of threat removal measures taken as a result of the incident. These undertakings are made to the other parties, although it is not they who directly benefit from them.

12–08 Threat removal measures are reasonable measures taken by any person (whether private or public) after an incident has occurred for the purpose of removing a grave and imminent danger of an escape or discharge of oil which, if it occurred, would create a serious danger of pollution damage. The coverage of the cost of such action is an incentive to both the shipowner and the potential victim to take immediate anti-pollution measures, before a spill occurs—an incentive which, it was seen in paragraphs 10–14 *et seq.*, the 1969 Liability Convention does not provide.[7]

12–09 Liability will not be assumed (either for pollution damage or threat removal measures) if the incident resulted from any of the circumstances described in Article III(2) of the Liability Convention,[8] or if it caused pollution damage anywhere in the world for any part of which the owner is liable under the terms of the Liability Convention. This latter provision means that the provisions of TOVALOP and the Liability Convention are mutually exclusive.

12–10 The second limb of TOVALOP, contained in Clause VI, ensures that the tanker owner is insured against the liabilities which he has voluntarily assumed, and also ensures that he can be reimbursed for any preventive measures taken after a spill and for any threat removal measures taken before it. Furthermore, the owner shall exercise his best efforts to take such measures.

---

[2] Any owner or bareboat charterer of a tanker may be a party.

[3] Defined in similar terms to those in the Liability Convention: see paras. 10–47 *et seq.*, above.

[4] Defined as any persistent hydrocarbon mineral oil, whether or not carried as cargo. Hence whale oil is (not unnaturally) excluded.

[5] Defined as "ship" in the Liability Convention but with the added words "whether or not it is actually so carrying oil": hence tankers in ballast are covered by TOVALOP but not by the 1969 Liability Convention (see paras. 10–07 *et seq.*, above).

[6] "Owner" is defined as in the Liability Convention (see para. 10–21, above) but with the proviso that, where a tanker is under bareboat charter, owner means the bareboat charterer.

[7] But the 1984 Liability Convention puts this right: see paras. 10–121 *et seq.*, above.

[8] See paras. 10–36 *et seq.*, above.

12–11   Each party must establish his financial capability to fulfil his obligations to the satisfaction of the International Tanker Owners' Pollution Federation Ltd., a company set up to administer the agreement. This may now be done by entering vessels in a specially formed mutual insurance association, the International Tanker Indemnity Association Ltd. (ITIA) or in a conventional Protection and Indemnity Club. The arrangement of such cover would have been impossible without limiting the liability of a party; now by Clause VII the limits roughly mirror those of the Liability Convention—U.S. $160 per limitation ton or U.S. $16.8 million (whichever the less).

12–12   A most important characteristic of the agreement is that the company set up to administer it ("the Federation") does not itself provide any insurance. Hence, when a person files a claim in response to being informed by the Federation that the tanker involved is subject to TOVALOP, any payment made comes from the insurer and not the Federation. The liabilities of parties are to each other, and consist of mutual promises. Insurance against the cost of having to fulfil them must be taken out by parties; the beneficiaries are the claimants and, in respect of their own voluntary clean-up costs, the parties themselves.

12–13   The main intention of the parties in making the fundamental changes in May 1978 was to provide in those areas not protected by the Liability Convention benefits and protection generally comparable with those available under the Liability Convention. Therefore, there was not felt to be a need to make further significant amendments to the agreement, although slight amendments were made in 1980 and 1981, the main one being an extension of the time limit to bring a claim from one year to two. As was seen in Chapter 10, in 1984 the Liability Convention underwent very significant change, and, as we shall see below in paragraphs 12–39 *et seq.*, these changes were anathema to a forceful part of the oil industry. At their instigation, CRISTAL Revised '85 was adopted, and a new form of TOVALOP, called the Pollution Liability Agreement Among Tanker Owners (PLATO) was proposed to go with it. Before analysing PLATO, however, it is appropriate to compare TOVALOP with the 1969 Liability Convention.

C. Comparison of TOVALOP and the Liability Convention

12–14   What, then, are the differences between TOVALOP and the Liability Convention? The main difference is that TOVALOP, being voluntary, covers only those tankers owned by the parties

307

thereto. It has just been seen that there are in fact very few tankers in the world not covered by TOVALOP, but it is always possible that a particular incident will be caused by a vessel owned by a non-party. For this reason it is often said that TOVALOP is *vessel specific*. The Liability Convention, being a legal regime, applies to all vessels within the scope of its provisions, although it is limited to pollution damage caused on, and measures taken to prevent or mitigate pollution damage in, the territory or territorial sea of a contracting state. Hence it is said by way of comparison that the Liability Convention is *territory specific*.

12–15 While such a distinction is inherent in the form of the two schemes (namely, a voluntary agreement amongst shipowners and an international agreement between sovereign states) there are differences not so caused. Perhaps the most important of these is the coverage by TOVALOP of the pure threat situation. TOVALOP allows a person who takes reasonable preventive measures while no oil has yet been spilled to recover his costs in many situations, whether he be the tanker owner or the potential victim, whereas the 1969 Liability Convention would neither reimburse such costs nor allow them to rank against a limitation fund set up because a spill subsequently took place.[9] The 1984 Liability Convention, however, would.[10]

12–16 Another important difference is that TOVALOP covers a spill from a tanker in ballast, whereas the 1969 Liability Convention does not. Again, the 1984 Liability Convention caught up with this development.[11] The third major difference from a legal point of view is that the limits to TOVALOP liability apply in all cases, not just those (as in the Liability Convention[12]) where the owner has not been guilty of actual fault or privity. The absence of such a restriction from TOVALOP, being a purely voluntary arrangement, is necessitated by insurance considerations.

12–17 A further significant difference is that TOVALOP covers the bareboat charterer, by providing that he shall be deemed to be the owner. This means that, unlike the Liability Convention regime,[13] there is no question of there being two (voluntary) payments where the tanker is subject to such a charter. However, it is not clear from Clause VIII(E) that a claimant could not accept a

---

[9] See paras. 10–15 *et seq.*, above.
[10] See paras. 10–121 *et seq.*, above.
[11] See paras. 10–118 *et seq.*, above.
[12] See paras. 10–72 *et seq.*, above.
[13] See paras. 10–21 to 10–29 above.

TOVALOP offer from the bareboat charterer and then sue the registered owner under national legislation. Lastly, TOVALOP's definition of pollution damage specifically excludes any loss or damage which is remote or speculative, or which does not result directly from the escape or discharge of oil. While the 1969 and indeed 1984 Liability Convention definitions probably have the same effect,[14] the point is clearer in TOVALOP.

12–18　There are also less important differences—for instance the limitation period under TOVALOP is shorter than that under the Liability Convention, disputes under TOVALOP must be settled by arbitration and TOVALOP does not cover whale oil—but these do not call for comment here. The striking conclusion to emerge from comparison of the two systems is that in many cases claimants are better protected by TOVALOP (when available) than by the 1969 Liability Convention. This was not always the case—before the changes made in May 1978, the Liability Convention was undoubtedly the more beneficial of the two from the claimant's point of view. TOVALOP was not then based on strict liability, its limits were lower and it applied only to governmental claimants. The scope of cover of TOVALOP after 1978, by providing a viable alternative to the 1969 Liability Convention, has probably had a considerable influence on governments against becoming party to that Convention.

12–19　One thing which the two systems have always had in common is the method of handling claims: in practice, the entity behind the owner will be a Protection and Indemnity Club (or ITIA) in both TOVALOP and the Liability Convention claims. Once the liability of the owner under either system is agreed, the only matter of contention is likely to be the reasonableness of the costs claimed. In the past negotiations on this question have almost invariably been handled on an amicable basis and a settlement reached without recourse to litigation or arbitration.

12–20　A government not yet party to the 1969 Liability Convention was and is, at least until 1986, faced with a choice of schemes, one purely voluntary which does not cover every tanker in the world but which provides very wide protection in respect of those vessels which it does cover, the other a legal scheme which does cover all laden tankers but which does not offer such wide protection in respect of them. How long that choice will remain depends

---

[14] See paras. 10–47 *et seq.* and 10–137 *et seq.*, above.

on whether PLATO—discussed in paragraphs 12–44 *et seq.*, below—enters into force.

3. CRISTAL

A. Introduction

12–21 Just as the 1969 Liability Convention was preceded by a tanker industry scheme to compensate the victims of oil pollution, so the 1971 Fund Convention was preceded by an oil industry scheme. The 1971 Fund Convention was adopted on December 18, 1971 and entered into force on October 16, 1978; the Contract Regarding an Interim Supplement to Tanker Liability for Oil Pollution (CRISTAL) was adopted on January 14, 1971 and came into effect on April 1 following, at which time oil companies receiving over 70 per cent. of the world's crude and fuel oil had become parties.[15] Currently this figure is still over 90 per cent. (comparable with TOVALOP). CRISTAL is a voluntary scheme adopted originally to compensate only the victims of oil pollution who had obtained insufficient compensation under existing laws or under TOVALOP, the compensation being provided within carefully defined limits by a fund, administered by an Institute set up under CRISTAL and contributed to by the oil company parties. As with TOVALOP, CRISTAL may be seen as an attempt by the oil industry to influence the then forthcoming Conference on the Fund Convention; as before, this met with mixed success. The latest amendments to CRISTAL, called CRISTAL Revised '85, must be seen in a similar light, but here the attempt is to jeopardise the entry into force of the 1984 Protocol to the Fund Convention.

B. The Pre-1985 Agreement

12–22 The original CRISTAL was limited to helping only the victims of oil pollution, and not shipowners. It was recognised soon after the adoption of the Fund Convention, with its shipowner relief provisions, that where a shipowner failed to take voluntary clean-up measures himself, the oil would be far more likely to cause damage compensable under CRISTAL; whereas if these measures had been taken, CRISTAL offered the shipowner nothing in the

---

[15] For comment, see G.L. Becker, "A Short Cruise on the Good Ships TOVALOP and CRISTAL" (1974) 5 J. Mar. L. & Comm., 609. The first version of CRISTAL is reproduced at 10 I.L.M. 137; the latest edition is obtainable from Marine Pollution Compensation Services Limited in London.

way of reimbursement. The expenses incurred by the shipowner were therefore borne by him, but they had the effect of reducing the liability of the CRISTAL fund. Shipowners and their insurers therefore entered into negotiations with the oil company parties to CRISTAL, as a result of which CRISTAL was changed, as from June 19, 1972, to offer shipowners reimbursement of a top slice of their voluntary clean-up costs; in return, the P. and I. Clubs formally undertook to encourage their members to take prompt and effective voluntary clean-up action. Hence, from that date, CRISTAL contained the two main elements of the 1971 Fund Convention: reimbursement of victims and shipowners.

12–23 When TOVALOP was amended as from February 20, 1973 to cover preventive measures taken prior to a spillage of oil upon the water, it became logical to make a similar amendment to CRISTAL. This was effected on May 25, 1973. So as from then, in contrast to the 1971 Fund Convention, CRISTAL covered the prespill situation in so far as measures taken by the shipowner to prevent pollution were concerned.

12–24 CRISTAL was amended again, this time fundamentally, in May 1978, so that with effect from June 1, 1978 it mirrored the provisions of the Fund Convention (which was soon to enter into force) as far as was felt desirable. Hence, since then, TOVALOP and CRISTAL have formed an integrated voluntary scheme close in scope to the 1969 Liability Convention and the 1971 Fund Convention, whereby the cost of a pollution incident is divided between the shipowning and the cargo interests in such a manner that the cargo interests only contribute when the shipowning interest's contribution is insufficient.

12–25 CRISTAL sets up a fund which is held by the Oil Companies Institute for Marine Pollution Compensation Limited, a company incorporated in Bermuda, and it is to this company and to each other that the parties owe their duties. The Institute collects the funds necessary to meet claims made under CRISTAL by raising a levy from the oil company parties based upon their receipts of crude and fuel oil. This fund will pay out to compensate any person (either private or governmental) who has suffered pollution damage or who has taken threat removal measures[16] as a result of an incident. There are, however, numerous prerequisites, exceptions and qualifications to this rule, all of which are contained in Clause IV.

[16] Defined as in TOVALOP.

12–26 It is a prerequisite to the Institute's liability that the oil involved in the incident was "owned" by an oil company party to CRISTAL. Clause V provides an extended definition of ownership so that, even if title was not in an oil company party at the time of the incident, the oil is nonetheless considered to be owned by such a party if the shipment in question has been sold by or is under contract to be sold to such a party, or if it is carried in a tanker owned by or chartered to such a party, and that party has elected in writing to have the shipment so considered. Such an extension of the concept of ownership enables a company which has a close connection with the shipment to ensure that it is covered by CRISTAL, even though it is not owned by a party thereto during all or part of its voyage by sea. It also enables such a company to eradicate the anomaly which would have arisen where oil is sold on c.i.f. or c. and f. terms; under such contracts, property in the oil usually passes on receipt of documents by the buyer, so that, if one was party to CRISTAL and the other was not, the cargo would have been covered by CRISTAL for part only of the voyage.

The other important prerequisite is that the tanker concerned is entered in TOVALOP. Hence, CRISTAL is both *vessel specific* and *cargo specific*.

12–27 The Institute is absolved from liability altogether if any of the events described in Article III(2) of the Liability Convention[17] caused the incident. Hence, CRISTAL follows the same policy as that followed in respect of shipowners' liability by the 1969 Conference, but diverges from that adopted in respect of the Fund Convention. It will be remembered that the Fund Convention does compensate where the incident was caused by "a natural phenomenon of an exceptional, inevitable and irresistible character" and it also does compensate where the incident was caused by an intentional act or omission on the part of a third person (*e.g.* terrorist action).[18]

12–28 CRISTAL is designed to compensate those who have suffered pollution damage only in so far as they have been unable to get adequate compensation from other specified sources; like the Fund Convention, it comes in on top of other provisions. Hence, it does not compensate a person in respect of pollution damage compensable under the Fund Convention itself[19]: one cannot prefer CRISTAL to the Fund. But this does not mean that CRISTAL will

---

[17] See paras. 10–36 *et seq.*, above.
[18] See para. 11–12, above.
[19] Clause IV(A)(1).

not pay up where the Liability Convention applies but the Fund Convention does not. Clause IV(E)(1) provides for the Institute's liability in such cases to the extent that the claimant has been unable to obtain full compensation by taking all reasonable steps to pursue the remedies available to him. Clause IV(F)(1) makes similar provision in respect of an incident to which the Liability Convention does not apply.

12–29 It has already been mentioned that the other major object of CRISTAL since June 1972 has been to compensate the shipowner himself. These provisions are now contained in Clauses IV(E)(2), (F)(2) and (G). The first limb is constituted by the provision that the Institute shall compensate the shipowner[20] for the cost of preventive measures[21] and threat removal measures[22] taken by him to the extent that he has been unable to obtain full compensation therefor without resorting to CRISTAL. The second limb is constituted by the provision that the Institute shall indemnify the shipowner for that portion of his liability to others which exceeds U.S. $120 per ton of the tanker's tonnage or $10 million (whichever the less) and which does not exceed $160 per ton or $16.8 million (whichever the less).[23] These limits are of course designed to mirror the equivalent provisions of Article 5 of the 1971 Fund Convention. It is noticeable, however, that under CRISTAL, while there is no proviso relating to breach of the various International Conventions relating to oil pollution and safety at sea, there is a proviso that "no such indemnity shall be paid to an Owner whose recklessness or wilful misconduct caused the Incident."

12–30 The second limb would still leave the shipowner bearing all the liability to others above $160 per ton or $16.8 million in cases where the legal regime relating to the incident did not entitle the shipowner to limit his oil pollution liability at that level or at all. Clause IV(H) therefore provided that in such a case, so long as the shipowner has not been guilty of causative wilful misconduct or privy to unseaworthiness, the Institute shall bear the excess of his liability over $160 per ton or $16.8 million (whichever is less). This clause led to claims in the *Kurdistan* case off Nova Scotia in 1979 and the *Princess Anne Marie* case off Cuba in 1980. The latter claim was expensive for CRISTAL (just over U.S. $26 million), and led to the deletion of Clause IV(H) as from June 1, 1984.

---

[20] Including in most cases the bareboat charterer.
[21] Defined as in the Liability Convention.
[22] Defined as in TOVALOP.
[23] Now the TOVALOP limit as well as the Liability Convention limit.

12–31   The deletion of Clause IV(H) is one sign of strain between the tanker owners and their P. & I. Clubs, on the one hand, and the oil companies on the other, over how the real burden of oil pollution should be divided between them. This has been exacerbated by prolonged depression in the oil tanker freight market since 1973, and by a predominant view among CRISTAL members that CRISTAL is an unwelcome, if necessary, addition to TOVALOP and the Liability Convention. A further sign of this strain is the cancellation as of the same date of an agreement between CRISTAL, ITIA and the International Group of P. & I. Clubs—called the Memorandum of Understandings—under which a minimum limit of U.S. $1 million was set for any ship, however small.

12–32   CRISTAL, of course, has to have a limit, and this was raised in May 1978 to be comparable at the time to that of the Fund Convention—U.S. $36 million. There is also provision for raising this limit up to $72 million if the Institute considers such an increase advisable in the light of experience. However, when the IOPC Fund raised its limit to 675 million gold francs (then about U.S. $54 million) in the wake of the *Amoco Cadiz* incident, the oil companies decided not to follow suit. This now creates a significant difference in the coverage of the industry and Conventional schemes.

C. Comparison of the Pre-1985 CRISTAL and the Fund Convention

12–33   It can therefore be seen that CRISTAL mirrors the Fund Convention—but how closely? Undoubtedly the most significant difference, from the point of view of coverage, is that CRISTAL has a lower limit than the IOPC Fund. Against this, it applies to pure threat situations whereas the 1971 Fund Convention does not. CRISTAL is, of course, limited to incidents where the oil concerned is "owned" by an oil company party to the contract, whereas the Fund is limited by reference to the territories and (in respect of shipowner relief) the ships of states party to the Convention.[24]

12–34   CRISTAL does not cover incidents caused by natural phenomena of an exceptional, inevitable and irresistible character,

---

[24] In practice this means that, until the Fund Convention is ratified by nearly all the states in the world, CRISTAL would on present membership levels of TOVALOP and CRISTAL cover more incidents than would the Fund Convention: but this is not a relevant consideration to a state contemplating ratification of the 1971 Fund Convention, since such ratification would bring with it the benefits of the Fund Convention to that state.

or those caused by intentional acts of third parties, whereas the Fund Convention does. In practice this is unlikely to be a significant difference, since few such incidents can be expected. The criteria for shipowner relief are in theory more generous in CRISTAL's case than in the Fund's, although again this is unlikely to be a serious difference.

12–35 Perhaps the most significant difference of all, greater even than that of the limits, is that CRISTAL is a scheme of last resort. Because of this, although it processes claims expeditiously, it only pays up when it can be shown that the claimant has exhausted all reasonable means of obtaining recovery from other sources. Hence, it sometimes takes a long time to pay out, for instance where litigation is pending to determine whether the owner will have a right to limit his liability, or in collision cases where the responsibility of the colliding vessel has yet to be determined. This may be illustrated by the fact that at any one time CRISTAL usually has about ten claims outstanding, some of which go back four or five years. In fact many of CRISTAL's disbursements have been made to governments or oil companies who have first paid the victims and taken upon themselves the right to claim from CRISTAL. The IOPC Fund, not being a last resort scheme, does not suffer from these limitations, although like CRISTAL its speed of response cannot be faster than the claimant's own ability to formulate his claim properly. The reality has therefore proved to be the reverse of the case which was expected in 1977 before the IOPC Fund came into existence; in the first edition of this work it was written: "Claims under CRISTAL are handled swiftly and with the minimum of bureaucracy; claims under the Fund Convention are bound to take longer. . . . "! As we have seen,[25] the IOPC Fund has completely disproved this prediction.

12–36 Another important distinction, from the point of view of a state, is that the state has a part to play in the approval of claims under the Fund Convention, but of course not under CRISTAL. This is so even where that state is the main claimant. The IOPC Fund is a public, international organisation and is accountable as such. The Institute is a private company run by its Board and shareholders in General Meeting; it determines its own policy and strikes its own bargains without reference to states or claimants.

12–37 The difference in limits and claims handling policy and procedure, and the cargo-specific nature of CRISTAL, will have

---

[25] See paras. 11–48 *et seq.*, above.

led to the conclusion for many states that on balance, the Fund Convention is a significant improvement on CRISTAL and so the better alternative. Others, such as those with high crude and fuel oil receipts but so far no major oil pollution incident in their waters, may have felt that CRISTAL is better.

12–38 However, the choice as discussed above is not going to remain any longer for on June 1, 1986, CRISTAL will either be superseded by CRISTAL Revised '85, or it will be terminated altogether.

4. PLATO AND CRISTAL REVISED '85

A. Introduction

12–39 Leading elements in that part of the oil industry which takes an active interest in oil pollution were deeply unhappy with the outcome of the 1984 Conference to revise the Liability and Fund Conventions.[26] Essentially, this was because they had failed to persuade the Conference to see it their way, despite a very determined effort. The Conference adopted limits for the 1984 Conventions which were much higher than they wanted, and divided the burden unfairly (as they saw it) between the shipowner and the oil industry. They were also unhappy with the way the Conference had dealt with the definition of *pollution damage*.[27] They therefore set themselves the task of providing an alternative to the 1984 Protocol to the Fund Convention, which is the instrument which effects the increase of the burden upon the oil industry, which might ensure that it never comes into force. The instrument designed for this purpose is CRISTAL Revised '85—an almost complete overhaul of the CRISTAL contract which was passed at a meeting of the shareholders of the Institute which administers CRISTAL on June 3, 1985.

12–40 This dissatisfaction is not a secret: the language of the Memorandum to Shareholders prepared by the Institute's Board and circulated prior to the meeting (the "Memorandum to Shareholders") is quite open about it:

> "The principal shortcomings of the Protocols are the failure of the [Liability Convention] to impose substantially higher tanker owner limits particularly for spills from small and medium size vessels; and

[26] See paras. 10–112 *et seq.* and 11–72 *et seq.*, above.
[27] See paras. 10–137 *et seq.*, above.

316

the failure to recognise the financial impact which this has on companies that receive oil in countries which are contracting states to the [Fund Convention]. While the Institute . . . supported a strengthening of a tanker owner's right to limit liability, this was on the assumption that revised limits would be substantially higher. The combination provided by the Protocols of (i) strengthened rights of a tanker owner to limit liability; (ii) expansion of the definition of pollution damage to include speculative costs for prospective restitution of the environment; and (iii) extension of the geographical scope to cover pollution damage in the Exclusive Economic Zone (EEZ) also results in an unacceptable increase in exposure for oil companies in [Fund Convention] countries."[28]

12–41 Nor is the real purpose of CRISTAL Revised '85 more than thinly veiled: the same Memorandum to Shareholders talks about the Protocols not forming a basis on which a "bridge" might be established to give effect, on a voluntary basis, to their terms prior to them entering into force; about the CRISTAL Review Committee being given a brief to develop an alternative plan more equitable than one based on the Protocols; about voluntary regimes probably representing the only truly practical way of providing adequate compensation; and about the concept of CRISTAL as an interim plan (*i.e.* one which essentially looks towards the implementation of an alternative legal regime) being abandoned. It further states:

> "It is not the purpose of this memorandum to speculate on the chances of the Protocols to [the Liability and Fund Conventions] achieving ratification. There are doubts that the [Fund Convention] Protocol, in particular, will receive sufficient support from governments. Certain governments which are major contributors to the IOPC [Fund] have publically expressed their dissatisfaction. It is for this reason that the Directors consider it to be no longer appropriate for CRISTAL to be an 'interim plan' and indeed this is further recognized in the recommendation that limits be increased in 1990 to ensure that the CRISTAL Revised '85 remains up to date."

If it was not the purpose to provide an alternative to the Fund Convention Protocol, why the need to increase the limits in 1990? Why abandon the interim concept? Why include this paragraph at all in the memorandum?

12–42 There is one other thread in the web of reasons for CRISTAL Revised '85, and that is dissatisfaction of many CRISTAL

---

[28] It is clear, however, that it is the division of the burden which really hurt, because the text of CRISTAL Revised '85 does not limit pollution damage to territorial waters, and the accusation that the 1984 definition of pollution damage allows speculative damage is, as we have seen in paras. 10–139 *et seq.*, entirely unfounded.

members with the existing CRISTAL. Because it does not apply in Fund Convention countries, CRISTAL members receiving crude and fuel oil in Fund Convention countries contribute to spills outside such countries, but CRISTAL members receiving crude and fuel oil outside Fund Convention countries do not contribute to spills inside them. Many members receiving contributing oil in Fund Convention countries accepted this situation only because CRISTAL was meant not to last forever. Under CRISTAL Revised '85, this situation would not apply—all oil company contributions to oil pollution funds would be mutualised.

12–43 But the oil industry realised that if it was to get a new CRISTAL off the ground, it would have to persuade the independent tanker industry (*i.e.* that part of the tanker industry not controlled by the oil companies) to go along with it and accept the liability below the threshold the oil industry wanted. Hence, the entry into force of CRISTAL Revised '85 is made dependent upon the entry into force of PLATO—a proposed agreement modelled upon TOVALOP, called the Pollution Liability Agreement Among Tanker Owners. This will happen on June 1, 1986 if Plato Limited—a company established to administer PLATO—determines that, as of March 31, 1986, PLATO has acquired 50 million gross registered tons of tanker tonnage as participating tankers. If this condition is not met, then on June 1, 1986, CRISTAL will cease to have effect except in relation to incidents occurring prior thereto, and it will eventually be wound up.

B. The PLATO Agreement

12–44 PLATO, which was signed on June 5, 1985, is a contract which has been modelled upon TOVALOP, and indeed the initial documentation, which includes an Explanatory Memorandum (the "Explanatory Memorandum") was mailed in mid-1985 to all existing members of TOVALOP. Hence, PLATO is a contract under which parties—who must be the owners or bareboat charterers of tankers—agree to take measures to prevent or abate pollution and to compensate pollution damage up to certain limits and to establish their financial responsibility to honour such undertakings. Parties become members of Plato Limited, a company set up to administer the agreement, but which may delegate its functions to the same company that now administers TOVALOP.

12–45 Because PLATO seeks to emulate TOVALOP, only those main features which differ from TOVALOP require mention here. Unlike CRISTAL Revised '85, which is specifically designed to replace CRISTAL, PLATO has no such provision with respect to TOVALOP and indeed there may be a period when the two co-exist. However, as was mentioned in paragraph 12–43 above, PLATO will only enter into force on June 1, 1986 if 50 million grt of participating tanker tonnage have become Participating Tankers by March 31, 1986.

12–46 One of the most striking changes from TOVALOP is that PLATO is intended to provide worldwide coverage, and so it applies to compensate persons even where the 1969 Liability and 1971 Fund Conventions are in force. Indeed, because the limits of PLATO exceed the 1969 Liability Convention (see Table 12.1 below) the IOPC Fund can become a claimant. Clause IV(B) provides that:

> "If an Incident occurs in a jurisdiction where the provisions of both the Liability Convention and the Fund Convention are in force, the Participating Owner shall . . . compensate—
> (1) the [IOPC] Fund in an amount equal to the amount that the [IOPC] Fund has paid or intends to pay as compensation as a result of the Incident, (irrespective as to whether said Participating Owner bears or would bear any liability under the Liability Convention with respect to the Incident); and
> (2) to the extent any Person remains uncompensated any Person who (i) sustains Pollution Damage or (ii) incurs Costs in taking Preventive Measures and Threat Removal Measures."

12–47 The words in brackets in Clause IV(B)(1) and the proviso to Clause IV(C) make it very clear that shipowners submit themselves to the same, more limited, defences as are available to the IOPC Fund in Fund Convention countries, *up to the limit of the IOPC Fund.*[29] Otherwise they have the usual TOVALOP and Liability Convention defences.[30] As the Explanatory Memorandum points out, a procedure will have to be worked out with the IOPC Fund relating to payments relating to incidents in Fund Convention jurisdictions. Not all incidents in such jurisdictions will involve the IOPC Fund: for instance, PLATO covers "pure threat" cases, whereas the IOPC Fund does not.

12–48 The other striking element of PLATO is its limits, which (as Table 12.1 shows) are higher than the 1984 Liability Convention and at 105,000 grt reach the current limit of the IOPC Fund, and rise still higher after December 31, 1989.

12–49 The definition of *pollution damage* adopted in both PLATO

---

[29] For the IOPC Fund's defences, see para. 11–12, above.
[30] See paras. 10–36 *et seq.*, above.

and CRISTAL Revised '85 has been changed from the old TOVA-LOP and CRISTAL, but except in relation to geographical scope it probably amounts to much the same thing as before. The new definition reads as follows:

> " 'Pollution Damage' means (i) physical loss or damage caused outside the Tanker by contamination resulting from the escape or discharge of Oil from the Tanker, wherever such escape or discharge may occur, including such loss or damage caused by Preventive Measures, and (ii) proven economic loss actually sustained as a direct result of contamination as set out in (i) above, including the Costs of Preventive Measures."

There is not as great a difference between this definition and the one adopted by the 1984 Protocols to the Liability and Fund Conventions as the Memorandum to Shareholders makes out, although of course there is no reference to impairment of the environment or to the cost of reinstatement "to be undertaken."[31] Also, although the Memorandum to Shareholders states that the extension of the Liability and Fund Conventions to the Exclusive Economic Zone is one of the factors resulting in an unacceptable increase in exposure for oil companies in Fund Convention countries, there is in fact no express limitation to damage being suffered on the territory or in the territorial waters of a state as there was in the old definition; therefore, as the Explanatory Memorandum states, the new definition recognises the right to recovery for damage within a state's Exclusive Economic Zone. Indeed, it is not even limited to that.

12–50 Participating Owners in PLATO need not have PLATO apply to all their tankers—only those they designate. This is a departure from the TOVALOP concept, where the scheme applies automatically to tankers owned by Participating Owners. Once a tanker is entered in PLATO, it must remain there for at least one year, so that if it is sold, the new owner must take on the obligation. Hence, PLATO is, like TOVALOP, *vessel specific*.

C. The CRISTAL Revised '85 Agreement

12–51 It is important to remember that the new agreement will not be applicable unless the requirements for its entry into force, discussed in paragraph 12–43 above, are fulfilled.

12–52 The basis of liability remains strict, with the same exceptions as CRISTAL and, indeed, the Liability Convention.[32] The

---

[31] *Cf.* paras. 10–137 *et seq.*, above.
[32] See paras. 10–36 *et seq.*, above.

scope of liability is, however, entirely different from CRISTAL. Probably the most dramatic element of the new agreement is that it seeks to re-establish the truly global applicability which CRISTAL enjoyed before the Fund Convention entered into force by covering spills in Fund Convention countries as well as others. In such a case, CRISTAL Revised '85 will supplement the cover provided by the IOPC Fund where the cargo was owned[33] by a member and where the other pre-requisites to liability are met. Chief among these are that the threshold amounts mentioned below have been paid by, or on behalf of, the shipowner.[34] Hence, while the new agreement remains *cargo specific*, it is no longer *vessel specific* in the way that CRISTAL was, in that there is no requirement that the ship be entered in PLATO. However, since the thresholds are above the 1969 and 1984 Liability Convention limits, the owner would only have paid such a threshold amount if either the ship was entered in PLATO, or the local legal regime was not based on either Convention, or because it is so based but the right to limit had been broken (in which case there probably would not be a claim on CRISTAL Revised '85 at all). Therefore, in most cases, CRISTAL Revised '85 will in fact be *vessel specific* to ships entered in PLATO.

12–53 In a case where the Fund Convention applies, CRISTAL Revised '85 provides for the reimbursement, within Cristal Limited's discretion,[35] of members' contributions to the IOPC Fund in respect of the same incident.[36] Thus, wherever the spill occurs, oil companies will pay their share of compensation due under CRISTAL Revised '85 and will share the cost in accordance with their contributing oil receipts. There are provisions in Clause IV(C)(7) to reinforce this mutuality principle which state that no payment may be made to a claimant who asserts or prosecutes a claim for Pollution Damage or the Cost of Preventive Measures or Threat Removal Measures against any fund (other than the IOPC Fund) established and/or maintained by means of assessments against Oil Companies.

12–54 The thresholds and limits of the new agreement are the

---

[33] The extended concept of ownership utilised under CRISTAL—see para. 12–28, above—is retained. The requirement of ownership by a member applies to all cases, not just to those in Fund Convention countries.

[34] Clause IV(C)(4).

[35] Cristal Limited will be the new name for the Institute.

[36] Clause IV(B)(1). It could also pay the IOPC Fund direct for such contributions, but it depends in practice on agreement being reached with the IOPC Fund, which may not be forthcoming.

other major feature, and these are best summarised in graphic form, so they are illustrated in Table 12.1. There are two sets of thresholds and limits—one applying after June 1, 1986 but before January 1, 1990, the other applying on or after January 1, 1990. The thresholds of CRISTAL Revised '85 are the limits of PLATO. The effect is striking. Shipowners bear a greater burden than they would under the 1984 Liability Convention, particularly in the highly important "small ship" end of the range.[37] Claimants bear a greater burden because the overall compensation available only rises to the level of the 1984 Fund Convention at 105,000 grt—the existing flat rate upper limit of CRISTAL is abandoned in favour of a tonnage–related limit. However, the amount available under CRISTAL Revised 85, like that under the 1971 Fund Convention in Fund Convention states, would be available from June 1, 1986 (rather than whenever the 1984 Fund Convention comes into force). Until January 1, 1990 CRISTAL Revised '85 offers increased cover over the current limit under the 1971 Fund Convention. The IOPC Fund might raise its limit to the maximum of 60 million SDRs (approximately US$ 60 million at present), which would put back the point at which CRISTAL Revised '85 improves on it to nearly 17,000 grt. More accurate direct comparison of the Convention and CRISTAL Revised '85 limits and thresholds is difficult over the long term because it is impossible accurately to predict the SDR/US dollar rate over the long term.

12–55 The "last resort" principle of CRISTAL noted above[38] is modified slightly,[39] but not significantly, so that in most, if not all, cases pollution claims are no more likely to be met soon after the spill than at present. There are, however, now three years in which to bring a claim, instead of the old two-year CRISTAL period.

12–56 The last feature worthy of mention is that the interim concept of CRISTAL has been abandoned, and CRISTAL Revised '85 is seen as a permanent alternative for states to the 1984 Fund

---

[37] This was a major objective of CRISTAL Revised '85—to make it apply far less often by making it come in for small ships only at a high threshold.

[38] See para. 12–36, above.

[39] Payment on behalf of members will be made to the IOPC Fund (assuming agreement is reached on that point) regardless of whether the claimant has pursued his remedies against third parties; claimants need not pursue their remedies against the tanker owner unless it can be demonstrated that the spill resulted from the shipowner's personal act or omission, committed with intent to cause such damage, or recklessly with knowledge that such damage would probably result (the same formula as used by Article V(2) of the 1984 Liability Convention—see paras. 10–147 *et seq.*, above); and a claimant can proceed against a national or local fund if or to the extent that payments under CRISTAL Revised '85 do not completely compensate the claimants.

TABLE 12.1
PLATO AND CRISTAL Revised '85
Compared with 1984 Liability and Fund Conventions

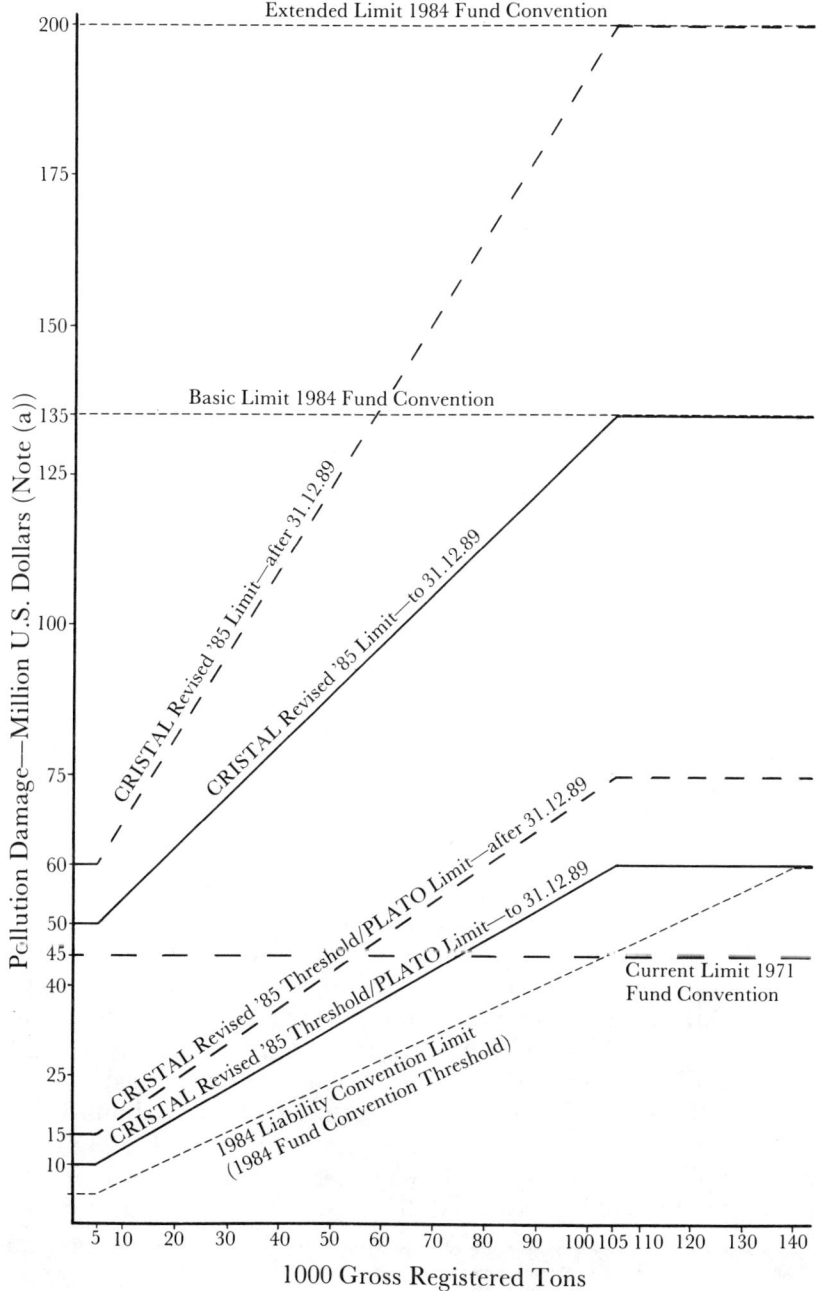

Notes to Table 12.1

(a) Liability and Fund Convention thresholds and limits have been converted from SDRs to US dollars at parity, which prevailed on the date Table 12.1 was compiled. The rate varies daily.

---

vention. The termination provisions (Clause III) are altered accordingly.

D. Conclusion

12–57   If independent tanker owners were to give in to oil company pressure and join PLATO, it may have much more to do with the current depressed state of the tanker market, giving rise to a need to be on goods terms with major oil company charterers, than anything else. Certainly, PLATO would be a millstone round the shipowners' neck once the 1984 Liability Convention gained widespread applicability. At the time of writing, it is not at all clear whether the 50 million grt hurdle will be met. The PLATO limits, which are also the thresholds for CRISTAL Revised '85, are higher than those of the 1984 Liability Convention, as Table 12.1 shows, but the really significant part of the difference is in the small ship end of the range, where a high proportion of incidents occurs. There is virtually nothing in the new scheme for independent tanker owners, except possibly that under this scheme, in a case within the CRISTAL Revised '85 limits, they are assured of the maximum liability under PLATO and claimants would have no incentive to try and make them pay for everything by attempting to break their right to limit their legal liability. But this would also be so under the 1984 Conventions. If and when the 1984 Liability Convention enters into force, their right to limit becomes virtually unbreakable.[40]

12–58   CRISTAL Revised '85 is a viable, workable alternative to the 1984 Fund Convention. It offers less cover earlier, whereas the 1984 Fund Convention offers more, later. The difference in limits is in the triangular area at the upper left hand part of the graph. CRISTAL Revised '85 has a limit which rises with tonnage, whereas the IOPC Fund 1984 has a flat limit irrespective of ship size. CRISTAL Revised '85 remains a scheme of last resort, whereas the IOPC Fund 1984 will not be. If the 1984 Fund Convention became universally ratified, the mutuality of CRISTAL Revised '85 would be achieved by the 1984 Fund Convention. The existence of CRISTAL Revised '85 must, however, act as a disin-

---

[40] See paras. 10–147 *et seq.*, above.

centive to at least some states to ratify the 1984 Fund Convention. We must wait until March 31, 1986 to see if PLATO attracts enough tonnage to bring CRISTAL Revised '85 into force. If it does not, CRISTAL will be terminated. Much will therefore depend on what pressure governments who wish to see the implementation of the 1984 Fund Convention are able to bring to bear on independent tanker owners.

PART III

SELECTED ISSUES IN UNITED KINGDOM LAW

13. Introduction and Overview of United Kingdom Law

13–01 The United Kingdom has adopted most of the international conventions applicable to oil pollution. The purpose of this Part of the book is to give some consideration to particular issues of United Kingdom law. As a preliminary, it is necessary to give a general introduction and so the aim of this Chapter is to indicate generally the area of application both of statute and the common law.

1. PUBLIC CONTROL OF OIL POLLUTION FROM SHIPS

13–02 The primary piece of legislation governing the control of oil pollution from ships is the Prevention of Oil Pollution Act 1971. This Act, as supplemented and amended by the Merchant Shipping (Prevention of Oil Pollution) Order 1983, made under section 20 of the Merchant Shipping Act 1979, together with the Merchant Shipping (Prevention of Pollution) (Intervention) Order 1980 and the Merchant Shipping Act 1974, provide the statutory framework for United Kingdom law designed to prevent oil pollution.

13–03 The Merchant Shipping (Prevention of Oil Pollution) Regulations 1983 implemented MARPOL 73/78 in the United Kingdom as from October 2, 1983. They apply to all UK registered ships and all ships in UK territorial waters, with some exceptions.[1]

A. Discharges[2] and Transfers

13–04 In principle, all discharges into the sea of oil or oily mixtures[3] are prohibited.[4] Tankers and non-tankers are dealt with separately, in terms of the allowable operational discharges. Thus,

---

[1] S.I. 1983 No. 1398. The exception includes warships—there is a power to exempt in Regulation 2.

[2] See also paras. 3–45 *et seq.* and paras. 14–38 *et seq.*

[3] "Oil" means petroleum in any form including crude oil, fuel oil, sludge, oil refuse, and refined products other than petrochemicals specified in Marine Shipping Notice No. 1077. "Oily mixture" means a mixture with any oil content: Reg 1.

[4] Regs. 12 and 13.

non-tankers may discharge so long as they are on a voyage, are more than 12 miles from land and do not discharge a mixture which has more than 100 parts per million of oil.[5] Such a ship must also comply with Regulation 14 which relates to the installation and use of oil discharge monitoring and control systems and oily water separating and oil filtering equipment. Otherwise, non-tankers may discharge only mixtures containing less than 15 parts per million of oil. Tankers may discharge only 50 miles from land when on a voyage and the instantaneous rate of discharge of oil content must not be more than 60 litres per mile and must not represent more than one thirty-thousandth[6] of the previous cargo carried. Tankers must also comply with Regulation 14. No discharges by any sort of vessel may be made in a "special area" of particular vulnerability.[7]

13–05 The Prevention of Oil Pollution Act 1971 prohibits the transfer of oil to or from any vessel at night.[8]

B. Construction[9]

13–06 Part IV of the 1983 Regulations deals with construction requirements for oil tankers. The basic requirement is that all vessels over a specified deadweight, both new and existing oil tankers,[10] must be fitted with segregated ballast tanks.[11] New crude tankers must in addition and existing crude tankers may as an alternative, be fitted with crude oil washing facilities.[12] Existing product carriers may operate either with segregated ballast tanks or dedicated clean ballast tanks.[13] New product carriers must have segregated ballast tanks.[14]

13–07 The operation of these systems is prescribed by the Regulations.[15] In addition, the equipment for pumping, priming and discharging the cargo of tankers is governed by Regulation 26, and Part V of the Regulations specifies standards of construction relat-

---

[5] Reg. 12.
[6] One fifty-thousandth for existing ships: for definition see Reg. 1.
[7] Reg. 16—the Mediterranean, the Baltic, the Black Sea, the Red Sea, the Gulf (the last two are not yet in force).
[8] s. 10.
[9] See further paras. 3–65 *et seq.* and 4–42 *et seq.*
[10] Reg. 17. The operative date defining "existing" and "new" tankers is, in essence, 1980.
[11] Reg. 18. Certain existing small tankers are permitted to operate for short interim periods with dedicated clean ballast tank systems—Reg. 18(9).
[12] Reg. 18(6).
[13] Reg. 18(10).
[14] Reg. 18(1).
[15] Regs. 19–24.

ing to sub-division and stability, designed to ensure that tankers are less likely to cause pollution if damaged by collision or stranding.

13–08   Regulation 25 requires all ships, tankers and non-tankers, to carry a residue or sludge tank adequate to receive oily residues prior to discharge to shore reception facilities.

13–09   Finally, more general construction requirements for tankers are to be found in the Cargo Ship Construction and Survey Regulations.[16]

C. Reception Facilities[17]

13–10   Harbour authorities whose facilities are used by tankers or other vessels carrying oil or oily residues or mixtures are empowered and required by October 1984, to provide reception facilities for the discharge of oily residues and mixtures adequate to meet the requirements of MARPOL 73/78.[18] The Secretary of State, to whom reports must be made, is empowered both to give directions and to grant exemptions.

D. Certificates and Records

13–11   Part II of the 1983 Regulations deals with survey and certification. All ships, tankers and non–tankers, must be surveyed and issued with a pollution prevention certificate. The survey is directed to ensuring that the vessel's construction complies with the requirements of the regulations. There are two varieties of certificate: an International Oil Pollution Prevention Certificate, (IOPP) for vessels trading to MARPOL countries, and a United Kingdom Oil Pollution Prevention Certificate (UKOPP) for others. If a vessel undergoes material change it must be re-surveyed. Otherwise the certificates last for five years, although a vessel holding an IOPP must be surveyed every year to ensure that the systems under review are still efficient.

13–12   Oil record books must be maintained by all ships, tankers and non-tankers, into which records of operations likely to cause pollution, such as ballasting or cleaning tanks, must be entered.

[16] S.I. 1981 No. 572 and S.I. 1984 No. 1219.
[17] See further paras. 3–90 *et seq.*
[18] Prevention of Pollution (Reception Facilities) Order 1984, (S.I. 1984 No. 862).

E. Enforcement[19-20]

13–13 There is a general power of inspection to ensure compliance with the regulations.[21] Vessels may be detained if the inspector is satisfied that the sailing of the vessel will constitute a threat to the marine environment. In addition, harbour authorities under Regulation 33 are empowered to refuse entry to ships on the same ground.

13–14 The Merchant Shipping Act 1984 establishes a system of inspectors who are empowered to issue Improvement and Prohibition Notices to ships. These notices, similar to those in use in shore-based industries under the Health and Safety at Work Act 1974 either, in the case of an Improvement Notice, specify a breach of appropriate regulations (which may include the 1983 Regulations) and require that it be remedied or, in the case of a Prohibition Notice, specify activities which in the inspector's opinion involve a risk of serious personal injury or serious pollution to navigable waters and require that they be discontinued. Notices may be appealed against to an arbitrator, but if not so appealed against, or if the appeal is unsuccessful, must be complied with on pain of a fine or, in the case of a Prohibition Notice, imprisonment.

13–15 In addition, straightforward prosecutions may be made under the Prevention of Oil Pollution Act 1971, s.19.

2. INTERVENTION[22]

13–16 The Torrey Canyon disaster led, amongst other things, to the 1969 Intervention Convention. This was enacted into United Kingdom law by the Prevention of Oil Pollution Act 1971, ss.12–16. These sections empower the Secretary of State to give directions whenever there has been an accident to, or on board, a ship and the Secretary of State is of the opinion that pollution on a large scale is likely on United Kingdom shores or in United Kingdom waters. The powers may, however, be used in respect of ships outside United Kingdom territorial waters and in respect of non-United Kingdom vessels where there is a threat to United Kingdom waters or shores.

[19-20] See also paras. 3–97 *et seq.*, and 4–21 *et seq.*, 4–38 *et seq.*, 4–58 *et seq.* and Chap 5.

[21] Reg. 32. See also Merchant Shipping Act 1974, s.27. If there has been a discharge of oil in United Kingdom territorial waters (which must be reported under Reg. 3) there may be an enquiry under the Merchant Shipping Act 1970, as amended.

[22] See also Chap. 6, above.

13–17   The directions that may be given appear to be unlimited as to their content. They may decline or require unloading the vessels or taking particular salvage measures. They may be issued to persons in charge of a ship, who may or may not be owners of the ship. The directions must be complied with and there are penalties for failure to comply.

13–18   Action which was not, on investigation, reasonably necessary or which did more harm than good, if taken in compliance with a direction issued under the Act, may give rise to a claim by a person damaged by such action against the Secretary of State. Intervention under the Act is, therefore, from the point of view of the Secretary of State not without risk. Action taken by agreement with the Secretary of State, as an alternative to the use of formal intervention powers, does not attract such consequential liability.

3. CIVIL LIABILITY

13–19   The Liability Convention[23] was enacted in the United Kingdom by the Merchant Shipping (Oil Pollution) Act 1971. This applies to ships carrying persistent oil in bulk as cargo and seeks to channel liability to the shipowner in respect of damage by contamination, the cost of measures reasonably taken after discharge for the purpose of preventing or reducing damage in the area of the United Kingdom and any damage caused by such measures.[24] The Act also provides for limitation of liability in accordance with the Convention.

13–20   The Act also introduces the system of compulsory insurance, evidenced by insurance certificates, and establishes a right of direct recourse against insurers.[25]

13–21   The Fund Convention[26] was implemented in the United Kingdom by Part I of the Merchant Shipping Act 1974. The IOPC Fund is designed to meet the costs of damage unmet by the Liability Convention. Thus the 1974 Act applies in respect of claims which cannot be met under the 1971 Act either because there is a valid defence under section 2 of that Act or because the person liable under that Act cannot meet his obligations or because the

---

[23] See Chap. 10, above.
[24] This is considered in greater detail in Chap. 15, below.
[25] ss.10–12.
[26] See Chap. 11, above.

damage exceeds the limits of liability under the 1971 Act. The 1974 Act applies only to pollution damage—defined in the same terms as are used in section 1 of the 1971 Act—in the United Kingdom or in United Kingdom territorial waters.

13–22 The 1974 Act also implements those provisions of the Fund Convention relating to partial indemnity of the shipowner's liability under the Civil Liability Convention and provides for subrogation rights in the IOPC Fund in respect of claims made against it.

13–23 Although the majority of United Kingdom law on oil pollution is contained within the statutes and regulations described above, the common law still has a part to play. First, oil pollution which causes damage not covered by the Liability Convention, or in United Kingdom terms, which is outside the provisions of section 1 of the Merchant Shipping (Oil Pollution) Act 1971, must be compensated, if at all, under common law rules. That requires that the claimant establish that the person responsible has committed a tort at common law. Such torts are likely to be trespass, public or private nuisance, or negligence. Each of these torts has its own conditions for recovery. The detail of this law is dealt with in Chapter 15.

13–24 Secondly, claims under the Merchant Shipping (Oil Pollution) Act 1971 are not dealt with in a comprehensive way by that statute. Questions of causation, quantification of damage and remoteness of damage still fall to be considered under common law rules. This, too, is considered in greater detail, in Chapter 15.

13–25 Finally, there may be some place for the operation of the common law in the public law sphere, although such place is not likely to be very large. A discharge from a ship in United Kingdom territorial waters, seriously affecting the coast and the rights of United Kingdom citizens connected therewith, may constitute a public nuisance. Public nuisances are crimes at common law and may be proceeded against as such.[27]

13–26 Oil pollution claims may therefore be brought both under the above-mentioned statutory provisions and under common law principles, even in respect of the same incident. The jurisdictional aspects of such claims, and issues relating to governing law, are the subject of Chapter 16.

---

[27] See paras. 15–25 *et seq.*

14. Prevention, Control and Clean-Up

SUMMARY

14–01 This chapter deals at the outset with the United Kingdom approach to international control of oil pollution from ships, including the global, regional and European Community instruments, and bilateral co-operation. It shows that the United Kingdom has a good record of ratifying such agreements. The provisions of United Kingdom domestic law reflecting these international obligations are also briefly examined.

14–02 United Kingdom pollution response organisation is examined, tracing the history from the *Torrey Canyon* period to recent major administrative reshuffles, following the *Amoco Cadiz* accident and the Eighth Report of the Royal Commission on Environmental Pollution. This oil pollution response is referred to both at the ministerial and local authority level.

14–03 The United Kingdom oil spill detection programme is reviewed, as well as the procedures relating to inspection of vessels, based on legislation which follows closely provisions of the IMO Conventions. Some examples are given of improved inspection procedures under the 1982 Paris Memorandum of Understanding. A further section analyses prosecution of pollution offences where the polluter has been identified. It also explains a relatively poor enforcement record, for reasons which are not endemic to the United Kingdom but are common to all other maritime states (difficulty in identifying the polluter and jurisdictional factors limiting the right of the port state to prosecute)

1. THE UNITED KINGDOM APPROACH TO POLLUTION CONTROL

14–04 The attitude of the United Kingdom towards the global international instruments on pollution control and prevention has generally been a positive one. As the host country of the International Maritime Organisation, the United Kingdom actively promoted the preparatory work of that Organisation for diplomatic conferences which adopted conventions analysed in Part II of this work.

14–05 Britain has been a keen supporter of the 1969 Bonn Agreement,[1] recently revised to include chemicals in addition to oil pollution and of the 1982 Paris Memorandum of Understanding.[2] Less enthusiastic support was given to the 1984 Bremen Conference on the North Sea, sponsored by the Federal Republic of Germany, whose aims included the possible designation of the North Sea as a special area within the meaning of MARPOL 73/78. United Kingdom Government officials took the view that, as MARPOL entered into force only recently, no further legal action was needed at present.

14–06 The attitude of the United Kingdom towards the ever-growing involvement of the European Community in the field of marine pollution has not been more than lukewarm, perhaps reflecting some doubts as to the extent of the competence conferred by the Treaty of Rome in respect of the environment. As far as marine pollution is concerned, especially ship-generated pollution, it is argued that the role of the IMO would be significantly weakened if the EEC spoke with one voice, precluding member states from individual participation. In particular, it was pointed out that other, perhaps less scrupulous, groupings would emerge, resulting in a break from the present practice of IMO, which is based on pragmatic rather than political concerns.

14–07 The United Kingdom holds strong views on the EEC practice of harmonising pollution rules, including the development of uniform emission standards, as opposed to environmental quality objectives.[3] The United Kingdom position is that certain discharge rules (especially relating to dumping and land-based pollution) need not be as stringent as those applicable to semi-enclosed seas of the European Community, such as the Baltic or the Mediterranean, where waters are shallower than around the United Kingdom, and the interchange of waters with oceans is slower, reducing the potential of the sea naturally to absorb the pol-

---

[1] Agreement for Co-operation in dealing with Pollution of the North Sea by Oil, 1969: See paras. 7–18 *et seq.*, above. The other parties are Belgium, Denmark, Federal Republic of Germany, the Netherlands, Norway and Sweden.

[2] The Paris Memorandum of Understanding on Port State Control was signed on January 26, 1982 and took effect on July 1, of that year. The other parties are Belgium, Denmark, Finland, France, The Federal Republic of Germany, Greece, Ireland, Italy, the Netherlands, Norway, Portugal, Spain and Sweden. See further para. 4–41, above.

[3] See, *e.g.* N. Haigh, *EEC Environmental Policy and Britain: An Essay and A Handbook*, (1984) at pp. 295–296.

lutants. Britain is a staunch opponent of unilateral measures taken by states, which depart from the established international practice relating to marine pollution. However, it actively supports bilateral agreements in appropriate cases, especially with its North Sea neighbours. Such agreements exist, *inter alia*, with France (mainly relating to ship-generated pollution) and Norway (principally in the area of offshore pollution). The so-called MANCHE plan has been in operation for several years, whereby the French and British engage in joint exercises testing their pollution plans (POLMAR) in France.

14–08 The safety of shipping in the Channel is dealt with by the Anglo-French Safety of Navigation Group (AFSONG). As a result of a meeting of AFSONG in September 1978, a voluntary reporting system was introduced for certain ships moving through the Channel from January 1, 1979. All loaded oil tankers and loaded gas and chemical carriers of 1,600 gross registered tons and over, and any vessel "not under command" or at anchor in a traffic separation scheme or associated inshore zone, was invited to report their presence to French shore stations at Ushant, Jobourg and Cap Griz Nez, and to the H.M. Coastguard Marine Rescue Co-ordination Centres at Brixham and Dover, when entering the traffic separation schemes or the associated inshore traffic zones off Ushant, the Casquets and in the Dover Strait.

2. UNITED KINGDOM POLLUTION RESPONSE ORGANISATION

14–09 In common with most maritime countries, the responsibilities for different aspects of marine pollution are shared in Britain amongst a wide range of ministries and other public bodies. Jurisdictional issues, including those connected with the United Nations Convention on the Law of the Sea, are the prime responsibility of the Foreign and Commonwealth office. The Secretary of State for Energy is in charge of control of pollution from offshore exploration and exploitation (*inter alia*, for environmental provisions in licences to drill for oil on the continental shelf). The Secretary of State for the Environment is responsible for overseeing control over discharges into the sea from land-based sources in England, whilst the Secretaries of State for Wales and Scotland have analogous responsibilities in Wales and Scotland. The Minister of Agriculture, Fisheries and Food has responsibilities for granting permits in respect of dumping of waste at sea, and also has more general responsibilities for marine pollution questions. The

Secretary of State for Transport has primary responsbility for legislation on oil pollution, especially from ship sources, and for clean-up action at sea. He also advises local authorities in England, Scotland and Wales on pollution response through the Marine Pollution Control Unit (MPCU). The Ministry of Defence assumes responsibility for pollution caused by naval vessels.

14–10   Although the involvement of a large number of ministries in different aspects of marine pollution would appear to be inevitable, many debates have taken place in Parliament on the question of better co-ordination of maritime policy. As a result in the early 1970s the government of the day conferred on the Lord Privy Seal the overall responsibility for maritime affairs. This was subsequently changed and the responsibility entrusted to the Secretary of State for Trade. The debate continues. Although the adoption of foreign models such as the appointment of a Minister for the Sea (as in Norway), or an interministerial committee on the sea (as in France, set up by President d'Estaing in August 1978) are not being considered, a maritime committee was set up in Parliament, with representatives from both the House of Commons and the House of Lords.

14–11   The two major reorganisations of government responsibilities relating to maritime pollution took place in 1978 and 1984. The first review was undertaken at the request of the then Prime Minister following the stranding of the *Amoco Cadiz*. An inter-departmental working group of officials was set up representing the Departments of Trade, the Environment, Energy, Agriculture, Fisheries and Food, Transport, Industry, the Scottish Office, the Welsh Office and the Department of the Environment for Northern Ireland. This body published its report on August 2, 1978, together with a report by a steering group under the chairmanship of Sir Leo Pliatzky: *Accidents at Sea Causing Oil Pollution: Review of Contingency Measures*. As a result of this report, the government decided to set up four inter-departmental groups to review (a) command, control and communications; (b) resources, research and development; (c) salvage, and (d) liability and compensation for marine pollution damage. The Government also agreed that the resources of the Department of Trade, which then had responsibility for pollution response, were over-stretched, and decided to establish, at the Department of Transport, a separate Marine Pollution Control unit (MPCU). Its principal terms of reference were as follows:

(1) to ensure that the arrangements for using the resources available for dealing with oil pollution are as effective as they can be;

(2) to develop a national plan, including measures to deal with other potential marine pollutants such as chemicals and other dangerous cargoes;

(3) to relate these plans to those of neighbouring countries so as to provide as much mutual support as possible;

(4) to take charge of operations at sea in the event of a marine pollution emergency in British waters.

The MPCU was also charged with reviewing the existing regional anti-pollution plans of other ministries and of offshore operators on the United Kingdom's continental shelf.

The second major administrative change took place in 1984, following the government review of the Eighth Report of the Royal Commission on Environmental Pollution: *Oil Pollution of the Sea,* presented to Parliament in October 1981.[4] The Commission criticised the then existing division of ministerial responsibilities to deal with oil spills, under which the Department of the Environment through local authorities was responsible for pollution response in respect of beaches and of the sea up to one mile from the shore, whilst the Department of Trade had responsibility for clean-up beyond the one-mile limit. The Royal Commission considered that this division of responsibilities was not conducive to an effective response in the case of an accident. The government accepted this criticism and transferred to the Department of Transport (which had earlier assumed all duties relating to marine pollution from the Department of Trade) the responsibility of dealing at sea with all oil spills, wherever they occur.

14–12   Civil and military aircraft and merchant and naval vessels are still advised to report sightings of oil to H.M. Coastguard. The coastguard then passes on this information to the MPCU which decides whether or not any response may be required. Contingency plans have already been developed to enable the MPCU to act at short notice should a major spill occur. The coastguard also has authorisation to initiate action in the case of a major incident. Aircraft are also being increasingly used for spraying dispersants. In 1981, six were in service for this purpose, whilst another two were being converted. Arrangements were made to distribute the aircraft at four bases, two in the north and two in the south. Fifteen operational bases were also established, from which all light aircraft could operate in an emergency. Dispersant stocks, which were originally ship orientated, have now been redistributed around the airfields to ensure that aircraft would not be inactive through lack of

---

[4] Cmnd. 8358.

dispersants in an emergency. These aircraft are on half-hour standby in daylight hours and on two-hour standby for relocation at night.

14–13 Port and harbour authorities deal with spills occurring within their area of responsibility; as the legal regime of many ports is regulated by legislation which is peculiar to each port, responsibilities, rights and duties vary widely from port to port. Major oil terminals, such as Milford Haven, have an elaborate system for dealing with spills. The latter also operates a special compensation scheme, applicable even when the polluter cannot be identified, thus going beyond the arrangements under the Liability Convention. Local authorities have no statutory duties to clean oil from beaches and adjacent waters, but have in practice assumed this responsibility. Coastal county councils in England and Wales have developed contingency plans to deal with oil pollution, in consultation with district councils. Pollution response plans vary from response to minor spills to dealing with major incidents. Both district and county councils now have designated officers (oil pollution officers) who may deal with small spills, oversee local contingency plans, exercise a supervisory and advisory function and co-ordinate the deployment of local authority resources and manpower, if required. The local authority pollution response plans are drawn up in consultation with local marine pollution control officers, the Department of Transport, the Nature Conservancy Council (NCC), local representatives of MAFF and neighbouring district and county councils.

14–14 In Scotland, regional and island councils draw up their pollution plans; they in turn consult district councils, and also the DAFS Fishery Inspectorate and Principal Officers of the Department of Transport.

14–15 In addition, an advisory Standing Committee on Pollution at Sea was set up in 1975 to advise the Secretary of State for Transport (formerly Trade) on measures necessary for dealing with oil and chemical pollution. It consists of representatives of most government departments, and selected representatives of the shipping industry, oil industry, local authorities, NCC, National Environment Research Council (NERC) and Advisory Committee on Pollution of the Sea (ACOPS). Following the strengthening of the role of the MPCU, the meetings of the Standing Committee have become less frequent.

14–16 No major spills have occurred in United Kingdom waters in recent years to test the existing response system, at either national or local level. As most incidents are not very large, it is especially important that local authority oil pollution plans are adequately designed and enforced. It might therefore be interesting to refer briefly to the contents of a typical plan, *e.g.* that of Hampshire County Council.[5] This plan, devised in 1978, shortly after the *Amoco Cadiz* accident, deals with a scheme for early warning reconnaissance, inshore spraying and clearance of pollution of the coastline, identifying the available infra-structure and quoting legal instruments empowering it to act (*i.e.* the Local Government Act 1972, s.138 and the Prevention of Oil Pollution Act 1971). The scheme also identifies the possible origin of the threat.

14–17　The first stage of the plan describes the action to be taken in case of minor pollution (routine beach clearance and minor pollution of coastline), including the setting in motion of the early warning system. The second stage of the plan deals with such spills as are likely to pollute any part of the Hampshire coastline. The plan then identifies the support that can be given by district councils within the County of Hampshire and lists the available clean-up equipment held by the County. Finally, warning is given regarding the health hazards and a list of specially scheduled sites of interest to the Nature Conservancy Council and the Ministry of Agriculture, Fisheries and Food is provided. The Plan's concluding sections deal with the role of the fire service, police, service personnel, and financing of operations. Provisions for disposal of oily residues are also made, and it is stipulated that the services of the Government Chemist are available for preliminary examinations of samples of oil. Telephone numbers of all responsible officials are listed in an Annex. It is interesting to note that no provisions are made for action to be taken in the case of a failure to act in accordance with the contingency plan.

14–18　The adequacy of contingency planning has for a long time been a concern of the European Communities which have commissioned several studies on the subject.[5a]

3. DETECTION OF SPILLS

14–19　Part of the United Kingdom response plan involves the initial detection of a spill. The organisation responsible for detections of spills in the United Kingdom is the Department of Trans-

---

[5] Hampshire County Council "County Oil Pollution Plan," June 1978.
[5a] See, *e.g.* budget item 6621 and proposal for a Council Directive on the Drawing Up of Contingency Plans **O.J.** C.215 of August 16, 1984.

port's Marine Pollution Control Unit. The Unit relies partly on reports received from passing ships, but in recent years has increasingly used aerial surveillance. This enables the spills not only to be sited, but also frequently helps authorities to identify the polluter—a task which can otherwise be difficult in such busy shipping lanes as the Channel. Aerial surveillance, which also helps in prosecution of offenders (as photographs taken provide the best evidence available) is, however, expensive. Ad hoc intensified "patrols" of selected areas have not so far persuaded the authorities that a greater percentage of spills is detected than during normal monitoring exercises.

14–20 Following recommendations by various Parliamentary and unofficial bodies, government-sponsored research is also being stepped up in the usage of infra-red detection (which can be operated both during the day and at night, although not in all weather conditions); the technique is used in conjunction with ultraviolet technique and radar, which can pick up spills at a greater range (of about 20 kilometres). These methods are important because of the large number of operational discharges which are carried out at night when spills cannot be detected with the naked eye.

14–21 Detection of spills remains a serious problem, a fact recognized by parties to the 1969 Bonn Agreement during their most recent meeting of experts. A recent research and development programme has been commissioned on the computer prediction of the fate of spilled oil, which is to be used in conjunction with surveillance flying to enhance the deployment of resources.

A. Oil record books

14–22 All ships to which the 1983 Merchant Shipping (Prevention of Oil Pollution) Regulations apply[6] must carry oil record books, if they are of 150 GRT and above (in the case of tankers) or 400 GRT and above (in the case of non-tankers). Non-tankers of this size must carry an oil record book in which operations affecting discharges of ballast or cleaning water from fuel tanks, machinery space bilge water, etc. must be recorded.[7] Tankers must also record fuller information relating to cargo operations, including details of loading, transfer of cargo between tanks, ballasting, cargo tank cleaning, discharge of ballast, etc.[8] Accidental discharges, even

---

[6] S.I. 1983 No. 1398. Reg. 2. Unless otherwise provided, the Regulations apply to U.K. ships, other ships within U.K. internal or territorial waters and Government ships registered or held in the U.K.

[7] Reg. 10(2)(a).

[8] Reg. 10(2)(b).

those covered by the exceptions to prohibited discharges, must be recorded.[9]

B. Reporting of discharges

14–23   There are in United Kingdom law specific obligations to report discharges of oil from a ship directly to the appropriate authorities. These requirements are also contained in the 1983 Regulations and apply to all ships within 200 miles of the United Kingdom regardless of nationality, to all United Kingdom ships which are within 200 miles of any land, and to all United Kingdom oil tankers when laden and all United Kingdom ships of 10,000 GRT and above (wherever they may be).[10]

14–24   The master of a ship to which the Regulation applies must make a report whenever any incident involves the discharge or probable discharge of oil or oily mixtures as the result of damage to the ship or its equipment or for the purpose of saving the ship or of saving life at sea. If the ship is within 200 miles of the United Kingdom, the report must be made to the controller of the coastguard, either directly or through the nearest United Kingdom coastal radio station.[11]

C. Reporting in Harbours

14–25   The Prevention of Oil Pollution Act 1971 requires that, before oil can be lawfully transferred to or from any vessel in a United Kingdom harbour between sunset and sunrise, notice must be given to the harbour master.[12] Such a notice must be given not earlier than 96 hours and not later than 3 hours before the transfer is to take place, but provision is made for the giving of general notices.[13]

14–26   Furthermore, owners or masters of vessels which discharge oil into the waters of a United Kingdom harbour or from which any oil is found to be escaping (or to have escaped) into such waters, are obliged to report the fact immediately to the harbour authorities, and failure to do so is itself an offence punishable by fine.[14]

---

[9] Reg. 10(3). Offences relating to failure to maintain oil record books and making false or misleading entries are dealt with in the Prevention of Oil Pollution Act, 1971, s.17(5).

[10] Reg. 31(1).

[11] Reg. 31(2), (4). For the details of the content of the report, see Reg. 31(5), (6). And see also para. 14–26, below.

[12] s.10(1).

[13] s.10(2), (3).

[14] s.11.

D. Monitoring

14–27 Monitoring of oil pollution is an important technique which enables authorities to quantify the problem, and to judge the success rate of various legal and technical rules to control pollution. The Department of Trade used to compile its figures on spills on the basis of reports which were received from the civil and military aircraft, and merchant and naval vessels. The Department of the Environment formerly compiled its own figures on the basis of reports from local authorities and published them in its *Digest of Environmental Pollution and Water Statistics*. Both statistical reports proved to be two narrowly based and the government has since accepted an independent survey compiled annually by the Advisory Committee on Pollution of the Sea (ACOPS), carried out in co-operation with local authorities, river purification boards in Scotland, ports and harbours, wildlife conservation bodies and the Department of Transport and Department of the Environment for Northern Ireland.

14–28 The total number of recorded spills in the last five years shows an encouraging trend: 1980—524; 1981—552; 1982—517; 1983—396; 1984—367. The number of spills does not, however, necessarily indicate the seriousness of the problem: the damage caused to the environment and the clean-up costs are not always commensurate with either the number or the size of spills.

14–29 The progressive entry into force of provisions of the Control of Pollution Act 1974 gave added responsibilities to the regional water authorities (England), the Welsh Water Authority and Scottish river purification boards. In particular, they are charged with monitoring the estuarial and coastal waters within the limit of Britain's territorial sea. The monitoring of activities of various bodies is co-ordinated by the Marine Pollution Monitoring Management Group, which was set up by the Department of the Environment in 1974.

4. CERTIFICATION OF SHIPS FOR OIL POLLUTION PURPOSES

A. Certificates and Surveys

14–30 Every United Kingdom tanker of 150 GRT or more and every United Kingdom ship of 400 GRT or more (not being a tanker) must carry on board the ship a certificate to the effect that the vessel complies with the requirements of the 1983 Regulations as to equipment for oil pollution prevention and control.[15] The certificates, which are valid for five years,[16] are of two types; one, the International Oil Pollution Prevention Certificate (IOPP), is issued to vessels trading between the United Kingdom and other states which are parties to MARPOL 73/78, while the other, the United Kingdom Oil Pollution Prevention Certificate (UKOPP) is issued to ships trading between the United Kingdom and states which have not acceded to MARPOL 73/78.[17]

Before a certificate can be issued, a ship must be surveyed to ensure that all the equipment required by the Regulations is "in good working order."[18] The procedure to be followed in carrying out the survey is set out in a Merchant Shipping Notice.[19]

Where an IOPP certificate has been issued to a United Kingdom ship, there is an obligation under MARPOL 73/78 to ensure that the ship is inspected from time to time to ensure its continued compliance. The United Kingdom has decided that, rather than subject United Kingdom vessels to unscheduled "spot-check" inspections, there shall be a mandatory annual survey of all United Kingdom ships holding an IOPP certificate, to be carried out within three months before or after the anniversary of the grant of the certificate.[20] Furthermore, within six months before or after the "halfway date of the period of validity" of the certificate, the ship must be submitted to an intermediate survey, which is of a more thorough-going nature.[21]

14–31 Ships which are the subject only of a UKOPP certificate are not subject to the requirement to submit to annual or intermediate surveys.[22]

B. Inspection of vessels

14–32 Inspection of ships in the United Kingdom is the responsibility of the Surveyor General, attached to the Department of Transport. Inspections relating to oil pollution prevention are carried out in the United Kingdom under the auspices of MARPOL

---

[15] Reg. 7(7).
[16] Reg. 7(5). The five-year period is a maximum but the certificate will cease to have effect if the vessel is significantly modified without approval, or if it changes flag. IOPP certificates also lapse if no intermediate survey is performed: Reg. 7(6).
[17] Reg. 7(1).
[18] Reg. 4.
[19] M. 1076. The Notice reflects the terms of IMO Resolution MEPC 11(18) of March 25, 1983.
[20] Reg. 5.
[21] Reg. 6.
[22] Merchant Shipping notice, M. 1076.

73/78 and of the Paris Memorandum of Understanding on Port State Control 1982.[23]

14–33 Any vessel to which the 1983 Regulations apply may be inspected at any time while it is in a United Kingdom port or terminal.[24] Initially, the inspector must confine himself to examining the IOPP or UKOPP certificate carried aboard the vessel, and he may not look behind the certificate unless there are "clear grounds" for believing that the condition of the ship does not match that suggested by the certificate. In such a case, the inspector must ensure that the ship does not sail, except for the nearest appropriate repair yard, until it can do so without presenting "an unreasonable threat of harm to the marine environment."[25]

Following the adoption of MARPOL 73/78, and examination of its various provisions with a view to their effective implementation by IMO's Marine Environment Protection Committee, it has become common practice for Member States to submit "MARPOL Enforcement Reports" to IMO. This applies to results of inspections carried out on one's own vessels following reports by third parties, and reports on alleged violations of MARPOL forwarded to flag states as a result of inspections carried out in home ports, or on the basis of reports passed on by aircraft or vessels on the high seas.

14–34 For example, in the most recent report sent to IMO, United Kingdom authorities specified that in the period between July 1, and December 31, 1984, reports of 10 incidents were submitted to flag states (two to Panama, and one each to Sweden, Greece, Italy, Singapore, Bahamas, Brazil, Argentina and Honduras). It is interesting to note that information on alleged violations in some cases was submitted by foreign ships, including the offending tankers.

14–35 In the same period, Britain investigated seven cases of alleged violations of MARPOL committed by British vessels, reported by other countries. In one case, the owner and master were prosecuted and fined; in another, formal warning was sent to the owners; in four cases, photographic evidence was awaited, whilst insufficient evidence was provided in the last case which was subsequently closed.

14–36 Inspections also take place under the terms of the Paris

---

[23] See note 2, above.
[24] In principle, of course, United Kingdom ships may be inspected elsewhere.
[25] Reg. 32.

Memorandum. Between July 1, 1983 and June 30, 1984, some 8,847 inspections were carried out on 7,350 ships. 1,844 of these vessels were inspected in the United Kingdom, and 25 of these were detained or delayed due to serious deficiencies.[26]

14–37 In addition, harbour masters have the power to board ships in other ports to inspect the ship or any machinery, boats, equipment or articles aboard, in the course of investigating discharges of oil or oily mixtures from the vessel into the harbour. They also have power to require production of the oil record book.[27]

5. DISCHARGES FROM SHIPS AT SEA OR IN UNITED KINGDOM HARBOURS

A. Discharge Standards and Offences

14–38 Criminal liability for oil pollution is now governed by the Merchant Shipping (Prevention of Oil Pollution) Regulations 1983. These Regulations, as mentioned above, introduce the provisions of MARPOL 73/78 into United Kingdom law, and they apply to all ships registered in the United Kingdom[28] wherever they may be and to foreign ships in United Kingdom territorial or internal waters.[29]

14–39 Discharges are subject to prohibitions and conditions which reflect those to be found in MARPOL 73/78.[30] Thus, oil tankers[31] may not discharge oil[32] or oily mixtures at all, unless certain conditions are satisfied. These conditions are that the tanker is proceeding on a voyage, is not within a special area,[33] is more than

---

[26] The Agreement, of course, is not confined to the implementation of pollution conventions, although many of the instruments covered (such as SOLAS and the STW Convention) are of great significance in reducing operational oil pollution. The vessels inspected represent 19% of those visiting the ports of parties (and 20% of the world fleet).

[27] Prevention of Oil Pollution Act 1971, s.18(6). The powers are also enjoyed by inspectors appointed by the Secretary of State under section 18(1).

[28] Within the meaning of the Merchant Shipping Act 1979, s.21(2). See Reg. 1.

[29] Reg. 2(1)(a), (b). The Regulations do not apply to public ships employed on government non-commercial service, Reg. 2(2), but see Reg. 2(1)(c) as to other government ships.

[30] See paras. 3–52 et seq., above.

[31] Vessels constructed or primarily adapted to carry oil in bulk in their cargo spaces, including combination and chemical carriers when actually carrying oil in bulk: Reg. 1.

[32] Petroleum in any form "including crude oil, fuel oil, sludge, oil refuse and refined products": Reg. 1.

[33] An area of sea where special mandatory controls over oil pollution are required by virtue of its oceanographic or ecological characteristics or of the character of the traffic in the area. Certain special areas are identified in Reg. 16, but Reg. 1 makes it clear that this list is not exhaustive.

50 miles from the nearest land, is discharging at a rate which does not exceed 60 litres of oil content per mile, has in operation an oil discharge monitoring and control system and slop-tank arrangement complying with the provisions of MARPOL 73/78, and does not discharge into the sea a total quantity of oil exceeding a certain fraction of the total quantity of the cargo of which the residues formed a part.[34] In the case of tankers regarded as "new" by the Regulations, this fraction is $\frac{1}{30,000}$. New tankers are those for which building contracts were placed after the end of 1975 (or if there is no such contract, whose keels were laid after June 30, 1976) or delivered after the end of 1979 or which have undergone a major conversion contracted for after the end of 1975, commenced after June 30, 1976 or completed after the end of 1979.[35] For all other tankers, the fraction is $\frac{1}{15,000}$. The only discharges from tankers to which these restrictions do not apply are those from machinery space bilges,[36] to which the "non-tankers" rules set out below apply, and discharges of clear or segregated ballast.[37]

14–40 Vessels other than tankers are subject to less stringent requirements. Such ships are prohibited from discharging oil or oily mixtures[38] into any part of the sea, unless they are proceeding on a voyage, not within a special area, more than 12 miles from the nearest land, are discharging in such a way that the oil content of the discharge is less than 100 ppm and have in operation a prescribed oil discharge monitoring and control system, oil-water separator, oil-filtering system or similar equipment.[39]

14–41 Both classes of ships are required to retain on board such residues of oil as they are unable to discharge lawfully under these provisions,[40] but Regulations require all tankers[41] over 150 GRT to have an adequate slop-tank arrangement,[42] and oil discharge control and monitoring systems.[43] Tankers smaller than 150 GRT

---

[34] Reg. 13(2).

[35] Reg. 1.

[36] Reg. 12(1)(a)(ii). Note that the stricter provisions of Reg. 13 apply to such discharges if oil cargo residues are present, and to cargo pump room bilge discharges.

[37] Reg. 13(3).

[38] Note that the Regulation excludes mixtures whose oil content (without dilution) does not exceed 15 ppm: Reg. 12(3).

[39] Reg. 12(2).

[40] Regs. 12(5), 13(5).

[41] Existing, *i.e.* not "new" tankers must comply by October 2, 1986.

[42] Equal to not less than 3% of the oil cargo capacity, unless the vessel has segregated or dedicated clean ballast tanks or uses crude oil washing, in which case 2% of oil cargo capacity is adequate: Reg. 15(1)–(3). Special provision is made for combination carriers and vessels not needing to supplement slop-tank capacity with further washing water.

[43] Reg. 15.

should in principle retain oil residues on board and discharge them into reception facilities,[44] and thus need not be so equipped.

14–42 Failure to comply with these Regulations constitutes a criminal offence. Discharges which do not meet the conditions described above and any discharge within a special area,[45] are punishable on summary conviction by a fine of not more than £50,000 and on indictment by a fine without upper limit.[46] The Regulations make it clear that the owner and the master may *each* be liable to prosecution for the same incident.[47] Other offences, although they may similarly attract an unlimited fine on indictment, are subject to a £1,000 maximum before the magistrates.[48]

14–43 Those accused of offences under the Regulations may avail themselves of a number of exemptions and defences. If it can be shown that the discharge was necessary to secure the safety of a ship (not necessarily the ship making the discharge) or to save life at sea, no criminal liability attaches in respect of it.[49] The same is true of discharges resulting from damage to the ship or its equipment (*e.g.* in a collision, or as the result of stranding or heavy weather), unless accompanied by a failure to take all reasonable precautions after the damage occurred or after the discovery of the discharge itself to stop or minimise the discharge or by action on the part of the owner or master taken with intent to cause damage or recklessly knowing that damage would probably result.[50] The broad definition of "oil," which includes refined products, makes it necessary to include a defence, or more properly an exemption, for the use of substances containing oil as defined in order to combat oil pollution damage, *e.g.* in a dispersant or coagulent.[51]

14–44 Furthermore, the Regulations provide that a person

---

[44] Reg. 15(4). Note that Reg. 15 also effectively provides that oil residues shall be retained on board tankers "engaged exclusively" on short-sea voyages of 72 hours maximum duration and remaining within 50 miles of the nearest land: Reg. 15(5).

[45] Note that, with a limited exception in respect of non-tankers of less than 400 GRT, all discharges (even those otherwise complying with Regs. 12 and 13) are prohibited within special areas: Reg. 16.

[46] Reg. 34(2).

[47] Thus overtaking the decision in *Federal Steam Navigation Co. Ltd.* v. *Department of Trade and Industry* [1974] 2 All E.R. 97 (H.L.)

[48] Reg. 34(1).

[49] Reg. 11(2).

[50] Reg. 11(b).

[51] Reg. 11(c). The exemption is subject to the approval of the government in whose jurisdiction the material is discharged.

charged thereunder may escape conviction if he shows that he took all reasonable precautions and "exercised all due diligence" to avoid the commission of the offence.[52]

B. Enforcement

14–45   The Prevention of Oil Pollution Act 1971[53] provides that prosecutions for oil pollution offences can only be instituted by the Attorney-General, a harbour authority (for offences committed in its harbours), the Secretary of State for Transport or his authorised officer, or a local fisheries committee (if the offence is committed within its district).[54] The number of prosecutions still lags considerably behind the number of likely offences committed. For example, in 1978 the British Transport Board and other port authorities prosecuted 22 British vessels, 16 foreign vessels and eight land installations, whilst the Department of Trade prosecuted one British ship only. In 1980, the Department of Trade prosecuted three British ships, whilst port authorities took action against 17 British vessels, 26 foreign ships and 11 land installations. In 1982, the Department of Trade took action against two British ships, and one foreign vessel, whilst port authorities instituted proceedings against 15 British ships, 25 foreign vessels and 16 land installations.[55] It is important to note that almost all prosecutions, especially by government agencies, resulted in convictions. This suggests that prosecutions are only brought when they are reasonably confident of a positive outcome.

14–46   There are three main reasons for the inadequate enforcement of pollution regulations which enables a considerable percentage of those in breach of those regulations to escape prosecution.

The first reason is the present concentration of enforcement responsibility on the flag state, for prosecutions may only be brought in the United Kingdom in respect of foreign vessels if the offence in question has been committed in United Kingdom inter-

---

[52] Reg. 34(3).

[53] Perhaps curiously, the Regulations do not make it clear that they are to be read subject to the enforcement and machinery sections of the 1971 Act, but the better view would seem to be that the Merchant Shipping (Prevention of Oil Pollution) Order 1983, (S.I. 1983 No. 1106) was only intended to give effect to MARPOL and, apart from the repeals necessary for this purpose, left the operation of the 1971 Act unaffected. For a discussion of the matter, see Bates, *United Kingdom Marine Pollution Law*, (London 1985) at pp. 43–44, where the point is elegantly and (it is submitted) correctly argued.

[54] Local fisheries committees are established under the Sea Fisheries Regulation Act 1966.

[55] Reports to Parliament by Secretary of State under Prevention of Oil Pollution Act 1971, s.26.

nal or territorial waters and if a suitable defendant is available.[56] Thus, the "passing ship" of foreign registry which does not touch at a United Kingdom port may avoid prosecution. To some extent, this problem would be mitigated by the wider recognition of "port state jurisdiction" on the model provided for in Article 218 of the United Nations Convention on the Law of the Sea.[57] That Article seeks to enable the authorities of a state in whose port the offending ship arrives to institute proceedings in respect of discharges contravening "applicable international rules and standards," regardless of where such discharge took place.[58] The port state may act either of its own motion or in response to a request from another state in whose waters[59] the discharge occurred or caused damage or the threat thereof.[60] The Convention is not yet in force, but even if it were, it would not necessarily solve all the problems associated with pollution prosecutions.

The second reason is a technical one; in many cases of deliberate discharges, it is difficult to identify the polluter, especially in busy shipping lanes.[61] In recent years, scientists have stepped up research in identifying polluters. A possible solution was developed in Sweden, using the so-called "tagging technique" whereby each cargo of oil is "labelled" in a different way by adding special metallic powders. The system has been in an experimental phase for many years, and was also introduced, on a trial basis, by the Helsinki Commission, which administers the 1974 Helsinki Convention. However, no competent international organisation, including the IMO, has so far found itself able to recommend the usage of this method, and opinions on its usefulness are divided, even in Sweden.

14–47 The third difficulty concerns the imposition of fines, which are regarded as being pitched at levels which are far from adequate

---

[56] See the Territorial Waters Jurisdiction Act 1878; in *Pinka* v. *R*. [1979] A.C. 107 it was held that summary jurisdiction does not extend beyond internal waters unless the statute specifically so authorises. As to timing, the general rule is that prosecutions must be commenced within six months of the commission of the offence, if the case is to be tried summarily: s.127 of the Magistrates' Courts Act 1980. The 1971 Act extends this, where the person to be charged is abroad, for two months after his next entry into the United Kingdom and dispenses with the usual restrictions on which court shall hear the case: s.19(4), (5).

[57] See paras. 5–54 *et seq.*, above. The United Kingdom has not signed the Convention.

[58] Unless it took place within the jurisdiction of another state, in which case the port state may only take action if requested to do so by that state, the flag state or a state damaged by the discharge, or if it is itself damaged or threatened thereby: Art. 218(2).

[59] Including the exclusive economic zone, as to which see Arts. 55–75 of the Convention.

[60] Art. 218(1), (3).

[61] A matter which is also significant in connection with civil actions based upon the Merchant Shipping (Oil Pollution) Act 1971.

to have any deterrent effect. On the face of the Regulations, the power to impose a fine of £50,000 on summary conviction and an unlimited fine on indictment seems commensurate with the gravity of pollution offences. In practice, fines have never approached this level, perhaps partly because the majority of prosecutions take place before magistrates who are unaccustomed to imposing fines so far in excess of the usual statutory maximum of £1,000. Thus, in 1978, a total of 46 successful prosecutions produced the sum of only £34,000; in 1980, the figure for 56 prosecutions was £70,000; and in 1982 the same number of prosecutions produced fines totalling £83,000. No doubt this stems in part from the reluctance of magistrates to regard deterrence as the guiding factor in determining the amount of any fine.[62] Indeed, at least in respect of foreign ships, the only person before the court is the master of the vessel, and in such cases the court is obliged to ignore the fact that the owner (or his insurers) may be prepared to stand behind the master and indemnify him for fines levied on him.[63] It has recently been suggested that, inter alia, the circumstances of the discharge and especially, where the discharge was intentional, the benefits which accrued from it, and the damage caused thereby, are factors which should influence the court when assessing the level of an appropriate fine.[64]

14–48 It should also be noted that, on occasions when fines imposed by magistrates have been appealed to higher courts, they have often been reduced[65] usually on the ground that they have not been commensurate with the ability of the person charged to pay so large a sum. There has been some suggestion that the recovery of fines from insurers should cease to be lawful, but the results have been inconclusive, perhaps in recognition of the necessity to protect seamen from unreasonable conduct on the part of foreign authorities.

Although the 1971 Act does provide that fines imposed for pollution offences may be paid to persons incurring expenses in clean-up or other remedial action,[66] it would appear that it is not a proper method of assessing the appropriate fine to base it upon the amount

---

[62] See e.g. s.35 of the Magistrates' Courts Act 1980, which requires a court to have regard to the means of the *person charged*.

[63] *R.* v. *Niel* (unreported), cited in Thomas, *Principles of Sentencing*, (1970), p. 222.

[64] In *Paderewski* v. *Tees & Hartlepool Port Authority*, an unreported case cited by Bates, *op. cit.* at p. 46, in which it was also held that the practice of owners of paying fines imposed on masters should be disregarded, for fear of encouraging the practice.

[65] See, *e.g. The Times*, May 15, 1976: fines of £30,000 on owner and £1,000 on master reduced to £20,000 and £750 respectively.

[66] s.20(2).

of the expenses to be incurred, without reference to other factors.[67] Frustrated by this relatively modest record of prosecutions, some local authorities have introduced novel systems of quasi-legal responses. For example, the Shetland Islands Council and the Sullom Voe Terminal, having introduced a 200-mile reporting system, decided to refuse entry into the port to such vessels as were caught discharging illegally their dirty ballast water outside Britain's three-mile territorial sea limit reasoning that, although the court was precluded from instituting legal proceedings (as discharge occurred outside Britain's jurisdiction, at least over a foreign vessel), economic sanction might prove to be an even more powerful deterrent.

6. CONCLUSIONS

14–49 The United Kingdom has been an active supporter of international control of marine pollution, hosting many of the international conferences which adopted conventions on vessel source discharges. It also supported regional co-operation in North Western Europe in this field, whilst some reservations remain on the role which the European Communities should play in environmental protection. Bilateral agreements were also concluded with countries with which the United Kingdom has continental shelf boundaries, such as Norway, or whose coasts are vulnerable to ship-generated pollution in the Channel (France).

14–50 The United Kingdom has structured its pollution response on the basis of experience gained following such accidents as the *Torrey Canyon* in 1967 and the *Amoco Cadiz* in 1978. Responsibilities are shared amongst various ministries, with overall responsibility being held by the Department of Transport. Local authorities have no statutory duties but have assumed the responsibility to clean up oil from beaches and adjacent waters. Port and harbour authorities deal with spills within their areas. Pollution contingency plans have also been drawn up, both at the national and local authority level. Co-ordination of Governmental response is entrusted to the Marine Pollution Control Unit, which is currently attached to the Department of Transport.

14–51 Detection of spills is monitored by the Department of Transport on the basis of reports submitted by passing ships and aircraft (both civil and military). Inspection of vessels is the

---

[67] *John* v. *Wright* (1980) S.L.T. 189.

responsibility of the Surveyor General. Reports are submitted to IMO on a regular basis, and also to the Secretariat which monitors implementation of the 1982 Paris Memorandum of Understanding on Port State Control. Inspection procedures are regulated by the Prevention of Oil Pollution Act 1971.

14–52 Prosecution for pollution offences is regulated by the 1971 Act and by the Merchant Shipping (Prevention of Oil Pollution) Regulations 1983. However, the number of prosecutions lags behind the number of offences committed. This is partly due to technical difficulties in identifying the polluter, and partly due to the lack of power of the port state to prosecute. This will change once the relevant provisions of the Law of the Sea Convention enter into force. For the time being, however, the United Kingdom has not even signed the Convention, supporting the stance taken by the United States. Fines have also proved to be an insufficient deterrent to would-be polluters. Magistrates' courts, which have jurisdiction in this matter, are not accustomed to imposing the maximum penalty provided by law (£50,000), and higher fines are often reduced on appeal. Also, insurance against fines is allowed, further reducing the deterrent impact of fines. Some local authorities, especially in the Shetland Islands, have thus started taking extra-legal action, by imposing sanctions over offending vessels where no jurisdiction can be exercised. This is mainly done by refusing the offending vessel right of entry into the port.

15. Principles of Civil Liability

SUMMARY

15–01   Civil claims for compensation for oil pollution in the United Kingdom will either be based upon the common law or upon the Merchant Shipping (Oil Pollution) Act 1971. This Act implements the Liability Convention in the United Kingdom, with some slight extension. It thus channels claims for compensation through the shipowner and provides for the Convention limitation provisions. It defines pollution damage, thus describing the claims that may be made under it. At common law, claims may be based upon trespass, public nuisance, private nuisance or negligence. These torts contain their own rules regarding types of recoverable damage and conditions of liability.

15–02   For claims under the Merchant Shipping (Oil Pollution) Act 1971 and the common law, the amount of damages recoverable are governed by similar rules. Questions of remoteness, causation and quantification are settled by the rules of common law, even in the case of claims under the Act, since the Act provides no special conditions.

15–03   When there is the possibility of multiple defendants, questions of joint and several liability are raised. The principles of United Kingdom Law are now primarily to be found in the Civil Liability (Contribution) Act 1978. This Act provides for contributions between joint tortfeasors and allows for recourse proceedings. Finally, when a claim is not based upon the Liability Convention, as enacted in the Merchant Shipping (Oil Pollution) Act 1971, the claim may be subject to global limitation under the Merchant Shipping Act 1894, s.503, (as amended principally by the Merchant Shipping (Liability of Shipowners and Others) Act 1957, implementing the Brussels Convention of 1956). However, the law of global limitation is to be substantially amended when the Limitation Convention 1976 comes into force. This convention has been enacted in the Merchant Shipping Act 1979 whose provisions in this respect have not yet been activated.

1. THE LIABILITY CONVENTION AND COMMON LAW

15–04   The common law can be applied to the issue of civil liability for oil pollution in two distinct situations: in common law countries where the Liability Convention has not been adopted and no special statutory regime has been introduced, and in those cases where the Liability Convention does not apply. In such cases the plaintiff must have recourse to the common law. In the United Kingdom claims covered by the Liability Convention are governed by the Merchant Shipping (Oil Pollution) Act 1971, as amended by the Merchant Shipping Act 1974. In view of the fact that these statutory provisions follow the provisions of the Liability Convention so closely, and of the fact that extensive analysis of the Liability Convention is contained in Chapter 10, the provisions of the 1971 Act will only be given space here where they depart from or supplement the Convention.

15–05   There is a surprisingly large variety of situations to which the Liability Convention does not apply, so that even in the United Kingdom, which is a party to the Convention, common law principles of liability have considerable practical importance. There are six main areas not covered by the Liability Convention. They are:

(1) Oil escaping from river and lake vessels, offshore installations, land installations and pipelines.[1]

(2) Oil escaping from dry cargo ships and tankers not carrying oil in bulk as cargo.

(3) Damage caused by non-persistent oils.

(4) Damage suffered by installations outside the territory or territorial sea of a party to the Liability Convention and all damage suffered on the territory or territorial sea of a non-party to the Liability Convention.

(5) Claims against salvors, bareboat charterers and certain other parties who may be involved in the management, operation, manufacture or repair of the ship.[2]

(6) Damage caused by oil spilling onto the sea and then catching fire.

Even when the Liability Convention does apply to a claim, some of the principles discussed in this chapter will be relevant to the

---

[1] In the U.K., s.30 of the Petroleum and Submarine Pipelines Act 1975, and s.11 of the Mineral Workings (Offshore Installations) Act 1971 give special actions in respect of breach of statutory duty, but they apply only to personal injury. Oil pollution rarely causes personal injury.

[2] Under U.K. and certain other municipal laws, salvors, bareboat charterers and others may be completely immune against proceedings for oil pollution damage claims: see below, paras. 15–97 *et seq.*

claim: for instance, those relating to causation, remoteness of damage and recoverable loss, and evidence.

15–06 In examining the principles of liability at common law, it is as well to remember that oil pollution not covered by the Convention can occur in a variety of different ways: for instance, it may occur as a result of a deliberate act such as the discharge of slops, oily bilges or cargo tank washings, or accidentally, as a result of collision, stranding, or leaks in bunker tanks or hoses. It might also occur if a bunker filling operation is negligently carried out, or during the transfer of crude oil at a single buoy mooring. In addition, pipelines may leak or burst for various reasons, and of course there will be other serious circumstances such as a blow-out from an offshore well. This Chapter concentrates on incidents involving ships.

2. PRINCIPLES OF COMMON LAW: ESTABLISHING BREACH OF A LEGAL DUTY

A. The Southport Corporation Case[3]

15–07 This case was the first case in the United Kingdom to be concerned with oil pollution damage. On that account alone it constitutes a most important authority; in addition it provided an opportunity for the discussion of most of the major issues of common law liability in such cases. A brief review of the facts of the case will now be given.

15–08 In December 1950 a small oil tanker of 680 tons gross, the *Inverpool*, was on a voyage from Liverpool to Preston. At the entrance to the Ribble estuary she encountered some very heavy seas, and for some unexplained reason her steering thereafter became erratic. The master decided to continue on course for Preston, although this involved the navigation of a narrow and shallow channel. She ran aground in this channel on a revetment wall. She was lying in a dangerous position, and was in danger of breaking her back. The safety of the ship and her crew were in peril. The propeller was found to be striking some object, and so the engines could not be used to get her off. The master decided to discharge about 400 tons of oil to lighten the ship, and this was carried by the wind and tide to the plaintiff's beach, which occasioned them considerable expense in the clearing-up.

---

[3] *Esso Petroleum Co. Ltd.* v. *Southport Corpn.* [1956] A.C. 218 at p. 222 (Devlin J.); *reversed* [1954] 2 Q.B. 182, (C.A.); *restored* [1956] A.C. 218, (H.L.).

15–09   The subsequent action was heard at first instance by Devlin J. (assisted after the hearing by one of the Elder Brethren), in the Court of Appeal by Singleton, Denning and Morris L.JJ., and in the House of Lords by Earl Jowett L.C. and Lords Normand, Morton, Radcliffe and Tucker. In the course of the litigation, causes of action in trespass, public nuisance, private nuisance and negligence were all discussed. We shall consider these causes of action in turn.

B. Trespass

DIRECT AND CONSEQUENTIAL INJURY

15–10   Trespass to land has been defined as "unjustifiable interference with the possession of land."[4] Even the slightest interference will suffice to ground an action,[5] but it seems to be agreed by writers[6] that the interference must be direct; if it is consequential, the appropriate remedy is nuisance. The distinction arises from the ancient forms of action: trespass lay only for direct injury, case for consequential injury.[7]

15–11   The distinction between direct and consequential interference is extremely difficult to perceive, and could be called a disgrace to our jurisprudence. Thus, it has been held that the roots of a tree on a defendant's land encroaching on a plaintiff's land is consequential, not direct,[8] but to throw stones upon a neighbour's land is direct injury.[9] Counsel for the defendants in the *Southport Corporation* case submitted that the interference by the oil was consequential, and not direct, and that therefore trespass would not lie.[10] Devlin J., while declining to go into the matter, thought that the injury probably was direct,[11] and at one point Morris L.J. seemed

---

[4] *Winfield and Jolowicz on Tort*, (12th ed.), p. 359.
[5] *e.g. Westripp* v. *Baldock* [1938] 2 All E.R. 779 (sand resting against the plaintiff's wall); *Gregory* v. *Piper* (1829) 9 B. & C. 591 (a stone rolling down from a pile onto the plaintiff's wall).
[6] *Winfield and Jolowicz on Tort*, (12th ed.), p. 364; *Salmond on the Law of Torts*, (17th ed.), p. 38; Harry Street, *The Law of Torts*, (17th ed.), p. 58; J. G. Fleming, *The Law of Torts* (6th ed.), pp. 15–17.
[7] Legal historians would not today lay the responsibility for this distinction at the door of mediaeval lawyers: it seems to have arisen relatively late in the day, to give some sort of order to the sprawling family of the "fertile mother of actions": see S.F.C. Milsom, *Historical Foundations of the Common Law*, p. 261.
[8] *Davey* v. *Harrow Corpn.* [1958] 1 Q.B. 60; [1957] 2 All E.R. 305.
[9] *Mann* v. *Saulnier* (1959) 19 D.L.R. (2d) 130, *per* West J.A., at p. 132 approving a passage from *Salmond on the Law of Torts*.
[10] [1956] A.C. 218 at p. 224.
[11] [1956] A.C. 218 at p. 225.

to agree,[12] but Denning L.J. was firmly of opinion that the distinction between direct and consequential injury mattered,[13] and that this was a clear case of the latter. He cited two ancient cases[14] and Viscount Simonds in *Read* v. *J. Lyons & Co.*[15] in support. Lord Radcliffe was hesitant: he was "not prepared to say that . . . the appellants action did constitute a trespass"[16] while Lord Tucker expressly agreed with Denning L.J.[17] None of the other judges offered an opinion on the point. The weight of opinion[18] and authority[19] is in favour of the distinction still being of importance, and of the view that even where oil is deliberately discharged on to the sea, the injury will be insufficiently direct,[20] on the basis that there rarely will be sufficient certainty that the spillage or discharge of the oil on to the sea will lead to the shore being contaminated— even in these days the wind and tide are not wholly predictable. However, a situation where the injury might be sufficiently direct is where the discharge takes place in harbour or at a terminal, and the harbour or terminal is affected. At any rate, it is quite clear that if oil spills onto the water and then catches fire, resulting damage would be consequential.

15–12 Does it make any difference if the vessel has accidentally spilled oil? Suppose, for instance, that there was an accidental spillage in circumstances such that the injury was direct—would trespass still lie?

15–13 There are two cases concerning trespass to the person which seem relevant. The first is *Fowler* v. *Lanning*.[21] The plaintiff

---

[12] [1954] 2 Q.B. 182 at p. 204.

[13] [1954] 2 Q.B. 182 at p. 195.

[14] *The Prior of Southwark's case* (1498) Y.B. Trin. 13 Hen 7, f. 26, pl. 4; and *Reynolds* v. *Clarke* (1725) 2 Ld. Raym. 1399.

[15] [1947] A.C. 156 at p. 166.

[16] [1956] A.C. 218 at p. 242.

[17] [1956] A.C. 218 at p. 244.

[18] See F. H. Newark, "The Boundaries of Nuisance" (1949) 65 L.Q.R. 480; R. W. M. Dias and B. S. Markesines, *The English Law of Torts*, (Brussels, 1976), pp. 95–96. Mr. Dias has suggested in a private communication that some forms of interference which *look* to be direct but which are held consequential (*e.g.* encroaching tree-roots) are so held because the action is invariably being brought in respect of the interference with use and enjoyment consequent on the encroachment.

[19] *Gregory* v. *Piper* (1829) 9 B. & C. 591, is an authority that deliberately to put matter where natural forces would take it onto the plaintiff's land is a trespass—and the need for a high probability in the effect of the natural forces was stressed. The interference relied on there was, *per* Bayley J., "a necessary or natural consequence" of the act. *Per* Parke J. "the defendant must be taken to have contemplated all the probable consequences of the act . . . the defendant [is responsible] for the necessary or natural consequences of his own act . . . ."

[20] C. J. Hamson takes a contrary view, [1954] C.L.J. 172, but he cites no authority and gives no supporting argument.

[21] [1959] 1 Q.B. 426; [1959] 1 All E.R. 290.

pleaded in his statement of claim that "the defendant shot the plaintiff." Diplock J. held that trespass to the person did not lie if the plaintiff's injury was caused unintentionally and without negligence on the defendant's part. He relied heavily on the 'traffic' cases discussed below in paragraphs 15–18 *et seq.*, and on two *obiter dicta* by Blackburn J. (also discussed below).

15–14   The second case is *Letang* v. *Cooper*,[22] in which the plaintiff sued the defendant in trespass and negligence, for running over her legs with his motor car while she was sunbathing in the grass car park of a hotel. It was clear that her action for negligence was time-barred, and the question was whether her action in trespass was also. The Court of Appeal held that it was, and in so doing Lord Denning M.R. and Dankwerts L.J. held that when the injury to a plaintiff is caused by the defendant's intended act, the cause of action is trespass; when unintended, negligence.[23] Diplock L.J. thought that no procedural consequences flowed from the choice by the pleader of describing a factual situation as either trespass or negligence[24] Lord Denning M.R., in a passage which seems totally to contradict what he said in the *Southport Corporation* case, said:

> "The truth is that the distinction between trespass and case is obsolete. We have a different sub-division altogether. Instead of dividing actions for personal injuries into trespass (direct damage) or case (consequential damage), we divide the causes of action now according as the defendant did the injury intentionally or unintentionally. If one man intentionally applies force directly to another, the plaintiff has a cause of action . . . in trespass to the person . . . . If he does not inflict injury intentionally, but only unintentionally, the plaintiff has no cause of action today in trespass. His only cause of action is negligence . . . .
>
> The modern law of this subject was well expounded by Diplock J. in *Fowler* v. *Lanning*, with which I fully agree. But I would go this one step further: when the injury is not inflicted intentionally, but negligently, I would say that the only cause of action is negligence, not trespass. . . ."[25]

15–15   Does this apply to trespass to land, even though both cases are expressly on trespass to the person? The arguments in the cases rely in part on trespass to land cases, and those parts directed towards the distinction between case and trespass are equally relevant to trespass to land. Professor Street's opinion is that while

---

[22] [1965] 1 Q.B. 232, [1964] 2 All E.R. 929.
[23] [1965] 1 Q.B. 232 at p. 240, *per* Lord Denning M.R., and at p. 242, *per* Dankwert L.J.
[24] *Ibid.* at p. 243.
[25] *Ibid.* at pp. 239–240.

Letang v. *Cooper* illustrates the current judicial tendency to avoid overlap of trespass and negligence, "it is too soon to conclude that trespass has no relevance when negligent conduct is relied on."[26] Perhaps a stronger conclusion is possible: that it is unlikely that trespass will be allowed when the oil spillage has been accidental.

15–16 It may therefore be concluded that trespass will not normally lie in oil pollution cases, either because the injury suffered will be held to be consequential, not direct, or because, in cases where the discharge was unintentional, the court is not likely to allow any other action than one in negligence.

15–17 These conclusions make consideration of possible defences to an action in trespass somewhat redundant, but since two special defences were argued in the *Southport Corporation* case, and since they are also relevant to the case in nuisance, they will be mentioned here.

The So-Called "Traffic" Rule

15–18 Devlin J. at first instance[27] regarded it as "well established that persons whose property adjoins the highway cannot complain of damage done by persons using the highway unless it is done negligently: *Goodwyn* v. *Cheveley*,[28] *Tillett* v. *Ward*,[29] and *Gayler and Pope Ltd.* v. *B. Davies & Son Ltd.*"[30] He then relied on two well-known dicta of Lord Blackburn[31] which indicate that the rule applies as much to navigable waterways as to highways on land. Morris L.J. agreed with him,[32] as did Lord Tucker,[33] the other judges not expressing an opinion on the point. That the rule is established for highways on land is undoubted: the question is, is it really established in relation to the owners of land abutting navigable waterways?

15–19 Despite the concurrence of three judges in the *Southport Corporation* case, there is considerable doubt that it is. The authority for the extension of the rule from land to sea rests on two *obiter dicta*; the judicial endorsement thereof in the *Southport Corporation* case is

---

[26] H. Street, *The Law of Torts*, (6th ed.), p. 16. A point of view which Lord Denning reiterated in *Miller* v. *Jackson* [1977] Q.B. 966 at p. 980.
[27] [1956] A.C. 218 at p. 226.
[28] (1859) 28 L.J. Ex. 298.
[29] (1882) 10 Q.B.D. 17.
[30] [1924] 2 K.B. 75.
[31] *Fletcher* v. *Rylands* [1866] L.R. 1 Exch. 265 at p. 286; *River Wear Comrs.* v. *Adamson* [1877] 2 App. Cas. 743 at p. 767.
[32] [1954] 2 Q.B. 182 at pp. 203–204.
[33] [1956] A.C. 218 at pp. 244–245.

weakened by the fact that Lord Tucker's remarks on the point were also *obiter*—he felt that there was no cause of action in trespass anyway. Devlin J. gave no policy reason for such an extension being desirable, and Morris L.J.'s concept of the basis for the rule seems off the mark: he thought (with Lord Blackburn in *Fletcher* v. *Rylands*) that "the circumstances were such as to show that the plaintiff had taken that risk upon himself, whereas it is plain from reading the traffic cases themselves that the basis is quite different." Martin B. in *Goodwyn* v. *Chevely*[34] said: "if a man . . . will not [fence his land] it seems to me he must put up with some of the inconveniences consequent upon it." In *Tillett* v. *Ward*[35] Stephen J. said it is "an exception which is absolutely necessary for the common affairs of life." The cases themselves concern straying cattle and bolting horses, against which eventuality it is at least technically possible to guard one's premises: it is impossible to guard one's coastline in such a direct way against oil pollution. It is further difficult to see how allowing the rule in oil pollution cases is absolutely necessary for the common affairs of life.

15–20 It is not absolutely clear from the report whether the traffic cases were cited by the defence at first instance in *The Wagon Mound* (*No.* 2),[36] but they probably were because Walsh J. expressly rejected Devlin J.'s reasoning in the *Southport Corporation* case: " . . . it cannot here be said that the spillage came about from the ordinary use of the harbour waters by the *Wagon Mound* and not from any unreasonable or excessive user, or that this was a risk which other users of the harbour must be regarded as having taken upon themselves. . . . "[37] On appeal to the Privy Council, counsel for the defendants cited the argument,[38] but the Board did not deal with it specifically in its advice.

15–21 For these reasons it seems wrong in law to apply the rule to oil pollution cases.[39] The trouble is that the only judicial authority positively against such an extension of the rule is Walsh J., cited above; and these things tend to be self-perpetuating. It would, therefore, be wrong to conclude that courts would not in the future

---

[34] (1859) 28 L.J. Ex. 298 at p. 302.
[35] (1882) 10 Q.B.D. 17 at p. 21.
[36] [1963] 1 Lloyd's Rep. 402.
[37] [1963] 1 Lloyd's Rep. 402 at p. 429.
[38] [1967] A.C. 617 at pp. 621–622.
[39] F. H. Newark, 17 M.L.R. 579 at p. 580, submits that the rule "is restricted to purely involuntary trespass," but gives no reason for such a submission. However, the idea is in accordance with the principle that trespass is an *intentional* tort. If the act were involuntary, there would be no trespass, and the plaintiff would have to show negligence.

continue to apply the rule to oil pollution cases: for the sake of principle, one can only hope they would not.

NECESSITY

15–22 The defence of necessity was argued in the *Southport Corporation* case, but is unlikely to be of much practical significance nowadays because it is only likely to apply where cargo is emptied onto the sea to save the ship. Such an occurrence is extremely unlikely these days: salvors and owners alike would always prefer to transfer the cargo to another tanker or a barge if this is at all possible. However, both in the *Torrey Canyon* case in 1967 and in the *Amoco Cadiz* case in 1978, cargo was deliberately released by bombing the wreck in order to obviate the possibility of its later escape. In all cases, it must be decided whether greater loss will be caused in oil pollution damage than will be suffered if the oil stays in the ship.

15–23 The plea also raises the issue of negligence: it was agreed by all the judges in the *Southport Corporation* case that the defence would not apply if the situation in which the vessel found herself was caused by her own negligence.[40]

For these reasons, the defence of necessity does not merit further consideration here.

CONCLUSION ON TRESPASS

15–24 Trespass will not normally lie where oil has been deliberately discharged to sea because the injury will rarely be sufficiently direct: in those rare cases, such as discharges in harbours or terminals, the spillage is in practice hardly ever deliberate. Where oil has been accidentally discharged to sea, trespass may well not lie on the grounds that the tort is only committed when the injury is intentional. Even if trespass does lie, the authority that a plaintiff adjoining a navigable waterway must prove negligence if he is to have a remedy must be overcome. In practice, the defence of necessity will rarely apply, and when it does, the issue of negligence is raised anyway.

---

[40] Although Walsh J., *obiter*, in *The Wagon Mound (No.* 2) [1963] 1 Lloyd's Rep. 402 thought the defence could be raised in an oil pollution case, this is in doubt, as is the question of whether compensation would be payable: see H. Street, *The Law of Torts*, (7th ed.), p. 3; J. G. Fleming, *The Law of Torts*, (6th ed.) pp. 87–91; F. N. Newark, 17 M.L.R. 579 at p. 581: "We may approve an act done to save life, but there is no more justice in charging up to a stranger the cost of saving your life than there is in requiring him to foot the bill for your daily keep."

C. Public Nuisance

PUBLIC NUISANCE AND CIVIL LIABILITY

15–25   Public nuisance is primarily a part of the criminal law. As such it may be prosecuted or, more usually, restrained by injunction sought in civil proceedings by the Attorney General or a local authority acting in exercise of powers conferred by the Local Government Act or other statute.[41] However, it is established law that an individual who suffers special damage as a result of a public nuisance may maintain civil proceedings for damage.[42] The standard judicial definitions of public nuisance, therefore, show the public law origin of the tort.

15–26   In *Attorney General* v. *PYA Quarries Ltd.*,[43] Romer L.J. produced the following definition of public nuisance, after a comprehensive review of the authorities: "It is . . . clear, in my opinion, that any nuisance is 'public' which materially affects the reasonable comfort and convenience of life of a class of Her Majesty's subjects. The sphere of the nuisance may be described generally as 'the neighbourhood,'[44] but the question whether the local community within that sphere comprises a sufficient number of persons to constitute a class of the public is a question of fact in every case."[45] There is always a danger in treating judicial definitions as statutory, but this particular definition has found such favour that it can be safely relied upon.[46]

15–27   It is immediately apparent that there is no problem here, as there is with trespass, of proving direct injury. Indeed, a nuisance may be constituted without proof of any actual injury at all: it is enough that the exercise of rights, whether specific or general, has been interfered with. However, since civil proceedings for public (as against private) nuisance require proof of special damage suffered by the plaintiff, in this case some injury must ensue. This may clearly be consequential.

---

[41] The so-called "relator" action: see, *e.g. Gouriet* v. *U.P.O.W.* [1978] A.C. 435. For statutory authority, see the Local Government Act 1972, s.222; and the Control of Pollution Act 1974, s.58(8).

[42] See *Winterbottom* v. *Lord Derby* [1867] L.R. 2 Exch. 316.

[43] [1957] 1 Q.B. 169, (C.A).

[44] See *Att.-Gen.* v. *Stone* (1895) 60 J.P. 168; *Att.-Gen.* v. *Cole & Son* [1901] 1 Ch. 205; *Att.-Gen.* v. *Corke* [1933] Ch. 89; all cited by Romer L.J.

[45] [1957] 2 Q.B. 169 at p. 184, closely echoing the words of Joyce J. in *Att.-Gen.* v. *Keymer Brick and Tile Co. Ltd.* (1903) 67 J.P. 434, which he quoted.

[46] *Winfield and Jolowicz on Tort*, (12th ed.), p. 379. Approved by Brown J. in *Att.-Gen. of British Columbia, ex rel. Eaton* v. *Haney Speedways and District of Naple Ridge Ltd.* (1963) 39 D.L.R. (2d) 48 at p. 54, where he held that seven families were sufficient to constitute a class.

15–28 It is also clear that if the area affected is too small, there can be no public nuisance; thus, as Denning L.J. said in the case, "when the nuisance is so concentrated that only two or three property owners are affected by it . . . then they ought to take proceedings on their own account to stop it and not expect the community to do it for them."[47] In the *Southport Corporation* case the oil was deposited with small breaks for a distance of seven and a half miles, it was one to three inches thick and varied in width from three to twenty feet, and doubtless this was in the same learned judge's mind when he held that there was a public nuisance in that case.[48] Unfortunately, his judgment on that point was not well reasoned; in the *Southport Corporation* case Lord Radcliffe simply conceded that it may "possibly" have been a public nuisance,[49] and Devlin J. seemed to think it was, but his remarks are not explicit.[50] It must always remain a possibility that the neighbourhood affected in a particular oil pollution incident is too small for the incident to constitute a public nuisance.

15–29 In addition, it must be possible that in a particular case, notwithstanding the fact that a neighbourhood has been affected, an insufficiently large number of persons to constitute a class of Her Majesty's subject will have been affected.

15–30 The matter is treated differently if there exists an identifiable public right. Interference with a public right, typically the blocking of a public right of way, always constitutes a public nuisance. Thus if, for example, there is a public right of way over the foreshore,[51] its obstruction by deposited oil might more easily be accounted a public nuisance. In such a case, it is of no significance that few persons were actually affected.[52] However, there is no general right of passage over the foreshore, nor a right of bathing.[53] Therefore, if the owner of the foreshore (usually, but not always the Crown[54]), or of the land above high-water mark habitually allows

---

[47] [1957] 2 Q.B. 169 at p. 191, citing *R. v. Lloyd* (1802) 4 Esp. 200.
[48] [1954] 2 Q.B. 182 at pp. 196–197; citing *R v. Mutters* (1864) Le. & Ca. 491 and *Scott v. Shepherd* (1773) 2 W. Bl. 892.
[49] [1956] A.C. 218 at p. 242.
[50] [1956] A.C. 218 at p. 225.
[51] The shore between the high and low water marks of ordinary tides.
[52] *Per* Denning L.J. in *Att.-Gen. v. P.Y.A. Quarries* [1957] 2 Q.B. 169 at p. 191: "Take the blocking up of a public highway or the non-repair of it. It may be a footpath very little used except by one or two householders. Nevertheless, the obstruction affects everyone indiscriminately who may wish to walk along it . . . ." It was accepted in *The Wagon Mound (No. 2)* [1967] A.C. 617, that the spillage of oil on to the waters of Sydney harbour was an interference with a public navigable waterway, and so a public nuisance.
[53] *Blundell v. Catterall* (1821) 5 B. & Ald. 268, followed by the Court of Appeal in *Brinckman v. Matley* [1904] 2 Ch. 313.
[54] *e.g.* the plaintiffs in *Blundell v. Catterall* and *Brinckman v. Matley* were private owners.

the public access to the foreshore, then the general test will have to be satisfied: if the foreshore is polluted, it will constitute a public nuisance only if a sufficiently large class of Her Majesty's subjects have been affected. If, however, the oil has interfered with a public right of fishing,[55] a public nuisance will have been committed.

15–31 It may therefore be concluded that in by no means all cases will oil pollution constitute a public nuisance. If even quite a large stretch of coastline (say two or three miles) owned by one person, and not habitually used for recreation by more than a few members of the public, were to be polluted, it is not certain that there would be a public nuisance. But where such a stretch is so used, for instance, harbours, public beaches and marinas, there probably will be a public nuisance. In such cases, the question of whether or not a plaintiff can recover damages will rest on whether he can show "special damage."

Special Damage

15–32 It is established law that a plaintiff can only recover damages in an action for public nuisance if he can show some damage peculiar or special to himself, which is in some way appreciably different from the annoyance and inconvenience suffered by the general public.[56] It is quite clear in oil pollution cases that the owner of land suffers special damage for the cost of clear-up[57] and damage to oyster or shellfish beds,[58] and indeed just the contamination of the land itself[59] is clearly different in nature and extent to that suffered by the general public, but the fisherman is in a different position.[60] The only judge in the *Southport Corporation* case specifically to mention special damage was Denning L.J., and he seems to have thought it beyond doubt that the plaintiffs there had suffered it.[61] One writer has suggested, however, that where the

---

[55] See below, para. 15–33 *et seq.*

[56] *Iveson* v. *Moore* (1699) 1 Ld. Raym. 486 (where colliery owner recovered for loss of profit occasioned by obstruction of the highway); *Rose* v. *Miles* (1815) 4 M. & S. 101 (where a carrier recovered from the expense of having to unload his barges and transport their cargo by land due to the defendant's obstruction of a creek); *Ricket* v. *Metropolitan Ry Co.* (1865) 5 B. & S. 156 (where a publican failed to recover loss of profit but where the principle for recovery was accepted); *Winterbottom* v. *Lord Derby* [1867] L.R. 2 Exch. 316 (passer-by failed to recover for obstruction of public way because no damage beyond delay and diversion, which was common to all).

[57] Now recoverable under U.K. law by virtue of the Merchant Shipping (Oil Pollution) Act 1971, s.15: see para. 15–116, *et seq.*

[58] See below, para. 15–116.

[59] See below, para. 15–117.

[60] See below, para. 15–125.

[61] [1954] 2 Q.B. at p. 197.

damaged shoreline is owned by the public, special damage cannot be shown because all members of the public have suffered alike.[62] Unfortunately this argument was not discussed in the case, but to lift the corporate veil here could lead to absurdity. If the owners of the *Corrimal* in *The Wagon Mound* (*No.* 2) case had not been private, but a state-owned trading company, to adopt the argument would be to deny the right of recovery in nuisance to her owners. The answer seems to be to treat the juridical person suffering loss as one juridical person.

FORESEEABILITY OF SPECIAL DAMAGE

15–33   The Privy Council held in *The Wagon Mound* (*No.* 2)[63] that for a plaintiff to recover in public nuisance, his damage must be of a foreseeable kind. As a decision of the Privy Council, strictly speaking it does not bind United Kingdom courts; but despite this fact and the somewhat doubtful grounds for the decision (Lord Reid, in delivering the advice of the Board, admitted that the authorities were inconclusive[64]), the case is likely to be followed in future.[65] It is noticeable that Parliament, while making preventive measures specifically recoverable in certain cases of non-Liability Convention liability[66] (and thus exceeding the United Kingdom's obligations under the Convention), refrained from specifically making damage by fire subsequent to the spillage automatically recoverable, thus retaining the *Wagon Mound* (*No.* 2) rule that it will be so only if it is foreseeable.

HAS THE TRADITIONAL TEST BEEN CHANGED?

15–34   In the *Wagon Mound* (*No.* 2) there was a great deal of discussion about the overlap of nuisance and negligence, and whether the latter must be shown to found an action in the former. This issue is discussed below, in paragraphs 15–44 *et seq.* after private nuisance has been mentioned.

IS DAMAGE BY FIRE FORESEEABLE?

15–35   This question is also relevant to an action in private nuisance, for as we shall see, foreseeability of damage is a condition of recovery there too.

---

[62] 104 L.J. 507 (J.B.M.).
[63] [1967] A.C. 617.
[64] *Ibid.* at p. 638.
[65] *Winfield and Jolowicz on Tort*, (12th ed.), p. 384; R. W. M. Dias, "Trouble on Oiled Waters" [1967] C.J. 62 at p. 65, n. 12. The principle has been applied in *Heaven* v. *Mortimore* (1967) 117 L.J. 326. (Karminski J.).
[66] Merchant Shipping (Oil Pollution) Act 1971, s.15.

15–36 It has been seen that damage caused by fire and explosion is specifically exempted from the Convention,[67] and so even if oil is discharged from a tanker carrying oil in bulk as cargo, the question of whether or not it is recoverable is governed by common law.

15–37 In *The Wagon Mound*[68] the Privy Council held that on the facts of the case, fire damage was *not* foreseeable[69]; in *The Wagon Mound (No. 2)* the same court,[70] in an action arising from the same incident, held that it was. Lord Reid, in giving the advice of the Board, purported to reach this opposite conclusion on the basis that the evidence before that court was "substantially different"[71] than the evidence before the earlier court. Regrettably, his subsequent explanation remains unconvincing.

15–38 Lord Reid's remarks on the subject indicate that the question of what damage, or of what kind of damage is foreseeable, depends upon the facts of each case before the court; this is supported by subsequent cases.[72] It is not, therefore, possible to say that in future damage by fire or explosion as a result of oil spilling on to water and subsequently igniting will always be held to have been foreseeable. However, such a finding must now be regarded as a serious possibility, particularly in the light of recent incidents, such as the *Betelgeuse* in Bantry Bay in 1979 and the *Castillo del Belver* off Cape Town in 1983, where spilled oil was ignited, thus illustrating the risk to all mariners.

D. Private Nuisance

Must the Defendant Own or use Land?

15–39 In the *Southport Corporation* case, Denning L.J. said: "In order to support an action on the case for a private nuisance the defendant must have used his own land or some other land in such a way as injuriously to affect the enjoyment of the plaintiff's land,"[73] and he cited Lord Wright in support.[74] Lord Radcliffe in

---

[67] See para. 10–49, above.
[68] [1961] A.C. 388.
[69] This conclusion was also reached by Walsh J. in *The Wagon Mound (No. 2)* [1963] 1 Lloyd's Rep. 402.
[70] [1967] 1 A.C. 617; Lords Reid and Morris both sat in both cases!
[71] [1967] 1 A.C. 617 at p. 640.
[72] *e.g. Bradford* v. *Robinson Rentals Ltd.* [1967] 1 W.L.R. 337 (*per* Rees J. at 345: " . . . what were the facts known to the defendants . . . which would reasonably lead them to foresee injury to the health of the plaintiff . . . ?"); *Cook* v. *Swinfen* [1967] 1 W.L.R. 457, (C.A.), especially at pp. 461–462, *per* Lord Denning M.R.; *Lamond* v. *Glasgow Corpn.*, 1968 S.L.T. 291 especially at pp. 292–293, *per* Lord Thomson; *Tremain* v. *Pike* [1969] 1 W.L.R. 1556.
[73] [1954] 2 Q.B. 182 at p. 196.
[74] In *Sedleigh-Denfield* v. *O'Callaghan* [1940] A.C. 880 at p. 903.

the House of Lords agreed with him.[75] But Devlin J. at first instance had thought that there was no principle that the nuisance must emanate from the defendant's close,[76] and Morris L.J. seemed to agree.[77]

15–40 Even if in 1953 there was ground for doubt on the matter, there is none now: for there is a line of cases which clearly establishes that a defendant can be liable for nuisance even though it did not emanate from his land, or from another's,[78] and Devlin J.'s approach has been expressly approved.[79] Opinion is also in favour of this rule.[80]

THE TRADITIONAL TEST OF PRIVATE NUISANCE

15–41 Private nuisance has been described as the unlawful interference with a person's use or enjoyment of land, or some right over or in connection with it.[81] Both Devlin J.[82] and Denning L.J.[83] in the *Southport Corporation* case agreed that for the plaintiff to succeed, he must have a proprietary interest in land which has been interfered with. This explains why the plaintiffs in the *Wagon Mound (No. 2)* could not sue in private nuisance[84]—their injury was to their ships—but the plaintiffs in *The Wagon Mound (No. 1)*,[85] whose wharf suffered damage, did.

15–42 Proof of the nuisance shifts the burden of proof that the interference is justifiable (*i.e.* reasonable) on to the defendant. It is

---

[75] [1956] A.C. 218 at p. 242.

[76] [1956] A.C. 218 at pp. 224–225.

[77] [1954] 2 Q.B. 182 at p. 204.

[78] *Smith* v. *Great Western Ry. and Anglo-American Oil Co.* (1926) 135 L.T. 112 (oil company liable in nuisance for leakage of oil from a railway tank on another's premises); *Hall* v. *Beckenham Corpn.* [1949] 1 K.B. 716 at p. 728, *per* Finemoore J., *obiter* (saying that a member of the public flying model aeroplanes in a park could be liable in nuisance); *Newman* v. *Conair Aviation Ltd.* (1972) 33 D.L.R. (3d) 474 (aviation company spraying crops held liable in nuisance); *Gertsen* v. *Municipality of Metropolitan Toronto* (1973) 41 D.L.R. (3d) 646 (corporation whose refuse lay on another's land and caused pollution by escaping methane gas held liable).

[79] *Per* Wilson C.J.S.C. in *Newman* v. *Conair Aviation Ltd.* (1972) 33 D.L.R. (3d) 474 at pp. 479–489; *per* Lerner J. in *Gertsen* v. *Municipality of Metropolitan Toronto* (1973) 41 D.L.R. (3d) 646 at pp. 682–683. Perhaps also *per* Veale J. in *Halsey* v. *Esso Petroleum Co. Ltd.* [1961] 1 W.L.R. 683 at p. 700.

[80] [1954] C.L.J. 172 at p. 173 (C. J. Hamson); *Winfield and Jolowicz on Tort*, (12th ed.), p. 400.

[81] *Per* the late Professor Winfield; the description has been judicially adopted—see the references in *Winfield and Jolowicz on Tort*, (12th ed.), p. 380.

[82] [1956] A.C. 218 at p. 224.

[83] [1954] 2 Q.B. 182 at p. 196.

[84] Walsh J. rejected the plaintiff's claim in private nuisance [1963] 1 Lloyd's Rep. 402 at p. 427.

[85] [1961] A.C. 388. The question of liability for nuisance was remitted to the Supreme Court of New South Wales for determination, but the action was dropped.

this aspect of the tort which has given it its traditional advantage over negligence, where the plaintiff must prove that the defendant's conduct was below that of the reasonable man. As one learned writer has said "the distinguishing aspect of nuisance, as compared with other heads of liability like negligence, is that it looks to the harmful result rather than to the kind of conduct causing it."[86] Prove your nuisance, and liability is established unless the defendant can exculpate himself. No defendant will be able to show that oil pollution is reasonable—it is simply not that kind of interference.

FORESEEABILITY OF DAMAGE

15–43 Although the *Wagon Mound* (*No.* 2) was a case in public nuisance, what it had to say about foreseeability of damage seems to apply equally to private nuisance, and there are dicta in the case to that effect.[87]

E. The Boundaries of Negligence and Nuisance

15–44 It is clear that as a matter of practice it will often be the case that facts which are in some way relevant to the proof or disproof of nuisance may also be relevant to the proof or disproof of negligence. At its simplest, nuisance shifts the burden of proof of the reasonableness of behaviour to the defendant, but the same facts may be at issue. Further, there may be special circumstances which negative a finding of nuisance, thereby raising the issue of negligence as a matter to be proved by the plaintiff. So, in the *Southport Corporation* case, all the judges except Denning L.J. were agreed that the question of liability depended upon whether or not the defendants had been negligent—because they held that either the traffic cases[88] applied or that the defence of necessity had been established.[89] But this intricacy makes for possible confusion.[90] In the *Wagon Mound* (*No.* 2), Lord Reid had much to say on the issue of when negligence might need to be proved in nuisance and this aspect of his judgment cannot be said to have clarified matters.

15–45 Walsh J., in the Supreme Court of New South Wales, had

---

[86] J. G. Fleming, *The Law of Torts*, (4th ed.), p. 338.
[87] [1967] A.C. 617 at pp. 636–637, 638, and 638–640, *per* Lord Reid.
[88] See above, paras. 15–18 *et seq.*
[89] See above, paras. 15–22 and 15–23.
[90] See R. M. Dias, "Trouble on Oiled Waters" [1967] C.L.J. 62. The classic academic investigations of the area are to be found in P. H. Winfield, "Nuisance as a Tort" [1931] C.L.J. 189, and F. H. Newark, "The Boundaries of Nuisance" (1949) 65 L.Q.R. 480.

held[91] that the fire damage to the plaintiff's vessels was not reasonably foreseeable by those for whose acts the defendants were responsible, but that this did not affect the defendants' liability in nuisance, because foreseeability of damage was not, in his view of the law, a prerequisite to recovery in that tort; however it was in negligence, and so he held that the defendants were not liable in negligence.

15–46 Now Lord Reid, delivering the advice of the Board, addressed himself to the question of whether or not foreseeability of damage was a prerequisite to recovery in nuisance.[92] After his own review of the cases,[93] he concluded: "In their Lordships' judgment the cases point strongly to there being no difference as to the measure of damages between nuisance and negligence *but they are not conclusive*"[94] (author's emphasis). So it was desirable to consider the question in principle, and it was in the course of so doing that the trouble began. It is important to realise that at this stage Lord Reid is still addressing himself to the question of foreseeability of damage. This question concerns not whether the plaintiff has to prove something about the actual quality of a defendant's conduct to succeed in nuisance, but whether, given the events which did take place, the damage actually suffered by the plaintiff ought to have been contemplated by a reasonable man.

15–47 Lord Reid said:

> "Comparing nuisance with negligence, the main argument for the respondent was that in negligence foreseeability is an essential element in determining liability and therefore it is logical that foreseeability should also be an essential element in determining the amount of damages: but negligence is not an essential element in determining liability for nuisance and therefore it is illogical to bring in foreseeability when determining the amount of damages. It is quite true that negligence is not an essential element in nuisance. Nuisance is a term used to cover a wide variety of tortious acts or omissions and in many negligence in the narrow sense is not essential. An occupier may incur liability for the emission of noxious fumes or noise although he has used the utmost care in building and using his premises. The amount of fumes or noise which he can lawfully emit is a question of degree and he or his advisers may have miscalculated what can be justified. Or he may deliberately obstruct the highway adjoining his premises to a greater degree than is permissible, hoping that no one will object. On the other hand the emission of fumes or

---

[91] [1963] 1 Lloyd's Rep. 402.
[92] [1967] 1 A.C. 617 at pp. 633–634.
[93] *Ibid.* at pp. 634–638.
[94] *Ibid.* at p. 638.

noise or the obstruction of the adjoining highway may often be the result of pure negligence on his part: there are many cases (*e.g. Dollman* v. *A. and S. Hillman Ltd.*[95]) where precisely the same facts will establish liability both in nuisance and in negligence. And although negligence may not be necessary, fault of some kind is almost always necessary and fault generally involves foreseeability, *e.g.*, in cases like *Sedleigh-Denfield* v. *O'Callaghan*[96] the fault is in failing to abate the nuisance of the existence of which the defender is or ought to be aware as likely to cause damage to his neighbour. (Their Lordships express no opinion about cases like *Wringe* v. *Cohen*,[97] on which neither counsel relied.) The present case is one of creating a danger to persons or property in navigable waters (equivalent to a highway) and there it is admitted that fault is essential—in this case the negligent discharge of the oil.

'But how are we to determine whether a state of affairs in or near a highway is a danger? This depends, I think, on whether injury may reasonably be foreseen. If you take all the cases in the books, you will find that if the state of affairs is such that injury may reasonably be anticipated to persons using the highway it is a public nuisance' (*per* Lord Denning M.R. in *Morton* v. *Wheeler*[98]).

So in the class of nuisance which includes this case foreseeability is an essential element in determining liability.

It could not be right to discriminate between different cases of nuisance so as to make foreseeability a necessary element in determining damages in those cases where it is a necessary element in determining liability, but not in others. So the choice is between it being a necessary element in all cases of nuisance or in none. In their Lordships' judgment the similarities between nuisance and other forms of tort to which *The Wagon Mound* (*No.* 1) applies far outweigh any differences, and they must therefore hold that the judgment appealed from is wrong on this branch of the case. It is not sufficient that the injury suffered by the respondents' vessels was the direct result of the nuisance if that injury was in the relevant sense unforeseeable."

15–48 In his consideration of principle, Lord Reid was seeking to find common ground between nuisance and negligence, so that he could conclude that, as foreseeability of damage was a prerequisite to recovery in the latter (see *The Wagon Mound* (*No.* 1)[99]), it ought to be in the former. He found the common ground in that "fault" is required to establish liability in each[1]: " . . . although negligence

---

[95] [1941] 1 All E.R. 355 (C.A).
[96] [1940] A.C. 880.
[97] [1940] 1 K.B. 229 (C.A.).
[98] (1956) January 31, (unreported), (C.A.) (No. 33 of 1956).
[99] [1961] 1 A.C. 388.
[1] [1967] 1 A.C. 617 at p. 639.

may not be necessary, fault of some kind is almost always necessary." Having done so, he completed his object by continuing " . . . and fault generally involves foreseeability."

15–49 It is perfectly clear that Lord Reid understood that not all fault is negligent fault: the sentence just quoted proves, as does his previous statement, that "it is quite true that negligence is not an essential element in nuisance."[2] The person who commits a nuisance may be legitimately described as being at fault, even though he has not been negligent—and indeed the traditional concept is that if he has committed a nuisance, proof that he has taken all reasonable care will not avail him, because his task is to show that the nuisance itself, the state of affairs complained of, is reasonable.

15–50 The difficulty arises only in that Lord Reid envisages that, for some types of nuisance, the "fault" encompassed therein involves foreseeability. Hence, in some cases (including "this case") "foreseeability is an essential element in determining liability."[3] And fault involving foreseeability may be difficult to distinguish from negligence.

15–51 There may be two ways of avoiding this apparent conflict. First, it can be argued that, to the extent that Lord Reid's comments are made in connection with the issue of the recoverability of damage and not the issue of the establishment of liability, they are *obiter* and, further, cannot stand with other explicit statements in the same judgments (notably "negligence is not an essential element in determining liability for nuisance").[4]

15–52 If, on the other hand, Lord Reid's remarks are to be taken as supporting the proposition that there are circumstances in which negligence is to be proved in establishing nuisance, then it may be that he means no more than that there are situations in which the facts which establish nuisance also happen to establish negligence. Thus, to quote Lord Reid once more:

> "emission of fumes or noise or the obstruction of an adjoining highway may often be the result of pure negligence on his part: there are many cases (*e.g. Dollman* v. *Hillman* [1941] 1 All E.R. 355) where precisely the same facts will establish liability in both nuisance and negligence."[5]

If that be so, the overlap is truly accidental.

---

[2] *Ibid.* at p. 639.
[3] *Ibid.*
[4] *Ibid.*
[5] *Ibid.*

15–53 If neither of the above arguments will convince, then it will be necessary to identify those circumstances in which, as a matter of law, the plaintiff needs to establish the negligence of the defendant in order to proceed successfully in nuisance. It is clear that in the *Wagon Mound* (*No.* 2) Lord Reid, who identifies the "fault" as "the negligent discharge of the oil," classifies the case as "one of creating a danger to persons or property in navigable waters (equivalent to a highway)"[6] It would seem that Lord Reid is doing no more than treating the so-called traffic rule (see above, paragraphs 15–18 *et seq*.), as a special case of nuisance requiring proof of negligence rather than a defence to an allegation of nuisance. In either case, the result is the same. If there are to be other analogous examples, then it will be necessary to generalise: but Lord Reid gives us no guidance. It may be that the principle lies in the fact that in so far as the defendant's conduct, which caused the damage complained of as a nuisance, consists of his exercising a public right, such as that of using a public highway, it is prima facie lawful. It would follow that proof of a further element of "fault," in Lord Reid's terms, would be necessary.

15–54 A generalisation at a further level of abstraction was provided by the learned authors of *Winfield and Jolowicz on Tort*[7]:

> "There are not two separate categories of nuisance, one fault-based and the other strict, but one principle that the defendant is liable if his interference with his neighbour's land is of sufficient gravity to constitute a nuisance in law and he is responsible for the interference in the sense that he knew or ought to have known of a sufficient likelihood of its occurrence to require him to take steps to prevent it."

15–55 It is respectfully submitted that, if there are cases of nuisance where negligence must be proved by the plaintiff as a necessary element in the tort, outside a novel formulation of the "traffic rule," then the above test forms the only rational basis for identifying such cases.

CONCLUSION ON PUBLIC AND PRIVATE NUISANCE

15–56 There is no binding precedent to suggest that the traditional basis of nuisance has been changed: the remarks of Lord Reid in *Wagon Mound* (*No.* 2) are either *obiter* and to be confined strictly to the context in which they were made, namely, to help establish the foreseeability of damage rule in nuisance, or are not to be taken as adding to the law of nuisance any further example

---

[6] *Ibid.*
[7] (12th ed.), p. 387.

beyond the "traffic rule" where negligence must be proved to establish liability. There is no evidence that the courts are willing to develop the law so as to bring together the torts of negligence and nuisance in any general way.

15–57 It is therefore reasonable to conclude that serious oil pollution incidents, including those which are in fact accidental, will be capable of founding actions for damages in public nuisance, without proof of negligence. Oil pollution incidents interfering with an interest in land appear capable of founding an action in private nuisance. However, it must remain possible that a court will follow the "traffic" cases and consequently hold in some cases that the plaintiff must prove negligence.

15–58 The barriers to a plaintiff trying to recover without proof of negligence are therefore considerable. Trespass to land is probably not available. In nuisance, he must fight arguments that the "traffic cases" apply, or that he must show that the defendant knew, or ought to have known that what he did was likely to cause the plaintiff actionable damage—or resort to negligence if he cannot.

F. Negligence

DUTY OF CARE

15–59 Traditionally, the tort of negligence is formulated in terms of a duty of care. Proof of liability in negligence requires proof of the existence of a binding legal duty, a failure in that duty and consequential damage. Duties of care exist in a wide variety of different circumstances: the widest description of those circumstances is to be found in the celebrated *obiter dictum* of Lord Atkin in *Donoghue* v. *Stevenson*,[8] the so-called "neighbour principle," the source of modern negligence. A duty of care will be owed to those whom one ought, as a reasonable man, to have in one's contemplation as being likely to be affected by one's acts or omissions. In short, the broadest principle of negligence would base liability upon foreseeability.

15–60 Since the decision of the Privy Council in the *Wagon Mound* (*No.* 1)[9] it has been clear that the allied question of remoteness of damage is also to be approached through the test of foreseeability. Thus, as a general rule, foreseeability of damage will establish both

---

[8] [1932] A.C. 562 at p. 580.
[9] [1961] A.C. 398.

liability and recoverability. However, the two issues are distinct. As we have seen in relation to liability in nuisance, in the judgment of Lord Reid in the *Wagon Mound (No. 2)*,[10] confusion can easily occur if the two issues are not kept distinct.

15–61 The broad neighbour principle has to be qualified. There are several examples of circumstances in which, although damage is foreseeable, yet there is no duty of care, or the duty must be formulated with qualifications or extensions. Thus, the occupier of land owes a duty of care to lawful visitors on his land, based upon foreseeability, but his duty towards trespassers cannot be so simply formulated, even when the presence of trespassers is entirely foreseeable.[11] There may be no duty owed to those outside the area of likely physical impact of a street accident who suffer physical consequences of the nervous shock induced by witnessing the accident.[12] The duty of care owed by those who make careless statements may be owed only to those in a relationship close enough to be described as "equivalent to contract" and the duty would seem to be subject to the right to exclude liability by the use of an appropriate disclaimer.[13] Finally, for many years it was thought that there was no duty to avoid purely economic loss.[14]

15–62 If the types of damage likely to be caused by an oil spill are considered under the four general headings of physical damage, environmental damage, economic loss and reinstatement costs,[15] then it is only the economic loss rule which is likely to have substantial importance. For that reason, that issue will be treated separately below. At this point, however, it is worth emphasising that in oil pollution cases there should be no duty of care problem in the case of physical damage to property, whether immovable property such as the foreshore or harbour installations, or movable property such as vessels, fishing gear, etc., or with the much rarer case of

---

[10] [1967] 1 A.C. 617: see above, paras. 15–44 *et seq.*

[11] See Occupiers' Liability Act 1957, s.1 and, for example, Infant trespassers and trespassers on railway premises seem to be treated differently: see *British Railways Board* v. *Herrington* [1972] A.C. 877. See now Occupiers Liability Act 1984, s.1.

[12] See, classically, the decision of the House of Lords in *Bourhill* v. *Young* [1943] A.C. 92, where a bystander in no physical danger from a street accident was not entitled to recover for nervous shock. The law has been substantially re-examined in *McLoughlin* v. *O'Brian* [1982] 2 W.L.R. 982, in which the mother of a family who suffered nervous shock as a result of witnessing the injuries to other members of her family caused by a road accident in which she was not involved did recover damages. The law, however, remains unclear.

[13] *Hedley Byrne and Co. Ltd.* v. *Heller and Partners Ltd.* [1964] A.C. 465.

[14] See below, paras. 15–63 *et seq.*

[15] A fuller analysis of types of loss which may be suffered in oil pollution cases is contained in paras. 10–42 *et seq.*

personal injury. Assuming the practical difficulties of proof can be met,[16] if the damage is reasonably foreseeable (as it generally will be) there is no reason why an action might not be brought. However, all the other types of loss or damage involve matters other than the physical damage to property owned or occupied by the plaintiff or personal injuries and must thus be classified, for the purposes of the common law, as economic loss.

PURE ECONOMIC LOSS

15–63 The classical objection to the imposition of a duty of care to avoid pure economic loss is that expressed by Cardozo C.J. in *Ultramares Corporation* v. *Touche*, that it would involve the creation of "liability in an indeterminate amount for an indeterminate time to an indeterminate class."[17] The objection came to be known as the "floodgates argument." However well or ill justified as a matter of judicial policy, it remained true that negligence causing purely pecuniary loss was seen as presenting particular problems. Even when the loss was clearly foreseeable and no causation problems were present, recovery was generally denied.[18]

15–64 The objection is not present in the same force in the case of economic consequences of physical damage. There, assuming that the economic consequences are not unforeseeable and therefore too remote,[19] there will generally be recovery. This aspect of the question was exhaustively examined by the Court of Appeal in 1973 in the case of *Spartan Steel and Alloys Ltd.* v. *Martin and Co. (Contractors) Ltd.*[20] Power supplies to the plaintiff's factory had been negligently interrupted by the defendants. As a result, their furnace had to be closed down and the "melt" then in progress dumped. The court allowed the recovery of the value of the dumped "melt" and the loss of expected profits on its sale, but disallowed a claim for loss of profit on four further "melts" which could not be carried out because the furnace was out of action. Economic loss immediately consequent upon physical damage was recoverable, but more remote economic losses were not. They were too far from the physi-

---

[16] See below, paras. 15–87 *et seq.*

[17] (1931) 255 N.Y. 170 at 179. The argument was seen as particularly apt in connection with negligent statements whose consequences are commonly purely financial. See *Hedley Byrne and Co. Ltd.* v. *Heller and Partners Ltd.* [1964] A.C. 465.

[18] See *Weller* v. *Foot and Mouth Disease Research Institute* [1966] 1 Q.B. 569; *Margarine Union GmbH* v. *Cambay Prince SS. Co. The Wear Breeze* [1967] 2 Ll. Rep. 315.

[19] See, in this connection *The Liesbosch Dredger (Owners)* v. *The Edison (Owners)* [1937] A.C. 449.

[20] [1973] Q.B. 27.

cal damage and, if treated independently, they fell foul of the principle denying a duty of care to avoid purely economic loss.[21]

15–65 In 1982 the House of Lords took the opportunity to consider the wider issue in *Junior Books Ltd.* v. *Veitchi Co. Ltd.*[22] The case concerned the negligence of a firm of specialist floor-layers who had so badly executed their contract with the head-contractor in the construction of a new building that the plaintiffs, for whom the building had been built, were put to expense in repair and replacement of the flooring. There was no physical damage, so the principle of *Spartan Steel* could not be called in aid. However, the plaintiffs succeeded. Clearly the decision establishes that there are cases in which foreseeable economic loss, without physical damage, is recoverable. It is less clear how those cases may be defined.

15–66 The major speech for the majority was delivered by Lord Roskill.[23] He rejected the "floodgates" argument and took as his starting point the broad *dictum* of Lord Wilberforce in *Anns* v. *Merton Borough Council*[24]:

> "The position now has been reached that in order to establish that a duty of care arises in a particular situation, it is not necessary to bring the facts of that situation within those of previous situations in which a duty of care has been held to exist. Rather the question has to be approached in two stages. First one has to ask whether, as between the alleged wrongdoer and the person who has suffered damage there is a sufficient relationship of proximity or neighbourhood such that, in the reasonable contemplation of the former, carelessness on his part may be likely to cause damage to the latter, in which case a prima facie duty of care arises. Secondly, if the first question is answered affirmatively, it is necessary to consider whether there are any considerations which ought to negative or reduce or limit the scope of the duty or the class of person to whom it is owed or the damages to which a breach of it may give rise."

15–67 In brief, reasonable foreseeability of damage creates a prima facie duty of care, which will lead to recovery unless there is

---

[21] Before *Spartan Steel* was decided, a general "threshold tort" theory attracted some support: that once *some* physical damage consequent on a negligent act was proved then all other foreseeable losses, even possibly suffered by other plaintiffs, which could be causally related to the same act were recoverable. The decision of the Court of Appeal in *Spartan Steel* clearly stopped the development of that idea, which would have had clear application to the problem of the "hotel keeper's lost profits" in an oil spill. The hotel keeper who has no interest in the beach that is fouled but who suffers loss of business as a result would have benefited from such a "threshold tort" argument.

[22] [1983] 1 A.C. 520.

[23] Lords Fraser and Russell agreed with Lord Roskill. Lord Brandon dissented. Lord Keith delivered a shorter concurring speech.

[24] [1978] A.C. 728 at p. 751.

a good policy reason for denying recovery.[25] Nonetheless, Lord Roskill, who throughout his judgment preferred the word "proximity" to the phrase "reasonable foreseeability," clearly felt that the question of the "prima facie duty of care" was not settled by a simple application of the old "neighbour principle" to the facts of the case. He found it significant that there was a degree of knowledge of the existence and identity of each other between the plaintiff and defendant and that the plaintiff had relied upon the skill of the defendant. So:

"Turning back to the present appeal I therefore ask first whether there was the requisite degree of proximity so as to give rise to the relevant duty of care relied on by the respondents. I regard the following facts as of crucial importance in requiring an affirmative answer to that question: (1) the appellants were nominated sub-contractors; (2) the appellants were specialists in flooring; (3) the appellants knew what products were required by the appellants [sic: surely respondents] and their main contractors and specialised in the production of those products; (4) the appellants alone were responsible for the composition and construction of the flooring; (5) the respondents relied upon the appellants' skill and experience; (6) the appellants as nominated sub-contractors must have known that the respondents relied on their skill and experience; (7) the relationship between the parties was as close as it could be short of actual privity of contract; (8) the respondents must be taken to have known that if they did the work negligently (as it must be assumed that they did) the resulting defects would at some time require remedying by the respondents expending money on the remedial measures as a consequence of which the respondents would suffer financial or economic loss."[26]

15–68 All that can be said with certainty is that there is no reason in principle why pure economic loss, without physical damage, may not be recovered in an action in negligence. Whether in a case arising out of an oil spill, for example the hotel keeper's claim for his lost profits resulting from the fouling of a beach which

---

[25] Doubt has been cast upon the validity of this approach by Oliver L.J. in *Leigh and Sillay Ltd.* v. *Aliakmon Shipping Co. Ltd.*, *The Aliakmon* [1985] 2 All E.R. 44 at p. 62, a case in which the Court of Appeal denied recovery to a cargo receiver without a bill of lading who sued the shipowner carrier for negligence, a loss which fell to be treated as pure economic loss since the plaintiff had no title to the goods, thus affirming the correctness of the decision in *Margarine Union GmbH* v. *Cambay Prince SS. Co.*, *The Wear Breeze* [1967] 2 Lloyd's Rep. 315 and overruling *Schiffart und Kohlen GmbH* v. *Chelsea Maritime Ltd*, *The Irene's Success* [1982] Q.B. 481. The Appeal Committee of the House of Lords has granted leave to appeal in *The Aliakmon:* it may be that the House will take the opportunity of re-examining both *Anns* v. *Merton London Borough* and *Junior Books* v. *The Veitchi Co.*

[26] [1983] 2 A.C. at p. 561. One is reminded of Lord Devlin's phrase "equivalent to contract" in *Hedley Byrne and Co. Ltd.* v. *Heller and Partners Ltd* [1964] A.C. 465, expressly relied upon by Lord Roskill in *Junior Books*. See now *The Aliakmon* [1985] 1 Lloyd's Rep. 199.

he does not own, the courts would find a duty of care remains in doubt. Such losses may easily be shown to be a foreseeable consequence of the negligent spill, but it is unlikely that as close a degree of proximity as was discovered in *Junior Books* would be found. Plaintiff and defendant are likely to be unknown to each other. There is unlikely to be reliance in any legally recognised sense of plaintiff on defendant for the safe carriage of the oil past the beach in question. Rarely will it be possible to say that the parties had any relationship at all, and never that it was "as close as it could be short of actual privity." It must be concluded that the existence of a prima facie duty of care to avoid purely economic loss in an oil spill case, although a possibility after *Junior Books*, remains unlikely.

15–69 Even if such a prima facie duty be held to exist, the plaintiff would still have to surmount the second of Lord Wilberforce's hurdles and show that there is no policy reason for denying liability. In this context, the rule of policy which was called in aid in the *Southport Corporation* case, the so-called "traffic rule" that the harmful consequences of the exercise of the public right of navigation should not be actionable, like the consequences of the exercise of the public right of use of the highway, might be seen to have a part to play.[27]

PURE ECONOMIC LOSS IN RELATION TO STATUTORY CLAIMS

15–70 Claims under the Merchant Shipping (Oil Pollution) Act 1971 may be made "for any damage caused in the area of the United Kingdom by contamination." Since, by section 20, the word "damage" includes "loss," it would seem that there is no statutory reason for excluding claims for pure economic loss. Such loss must, however, result from contamination caused by the spill, which raises a question of causation not specifically answered in the statute itself. In the absence of precedent, the question whether claims such as that for the hotelier's lost profits or the holiday-maker's loss of holiday value must be regarded as open. Clearly it would be open to a court to hold that the hotelier's losses were not "caused by" the contamination: it is, however, submitted that it is unlikely that it would do so. The same principles must apply to claims for damage resulting from preventive measures, since the same language is used: "any damage caused in the area of the United Kingdom by any measures so taken."

---

[27] See *Esso Petroleum* v. *Southport Corporation* [1956] A.C. 218; *Overseas Tankship (U.K.) Ltd.* v. *Miller SS. Co. Pty. Ltd, the Wagon Mound (No. 2)* [1967] 1 A.C. 617, and above, paras. 15–07 *et seq.*

15–71 Claims for the cost of preventive measures are also permissible under the Act. Such claims are clearly claims for pure economic loss.

G. Causation

15–72 As has been indicated, the question of causation arises both in statutory and common law claims. Under the Merchant Shipping (Oil Pollution) Act 1971, the damage must be "caused" by the contamination, etc. At common law, the loss or damage must be caused by the tort, the trespass, nuisance or act of negligence.

15–73 Assuming the existence of a duty of care, liability will depend upon proof that the allegedly negligent act has caused the loss in question. Causation is a complex topic in the common law, and little attempt has ever been made to evolve a coherent theory of causation, the courts in general preferring the flexible imprecision of "effective cause" to the continental subtleties of "*causa causans*" and "*causa sine qua non.*"[28]

15–74 In oil-spill cases, it is possible for the argument to be raised that the spill was not the true cause of the pollution: some other factor, such as the failure by some other person or authority to mount effective clean-up operations or to deal with adequate celerity with the threat of the spill after the shipping casualty that preceded it may have a role. It is now plain, however, that in law an event can have more than one cause,[29] and so the attractiveness which this argument may have had in the past for judges who adhere to the sole or monistic theory of causation[30] can no longer be relied upon. No attempt will be made here to assess the chances which the argument might have in an action today, for, as one writer has put it, "the whole approach to the concept of causation is unfettered and permits infinite variation with freedom to choose one or more causes out of a selection of factors which contributed towards the actual happening of the event."[31] It is sufficient to

---

[28] See H. L. A. Hart and A. M. Honore, *Causation in the Law*, (Oxford, 1959), also substantially printed at (1956) 72 L.Q.R. 58, 260, 398. For a typical "common-sense" approach to causation in the House of Lords in an analogous situation, see *Yorkshire Dale SS. Co. Ltd.* v. *Minister of War Transport* (*The Coxwold*) [1942] 2 All E.R. 6.

[29] See, *e.g.* the majority decision in the House of Lords in *Stapley* v. *Gypsum Mines Ltd.* [1953] A.C. 663, and the unanimous decision of the House of Lords in *Baker* v. *Willoughby* [1970] A.C. 467.

[30] Notably Lord Wright: see "Notes on Causation and Responsibility in English Law," [1955] C.L.J. 163, emphasising the importance of this theory.

[31] R. A. Percy (ed.), *Charlesworth on Negligence*, (5th ed.), para. 77.

point out that, should a modern bench be inclined to take the view urged by the argument, there are precedents to support such a view.[32]

15–75 A second argument which invites a similar view of the events but which is quite opposite in its approach, is that the pollution was the result, not of the failure to keep the oil in, but of the stranding; and that, while the failure to keep the oil in may or may not be admittedly negligent, the stranding itself was not negligent. This argument will be particularly attractive to the defendant where the ship has gone aground as a result of the negligence of another vessel, as may be the case when, for instance, the stranding has resulted from a course alteration made to avoid another ship proceeding in the wrong direction down a traffic separation scheme.[33]

15–76 Another argument may be stated in this way: that the casualty and the spillage must be regarded as a single, continuing event, and that, because all reasonable efforts were made after the spillage to contain the oil, there has been no breach of the duty of care. This argument implicdly formulates the duty of care for breach of which the plaintiff must sue as being a duty to take reasonable care to stop any spilled oil reaching a place where it affects a plaintiff, rather than being a duty to take care that oil does not spill at all.

15–77 Naturally, the chances of any one of these arguments succeeding in any given case will depend heavily on the individual circumstances which pertain. In addition, it may be difficult to persuade the court to accept the formulation of duty and responsibility relied upon. This may be particularly true of the third argument described above. They do, however, illustrate some of the legal difficulties which may be encountered in proving negligence in an oil pollution case.

H. Proof of Negligence: Res Ipsa Loquitur

15–78 As a general rule in negligence, the burden of proving that the defendant's behaviour fell below the standard of care imposed

---

[32] *e.g. Radley* v. *London and North Western Ry.* (1876) 1 App. Cas. 754 (H.L.); *British Columbia Electric Ry. Co.* v. *Loach* [1916] 1 A.C. 719, (P.C.); *Swadling* v. *Cooper* [1931] A.C. 1 (H.L.); *Quinn* v. *Burch Bros (Builders) Ltd.* [1966] 2 Q.B. 370 (C.A.); *Norris* v. *W. Moss & Sons Ltd.* [1954] 1 W.L.R. 346 (Vaisey J.).

[33] This is thought to have been the cause of the collision between *H.M.S. Achilles* and the *Olympic Alliance* in the Dover Strait on November 12, 1975. See *The Times*, November 13, 1975, p. 1; *The Guardian*, November 13, 1975, p. 1; *The Times*, February 19, 1976, p. 4.

upon him by the legal duty of care rests with the plaintiff. He who alleges must prove. However, there are exceptional circumstances in which the facts of the damage suffered point inexorably to negligence: where it seems unlikely that the event could have occurred without negligence on the part of the defendant or some person for whom he was responsible. In such a case, the burden of proof is shifted to the defendant and it rests with him to satisfy the court that, despite appearances, the event did occur without negligence for which he was responsible.

15–79   The classic statement of this principle, *res ipsa loquitur*, the thing (or the facts) speaks for itself, is to be found in the old case of *Scott* v. *London and St. Katherine's Docks Co.*,[34] where a passer-by was injured by six bags of sugar which fell upon him from a warehouse teagle opening. His action succeeded and Erle C.J. said[35]:

> "There must be reasonable evidence of negligence. But where the thing is shown to be under the management of the defendant or his servants and the accident is such that in the ordinary course of things does not happen if those who have the management use proper care, it affords reasonable evidence, in the absence of explanation by the defendant, that the accident arose from want of care."

15–80   The principle requires, as one modern authority has it "that the mere fact of the accident happening should tell its own story and raise the inference of negligence so as to establish a *prima facie* case against the defendant."[36] In detail, the circumstances giving rise to the event complained of must be in some sense under the control of the defendant, the accident must be such as not to admit of a non-negligent explanation, and no such explanation must be offered or be available.

15–81   Collisions and strandings of ships quite often seem to fit these conditions. The defendant shipowner is in control, through his servants, of the ship whence the oil emanated. Well-navigated, well-maintained, well-run ships do not collide with other ships or go aground. In the absence of explanation, such as the stress of very heavy weather, the prima facie case would seem to be made. And, particularly in cases of jammed steering-gear or other mechanical failure aboard the accused vessel, the principle can often be applied.[37]

---

[34] (1865) 3 H. & C. 596.
[35] *Ibid.* at p. 601.
[36] See *Winfield and Jolowicz on Tort*, (12th ed.), p. 108.
[37] See for example *The Merchant Prince* [1892] P. 179, much relied upon in the *Southport Corporation* Case.

15–82 The argument was raised in the *Southport Corporation* case.[38] The vessel had gone aground as a result of a defect in her steering brought about by a fractured stern frame. The allegation of negligence was, however, made against the master in respect of the discharge of the oil after the grounding, and against the owners only in respect of their legal responsibility for the negligence of their servant, the master. At first instance, Devlin J. refused to apply the maxim so as to impose upon the owners the heavy duty of showing that their maintenance and management of the ship was without negligence. There had to be a primary allegation of negligence and the principle *res ipsa loquitur* was to be applied only within the bounds of that allegation. In this, he was upheld in the House of Lords.

15–83 This decision, by concentrating upon the primary require-ment that the plaintiff raise some particular question of negligence before the principle can apply, had the effect of reducing the practi-cal usefulness of the doctrine. It is insufficient to point to the oil spilt on the foreshore. It is necessary to identify some circumstances under the control of the defendant which colourably gave rise to the spill before the defendant can be called upon to justify his behav-iour.

I. Negligence and the Navigation Regulations

15–84 The current navigation regulations are the International Regulations for Preventing Collisions at Sea 1972 as amended by Resolution A464(XII) of IMCO. They were brought into effect in the case of United Kingdom registered vessels on July 15, 1977 and, in respect of other vessels the date of entry into force of the regulations in the flag state. As a matter of English law the regula-tions currently have force in virtue of the Merchant Shipping (Dis-tress Signals and Prevention of Collision Regulations, 1983)[39] which came into force on June 1, 1983.

15–85 The regulations provide rules and guidance in steering and sailing rules, lights and shapes, distress signals and other mat-ters. By Rule 10 vessels are obliged to obey the rules of traffic separ-ation schemes, and by Regulation 1(2)(b) a traffic separation scheme within the rules is one adopted by IMO and declared as so adopted by a Notice to Mariners.

15–86 Before 1911, the rule was that a breach of the Regulations

---

[38] [1956] A.C. 218. See above paras. 15–07 *et seq.*
[39] S.I. 1983 No. 708, made under s.21 of the Merchant Shipping Act 1979.

raised a presumption of causative fault; section 419(4) of the Merchant Shipping Act 1894 enacted that "where, in a case of collision, it is proved to the court before which the case is tried that any of the collision regulations have been infringed, the ship by which the regulation has been infringed shall be deemed to be in fault, unless it is shown to the satisfaction of the court that the circumstances of the case made departure from the regulation necessary." This rule has now been abolished by section 4 of the Maritime Conventions Act 1911, and so the question nowadays is, did the failure to obey the regulations contribute to the damage suffered?[40] If not, the failure will not affect the question of civil liability; if so, the question will be whether the failure was negligent. Hence, the issue of negligence must be met at common law without the assistance of presumptions, and a plaintiff will be unable to avail himself of the fact that in almost every casualty some breach of the regulations (albeit usually a minor one) can be shown.

J. Evidence: the Marine Enquiry

15–87 Enquiries into maritime casualties are now governed by sections 55 to 58 of the Merchant Shipping Act 1970, as amended by section 32 of the Merchant Shipping Act 1979.[41] These provisions replaced the law contained within Part VI of the Merchant Shipping Act 1894. The new law defines shipping casualty so as to include "the loss or presumed loss, stranding, grounding, abandonment of or damage to a ship; or loss of life caused by fire on board or by any accident to a ship . . . or by any accident occurring on board . . . ; or any damage caused by a ship," if the ship is United Kingdom registered or in United Kingdom waters.[42] The 1894 Act had required that the casualty occur "on or near the coasts of the United Kingdom."[43] In addition the Merchant Shipping (Prevention of Oil Pollution) Order 1983[44] made under the Merchant Shipping Act 1979, applies the regime of marine enquiries to oil discharges made in contravention of the Prevention of Oil Pollution Regulations,[45] which gives effect to the 1973 International Convention for the Prevention of Pollution from Ships.

---

[40] See *The Heranger (Owners)* v. *The Diamond (Owners)* 1939 A.C. 94 for a classic exposition of the position.

[41] The 1979 Act amended the 1970 provisions so as to include so-called "near misses" which were defined as "events . . . of a kind likely to cause events which, if they occurred, would constitute a shipping casualty." The 1970 legislation was not brought into force until July 1, 1983 (S.I. 1982 No. 1617).

[42] Merchant Shipping Act 1970, s.55.

[43] Merchant Shipping Act 1894, ss.464–465.

[44] S.I. 1983 No. 1106, Reg. 5 in force October 2, 1983.

[45] S.I. 1983 No. 1395.

In relation to civil proceedings, the question arises as to how far the results of the marine enquiry may be utilised. The position seems to be as follows.

15–88 The findings of a court of inquiry expressed in documentary form, or the oral testimony at subsequent civil proceedings of the officer who conducted the inquiry, would be regarded at common law as opinion evidence or hearsay or both. A finding that ship X failed to alter course to starboard would be opinion evidence, as would a finding that this constituted negligence; the recorded evidence of an eye-witness to the effect that ship X failed to alter course to starboard would be hearsay at the subsequent civil proceedings.

15–89 Many of the more valuable parts of the report are, therefore, likely to be opinion evidence in subsequent civil proceedings, and as such are excluded at common law unless (a) the evidence concerns matters which call for the special skill or knowledge of an expert and the witness is an expert on such matters, or (b) the facts upon which that opinion is based cannot be stated without reference to the opinion in a manner equally conducive to the ascertainment of the truth.[46] This latter exception is given statutory clarification by section 3(2) of the Civil Evidence Act 1972:

> "It is hereby declared that where a person is called as a witness in any civil proceedings, a statement of opinion by him on any relevant matter on which he is not qualified to give expert evidence, if made as a way of conveying relevant facts personally perceived by him, is admissible as evidence of what he perceived."

15–90 These common law exceptions do not cover findings such as the ones mentioned above, and the formulation in section 3(2) of the Civil Evidence Act 1972 does not cover them because the officer who conducted the inquiry did not personally perceive them: although an eye-witness might be permitted to give oral testimony at the trial of the civil action as to what was the speed of ship X, or her course, a finding of the court of inquiry at which such evidence had already been given, to the same effect, would be inadmissible.

This exclusion is not affected by section 1 of the Civil Evidence Act 1972, for, although it permits hearsay evidence of opinion to be given by extending the provisions of sections 2 and 4 of the Civil Evidence Act 1968, it is necessary to start the chain of record with a matter of which direct oral evidence would be admissible—and as

---

[46] *Sherrard* v. *Jacob* [1965] N.I. 151 at pp. 157–158, *per* Lord Macdermott, cited by Sir Rupert Cross, *Evidence*, (4th ed.) 381.

has been seen, direct oral evidence by the officer who held the inquiry would be inadmissible.

15–91 It therefore appears that the findings of the marine inquiry, which are potentially so valuable to prospective parties to a civil action for oil pollution damage, would be inadmissible in those proceedings.

15–92 Certain perhaps less valuable elements of the report will not be opinion evidence, but will be hearsay—for instance, the recorded evidence of eye-witnesses as to what actually took place. This too will be excluded at common law unless it can be fitted into one of the recognised exceptions to the hearsay rule. The question arises, therefore, whether statements of this nature contained in the report are within the "public documents" exception to the hearsay rule.

15–93 There is a longstanding exception, retained for civil proceedings by the Civil Evidence Act 1968[47] that public documents, including inquisitions, surveys, assessments, reports and returns, are admissible but not conclusive as to their contents when made under public authority. For a document to qualify under this exception, it must be one produced as a result of "a judicial or quasi-judicial duty to inquire by a public officer."[48] Further, it will only be a public document if it is made for the purpose of the public making use of it, and to enable the public to refer to it.[49]

15–94 Despite the formal nature of the inquiry and the statutory basis thereof, it is difficult to find this object in the proceedings. Section 56(5) of the 1970 Act requires that, in the usual case, the report shall be made "to the Board of Trade" (although copies must be given to parties to the proceedings on application[50]). The lack of conclusive evidence as to the purpose of these inquiries could prove fatal to a party seeking to adduce a report produced by one. Of course, in other countries the basis may be more clearly established as being public in the required sense.

15–95 In view of the doubtful admissibility of these parts of the report at common law, it becomes specially relevant to consider the

---

[47] s.9(2)(c). By s.9(6), the words used in s.9 do not alter the common law rules, but merely identify them.

[48] *Phipson on Evidence*, (12th ed., 1976), para. 1099, citing Lord Blackburn in *Sturla* v. *Freccia* (1880) 5 App. Cas. 623. Remarks of Lord Blackburn applied in *Thrasyvoulos Ioannou* v. *Papa Christoforos Demetriou* [1952] A.C. 84 [1952] A.C. 84 (P.C.).

[49] *e.g.* in *Sturla* v. *Freccia* (1880) 5 App. Cas. 623, and *Ioannou* v. *Demetriou* [1952] A.C. 84 (P.C).

[50] Shipping Casualties and Appeals and Re-hearing Rules S.R. & O. 1923 No. 752, r. 18.

statutory position. There are provisions in the Merchant Shipping Act itself, and in the Civil Evidence Acts of 1968 and 1972.

Section 695(1) of the Merchant Shipping Act 1894, provides:

> "Where a document is by this Act declared to be admissible in evidence, such document shall, on its production from the proper custody, be admissible in evidence in any court or before any person having by law or consent of parties authority to receive evidence, and, subject to all just exceptions, shall be evidence [and in Scotland sufficient evidence][51] of the matters stated therein in pursuance of this Act or by any officer in pursuance of his duties as such officer."

It is noticeable that, while section 484(2) makes the report of a naval court[52] admissible, section 466 makes no similar provision for non-military inquiries. Such omission is in line with earlier decisions on the Merchant Shipping Act 1854, which Act had contained wider rules on admissiblity which were restricted by the courts.[53] It may therefore be firmly concluded that section 695(1) of the 1894 Act does not affect the common law rules discussed above in so far as reports of civil marine inquiries are concerned.

The Civil Evidence Act 1968 is of much greater use, for it appears to allow in evidence those parts of the report which are hearsay only, *e.g.* recorded evidence given to the inquiry, such as where A/B Smith states that "ship X sounded three blasts at 0842 hrs." It seems likely that such evidence would be admissible either under section 2(1) or section 4(1),[54] although it may need to be proved in a manner authorised by the court, for the proviso to section 2(3) requires this where "the statement in question was made by a person while giving oral evidence in some other legal proceedings"—and a Merchant Shipping Act Inquiry may well be a "legal proceeding" within the definition of section 18(2).[55] This admissibility is extended to the record of original opinion evidence to the inquiry by section 1 of the Civil Evidence Act 1972, if the opinion

---

[51] Added by the Merchant Shipping Act 1970, s.100(1), Sched. 3, para. 3.

[52] As to the constitution, etc., of naval courts, see the Merchant Shipping Act 1894, ss.480–485. These provisions are prospectively repealed by the Merchant Shipping Act 1970, s.100(1) and Sched. 5, on the recommendation of the Pearson Report, Cmnd. 3211, para. 183, and are not re-enacted.

[53] See *Nothard* v. *Pepper* (1864) 17 C.B.N.S. 39; *The Little Lizzie* (1870) L.R. 3 A. & E. 56; *McAllum* v. *Reid* (1870) cited in L.R. 3 A. & E. 57; *The Henry Coxon* (1878) 3 P.D. 156.

[54] In *Taylor* v. *Taylor* [1970] 1 W.L.R. 1148, the Court of Appeal allowed the transcript of evidence given in a criminal prosecution to be given in evidence at a subsequent civil trial, on the grounds that both sections allowed it.

[55] s.18(2): "In this Act . . . 'legal proceedings' includes an arbitration or reference, whether under an enactment or not." *Cf.* s.18(1): "In this Act 'civil proceedings' includes . . . (b) an arbitration or reference, whether under an enactment or not, but does not include civil proceedings in relation to which the strict rules of evidence do not apply."

evidence was within the common law exceptions discussed above or section 3(2) of the Civil Evidence Act 1972.

CONCLUSION ON EVIDENCE

15–96   The most valuable parts of a marine inquiry's report, namely the findings of that inquiry, are probably inadmissible in subsequent civil proceedings in the United Kingdom and in jurisdictions where the common law rules on opinion evidence are unaltered by legislation. However, those parts of the report which record evidence given to the inquiry may in certain jurisdictions be admissible in subsequent civil proceedings under the "public documents" exception to the hearsay rule. Where legislation akin to the United Kingdom Civil Evidence Acts 1968 and 1972 has been passed, such parts of the report will be definitely admissible.

K. Persons Exempt from Liability

15–97   The Merchant Shipping (Oil Pollution) Act 1971, which gave effect in the United Kingdom to the Liability Convention, also introduced the provisions in the Convention relating to exemption from liability and channelling of liability.[56] Section 1 of the Act makes the owner liable. Section 20 defines "owner" as "the person registered as owner," with the saving that if a ship is owned by a state and operated by a person registered as ship's operator, then the ship's operator is treated as owner.

The purpose of the Convention is to channel liability through the owner. To that end Article II(4) restricts liability of the owner for pollution damage to that provided for in the Convention and excludes the liability of servants or agents of the owner. That does not deal with the position of salvors or bareboat charterers. The Merchant Shipping (Oil Pollution) Act goes further. Section 3 provides:

> "Where, as a result of any occurrence taking place while a ship is carrying a cargo of persistent oil in bulk, any persistent oil carried by the ship is discharged or escapes then, whether or not the owner incurs a liability under section 1 of this Act—
> (a) he shall not be liable otherwise than under that section for any such damage or cost as is mentioned therein; and
> (b) no servant or agent of the owner nor any person performing salvage operations with the agreement of the owner shall be liable for any such damage or cost."

15–98   Thus, where the Act applies (and it does not apply to

[56] See above, para. 10–21 *et seq.*

those situations which are outside the Convention),[57] neither servants nor agents nor salvors are liable. However, in the last case, the salvor must be performing salvage operations with the *agreement* of the owner. Clearly, salvors operating under Lloyd's Open Form, or some similar salvage contract, are within the exemption. It is however conceivable that some voluntary salvors would not be. It is a rare but possible case that a salvor might render services to an abandoned vessel, without the prior knowledge of the owner. To such a person the exemption would not seem to apply.[58]

15–99 Charterers are also specially dealt with. Section 7 of the Act provides:

> "Where, as a result of any discharge or escape of persistent oil from a ship, the owner of the ship incurs a liability under section 1 of this Act and any other person incurs a liability, otherwise than under that section, for any such damage or cost as is mentioned in subsection 1 of that section, then, if—
>> (a) the owner has been found, in proceedings under section 5 of this Act, to be entitled to limit his liability to any amount and has paid into court a sum not less than that amount; and
>> (b) the other person is entitled to limit his liability in connection with the ship by virtue of the Merchant Shipping (Liability of Shipowners and Others) Act 1958;
>
> no proceedings shall be taken against the other person in respect of his liability, and if any such proceedings were commenced before the owner paid the sum into court, no further steps shall be taken in the proceedings except in relation to costs."

The provision applies only in respect of a spill that is within the Act, and therefore within the Liability Convention. Only the owner is liable under the Convention. Therefore if the charterer is liable for the same spill, he must be liable otherwise, for example under the common law principles discussed above. If both are able to limit their liability, the owner under the Convention and the Charterer generally under the 1957 Brussels Limitation Convention, as enacted in the United Kingdom, the charterer is granted immunity. If either is unable to limit, or if the owner is unavailable for legal proceedings, then the regimes applicable to each apply and both actions may continue.

---

[57] See above para. 15–05.

[58] The more likely, but more difficult, case of a salvor with the *consent* but not the prior *agreement* of the owner would be deprived of immunity only by the narrowest and most literal interpretation of the section. On the facts, it would seem likely that a court would be able to discover sufficient evidence of at least an informal agreement. Thus the voluntary salvor would be exempt.

It is to be noted that these exemptions apply only in respect of claims within the Convention.

L. Damages Recoverable

15–100   The types of damage that may be caused by an oil-spill are varied. The question of the extent to which they may be compensated at common law and in statutory claims substantially depends upon the application of rules of remoteness of damage, which are applicable to both types of claim. It is first desirable to discuss the types of damage that may be suffered as a result of an oil spill.

15–101   First, there is physical damage. The spilled oil may foul vessels and their gear, harbours and harbour equipment and contaminate beaches and coastlines. Such damage will generally occur on land or within territorial waters, save only that vessels may suffer fouling on the high seas. More rarely, there may be cases of personal injury. Skin conditions such as eczema may be brought about by contact with oil or more remotely persons may suffer injuries through slipping on oil-covered rocks, but such claims may well be difficult to establish.

15–102   Then there is environmental damage. Wild birds, fish and all types of marine flora and fauna may be killed and the stocks seriously depleted. Such damage may well be regarded as the most important, but it is also extremely difficult to quantify and presents substantial problems of *locus standi* to the common law.

15–103   The third main type of damage may be described as economic loss: for instance, loss of amenities (such as beaches, marinas, even harbours which must be closed for cleaning), loss of profits by hoteliers, publicans and those in the tourist industry generally, as well as loss of holiday value by holidaymakers themselves. In addition, much of the environmental damage referred to above can have an economic effect: for instance, fishermen may be unable to fish where they intended to fish or they may find that their catch is less saleable because of its actual or possible tainted condition.

15–104   The fourth type of damage is strictly only a subspecies of the third: the costs of reinstatement of the environment. The costs of preventive measures and the costs of cleaning the sea and the coast: the use of booms and other mechanical means for the retention or removal of spilled oil, the cost of detergent, the expenses of recovering and cleaning affected seabirds, etc., can all be regarded as economic consequences of an oil-spill. It is noticeable that these expenses may be incurred by a variety of different persons or bodies acting under a variety of powers, obligations and motives, from clean-up expenses incurred by national government acting under statutory powers to the costs of voluntary organisations devoted to the protection of some aspect of the environment.

15–105 The differing types of damage have been a matter for discussion in the international debate. The definition of pollution damage in the Liability Convention and its application to these questions as well as the expansion of that definition in the 1984 Protocol have been discussed already.[59] Our concern at this point is to consider how far the common law rules and the provisions of the Merchant Shipping (Oil Pollution) Act 1971 address these matters.

DAMAGES RECOVERABLE UNDER THE MERCHANT SHIPPING (OIL POLLUTION) ACT 1971

15–106 Section 1(1) of the Merchant Shipping (Oil Pollution) Act 1971 enacts that:

> "Where, as a result of any occurrence taking place while a ship is carrying a cargo of persistent oil in bulk, any persistent oil carried by the ship (whether as part of the cargo or otherwise) is discharged or escapes from the ship, the owner of the ship shall be liable, except as otherwise provided by this Act[60]—
>
> (a) for any damage[61] caused in the area of the United Kingdom by contamination resulting from the discharge or escape; and
>
> (b) for the cost of any measures reasonably taken after the discharge or escape for the purpose of preventing or reducing any such damage in the area of the United Kingdom; and
>
> (c) for any damage caused in the area of the United Kingdom by any measures so taken."

This definition follows closely that used in the Convention. Articles I (6) and I (7) read:

> " 'Pollution damage' means loss or damage caused outside the ship carrying oil by contamination resulting from the escape or discharge of oil from the ship . . . and includes the costs of preventive measures and further loss or damage caused by preventive measures.
>
> 'Preventive measures' means any reasonable measures taken by

---

[59] See above, paras. 10–47 to 10–58, 10–137 to 10–140.

[60] A reference to the exemptions from liability described above and contained in s.3 and s.7 of the Act and to the defences permitted under s.2—war, malicious act of a third party and negligence on the part of governmental authority in the maintenance of navigational aids.

[61] By s.20, "damage" includes "loss".

any person after an incident has occurred to prevent or minimise pollution damage."

15–107 It follows that the damages recoverable under the Merchant Shipping (Oil Pollution) Act 1971, in English law closely follow those recoverable generally under the Convention.[62] Since in the Act "damage" includes "loss"[63] it would follow that, at least in principle, economic loss directly related to the contamination caused by the oil spill (in contradistinction to other damage, such as fire) would be recoverable. However, as was pointed out above in the context of considering the Convention,[64] problems of causation,[65] quantification and remoteness are dealt with neither by the Convention nor by the Act. These matters must therefore be dealt with by the general law. Clearly, problems of causation and remoteness become more difficult and less easily soluble the further away from the physical damage the loss is to be found. The owner of a seafront hotel with a private beach, contaminated by spilt oil, is in a stronger position with regard to lost profits than is the hotel-keeper the other side of town or the operator of sightseeing motor-coach tours. These matters are therefore discussed further below in connection with common law claims, where the same issues arise.

15–108 The costs of reasonable preventive measures and any damage caused by those measures are specifically mentioned in the Act. Issues of remoteness and whether there is a duty of care in respect of such matters are not therefore raised, although questions of causation and quantification are.

15–109 As at the time of writing, there have been no reported nor available unreported cases on the Merchant Shipping (Oil Pollution) Act 1971.

DAMAGES AT COMMON LAW: REMOTENESS

15–110 It is clear from the two *Wagon Mound* cases,[66] and from their subsequent development, that a defendant in both negligence and nuisance is liable for the full extent of the foreseeable kind of damage caused by his act or omission. This leads to the question of what kinds of damage oil pollution can cause, and to the further question of whether any of these kinds of damage are irrecoverable at common law. The common law rules on the subject also apply to

---

[62] See above, para. 10–47.
[63] s.20.
[64] See above, paras. 10–50 *et seq.*
[65] See above, paras. 15–72 *et seq.*
[66] (No. 1) [1961] A.C. 388; (*No. 2*) [1967] 1 A.C. 617.

statutory liability, so that, unless the statute can be construed as making an exception, damage which is irrecoverable at common law will also be irrecoverable in an action for breach of statutory duty.[67] The significance of this point is seen when it is appreciated that the Liability Convention does not make extensive provision for exceptions to rules of remoteness.

15–111 Remoteness of damage contains one potent source of confusion. Pure economic loss, as opposed to economic consequences of physical damage, may well be remote under the ordinary rules of foreseeability. As has been discussed above[68] pure economic loss also raises the more fundamental question if the defendant is sued in negligence of whether there is a duty of care. Such considerations do not, of course, apply to actions on the Merchant Shipping (Oil Pollution) Act 1971 or actions in trespass actions in nuisance, although other rules, such as that requiring an interference with ordinary use of land for there to be an action in nuisance tend to cover similar ground. Thus, outside negligence, the matter of pure economic loss is solely a question of remoteness—whether such loss is or is not a foreseeable consequence of the tort. If negligence is alleged, the matter has two distinct, but hard to distinguish, aspects.

15–112 It must be emphasised that what follows relates only to the issue of remoteness.

Physical Damage and Environmental Damage

15–113 Physical damage is central to the law of tort. Provided there are no problems of causation, physical damage caused by contamination by oil is in practice always foreseeable and always recoverable: theoretical problems of unforeseeable types of physical damage are hard to envisage.

15–114 To the general principle, one exception must be made. In the first *Wagon Mound* case,[69] the fire damage caused by the ignition of the oil floating on the surface of Sydney Harbour was held to be too remote a consequence of the spill. On the facts as proved to the court, the chain of causation which had possibly led to the fire had too many chance occurrences to qualify as reasonably foreseeable.

---

[67] This follows from *Danaghey* v. *Boulton and Paul Ltd.* [1968] A.C. 1, where the House of Lords held that a plaintiff can recover for the kind of damage which the statutory duty is designed to prevent, even though it occurred in a way not apparently contemplated by Parliament. The interpretation of what constitutes the kind of damage contemplated was wider than in *Gorris* v. *Scott* (1874) L.R. 9 Exch. 125, and this development may continue.

[68] See above, paras. 15–62 *et seq.*

[69] [1961] A.C. 388.

However, on proof of a slightly different set of facts, relating to a different plaintiff damaged by the same fire deriving from the same spill, in *Wagon Mound (No. 2)*[70] such loss was held to be reasonably foreseeable and therefore not too remote. The two cases together provide an instructive example supporting the principle that reasonable foreseeability, and therefore remoteness, is a question of fact in a particular case.

15–115 It must also be noted that actions for breach of statutory duty under the Merchant Shipping (Oil Pollution) Act 1971 cannot include fire damage, since the damage must be "caused by contamination."[71]

15–116 At common law, and under statute[72] the cost of cleaning up land fouled by oil will be recoverable as will the cost of cleaning, repair or replacement of contaminated chattels, such as boats or fishing gear.[73] If profit-earning chattels are affected, then the loss of profits reasonably attributable to them will also be recoverable. There will however be no recovery of unexpectedly high losses deriving from the loss of an unusually lucrative charter of a ship, or even from the extra costs caused to a plaintiff on account of his extreme impecuniosity.[74]

15–117 Pure environmental loss presents particular problems. The base of the damage suffered is physical: injury to flora and fauna in the wild state. There is a problem of the plaintiff's *locus standi*. There is no property in wild animals, plants, etc., until they are reduced to possession. Then they become the property of the possessor.[75] Therefore no tort claim based upon damage to the environment alone can be based upon physical damage to an interest of the plaintiff that has been invaded by the oil. The plaintiff has no sort of title and cannot complain of the physical damage. It therefore follows that any such claim can at best be regarded as a species of claim for economic loss—although in most cases such a loss will be hard to quantify. As we have seen, the establishment of

---

[70] [1967] 1 A.C. 617.

[71] s.1(1): see above, para. 15–106.

[72] The Act specifically deals with clean-up costs in s.1(1)(*b*).

[73] *Rust* v. *Victoria Graving Dock Co. and London and St. Katharine Dock Co.* (1887) 36 Ch.D. 113 (C.A.) (where the cost of repairing houses damaged by flood was allowed in a nuisance action).

[74] See, on the last point, *The Liesbosch Dredger (Owners)* v. *The Edison (Owners)* [1933] A.C. 448 and, generally, *Winfield and Jolowicz on Tort* (12th ed.), p. 137 *et seq.*

[75] *R.* v. *Mallison* (1902) 86 L.T. 600 (where fish caught on the high seas were held to be the property of the owner of the smack by which they were taken); it would appear, however, that there is no property if the fish are merely enclosed in a net: *Young* v. *Hichens* (1844) 6 Q.B. 606. On birds see *The Case of Swans* (1592) 7 Co. Rep. 15.

a duty of care in such circumstances may well be problematical. But assuming that this be possible in negligence, or that the facts permit the use of some other tort, then the question still remains as to whether such losses are too remote. The losses suffered by environmental groups, users of the water or countryside seem likely to be regarded as insufficiently precise to be reasonably foreseeable.

15–118 However, shellfish are different, being specially provided for both at common law and now by statute. Not only are shellfish capable of being owned while in a bed,[76] but they are specially protected from interference.[77] Hence, damage to owned shellfish is a recoverable head of damage in an action in trespass, nuisance, negligence or breach of statutory duty.[78]

Cost of Preventive Measures

15–119 The cost of preventive measures is specifically dealt with under the Act.[79] The extent of recoverability is, subject to questions of remoteness and quantification, that which is recoverable under the Convention.[80]

15–120 At common law, the preliminary question is whether the claim amounts to a claim for purely economic loss. If it is merely the quantification of damage to a piece of property, the foreshore or a fishing boat, then no problem of duty of care arises. If, however, preventive measures are taken by a person or body with no property interest at stake, the claim is one for economic loss alone. So, if based in negligence, a duty has to be established which, even after the decision in *Junior Books*[81] may be problematic. Claims in negligence, without an interest in land, also present some problems.[82]

---

[76] Sea Fisheries (Shellfish) Act 1967, s.7(1), (2) and (3). By s.22(2), "Shellfish" includes "crustaceans and molluscs of any kind, and includes any part of a shellfish and any (or any part of any) brood, ware, half-ware or spat of shellfish, and any spawn of shellfish, and the shell, or any part of the shell, of a shellfish."

[77] Sea Fisheries (Shellfish) Act 1967, s.7(4). This provision makes it an offence for any third party to, *inter alia*, knowingly deposit any ballast, rubbish or other substance within the limits of the several fishery, and it further makes such person liable to the owner or grantee of the fishery for all damage sustained by reason of the unlawful act whether or not a prosecution or conviction has been obtained. The section is ambiguous in that it is not clear whether it is necessary for the offender to know of the whereabouts or existence of the shellfish, or whether it suffices that he knows of the act he is doing.

[78] In *Foster* v. *Warblington U.D.C.* [1906] 1 K.B. 648 the plaintiff succeeded in nuisance for the pollution by sewage of oysters owned by him, even though he had no title to a several fishery, on the basis of his mere occupation of the beds; and in *Nicholls* v. *Ely Beet Sugar Factory Ltd.* [1936] Ch. 343, the Court of Appeal held that it is not even necessary to prove damage to sue for infringement of a several fishery although causation must be shown.

[79] s.1(1)(*b*).

[80] See above, para. 10–48.

[81] [1983] 1 A.C. 580; see above, paras. 15–65 *et seq.*

[82] See above, paras. 15–62 *et seq.*

15–121 Assuming a duty, or that a tort can be established in some other way, the questions that remain are questions of causation and remoteness. An example might be useful.

Suppose that a ship has gone around 20 miles from the coast; oil is seeping from her tanks and is being carried by wind, tide and current towards the shore. The local authority sends out a small armada of boom-laying boats, detergent spraying boats and slick-lickers, and actually succeeds in cleaning up all the oil before it reaches the shore. Are the costs involved a recoverable head of damage at common law?

15–122 One argument which might be taken by the defendant is that, because the oil never reached the shore, no actionable damage has been suffered: only trespass is actionable *per se*,[83] but for nuisance and negligence damage must be shown. The effrontery of this argument is mitigated when it is appreciated that what lies behind it is the argument that the oil might never have actually reached the shore. However, a court should have little difficulty in rejecting the argument where the circumstances show that the measures taken were reasonable: if the defence of *novus actus interveniens* can be rejected so often on the grounds that the event relied on as breaking the chain of causation was reasonably foreseeable,[84] then this argument can be rejected on the grounds that the taking of reasonable preventive measures was foreseeable. Of course, the action taken by the plaintiff must be weighed in the balance with the threat posed, in order to decide what is reasonable in these cases[85]; and so it is not difficult to imagine circumstances in which preventive measures might well be irrecoverable. For instance, suppose the oil is spilled 100 miles out to sea, and starts to blow towards the shore, at which point the armada sets off; the wind changes direction and blows it out to sea where it is broken up by natural forces; here, the cost of any detergent spraying so far offshore would probably be too remote.[86]

15–123 On principle it would seem likely that it does not matter that the person who has incurred the expense of taking reasonable

---

[83] Trespass will very rarely be available, and probably never in the supposed situation; see above, paras. 15–10 *et seq.*

[84] *D'Urso* v. *Sanson* [1939] 4 All E.R. 26; *Steel* v. *Glasgow Iron and Steel Co.* 1944 S.C. 237; *Hyett* v. *Great Western Ry Co.* [1948] 1 K.B. 345.

[85] *Sayers* v. *Harlow U.D.C.* [1958] 1 W.L.R. 623 (C.A).

[86] A not unsimilar case occurred following the collison between the *Olympic Aliance* and *H.M.S. Achilles*, on November 12, 1975. The Dutch spent £16,000 spraying the slick with detergent while it was proceeding parallel to the coast and then out to sea. See IMCO Doc. MEPC/Circ. 31. The United Kingdom and French reports on the same incident are contained in MEPC/Circ. 32 and 33 respectively.

preventive measures is not the person whose land might have been oiled had the measures not been taken.

15–124   It must nonetheless be admitted that the conclusion that reasonable preventive measures are recoverable rests on reasoning by analogy: there is a paucity of precedent. Perhaps for this reason the United Kingdom law was confirmed by statute in 1971. Section 15(1) of the Merchant Shipping (Oil Pollution) Act 1971 provides that:

> "Where—
> (a) after an escape or discharge of persistent oil from a ship, measures are reasonably taken for the purpose of preventing or reducing damage in the area of the United Kingdom which may be caused by contamination resulting from the discharge or escape; and
> (b) any person incurs, or might but for the measures have incurred a liability, otherwise than under section 1 of this Act[87] for any such damage;
> then, notwithstanding that subsection (1)(b) of that section does not apply, he shall be liable for the cost of the measures, whether or not the person taking them does so for the protection of his interests or in the performance of a duty."

The Fisherman's Lost Profits

15–125   In *Attorney-General for British Columbia* v. *Attorney-General for Canada*,[88] Viscount Haldane L.C., giving the advice of the Privy Council, reviewed the authorities on fishing rights of all kinds[89] and concluded that " . . . the subjects of the Crown are entitled as of right not only to navigate but to fish in the high seas and tidal waters alike,"[90] and that this right can be taken away only by competent legislation.[91] The fact that for over a century there have been United Kingdom regulations on how sea fishing may be conducted, and where,[92] does not detract from the existence of the right. It would therefore appear that an oil pollution incident which prevents a member of the public from exercising this right consti-

---

[87] s.1 of the Act provides for strict liability where a tanker carrying oil in bulk as cargo has spilled oil.

[88] [1914] A.C. 153.

[89] Lord Hale, *De Jure Maris: Neill* v. *Duke of Devonshire* (1882) 8 App. Cas. 135; *Malcomson* v. *O'Dea* (1863) 10 H.L. Cas. 593.

[90] [1914] A.C. 153 at p. 169.

[91] *Ibid.* at p. 170. See also *McRae* v. *British Norwegian Whaling Co. Ltd.* [1927–31] Nfld. L.R. 274, declaring the right.

[92] The legislation is voluminous and complicated, but the main instruments are the Sea Fish (Conservation) Act 1967 and the Sea Fisheries Act 1968, and subordinate legislation made thereunder.

tutes a public nuisance, so long as the criteria discussed above, under the heading "Special Damage," are fulfilled.[93]

15–126 However, for a fisherman to have an action for damages in public nuisance he must show special damage—see above, under the heading "Foreseeability of Special Damage." Where a public right of fishery has been interfered with, all members of the public alike suffer interference, and it is difficult to see how a professional fisherman could show that element of discrimination or peculiarity needed, unless perhaps he was actually fishing at the time of the spill, and had to stop as a result.[94] In the leading case of *Rose* v. *Miles*,[95] where a carrier brought an action on the case to recover the expense of having to unload his barges and transport their cargo by land due to the defendant's obstruction of a public navigable waterway, Lord Ellenborough C.J. said:

> "In *Hubert* v. *Groves* (1794) 1 Esp. 147, the damage might be said to be common to all, but this is something different, for the plaintiff was in the occupation, if I may so say, of the navigation; he had commenced his course upon it, and was in the act of using it when he was obstructed. It did not rest merely in contemplation. Surely this goes one step farther; this is something substantially more injurious to this person, than to the public at large, who might only have it in contemplation to use it. And he has been impeded in his progress by the defendants wrongfully mooring their barge across, and has been compelled to unload and to carry his goods over land, by which he has incurred expense, and that expense caused by the act of the defendants. If a man's time or his money are of any value, it seems to me that this plaintiff has shown a particular damage."[96]

15–127 It is, perhaps, just as well that this is so, for if it were otherwise, every fisherman within reach of the spill area would claim that he was going to have fished there but for the spill, and thus "cash in" on the course, so that the genuine fisherman who really has had his habitual ground seriously affected would go without a remedy.

15–128 Would it make any difference if, on the morning follow-

---

[93] *McRae* v. *British Norwegian Whaling Co. Ltd.*, above, note 91, (waters polluted by whaling factory preventing exercise of public right of fishing held to be a public nuisance); *Fillion* v. *New Brunswick International Paper Co.* [1934] 3 D.L.R. 22 (almost identical facts as in *McRae*—held a public nuisance); *Hickey* v. *Electric Reduction Co. of Canada Ltd.* (1970) 21 D.L.R. (3d) 368 (discharge of poisonous materials into Placentia Bay affecting public right of fishing held on a preliminary point of law to be a public nuisance).

[94] In all three cases of *McRae, Fillion and Hickey*, above, the claims of fishermen were rejected as being too remote, for special damage could not be shown.

[95] (1815) 4 M. & S. 101.

[96] *Ibid.* at p. 103.

ing the casualty, a fisherman proceeds deliberately to the fishing ground polluted by oil, and there catches a tainted catch? While this might constitute special damage on the above principles, the conclusion that the plaintiff has been 100 per cent. contributorily negligent seems inescapable. Alternatively, the point may be phrased in causative terms by saying that the cause of the plaintiff's loss is not the defendant's act, but his own.

The Hotelier's Lost Profits

15–129   The question of the hotelier's lost profits will be almost exclusively decided in terms of whether there is a duty of care to avoid this type of pure economic loss, except in the rare case where the fabric of the hotel or its land or chattels have been contaminated by the oil-spill. As such, cases will fall to be decided in accordance with the principles discussed in paragraphs 15–63 *et seq.*, above.

15–130   If a duty is held to exist, the question remains whether loss of business is or is not too remote. On principle, it would seem that ordinary loss of business (the difference between reasonably expected trade and the business actually done) would be reasonably foreseeable and not too remote.

M. Joint and Several Liability

15–131   By section 1(3) of the Merchant Shipping (Oil Pollution) Act 1971, enacting Article IV of the Convention:

> "Where persistent oil is discharged or escapes from two or more ships and—
> (a) a liability is incurred under this section by the owner of each of them; but
> (b) the damage or cost for which each of the owners would be liable cannot reasonably be separated from that for which the other or others would be liable;
> each of the owners shall be liable, jointly with the other or others, for the whole of the damage or cost for which the owners together would be liable under this section."

The situation clearly envisaged by Article IV is where two tankers, A and B, collide, and their cargoes spill into one slick, which causes P damage. The fact that P may be unable to show that x per cent. of his damage is due to A and the remainder to B is irrelevant because he is entitled to judgment for all his damage against both A and B; subject to A or B limiting their individual liabilities, P could recover in full from either of them. This, it may be noted, is identical to the English common law and Admiralty rule in collision cases where vessel C is damaged by a collision between vessels A and B for which they are both to blame.[97]

15–132 But this does not mean that in common law or other countries Article IV will be superfluous, for there would without it be no guarantee that the rule would apply in the unusual situation dealt with by the Convention. For instance, in the United Kingdom several defendants have been held severally liable for all P's damage which is reasonably separable in some cases,[98] but not in others.[99] Article IV ensures that the rule will apply.

15–133 Even where the section applies, several questions are left for resolution by the common law: they include separation, the effect of limitation and the principles upon which separation of damage is to be made.

15–134 The common law rule was that there should be no contribution between joint or concurrent tortfeasors.[1] This position was changed first by the Law Reform (Married Women and Joint Tortfeasors) Act 1935, which itself was amended and extended in this regard by the Civil Liability (Contribution) Act 1978. This provides, in section 1, that any person liable for any damage may seek a contribution from any other person liable in respect of the same damage, whether jointly or severally liable and whether or not the basis of the liability is the same. This has the effect of placing the matter of apportionment between concurrent tortfeasors on the same basis as that which pertains between plaintiff and defendant under the Law Reform (Contributory Negligence) Act 1945. That would seem to imply that both fault and culpability must be taken into account.[2] The provisions of section $2(3)(a)$ and $2(3)(c)$ would seem also to ensure that if the owners of one vessel involved in a collision giving rise to an oil spill limits liability, then that vessel may not be required to contribute more than the limitation amount.

---

[97] *The Devonshire* [1912] A.C. 634; and see G.L. Williams, *Joint Torts and Contributory Negligence* (London, 1951), pp. 23–32.

[98] *e.g. Grant* v. *Sun Shipping Co. Ltd.* [1948] A.C. 549, (H.L.); see especially *per* Lord Parcq at p. 563: " . . . when separate and independent acts of negligence on the part of two or more persons have directly contributed to cause injury and damage to another, the person injured may recover damages from any one of the wrongdoers, or from all of them"; and see also *Fleming* v. *Gemmill* 1908 S.C. 340 (several defendants contributing to sewage pollution of a stream jointly and severally liable in nuisance).

[99] Notwithstanding the indivisibility of damage, liability was apportioned as between several defendants in *Croston* v. *Vaughan* [1938] 1 K.B 540, (C.A.), (road accident) and in *Pride of Derby and Derbyshire Angling Association Ltd.* v. *British Celanese Ltd.* [1953] Ch. 149, (river pollution).

[1] Merryweather v. *Mixan* (1799) 8 T.R. 186.

[2] See *Winfield and Jolowicz on Tort*, (12th ed.), p. 608.

15–135 If damage is not reasonably separable, and therefore section 1(3) of the Merchant Shipping (Oil Pollution) Act 1971 cannot apply, the question of whether the two shipowners are joint or several tortfeasors falls to be decided by the general law. The provisions of section 1(1) of the Civil Liability (Contribution) Act 1978 however, would now seem to render that question insignificant. The Act applies where any person is "liable in respect of the same damage (whether jointly with him or otherwise)." Thus the court can order contribution in all cases.

N. Limitation of liability

15–136 For oil spills within the Convention, in virtue of the Merchant Shipping (Oil Pollution) Act 1971, s.4, limitation proceedings may be taken within the special terms of the Convention.[3] Outside the Civil Liability Convention limitation is governed by other Conventions: the two Brussels Conventions of 1924 and 1957 and the new London Convention of 1976.[4] The first two are in force and enacted as part of United Kingdom law. The last is contained within the Merchant Shipping Act 1979, to be activated when the Convention obtains the necessary number of ratifications.

As a matter of domestic law, limitation of liability is governed by the Merchant Shipping Act 1984, s.504, as amended, substantially by the Merchant Shipping (Liability of Shipowners and Others) Act 1958.

15–137 It is clear that the right to limit in United Kingdom law is available to shipowners and ship operators in respect of oil spill. The claims covered include loss of or damage to property outside the limiting ship and "infringement of any rights" so long as this loss damage or infringement is caused by "the act, or omission of any person (whether on board the ship or not) in the navigation or management of the ship or in the loading, carriage or discharge of its cargo . . . ."[5] Such language seems apt to cover not only claims in respect of physical damage, but also (in so far as they can be made in United Kingdom law) claims for economic loss or environmental damage.

15–138 The amount of limitation is fixed by reference to limitation tonnage, as derived from the Merchant Shipping Act 1984,

---

[3] See above, paras. 10–59 to 10–78, 10–141 to 10–154. s.8A of the 1971 Act preserves the limitation rights of ships registered in states party only to the 1957 Brussels Convention.

[4] See above, paras. 9–29 *et seq.*

[5] See s.2(1) of the 1958 Act: note the differences in phrasing between the Act and Art 1(1) of the 1957 Brussels Convention which the Act was designed to enact.

s.12, and is based upon register tonnage. The limitation fund used to be expressed in gold francs but, in virtue of the Merchant Shipping Act 1981, is now[6] to be calculated in Special Drawing Rights, in order to give effect to the 1979 Protocol to the Brussels Convention. The number of Special Drawing Rights allocated are: in respect of the larger fund, to be established in case of person injury under section 503(1)(i) of the Merchant Shipping Act 1894, 206.67 Special Drawing Rights in place of 3,100 gold francs; in respect of the smaller fund, to be established when there is no personal injury claim under section 503(1)(ii) of the Merchant Shipping Act 1894, 66.67 Special Drawing Rights in place of 1,000 gold francs. A special drawing right is a unit defined by the International Monetary Fund and a sterling equivalent is available on a daily basis. There is therefore no further need for regular Sterling Equivalent Orders, used heretofore in respect of gold francs.

15–139 The right to limit is available under the present law only when the loss or damage has taken place without the defendant's "actual fault or privity."[7] There are clearly two aspects to this phrase. The owner will lose his right to limit if the fault giving rise to the claim, being one for which he was legally responsible, was in fact the fault of another, but he, the owner, was *privy* to it. In *The Eurysthenes*,[8] an insurance case on the analogous question of whether a ship had been sent to sea in an unseaworthy state with the privity of the assured, the Court of Appeal was prepared to hold an owner privy to the acts of his servants if he knew of the fault, or knew of facts which pointed inexorably to the fault, and did nothing about the matter. The Court would not, however, extend the notion so as to apply an objective standard and hold an owner privy because he ought to have known, as a reasonably prudent owner, of the fault.

While privity deals with the fault of another, "actual fault" deals with the fault of the owner itself. The practical problems here centre on the matter of delegation. A shipowning company will delegate authority to appropriate directors or managers. On ordinary principles, faults which can be brought home to the company directors are clearly the fault of the company. In *The Lady Gwendolen*,[9] the Court of Appeal was prepared to deprive a shipowning company of the right to limit on the basis of the fault of the company's marine superintendent. If the delegation was reasonable

---

[6] In force from November 29 1984.
[7] 1894 Act, s.503(1).
[8] [1976] 21 Lloyd's Rep. 171.
[9] [1965] 1 Lloyd's Rep. 335.

and genuine, then the company would not avoid responsibility for the fault of the delegate.

Another way of raising the same issue is to argue that the *system* of delegation itself amounts to a fault. This raises a more general issue of fault in the management of the ship. In *The Alletta*,[10] fault lay in failing to supply copies of relevant bye-laws.

15–140   Much of this law was recently re-examined by the House of Lords in *Grand Champion Tankers Ltd.* v. *Norpipe A/S, The Marion*.[11] The case concerned the failure to supply up-to-date charts: the vessel, relying an old charts, anchored above an oil pipeline and caused much damage by rupturing it. The House appeared to apply both the "delegation" approach and the "managerial fault" approach. The shipowners had utilised a separate managing company which worked under contract with them. That company had introduced a system of control in the matter which had failed. It failed for a complex variety of reasons, chief among which was that the system was not itself subject to effective monitoring by the managing director of the management company who had instituted it. In this lay the fault. It is to be noted that the fault lay in failure in control and at the same time was the fault of a delegate of the owners, the employed (and independent) managing company.[12] The *Marion* seems to represent a policy restrictive of the right to limit.[13]

3. CONCLUSION ON ENGLISH LAW

15–141   The benefit of the introduction of the Liability Convention into English law by the Merchant Shipping (Oil Pollution) Act 1971 lies mainly in the realm of liability. At common law there are great difficulties in the way of establishing that a tort has been committed. Trespass is available only to occupiers of land whose property has been physically invaded. Nuisance contains similar doubts and problems relating to the standing of a plaintiff as well as being restricted by the availability of the defence of necessity and the so-called "traffic rule." Negligence suffers from two main defects: first, it may be hard to establish failure aboard the vessel to take due care; secondly, the recoverability of pure economic loss is problematic.

[10] [1973] 1 Lloyd's Rep. 375.
[11] [1984] A.C. 563.
[12] See also *The Norman* [1960] Lloyd's Rep. 1 (H.L.).
[13] It is to be noted that the 1976 Convention replaces "Actual fault or privity" with "personal act" of an intentional or reckless kind, thus killing the "delegation" argument.

15–142 Strict liability, or its close cousin that appears in the convention, is of enormous value to claimants. The burden of proof of the limited defences available lies firmly on the shipowner. The way of the claimant is much clearer.

15–143 Some questions remain unresolved. The difficult matters of more remote consequences of oil spills, economic loss and environmental damage are not specifically dealt with in the United Kingdom legislation. At least in part, therefore, they fall to be decided by the rules of common law. Those rules are not entirely clear and to some central questions somewhat tentative answers have to be offered.

16. Jurisdiction, Governing Law and Enforcement of Judgments: the Treatment of Foreign Elements*

SUMMARY

16–01 This chapter begins by elaborating the legal principles upon which an English court will exercise its jurisdiction over a defendant in respect of oil pollution claims. The rules have become unduly complicated because successive sets of special provisions have been enacted in response to particular International Conventions, so that in any one case the plaintiff must now consider both the juridical basis for his claim and considerations relating to the defendant before identifying the relevant jurisdictional provisions. He must also consider the facts of the case—for instance whether or not it arose out of a collision—the type of relief sought (interim or final) and mode of proceeding (*in personam* or *in rem*). This chapter also deals, albeit more briefly, with the question of the law which before an English court will govern an issue in a case with a foreign element; and finally with the enforceability in England of a judgment given in another country. This chapter does not consider procedural issues in Scotland or Northern Ireland.

1. INTRODUCTION

16–02 An action for damage or loss suffered as a result of oil pollution caused by a ship (called in this chapter an "oil pollution claim") brought in an English court will often involve a foreign element. That "element" may be a party to the proceedings, any ship concerned in the pollution, the place of the event causing the pollution or any transaction (such as a salvage agreement or an indemnity clause) relevant to the pollution. It is only if *all* such elements are English that there is no question either of the ability to bring the dispute to an English court or whether the dispute should

---

* This chapter is written as if the Civil Jurisdiction and Judgments Act 1982 had entered fully into force. In fact, this is not expected to occur until 1986. As to the Act, see paras. 16–14, 16–23 to 16–33 and 16–52, below.

be adjudged by English law. Where a foreign element does exist, the English court must first decide whether it has the power to hear the proceedings connected with the dispute and then whether it will exercise that power by hearing the proceedings. Other questions the court may have to consider are whether English or a foreign law should govern the resolution of the dispute, and which remedy to grant—relating to any interim relief requested prior to judgment and relief following judgment on the merits. Where a judgment in an oil pollution claim has been obtained in a foreign court, the plaintiff may wish to enforce it in England where the defendant has assets more easily available than in the state in which judgment has been obtained. Conversely where a claim in a foreign court has been rejected a defendant may wish to plead in any claim in an English court that the foreign judgment should be recognised in England. All these issues are considered below.

16–03 The questions of jurisdiction, governing law and recognition and enforcement of foreign judgments therefore underlie the whole of the legal process relevant to oil pollution claims.[1] Although it might be felt that all oil pollution claims involving a foreign element should be treated in the same manner from the procedural viewpoint, this is not in fact the case: the rules relevant to the presence of a foreign element differ according to the legal basis of the claim (*e.g.* contract, tort or statute) and the type of proceedings (*i.e.* action *in personam* or action *in rem*) by which the plaintiff seeks to enforce the claim. The reasons, as so often is the case with procedural questions, are partly historical; and superimposed upon the historical pattern are provisions enacted to give effect to international conventions dealing with specific issues such as collision, arrest of ships and harmonisation of laws within the European Economic Community.

A. Procedurally Relevant Concepts

CONVENTION CLAIMS

16–04 An oil pollution claim will normally be based either on a statutory provision or upon tort.[2] The statutory provisions are contained in the Merchant Shipping (Oil Pollution) Act 1971 and the Merchant Shipping Act 1974. The Merchant Shipping (Oil Pollu-

---

[1] As to enforcement of maritime claims generally see D. Jackson, *Enforcement of Maritime Claims*, (Lloyd's of London Press Ltd., London, 1985).

[2] As to tort liability see Chapter 15 above. It is possible for a claim in contract to be based on oil pollution damage (for instance indemnity claims, as to which see, *e.g. The Jade* [1976] 1 All E.R. 920 (H.L.).

tion) Act 1971 enacts into United Kingdom law the International Convention on Civil Liability for Oil Pollution Damage 1969 (the Liability Convention)[3] and also provides in section 15 for an action to recover the cost of preventive clean up measures. The Merchant Shipping Act 1974 enacts the International Convention on the Establishment of an International Fund for Compensation for Oil Pollution Damage 1971 (the Fund Convention).[4] In this chapter any claim based on either the Liability or Fund Convention will be referred to as a "Convention claim."

Collision Claims

16–05 Whether the legal basis of the claim be statute, tort or any other the ability to bring proceedings in England will be limited by relevant statutory restrictions. In particular one of a number of specified links between the dispute and England is required to bring a claim in England if that claim is based on failure to comply with the Collision Regulations or on the carrying out of, or failure to carry out, a manoeuvre in the case of one or more ships. These links are required because the United Kingdom is a party to the International Convention on Certain Rules concerning Civil Jurisdiction in Matters of Collision, 1952 ("the Collision Convention") and this Convention requires them. Such claims are referred to in this Chapter as "Collision claims."[5]

Claims in personam and Claims in rem

16–06 An oil pollution claim is enforceable, as is any other claim in English law, by an action *in personam*—that is, an action against the person liable for the damage. However, most oil pollution claims will fall within the Admiralty jurisdiction of the High Court[6] and within the heads of claim specified in the Supreme Court Act 1982 as enforceable not only by action *in personam* but also by action *in rem*.[7] An action *in rem* is one where the plaintiff sues property— usually a ship—rather than a person. The choice of action *in personam* or action *in rem* affects the procedure in an English court as well

---

[3] See Chap. 10, above.
[4] See Chap. 11, above.
[5] See Supreme Court Act 1981, s.22, enacting provisions of the International Convention on Certain Rules concerning Civil Jurisdiction in Matters of Collision, 1952 (the Collision Convention). See further para. 16–21, below. As to the effect on Collision claims of the Civil Jurisdiction and Judgments Act, 1982 see paras. 16–25 *et seq.*
[6] The Admiralty jurisdiction of county courts is set out in the County Courts Act 1984, ss.26–31 and County Court Rules, Order 40. In general, claims for a sum must be below £5,000 (or in the case of salvage the property salved must not exceed £15,000).
[7] See Supreme Court Act 1982, ss.20, 21.

as the choice of the court in which the claim may be heard. Actions *in rem* may only be brought in the Admiralty court.

16–07 The action *in rem* has three fundamental advantages over the action *in personam*. First, a claimant may through arrest or the threat of arrest of property obtain security for his claim; secondly the issue of a writ *in rem* makes the claimant a secured creditor in respect of the property in relation to which the writ is issued, and thirdly the property may be sold by the court with a clear title to the purchaser (the claim *in rem* being transferred to the proceeds).

MARITIME LIENS

16–08 Most oil pollution claims will fall within the rubric "damage done by a ship" as provided in section 20(2)(*e*) of the Supreme Court Act 1981.[8] This means they fall within the Admiralty jurisdiction of the High Court, and can be enforced by action *in rem* under section 21(4). Section 20(5) specifically provides that Convention claims are included in the rubric. However, added security is provided in the case of any claim based other than by statute on damage done by a ship (physical injury or damage to property) in that this is one of the few claims which attracts a maritime lien. As such, a lien attaches at the moment of damage and confers security in the property (usually a ship) in relation to which the claim is brought. In addition it has the highest priority, after existing possessory liens, as against other claims *in rem*.[9] An oil pollution claim other than a Convention or other statutory claim stemming from a collision would therefore attract a maritime lien insofar as any damage would so attract it. However, it is fairly clear

---

[8] There is no reason to limit the scope of "damage" in this statutory head of claim so as to exclude any damage recoverable *in personam*. So an oil pollution claim founded on purely economic loss should be enforceable in Admiralty if it meets the statutory criteria set out in the Supreme Court Act 1981 *and* if it would be enforceable as a tortious claim generally. However, if the economic loss (whether through indemnity or directly suffered) is caused through a statutory liability imposed on the defendants (particularly without fault) it is arguable that such loss was not caused by the ship but arises through the statute: see, *e.g.* the reasoning in respect of actions under the Fatal Accidents Acts in *The Vera Cruz* (1884) 10 App.Cas. 59. This restrictive approach to causation would exclude the claims under the Liability or Fund Convention from the head of claim but for express statutory provisions that such claims are within it. See also Supreme Court Act 1981, s.20(2)(*d*), "damage received by a ship"; s.20(2)(*j*), any claim in the nature of salvage. For damage to be done by a ship physical contact is not required but the ship must be the "instrument of mischief": See, *e.g. The Jade* [1976] 1 All E.R. 920.

[9] A maritime lien has advantages over an action *in rem* of itself in that, (i) it attaches when the cause of action arises; (ii) there is no "*in personam*" liability requirement when the action is brought; (iii) in some circumstance there is no requirement of "*in personam*" liability when the cause of action arises; (iv) there is priority over other claims *in rem*. Other claims attracting maritime liens are salvage, masters' and seamen's wages, masters' disbursements and bottomry. As to maritime claims generally see Jackson, *op cit.*, Chap. 12.

that a claim stemming from a collision but based on the Liability or Fund Convention does *not* attract a maritime lien: such a claim is dependent on a statutory liability removed from the concept of fault liability to which the lien attaches and has been added to, rather than incorporated in, the concept of "damage" within the relevant phrase.

16–09 A claimant suing in England may therefore use an action *in personam* or, if the claim falls within the statutory framework, an action *in rem*. An action *in personam* will found the power to issue a so-called Mareva injunction discussed below in paragraph 16–14: an action *in rem* carries with it the power of arrest, the status of preferred creditor at the latest as from the issue of the writ and the ability to enforce the claim against a purchaser of a ship. In an action *in personam* a claimant may be able to serve a writ out of England; and the Civil Jurisdiction and Judgments Act 1982 introduces a European framework for service of a writ out of the jurisdiction without leave in an EEC member state in respect of the matters within its scope (see paragraph 16–23, below).

2. INTERIM RELIEF IN AN OIL POLLUTION CLAIM

16–10 A plaintiff may seek relief prior to and pending judgment because of the need to preserve evidence or property at issue, or to reduce the risk of judgment being unsatisfied or because of the nature of the claim made. The relief will normally take the form either of an injunction against the defendant prohibiting acts disturbing existing circumstances, or of seizure of an asset of the defendant pending judgment. In respect of an oil pollution claim the most important forms of interim relief are (i) arrest of ship (or alternative security) and (ii) a so-called Mareva injunction restraining removal of assets from the jurisdiction or dissipation of assets within the jurisdiction.[10]

A. Arrest of Ships

16–11 In English law, arrest of property is restricted to claims enforceable by an action *in rem*: the property which may be arrested is therefore restricted to property in relation to which an action *in rem* may be brought. In respect of most claims—and probably most oil pollution claims—this is restricted either to the ship in respect of

---

[10] For details of these remedies see Jackson, *op cit.*, Chaps. 9 and 10.

which the action is brought or (in some cases) a "sister ship."[11] Arrest is based on the issue and service of a warrant which in turn depends on the issue of a writ *in rem* by the claimant. The power to arrest cannot therefore be exercised save through and following the commencement of proceedings on the merits.

16–12 The United Kingdom has ratified the International Convention for the Unification of Certain Rules Relating to the Arrest of Sea Going Ships, 1952 (discussed in Chapter 5 and known as the Arrest Convention)[12]: and through the Administration of Justice Act 1956 and its successor the Supreme Court Act 1981,[13] most (if not all) the Convention provisions are enacted into English law. However, because of the English view that arrest is dependent on the issue of a writ in an action *in rem* on the merits (and is not a question of relief prior to the merits action) the Convention provisions are couched in the Act in terms of jurisdiction on the merits of the claim.

16–13 An oil pollution claimant who wishes to arrest a ship therefore may do so only as part of his substantive action[14]: and other claimants *in rem* may also lodge claims against the same ship. The defendant may lodge bail or a bank, P. and I. Club or other insurer's guarantee to prevent or obtain release from arrest of the ship. Unlike property under arrest such bail or guarantee is limited only to the claim for which it is initially provided, and there is therefore no question of priority of claims in respect of the same security.

B. Injunction

16–14 Although an action *in rem* and therefore arrest is available in most cases only against a ship, a remedy almost as powerful may be available to an oil pollution claimant in respect of other assets of a defendant. This is an injunction—an order enjoining a defendant or potential defendant from removing assets specified in the order[15] from England or from dissipating them in England. Such injunc-

---

[11] In respect of maritime liens (see para. 16–08, above) cargo or freight may be the subject of a claim *in rem* and where this is so, may be arrested—but such a lien does not extend to a sister ship.

[12] The Convention provides a framework for (i) arrest as interim relief (Arts. 2–6) and (ii) jurisdiction to hear the suit on the merits based on arrest (Art. 7).

[13] As from January 1, 1982.

[14] But once having initiated the action the ship may be detained under arrest if the proceedings are stayed. See Civil Jurisdiction and Judgments Act 1982, s.26 (in force from November 1, 1984) and *The Tuyuti* [1984] Q.B. 838.

[15] Or up to the amount specified in the order.

tions are known as Mareva injunctions, after an early leading case on their award.[16] Before the Civil Jurisdiction and Judgments Act 1982, ss.24 and 25, the validity of such an injunction was ultimately dependent upon the jurisdiction of the English court in an action *in personam* on the merits against the defendant, but now the injunction may also be awarded where proceedings have been begun in another EEC member state.[17] The injunction is an order against a defendant. Its practical effect is normally to freeze the assets specified in the order until judgment (or execution of the judgment) in the action in being or to be started.[18] It does not confer any preference on the claimant in respect of the claim nor impose any charge on the assets.[19] Failure of the defendant or a third party who knows of it to comply with the injunction is contempt of court, which carries a sanction of fine or imprisonment.

16–15 English law therefore provides an oil pollution claimant with sound opportunities for interim relief. Although the Mareva injunction confers no priority—as does an arrest following the issue of a writ *in rem*—it is not limited in scope to the value of the ship. The two remedies may be applied for by the same plaintiff in respect of the same claim, which the plaintiff might want to do, for instance, where his claim exceeds the value of the ship. Given the high incidence of oil pollution claims in EEC waters, the extension of the injunctive relief by the 1982 Act is particularly significant.

3. JURISDICTION ON THE MERITS

A. Range with Respect to Types of Claim and Persons

16–16 Before the advent of the special rules created by the 1969 Liability Convention, the 1971 Fund Convention, the 1952 Collision Convention and the 1968 Brussels Convention on Civil Jurisdiction and Enforcement of Judgments, the English rules of jurisdiction on the merits related solely to the presence of the

---

[16] *Mareva Compania Navieva A.A.* v. *International Bulk Carriers S.A.* [1975] 2 Lloyd's Rep. 509.

[17] See *The Siskina* [1979] A.C. 710. The statutory provisions enable a court to grant a Mareva injunction in respect of proceedings in civil or commercial matters in a court of any EEC member state. An oil pollution claim is within the scope of the provisions. This will considerably strengthen the hand of oil pollution claimants in EEC member states.

[18] An injunction should not be granted if it will interfere substantially with the trading activities of an innocent third party (as, *e.g.* a shipowner when cargo in the ship is the subject of attack through a Mareva injunction: See *The Eleftherios* [1982] 1 W.L.R. 539).

[19] *Cretanor Maritime Co. Ltd.* v. *Irish Marine Management Ltd.* [1978] 1 Lloyd's Rep. 425. Lord Denning has contended that the injunction operated "*in rem*" (see, *e.g. Z. Ltd.* v. *A.* [1962] Q.B. 558 at p. 573)—but with respect this seems clearly wrong.

defendant or his arrestable property within the jurisdiction. But now, statutory provisions based upon (and in some cases extending) these Conventions provide substantial exceptions to the old rules. This leads to a very complex web of rules in which there is no particular coherence. The exceptions can only be understoood in the context of the Conventions they implement.

16–17 A general rule can be stated as follows. Subject (a) to the Merchant Shipping (Oil Pollution) Act 1971 and the Merchant Shipping Act 1974, which are applicable to claims based on the Liability and Fund Conventions; (b) to section 22 of the Supreme Court Act 1981 which now enacts the provisions based on the Collision Convention; and (c) to the statutory framework of the Civil Jurisdiction and Judgments Act 1982 which implements the provisions of the Brussels Convention, jurisdiction in an oil pollution claim depends—

- (i) in an action *in personam* either on submission of the defendant[20] or service of the writ *in personam* on the defendant or a substitute as specified by the Rules of Court, 1965 (as amended)[21];
- (ii) in an action *in rem* on service of the writ *in rem* on the property in relation to which the action is brought or a substitute as specified by the Rules of Court 1965 (as amended).[22]

16–18 Convention claims—that is, oil pollution claims based either on the Liability or Fund Convention—may be brought in an English court under legislative provisions reflecting those of the Conventions.[23] In the case of Liability Convention claims, the relevant provision is the Merchant Shipping (Oil Pollution) Act 1971, s.1 and in the case of the Fund Convention it is the Merchant Shipping Act 1974, s.4. Actions in the High Court brought under these provisions are Admiralty proceedings[24] and must therefore be brought in the Admiralty Court. Subject to restrictions in the enforcement of Collision claims,[25] Convention claims may be pursued by action *in personam* or by action *in rem*.

---

[20] A defendant may submit by agreement or through participation in proceedings on the merits. See, *e.g. The Messiniaki Tolmi* [1984] 1 Lloyd's Rep. 266. A defendant to an action *in rem* will submit to jurisdiction *in personam* through participation in the *in rem* proceedings.

[21] As to service and deemed service through acknowledgment or endorsement of the writ, see R.S.C. Order 10, Rule 1.

[22] See R.S.C. Order 75, Rule 8. This jurisdiction has its international root in Art. 7 of the Arrest Convention, 1952 which permits jurisdiction on the merits in the country of arrest if its national law permits it.

[23] See paras. 10–84, *et seq.* and para. 11–43 *et seq.* above.

[24] R.S.C. Order 75, Rules 2 and 2A.

[25] As to which see para. 16–21.

16–19 The Merchant Shipping (Oil Pollution) Act 1971 creates rights in furtherance of the Liability Convention in respect of oil pollution which at least in part occurs in the United Kingdom.[26] In addition (and outside the Convention), it provides for a claim for the cost of measures reasonably taken to reduce or prevent damage in the United Kingdom where liability arises otherwise than under the Convention (for instance in the tort of nuisance).[27] In respect of a Convention claim no action may be brought based on damage occurring in another Convention country because, as was seen in paragraphs 10–84 *et seq.*, the scheme of the Liability Convention is that actions under it be brought in a place where the damage occurs.[28] The claim itself is therefore restricted *by definition* to one based on damage connected with the United Kingdom.

16–20 The substantive provisions of the Merchant Shipping Act 1974, like those of the Merchant Shipping (Oil Pollution) Act 1971, are based on pollution damage in the United Kingdom, but they also provide for actions for pollution damage in a Fund Convention country while (as now) the headquarters of the Fund is in the United Kingdom.[29] Once again the restriction is imposed by the limits of the claim and it follows the provisions of the Convention enacted by the statute.

16–21 There is something of a conflict between the provisions of the Liability and Fund Conventions as enacted by the 1971 and 1974 Acts, and the provisions of the 1952 Collision Convention[30] as enacted by the Supreme Court Act 1981. Section 22(1) of the Supreme Court Act 1981 restricts the bringing of actions *in personam* based on a collision claim to circumstances where there is a substantive contact with England. The statutory provision reads:

> "**22.**—(1) This section applies to any claim for damage, loss of life or personal injury arising out of—
>
> (a) a collision between ships; or
> (b) the carrying out of, or omission to carry out, a manoeuvre in the case of one or more of two or more ships; or
> (c) non-compliance, on the part of one or more of two or more ships, with the collision regulations.
>
> (2) The High Court shall not entertain any action *in personam* to enforce a claim to which this section applies unless—

---

[26] s.1. The restriction is to the United Kingdom and therefore damage in Scotland or Northern Ireland will found an action in England. The Act extends to Scotland and N. Ireland.
[27] s.15.
[28] s.13(2).
[29] s.11.
[30] See note 5, above.

   (*a*) the defendant has his habitual residence or a place of business within England or Wales; or

   (*b*) the cause of action arose within inland waters[31] of England or Wales or within the limits of a port of England or Wales; or

   (*c*) an action arising out of the same incident or series of incidents is proceeding in the court or has been heard and determined in the court."

16–22 Hence, where a Liability or Fund Convention claim arises out of collision it is arguable that in these circumstances the claim is subject also to the Collision Convention restrictions. As a consequence, a plaintiff might be deprived of a forum altogether; and a court may therefore be encouraged to hold that a Convention claim does not "arise out" of the collision or other act or omission within the Collision Convention. However, it is certain that any oil pollution claim the basis of which lies outside the Liability or Fund Convention (*e.g.* in the tort of nuisance or negligence) is so restricted by the Collision Convention.

16–23 The jurisdiction of an English court to hear an oil pollution claim will be further affected as regards any defendant domiciled in an EEC Member State by the coming into force of the Civil Jurisdiction and Judgments Act 1982. The Act enacts into English law the Brussels Convention on Jurisdiction and the Enforcement of Judgments in Civil and Commercial Matters 1968, as amended on the accession of Denmark, Ireland and the United Kingdom to the Convention[32] (hereafter the Jurisdiction and Judgments Convention). An oil pollution claim is within the scope of the Convention.

16–24 By the Jurisdiction and Judgments Convention, subject to exceptions, a defendant domiciled in an EEC State may be sued in that State in respect of any matters within the scope of the Convention. Domicile is left for definition to national laws and the 1982 Act sets out detailed rules for the ascertainment of domicile in the United Kingdom. It provides in essence that an individual is domiciled in the United Kingdom if he is resident there and his residence indicates a substantial connection with the United Kingdom: and a corporation is domiciled where it has its seat, *i.e.* if it is incorporated or formed under the law of the United Kingdom and it has its registered office there or its central management and control is

---

[31] Inland Waters excludes territorial waters.

[32] The Act also provides for the allocation of jurisdiction as between England, Scotland and Northern Ireland as parts of the United Kingdom, a new jurisdictional framework for Scotland, recognition and enforcement of judgments and miscellaneous rules (*e.g.* relating to interim relief).

exercised there.[33] Exceptions to the general rule that jurisdiction depends on the domicile of the defendant are as follows:

  (i) where there is a jurisdictional agreement within Article 17;
  (ii) where the defendant appears other than to contest jurisdiction (Article 18);
  (iii) where the claim is an insurance claim (Articles 7–12A);
  (iv) where the "exclusive jurisdiction" provisions apply (in particular, where the proceedings concern the enforcement of a judgment, the court of the EEC State in which the judgment is to be enforced has exclusive jurisdiction) (Article 16); and
  (v) where the Convention provides for "special jurisdiction"—in respect of oil pollution claims the primary grounds for special jurisdiction are:

    (a) in tortious matters, the courts of the place of the harmful event (Article 5)[34];

    (b) in claims connected with the use or operation of a ship, jurisdiction over limitation of liability following jurisdiction over liability (Article 6A).

16–25   Further, Article 57 provides that:

> "This Convention shall not affect any conventions to which the Contracting States are or will be parties and which, in relation to particular matters, govern jurisdiction or the recognition or enforcement of judgments.
>
> This Convention shall not affect the application of provisions which, in relation to particular matters, govern jurisdiction or the recognition or enforcement of judgments and which are or will be contained in acts of the Institutions of the European Communities or in national laws harmonised in implementation of such acts."

The article thereby preserves the jurisdictional provisions of the Liability and Fund Conventions and of the Collision Convention but the extent to which such provisions are thereby incorporated into the jurisdictional scheme of the 1982 Act is unclear. Article 57 enables States to maintain any jurisdiction on the merits based on arrest by becoming parties to the Arrest Convention 1952.[35] The effect of Article 57 in oil pollution cases is, therefore, that actions to enforce claims based on liability within the Liability and Fund Conventions or the Collisions Convention are governed by the 1982 Act, and are also subject to the principles discussed in paragraphs

---

[33] See ss.41–46 of the 1982 Act.
[34] Construed by the European Court in *Bier* v. *Mines de Potasse d'Alsace* [1976] 2 E.C.R. 1735 to mean place of the event causing the harm or the place of the harm.
[35] See note 22, above. See also Art. 25(2) of the Accession Convention spelling out in more detail continued application of Conventions on particular matters.

16–18 to 16–22 above. However it would appear that this does not present a problem since an action under the Liability or Fund Convention would satisfy the criteria of the Jurisdiction Convention, for damage must have occurred in the United Kingdom or, in respect of an action against the Fund, the Fund is "domiciled" in the United Kingdom. In relation to a claim within the Collision Convention, it may be necessary to satisfy the criteria of the Convention *and* the Jurisdiction Convention. However, such a requirement would hardly be imposed so as to deprive a plaintiff of any forum at all.

B. The Establishing of Jurisdiction

A Claim in personam

16–26 In addition to any statutory requirement (such as for Collision claims) jurisdiction over a claim *in personam* is established by the service of a writ *in personam*. Unless the defendant submits to the jurisdiction the writ must be served on the defendant whether he is within or outside England. If the claim is not within the Civil Jurisdiction and Judgments Act 1982, service outside England normally requires leave of the Court. Leave is permitted in respect of specified categories of claim on the establishing of specified connections with England (as, for example, place of the tort). In addition to the fitting of a claim within such an appropriate category, a plaintiff must persuade the court to exercise its discretion in his favour.

16–27 The framework applicable to the granting of leave is set out in the Rules of the Supreme Court 1965 (as amended) and in particular in Order 11, Rules 1 and 2 and in Order 75, Rules 2 and 4. The rules are drafted so as to take into account statutory provisions[36]; and in particular they provide that a writ *in personam* issued in respect of a claim enforceable in an English court under the Civil Jurisdiction and Judgments Act 1982 may be served out of England *without* leave: this reflects the basis of jurisdiction under the Brussels Convention, discussed above. The restrictions on jurisdiction over Collision claims are reflected in the rules that a writ *in personam* in such cases may be served out of the jurisdiction only if the court would have jurisdiction if the writ were served in England.[37]

---

[36] See, *e.g.* R.S.C. Order 75, Rule 4(1) (collision claims).
[37] As to which see para. 16–21, above. The restrictions also apply to limitation actions, *i.e.* actions by a shipowner or other person for limitation of liability in respect of a ship under the Merchant Shipping Acts 1894–1984.

16–28   A writ issued to enforce a Convention claim may be served outside England with leave of the court.[38] While the IOPC Fund has its headquarters in England, the need for service out of the jurisdiction will not arise in connection with any claim against it.

16–29   Claims within the Civil Jurisdiction and Judgments Act 1982 are governed by the framework of the Jurisdiction and Judgments Convention. While in respect of claims within the Act service of a writ remains a procedural necessity, such service does not provide a jurisdictional base. An English court therefore cannot in respect of such a claim within the Act assert the power to hear a case simply on the ground that the defendant was served with a writ in England; and service of a writ outside England is controlled by the provisions of the Act providing jurisdictional bases, rather than by leave of the court.

16–30   Where the oil pollution claim is not within the Merchant Shipping (Oil Pollution) Act 1971, the Merchant Shipping Act 1974 or the Civil Jurisdiction and Judgments Act 1982, the claim will probably be based on tort. An example would be where the ship was not carrying oil in bulk as cargo and the defendant is not domiciled within the EEC. If the claim is a Collision claim, it is governed by R.S.C. Order 75, Rules 2 and 4; otherwise it must fall within R.S.C. Order 11, Rule 1(1)(*f*). Rule 1(1)(*f*) permits service of a writ out of the jurisdiction if "damage was sustained or resulted from an act committed, within the jurisdiction."[39] Alternatively leave may be sought under Order 11 on the basis that the defendant is domiciled in England or that the action is brought against a person duly served within or out of the jurisdiction and that the person out of the jurisdiction is "a necessary or proper party thereto."[40]

16–31   Whatever the category to qualify for consideration, the claimant relying on Order 11 must advance a good arguable case that the claim is within the paragraph.[41] The wording of the rules prior to the coming into force of the Civil Jurisdiction and Judgments Act 1982 necessarily posed the issue of the place at which the tort was committed with judicial emphasis sometimes on the event

---

[38] R.S.C. Order 75, Rule 4(1A). See also R.S.C. Order 75, Rule 4(4) which apparently has the effect of requiring leave in all Admiralty proceedings. The Court has a discretion whether to grant leave. *Cf.* para. 16–33, below.

[39] It seems fairly clear (and logical) that "jurisdiction" includes land and territorial waters. Other categories include contract and security over movable property situate in the jurisdiction (see paras. (d), (e), (1)).

[40] R.S.C. Order 11, Rule 1(1)(a), (c).

[41] See *The Al Wahab* [1983] 2 Lloyd's Rep. 365 (H.L.).

and sometimes on the damage sustained.[42] The amended wording makes it clear that the occurrence of either the event or the damage within England can be found the power to grant leave.[43]

16–32 Traditionally English courts have expressed reluctance in exercising discretion under Order 11 to grant leave, viewing the service of a writ in a foreign country as an interference with the exclusive jurisdiction and sovereignty of the foreign country where service is to be affected.[44] In truth the service of a writ is of far less moment as an act of interference than, for instance, according jurisdiction over a person having no connection with England other than transitory presence here.[45] In any event the extent to which the expressed reluctance has been reflected in practice is uncertain. Despite occasional judicial references to foreign sovereignty it seems as if the discretion is now viewed (as, it is suggested, it should be viewed) as a variant of the principle of the appropriate forum.[46]

16–33 The granting of leave in an action based on a Collision claim is governed by R.S.C. Order 75, Rules 2 and 4. Discretion exercised under this rule will not necessarily turn on factors identical to those seen as relevant to leave in tort cases under R.S.C. Order 11. The court will no doubt exercise its power in the light of the principles of the Collisions Convention.[47]

A CLAIM IN REM

16–34 Jurisdiction in an action *in rem* is established either through service of a writ *in rem* or, it would seem, arrest of the property in relation to which the action is brought. The writ is normally required to be served on ship or cargo[48] but acknowledgment of service by a defendant or endorsement of the writ by his solicitor

---

[42] *i.e.* (in Rule 11(1)(h)) "if the action begun by the writ is founded on a tort committed within the jurisdiction."

[43] As similar jurisdiction under the Civil Jurisdiction and Judgments Act 1982 must be interpreted (see note 34, above).

[44] See generally Dicey and Morris, *The Conflict of Laws* (10th ed., 1980) p. 197.

[45] Or jurisdiction in an action *in rem* through service of a writ on property in England—rules of exhorbitant jurisdiction outlawed by the Jurisdiction and Judgments Convention in respect of matters and defendants within its scope: see Art. 3. As to the saving of other Convention rules see Art. 57 and para. 16–25, above.

[46] As to which see para. 16–41, below.

[47] For an example of the nature of the discretion see *The Aegean Captain* [1980] 1 Lloyd's Rep. 617.

[48] A writ in an action aimed at freight may be served on ship or cargo (R.S.C. Order 75, Rule 8); when aimed at proceeds of sale the writ is filed in the Admiralty Registry (Order 75, Rule 14).

will render such service unnecessary.[49] A writ *in rem* may not be served out of the jurisdiction.

16–35   An action *in rem* is geared either to the property[50] in relation to which it is brought (*e.g.* in an oil pollution claim the ship causing the pollution) or, in a limited class of cases, a "sister ship." An action *in rem* may be brought only in relation to one ship although writs *in rem* relating to the same claim may be issued in respect of a number of ships.[51] A claimant may therefore issue a writ (and become a preferred creditor) and wait for the ship having the greatest value to come into the jurisdiction. The claim is then enforceable against that ship and it is in relation to that ship that the claimant is a secured creditor.

16–36   The conditions for the bringing of an action *in rem* are set out in the Supreme Court Act 1981, ss.20 and 21 which as from January 1, 1982 replaced the Administration of Justice Act 1956, ss.1 and 3. The Administration of Justice Act 1956 enacted provisions of the Arrest Convention 1952 into English law and represented an amalgam of historical English Admiralty development with the principles of the Arrest Convention. In the main the Supreme Court Act 1981 re-enacts the Administration of Justice Act 1956 but extends the scope of the action *in rem* by making it available in certain circumstances against demised chartered ships[52]: an action *in rem* will lie against a ship[53] if (i) the claim attracts a maritime lien (as does a non-Convention oil pollution claim based on damage done by a ship) or (ii) is one for which the Act provides that an action *in rem* will lie (as is damage done by a ship) and the person who would if sued be liable *in personam* was (a) at the time the cause of action arose the owner, charterer or person in possession or control of the ship and was (b) at the time of issue of the writ the owner or demise charterer.

16–37   Unless the claim attracts a maritime lien an action *in rem* can only be brought if at the time the action is brought the shipowner or demise charterer would if sued be liable *in personam*. It is imperative therefore to issue the writ as soon as possible so that a transfer of ownership of the ship cannot remove it from the scope of liability *in rem*. The risk of a transfer resulting in an escape from

---

[49] R.S.C. Order 75 Rule 8(3), Order 10 Rule 1(4), (5).

[50] *i.e.* ship, cargo or freight (but usually a ship).

[51] See Supreme Court Act 1981, s.21(8). Maritime lien claims are limited to the ship involved in the claim. As to maritime liens in general see Jackson, *op. cit.*, Chaps. 2, 4 and 12.

[52] Supreme Court Act 1981, s.21(4)(*a*).

[53] Or in the case of some maritime liens, cargo or freight.

liability *in rem* is particularly marked in the case of a one ship company. In most cases an English court will not probe a company structure to trace ownership of a ship through subsidiary companies to a holding company. It will not pierce the corporate veil.[54]

16–38   The Arrest Convention added significantly to the scope of arrest in English law through the introduction of arrest of a ship not concerned in a claim: the one ship company may be a retort to this development. Translated into English law therefore an action *in rem* for an oil pollution claim may be brought in respect of a claim in relation to one ship (the particular ship) against another if:

(a) at the time the cause of action arose the owner or charterer of the particular ship or the person in possession or control of it would if sued be liable *in personam* and

(b) at the time the action was brought that person owns the other ship.[55]

Sovereign Immunity and Forum Non Conveniens

16–39   Even if an English court has power to adjudicate on the merits, a claimant may be met by a plea that the power should not be exercised. Of the grounds on which such a plea may be based[56] the most relevant to an oil pollution claim are (i) sovereign immunity and (ii) the forum is inappropriate, usually called a plea of "*forum non conveniens.*"

16–40   As to sovereign immunity, England has finally succumbed to the restrictive immunity theory. The State Immunity Act 1978 now enacts this principle in adopting the provisions of the European Convention on State Immunity 1972 and of the International Convention for the Unification of Certain Rules relating to the Immunity of State-owned Vessels 1926 (the Brussels Convention) and its Protocol of 1934. Under the State Immunity Act, a foreign government is entitled to immunity from suit unless it has submitted to English jurisdiction or is engaged in a commercial activity.[57] In the context of commercial activity, there is no immunity from jurisdiction over a claim in connection with a ship if

---

[54] Compare *The Maritime Trader* [1981] 2 Lloyd's Rep. 153 with *The Saudi Prince* [1982] 2 Lloyd's Rep. 255.

[55] s.21(4)(*b*) of the Supreme Court Act 1981.

[56] Other grounds included are arbitration and jurisdictional agreements. *Cf.* Jackson, *op. cit.*, Chap. 6.

[57] State Immunity Act 1978, ss.1–3. But as regards states parties to the Brussels Convention see s.10(6).

when the cause of action arose, the ship was in use or intended for use for commercial purposes.[58]

16–41 As to the appropriate forum[59] the principle has developed in substance since the decision of the House of Lords in 1973 in *The Atlantic Star*,[60] a case concerning a collision between Belgian and Dutch ships in the River Scheldt in which the House declared Belgium to be the more appropriate forum in that case. The principle, as developed and reiterated most recently by the House of Lords in *The Abidin Daver*,[61] requires a defendant who wishes to obtain a stay of English proceedings to show that England is not the natural forum and that there is a natural forum in which justice can be done at substantially less inconvenience and expense. Any stay must not deprive the plaintiff of a "legitimate juridical advantage."[62] Once the defendant has established that there is a more appropriate forum, it seems as if it is for the plaintiff to establish the advantage which brings him to England.[63] The issue is at the discretion of the trial judge and an appellate court will interfere with that discretion only if its exercise is plainly wrong, or relevant factors have not been taken into account or irrelevant factors have been taken into account.[64] The role of the principle in cases within the Civil Jurisdiction and Judgments Act 1982 is open to doubt. It is strongly contended by some that within the jurisdictional framework of the Jurisdiction and Judgments Convention there is no room for a qualification of jurisdictional competence on the basis of appropriateness.[65]

4. GOVERNING LAW

16–42 Once the English court has jurisdiction over the defendant the question arises whether any law apart from English law is to be

---

[58] *Ibid.*, ss.10(1), (2). Where the action *in rem* is against a "sister ship" both ships must be "in use or intended for use for commercial purposes" (s.10(3)). As to claims against cargo, see s.10(4).

[59] An English court has power to order a party to foreign proceedings to desist from those proceedings. It will rarely be exercised but the applicable general principles are those governing the stay of English proceedings. See *Castanho v. Brown and Root Ltd.* [1981] A.C. 557. (H.L.) .

[60] [1974] A.C. 436.

[61] [1984] 1 All E.R. 470.

[62] See *ibid.* at p. 484.

[63] See *ibid.* at p. 479 (Lord Keith); and at p. 476 (Lord Diplock). But it is not entirely clear. See *European Asian Bank A.G. v. Punjab and Sind Bank* [1982] 2 Lloyd's Rep. 356.

[64] See *The Abidin Daver* [1984] 1 All E.R. 470 at p. 482 (Lord Brandon).

[65] See Schlosser Report on the Accession Convention, *Official Journal of European Communities* (5.3.79) No. C59/97–98.

applied to the case. More specifically, four situations may be supposed, and the problem of governing law considered in respect of each:

(1) A ship within English territorial waters spills oil which causes damage within English territorial waters;

(2) A ship on the high seas and outside the territorial waters of any state spills oil which causes damage within English territorial waters;

(3) A ship in foreign territorial waters spills oil which causes damage in English territorial waters;

(4) A ship in foreign territorial waters or on the high seas spills oil causing damage outside English territorial waters.

16–43   Even if foreign law is applicable in English proceedings by governing law rules, its actual application in a case is within the control of the respective parties; for unless pleaded and proved foreign law is deemed in an English action to be identical to English law. English law applies to all procedural matters, including under that rubric remedies and the quantification of damages.[66] Time bar questions will cease to be procedural once the Foreign Limitation Periods Act 1984 comes into force.

16–44   An oil pollution claim will be based either on the statutory provisions enacting the Liability and Fund Conventions, or on tortious liability coupled with the statutory provision of section 15 of the Merchant Shipping (Oil Pollution) Act 1971, or purely on tortious liability. No question of governing law arises in an action based on the Merchant Shipping (Oil Pollution) Act 1971, whether the claim is a Convention claim or is based on section 15; nor does it arise in an action against the IOPC Fund under the Merchant Shipping Act 1974. In so far as the causes of action under either statute are based on damage and reflect the Liability and Fund Conventions, they are restricted primarily to damage occurring in the United Kingdom, or in the case of the Fund Convention, occurring in a Fund Convention country. The Acts apply to all parts of the United Kingdom so there is no potential internal conflict between those parts and the statutes provide liability criteria which are clearly controlling and mandatory. Foreign law has no place in such a framework.[67]

16–45   The issue of governing law therefore arises in oil pollution

---

[66] Also, apparently, it applies to priority between claims, and (to some) the heads of claim (such as pain and suffering). See *Chaplin* v. *Boys* [1971] A.C. 356.

[67] Similar reasoning applies to an action in respect of clean up measures under the Merchant Shipping (Oil Pollution) Act 1971, s.15.

claims generally only in relation to those based on tortious principles rather than on statutory enactments of international Conventions. Apart from the question of tort on the high seas, the general tortious choice of law rules will apply to an oil pollution claim. They are:

   (i) English law applies to any tort committed in England[68] and (as the law of the forum) to any matter of procedure.[69]

   (ii) Subject to (iii), in respect of any tort committed on the land territory of a state other than England the act must be actionable according to the law of the place where the tort is committed and *also* according to English law (assuming that it had been committed in England).[70]

   (iii) Where in respect of a substantive, as distinct from a procedural, issue the foreign law applicable differs from English law, and the issue can be segregated from the other issues in the case, it may be governed by the "proper law" *i.e.* the law which reflects the "centre of gravity" of the event to be decided in the context of the policy of that law.[71]

16–46 The governing law rules applicable to torts committed on the high seas are less clear but may be adapted from those generally applicable. However a tort on water raises a question not applicable to a tort on land—that the high seas are not the territory of any state. In that context, on occasion, English courts have applied the "general maritime law"—by which it seems that English law is meant. Where a tort occurs within the territorial waters of a state it seems that the same governing law rule will apply as if it had occurred on land.[72] If it occurs on the high seas and involves one ship only it is suggested that the ship should be treated as the territory of the State of the ship's flag for the purpose of applying the

---

[68] The authority for this proposition rests in a defamation case: See *Szalatray Stacho* v. *Fink* [1947] K.B. 1.

[69] Including quantification of damages.

[70] Prior to August 24, 1982 when section 30 of the Civil Jurisdiction and Judgments Act 1982 came into force, an English court would be likely to decline jurisdiction over any claim based on damage to foreign immovable property. Since that date the prohibition applies only where the proceedings are principally concerned with title to or possession of immovable property.

[71] See *Chaplin* v. *Boys* [1971] A.C. 356 at pp. 391–392 (*per* Lord Wilberforce). But the case is unsatisfactory because of the divergent approaches of the members of the House. Lord Wilberforce addresses his remarks only at the possible inapplicability of the foreign law. In *Coupland* v. *Arabian Gulf Petroleum* [1983] 2 All E.R. 434 at p. 446, Hodgson J. treated it as a formula for replacing English law. The Court of Appeal did not deal with the issue (see [1983] 3 All E.R. 226).

[72] Although some might argue that the analogy of territorial waters to land is less apt than to the high seas.

general rules.[73] If it occurs on the high seas and extends to or involves more than one ship it seems that English law governs in its role as the "general maritime law."[74]

16–47 The automatic application of English law to any such incident on the high seas is difficult to defend but may be a refuge from problems caused by the possibility of two or more foreign laws being potentially applicable. Although it is somewhat insular, the one virtue of asserting the applicability of English law is certainty. However, it does little to discourage forum shopping: few states are likely to recognise the claim of English law as the "general maritime law."

16–48 There are two major defects to the basic tort governing law rules. The first is the need for a claimant to satisfy two systems of law—precisely the consequence a principle of governing law is intended to prevent. Secondly, the rule depends on the "place" of the tort—posing problems where an act takes place in one state (or outside the jurisdiction of any state) and damage is sustained in another.

16–49 Most of the English cases on "place" of a tort have been decided in the context of whether the court has jurisdiction to hear the case, rather than on which law governs, and have tended to emphasise the "completion" of the tortious act. The amended rule of court coming into force[75] with the major part of the Civil Jurisdiction and Judgments Act 1982 provides a foundation of jurisdiction where either the act or damage occurred in England and this may lead to a change of attitude by English courts.

16–50 It must be remembered (and the amendment to the rules of court emphasises this) that in a jurisdictional context the concern is not to decide *a* governing law but simply whether the governing law is English. In a governing law context a choice must be made so as to point to one law. Although it is a matter of fact and degree it is suggested that prima facie the place of the tort is where it is completed so that where oil pollution damage is suffered in England, English law would govern wherever the cause occurred. If the damage was suffered on land outside England, actionability under both English and the foreign law would be

---

[73] And therefore that law would be the foreign law in the context of the rules requiring actionability according to English and foreign law.

[74] See, *e.g. The Leon* (1881) 6 P.D. 148 and the necessary implication of *The Esso Malaysia* [1975] Q.B. 198.

[75] Now R.S.C. Order 11, Rule 1(1)(f)—this falling into line with the interpretation placed on "place" of the tort by the European Court in construing a similar provision in the Jurisdiction and Judgments Convention (see note 34, above).

required for liability in English proceedings. If the damage was suffered on the high seas as the result of the involvement of one ship, the law of the flag of that ship would form the foreign law for this "double barrelled" rule requiring foreign and English actionability. Where more than one ship is involved, or the act and damage are not restricted to one ship it seems that English law will apply.[76]

16–51 An oil pollution claimant may of necessity or because of obvious choice (such as where the defendant's assets are in England) seek to bring proceedings in an English court. To do so jurisdiction must be both established and exercised. In other cases, there may be a choice to be exercised after considering a number of factors including the chances of success, the availability of interim relief (and its effectiveness) and the likelihood of establishing jurisdiction. Whether the proceedings are brought by choice or by necessity, the issue of the governing law must also be borne in mind, particularly the hurdle in tort cases of establishing liability in English and foreign law. The tactical question remains of whether a claimant should seek to prove the foreign law or rely on the presumption that until proved otherwise it is assumed to be the same as English law. As to this, however, the choice may be made by the opposing party and control is therefore not entirely vested in the claimant.

5. RECOGNITION AND ENFORCEMENT OF FOREIGN JUDGMENTS

16–52 So far in this chapter the issue has been the effect of a foreign element in English proceedings. The remaining question concerns the effect in English law of the judgment of a foreign court. To what extent will such a judgment be recognised so as to prevent relitigation of the same issue, and to what extent will a judgment be enforced so as to allow recovery by a claimant relying on it? First, a cause of action *in personam* will merge into a judgment if the judgment be of an English or (where the judgment is recognised or enforceable in England) a foreign court, and consequently the claimant can only rely on the judgment and not the claim.[77] A judgment in an oil pollution claim given by a court of an EEC member state which is a party to the Jurisdiction and Judgments

---

[76] But only to the extent to which that law permits—so, *e.g.* Convention claims generally require the damage to have occurred in the United Kingdom.
[77] Civil Jurisdiction and Judgments Act 1982, s.34.

Convention will be recognised and enforced in England through registration in accordance with that Convention.[78] No review of substance is permitted and only factors specified in the Convention (such as English public policy or irreconcilability with a previous judgment between the parties) can be pleaded against recognition or enforcement.[79]

16–53 A judgment outside the scope of the Jurisdiction and Judgments Convention, if for a sum of money, may be enforced either through registration where this procedure is made available or, where not, by an action on the judgment. Such judgments *in personam* given by a court of a country to which either the Administration of Justice Act 1920 or the Foreign Judgment (Reciprocal Enforcement) Act 1933 has been extended may be registered and are then enforceable as the judgment of an English court. The Act of 1933 is applied to judgments on claims made under the Liability or Fund Conventions in courts of Convention countries.[80] The prerequisites for registration are similar to the qualifying and disqualifying elements relevant to enforcement through an action on the judgment.[81]

16–54 When registration is not available the ability to enforce or obtain recognition of a judgment by action is subject to a number of factors. The judgment must not constitute a fine or a penalty, nor must it be contrary to natural justice, obtained by fraud or be contrary to a jurisdiction or arbitration clause[82]; nor may it be contrary to public policy. Finally the foreign court must have had jurisdiction. In an action *in rem* this requires that the "*res*" (*i.e.* the ship) at which the action is aimed is in the territory of the court. In an action *in personam* the defendant must have been either resident in that territory or have submitted to the jurisdiction. Subject to these requirements, a foreign judgment will be enforced and recognised in an English court.

16–55 An oil pollution claimant seeking to enforce a Convention claim will have little choice of forum because of the link between the place of the damage or the place of the Fund and the claim. To

---

[78] Arts. 26, 31.

[79] See Arts. 28, 29. A further disqualifying element is the lack of service of the document instituting the proceedings where the judgment is by default.

[80] Merchant Shipping (Oil Pollution) Act 1971, s.13(2); Merchant Shipping Act 1974, s.6(4).

[81] See para. 16–55, below.

[82] As to the effect of jurisdiction and arbitration clauses, see Civil Jurisdiction and Judgments Act 1982, s.32—subject to any obligation under the Jurisdiction and Judgments Convention.

such a claimant, the ability to enforce a judgment in a state other than that in which it is given is extremely important. Such enforcement in the United Kingdom is made available both directly through the legislation enacting the Liability and Fund Conventions and, where applicable, in relation to EEC Member States, the Civil Jurisdiction and Judgments Act 1982. Relevant to the choice of forum for an oil pollution claimant wishing to enforce a non-Convention claim will be not only such factors as the procedure and assets available but also potential enforceability of any judgment in other states. As has been seen, English law provides for the enforcement of a foreign judgment either through registration, or, where registration is not applicable, an action on such a judgment.

6. CONCLUSIONS

16–56 By virtue of the nature of an oil pollution claim, it will tend to pose for the claimant questions of selection of the forum, the law to govern the claim and possibly enforcement in one state of a judgment obtained in another. An English Court can only hear a case and award a remedy under its jurisdiction. An oil pollution claimant must therefore appreciate the basis of that jurisdiction and the modes of its exercise. This chapter has dealt with the pre-requisites for and method of enforcement of an oil pollution claim in England, the law which will govern the claim and the enforcement in England of a judgment obtained elsewhere. The extent to which an oil pollution claimant will be able to opt for England as a matter of choice or be faced with a question of the governing law will vary according to whether the claim is within or outside the Liability or Fund Convention.

A. Convention Claims

16–57 A Convention claim in England must be linked to damage occurring in England or, if under the Fund Convention be against the Fund as it is situate in England. Further, the statutory specification of liability criteria leaves little room for any argument on governing law and provision is made for enforcement in England of judgments obtained in other Convention countries. Nevertheless questions of jurisdiction do remain for English law: first, because of the possible applicability of the jurisdictional frameworks of the Collision Convention and the EEC Jurisdiction and Judgments Convention; secondly, because of the availability of twin modes of enforcement through the action *in rem* and action *in personam*; and thirdly because of the undoubted relevance of security for a claim.

B. Non-Convention Claims

16–58 An oil pollution claimant's ability to select a forum will depend on the nature of the claim. Unless restricted (as with Collision claims) the option is there, and will be exercised no doubt bearing in mind the presence of the defendant or his assets and procedure. It is therefore essential for an oil pollution claimant to know not only the jurisdictional framework of any state in which an action may be brought but also the law which may be applied by the courts of that state, the modes of enforcement and the availability of security for the claim. This chapter has dealt with the rules of English law relevant to this necessary inquiry. Whatever the degree of choice of forum, an oil pollution claimant must appreciate the means available by which the end (of enforcement of the claim) may be reached.

PART IV

SELECTED ISSUES IN UNITED STATES LAW

17. The Framework of United States Legislation

SUMMARY

17–01 United States legislation in the vessel oil pollution field has its beginnings in general maritime legislation which sought to encourage growth in the shipping industry, perceived to be crucial to American economic development, and which subsequently sought to regulate conduct in that same industry when it became evident that such economic encouragement did not provide incentives in the area of environmental control. This chapter presents an historical overview of vessel oil pollution legislation in the United States leading to an outline of the existing statutory framework of vessel-related oil pollution law.

17–02 As may be seen in the chapters on international law, development in the oil pollution liability area has tended to be consensual and comprehensive. While the nature of consensus requires resolution of diverse and often conflicting positions, the compromise which results is usually sufficiently responsive to each participant's needs to ensure broad international support. United States legislation in the oil pollution liability area has deviated substantially from the international efforts towards consensus and uniformity found in the existing international conventions to which the United States is not a party. Rather, the United States has chosen to build upon domestic legislation as and when new areas for governmental intervention are identified. The result has been that legislation has developed in a piecemeal fashion and reforms have been limited to remedial legislation consistent with identified needs. Oil pollution liability law in the United States is, therefore, impossible to treat in the same manner as in the international law sections since, as discussed below, there is not one, but there are several statutes and amendments thereof dealing with oil pollution liability and these statutes have evolved from laws rooted in the nineteenth century. For this reason, this chapter will consider the historical background of United States pollution liability law and proceed briefly to describe the relevant statutes.

433

1. HISTORICAL OVERVIEW OF UNITED STATES LEGISLATION

17–03 Any discussion of this area must begin with the United States Limitation of Liability Act.[1] The Limitation Act was adopted into law in 1851 after courts had refused to accept the common law principle of limitation in the general maritime law based on abandonment of the owners' interest in the vessel and pending freight.[2] Because the growth of international trade in the United States had led to development of a strong indigenous merchant fleet, it was the desire of Congress to put the domestic fleet on an equal footing with the English merchant fleet which had the benefit of limited liability.[3] United States courts, in giving a liberal effect to such policy, noted that the purpose of the legislation was "to promote the building of ships and to encourage persons engaged in the business of navigation,"[4] and "to put American shipping upon an equality with that of other maritime nations."[5]

17–04 The Limitation Act, which is still in force, provides that the liability of a shipowner for any loss, damage or injury for any act done or loss sustained "without the privity or knowledge of such owner . . . shall not . . . exceed the amount or value of the interest of such owner in such vessel, and her freight then pending."[6] Unlike the 1957 Brussels Convention on Limitation, to which the United States is not a signatory, the Limitation Act permits the registered or demise owner of a vessel to establish a fund based on vessel market valuation at the conclusion of the casualty voyage, not on the Brussels Convention concept of registered tonnage.[7] The current incongruity of the Act was dramatically demonstrated in the *Torry Canyon* case in 1968, where the Federal Court in New York approved a $50 limitation fund based on the market valuation of one salved lifeboat when the *Torrey Canyon* was herself lost after her 1967 stranding and sinking off the coast of England, resulting in

---

[1] Act of March 3, 1951, 9 Stat. 635; 46 U.S.C. §181–189. It is not the purpose of this treatise to present a detailed examination of the U.S. Limitation Act, but only to describe its significance in terms of stimulating the need for congressional reform in the oil pollution field. An excellent and detailed discussion of the U.S. Limitation Act may be found in 3 *Benedict on Admiralty* (1983) and Gilmore & Black, *The Law of Admiralty*, 818–957 (2nd ed., 1975).

[2] See, *e.g. The Rebecca*, 20 Fed. Cas. 373 (D.Me. 1831); *The Maine* v. *Williams*, 152 U.S. 122 (1894).

[3] See 3 *Benedict on Admiralty* (1983) §6 for a discussion of Congressional policy.

[4] *Moore* v. *American Transportation Co.*, 650 U.S. (1860).

[5] *The Maine* v. *Williams* 152 U.S. 122 (1894).

[6] 46 U.S.C. §184. Other provisions provide for an additional separate fund for personal injury and death claimants: 46 U.S.C. §183(b).

[7] See paras. 9–34 *et seq.*, above.

pollution expenses of some $15,000,000.[8] Although the *Torrey Canyon* dispute was settled for the more substantial, but relatively modest, amount of $3,000,000, the public outcry that resulted from the prospect that a shipowner could escape liability in the United States for such a massive pollution disaster focused attention on the need for congressional reform.[9]

17–05   The Limitation Act, although applying to pollution liability, was not enacted specifically to address pollution liability. The first specifically pollution-related legislation was not long in coming, however. In 1886 the first federal effort to control water pollution from vessels was passed—the New York Harbor Act of 1886.[10] Annexed to a general harbor appropriations bill, the New York Harbor Act made it unlawful 'to cast, throw, empty, or unlade, or cause, suffer or procure to be cast, thrown or emptied . . . from or out of any ship, vessel, lighter . . . or other craft . . . or from the shore . . . any ballast, stone, slate, gravel . . . or other refuse or mill waste of any kind into New York harbor." There were no enforcement or penalty provisions until 1888 when new legislation added the ports of Hampton Roads and Baltimore and made violation a criminal misdemeanor punishable by fines of between $500 and $2,500 plus imprisonment.[11] In 1894, Congress enacted legislation which provided for penalties against masters, pilots, engineers or others onboard, and against the offending vessels who deposited refuse in navigable federally improved waters.[12]

17–06   The first significant general federal water pollution legislation was enacted in 1899, and became popularly known as The Refuse Act.[13] The Refuse Act enlarged the scope of the New York Harbor Act and superceded the existing legislation by making it unlawful "to throw, discharge, or deposit . . . any refuse matter of any kind or description whatever . . . into any navigable water of the United States." The enforcement provisions were penal in nature and made it a misdemeanor for violation. The Act imposed fines upon persons and corporations which violated or knowingly aided, abetted, authorized, or instigated a violation of from $500 to $2,500, or imprisonment of from 30 days to one year, or both.[14] A

---

[8] *In re Barracuda Tanker Corpn.* 281 F.Supp. 228 (S.D.N.Y. 1968).

[9] See, *e.g.* Gilmore & Black, *The Law of Admiralty*, 824–834, 840–846 (2nd ed., 1975).

[10] 24 Stat. 329, See, Healy & Sharpe, *Admiralty*, 732 (1974).

[11] 25 Stat. 209, 33 U.S.C. §441. Of course, these were very substantial amounts in 1888. Today they seem relatively insignificant.

[12] 28 Stat. 363.

[13] 33 U.S.C. §407, 30 Stat. 1152.

[14] 33 U.S.C. §411, 30 Stat. 1153.

435

unique aspect of the Act was that the court must award to an informer one-half of the fines recovered.[15] Moreover, any master, pilot or engineer who knowingly engaged in unlawful discarges described in the Act would be subject to the fines imposed and "shall also have his license revoked or suspended for a term to be fixed by the judge before whom tried and convicted."[16] Vessels used or employed in violating the Act were liable *in rem* for the penalties.[17]

17–07 The Refuse Act remedies are purely criminal in nature, so that the United States may not recover civil damages under it and private individuals have no right of recovery by virtue of the Act.[18] Despite its ambiguity, and despite the coexistence of new legislation specifically aimed at discharges of oil from ships, the general provisions of the Refuse Act are still used to prosecute offenders in most oil pollution cases in which there is substantial governmental intervention in clean-up. For instance in 1981, in *United States* v. *Big Sam*,[19] the owner of a tugboat and the tugboat *in rem*, which was in collision with an oil barge on the Mississippi River causing an oil spill, were penalized for discharging "refuse" into the navigable waters of the United States.[20]

17–08 Legislation specifically directed to oil discharges from vessels into navigable waters, as opposed to pollution in general, had to wait until the passage of the Oil Pollution Act of 1924.[21] This Act made it unlawful to discharge oil upon the coastal navigable waters

---

[15] This was held to be non-discretionary in the event of a single court conviction: *U.S.* v. *Anaconda Wire & Cable Co.*, 342 F.Supp. 1116 (S.D.N.Y. 1972).

[16] 33 U.S.C. §412, 30 Stat. 1153.

[17] *Ibid.*

[18] Private persons have attempted to use the informer's reward as a basis to institute a civil action or to compel governmental action under the Refuse Act; see *e.g.*, *Bass Anglers Sportsman's Soc. of America* v. *U.S. Plywood-Papers Inc.*, 324 F. Supp. 302 (D.C. Tex. 1971), but these efforts have not been successful and, in any event, would be of limited utility with respect to the typical case of vessel oil pollution under the Refuse Act. Although criminal in nature, it has been held that neither *scienter* nor *mens rea* are necessary to be proved to support a conviction for violation of the Refuse Act. *U.S.* v. *White Fuel Co.*, 498, F.2d 619 (1st Cir. 1974); *U.S.* v. *American Cyanamid Co.*, 345 F. 2d 1202 (S.D.N.Y. 1973), *aff'd*, 480 F.2d 1132. It should also be noted that the Refuse Act has been used as the basis for establishing a common law right of recovery for clean-up costs by the United States. See, *e.g.* *Matter of Oswego Barge Corp.*, 664 F. 2d 324 (2d Cir. 1981); *Wyandotte Transp. Co.* v. *U.S.*, 389 U.S. 191 (1967).

[19] 505 F. Supp. 1029 (E.D. La. 1981).

[20] This was not the first such case to hold that the Refuse Act encompassed oil pollution. In *U.S.* v. *Standard Oil Co.*, 384 U.S. 224 (1965), the Supreme Court, in a split 5–3 decision, held that "refuse-matter" included oil. That case involved oil pollution of a navigable river by a refinery when a shut-off valve had been accidentally left open. In *Matter of Oswego Barge Corpn.*, 664 F. 2d 327 (2d Cir. 1981), the Second Circuit found that "refuse" included oil from a massive spill resulting from the grounding of an oil barge in the St. Lawrence Seaway.

[21] 43 Stat. 604.

of the United States, "except in case of emergency imperiling life or property, or unavoidable accident, collision, or stranding." The Oil Pollution Act of 1924 was also criminal in nature, making violation punishable by fines of between $500 and $2,500 and imprisonment of between 30 days and one year. Vessels were liable for the pecuniary penalty *in rem* and United States masters were subject to licence revocation for violations. The Act was amended in 1966 to include a *mens rea* requirement by defining "discharge" as "any grossly negligent or wilful spilling, leaking, pumping, pouring, emitting, or emptying of oil." Violators were subject to fines and imprisonment and vessels from which oil was discharged were subject to fines of up to $10,000.[22] These amendments introduced for the first time an additional duty which is still (in modified form) a feature of the current legisation, namely that violators were required immediately to remove oil discharged into navigable waters and upon adjoining shorelines, failing which, the Secretary of the Interior was entitled to remove the oil and seek reimbursement from the violator.

17–09 These amendments were short lived since the Oil Pollution Act of 1924, as amended in 1966, was repealed with the passage of the Water Quality Improvement Act of 1970.[23]

17–10 Between the enactment of the Oil Pollution Act of 1924 and its repeal in 1970 came the enactment of the historic Federal Water Pollution Control Act of 1948 ("FWPCA") which remains, with amendent, as a cornerstone of United States law in the vessel oil pollution field.[24] The FWPCA was enacted for the purpose of "restor[ing] the chemical, physical and biological integrity of the Nation's Waters." As a declaration of National Policy the following goals, *inter alia*, were set: (1) the discharge of pollutants into navigable waters is to be eliminated by 1985; and (2) the interim goal of water quality which provides for the protection of fish, shellfish and wildlife and provides for water recreation is to be achieved by July 1, 1983. The Act contains a comprehensive federal water pollution plan which covers such diverse areas as research, investigation and training programs, sewage treatment, solid waste disposal, agricultural pollution, thermal discharges, mine water pollution and water

---

[22] 80 Stat. 1252; this was a rider to the Clean Water Restoration Act of 1966, 80 Stat. 1246–53.

[23] 84 Stat. 91.

[24] 62 Stat. 1155, originally 33 U.S.C. §§466(a)–466(j); renumbered as 33 U.S.C. §§1151–1173 in 1970 under the Water Quality Improvement Act; substantially amended and renumbered as 33 U.S.C. §§1151–1376 in 1972 under the Federal Water Pollution Control Act Amendments of 1972, other significant amendments were adopted in 1977 under the Clean Water Act.

treatment. The provision which is relevant to vessel oil pollution is section 311 of the Act, now codified as 33 U.S.C. §1321. Section 311 declares as policy that "there should be no discharges of oil or hazardous substances into or upon the navigable waters of the United States, adjoining shorelines, or into or upon the waters of the Contiguous Zone, or in connection with activities under the Outer Continental Shelf Lands Act or the Deepwater Port Act of 1974."[25] Section 311, which will be described more fully in Chapter 19, is both prohibitive and curative in nature, in that it deals both with prevention and liability. It prohibits the discharge of oil in harmful quantities and requires the owner to notify the appropriate agency of relevant discharges.[26] The Act authorizes the President to remove oil discharged into navigable waters unless he determines that the owners will take proper steps to remove it themselves.[27] Any such efforts entitle the government to reimbusement for actual costs of removal.[28]

2. THE CURRENT POSITION

17–11 In addition to FWPCA, there are three statutes concerned with problems of vessel oil pollution only in connection with discrete federal projects: the Trans-Alaska Pipeline Authorization Act (1976),[29] the Deepwater Port Act (1976)[30] and the Outer Continental Shelf Lands Act amendments (1978).[31] Each of these Acts, which will be discussed in more detail in Chapter 20, is concerned with liability, clean-up and response to vessel oil pollution arising out of activities related to a specific geographic area and activity. Thus, whereas FWPCA applies generally to oil pollution affecting the navigable waters of the United States, it will be observed that these three statutes have not been applicable to the vast majority of vessel oil pollution matters that have arisen or that can be expected to arise in the near future.[32]

---

[25] 33 U.S.C. §1321(b)(1).

[26] 33 U.S.C. §1321(b)(3)(5).

[27] 33 U.S.C. §1321(c)(1).

[28] 33 U.S.C. §1321(f)(1).

[29] 43 U.S.C. §§181–195. This statute is applicable to vessel activity between pipeline terminals in Alaska and United States ports.

[30] 33 U.S.C. §§1501–1524. This statute is applicable to the present and future development of deepwater ports beyond the three-mile U.S. territorial sea for deep laden tankers, of which there is presently only one.

[31] 43 U.S.C. §§1331–1335. This statute is applicable to vessels carrying oil from offshore facilities on the Outer Continental Shelf or adjacent seas.

[32] Of course, there has been significant vessel activity falling within the jurisdiction of the Trans-Alaska Pipeline Authorization Act.

17–12   Apart from the liabilities and obligations which arise from the above legislation, there is a separate and further area of governmental regulation in the field of oil pollution *prevention* where, unlike the law and policy relating to oil pollution *liability*, the implementing legislation has been in response to standards and rules generally developed in international agreements. The Protocol of 1978 Relating to the International Convention for the Prevention of Pollution from Ships, 1973 (MARPOL 73/78) entered into force for the United States on October 2, 1983,[33] and the 1969 Convention Relating to Intervention on the High Seas in Cases of Oil Pollution and the 1973 Protocol Relating to Intervention on the High Seas in Cases Other than Oil Pollution have been in force in the United States since February 1974 and March 1983, respectively. Since international treaties and agreements are not self-effectuating in the United States, implementing legislation is required to make these agreements part of United States domestic law. The implementing legislation and federal regulations relating to these agreements will be discussed in Chapter 18.[34]

17–13   Hence, as may be seen from this brief discussion, the implementation of United Stated policy on oil pollution liability, as distinguished from oil pollution prevention, has largely been a patchwork affair, with certain overlapping statutory coverage. Legislation has developed both in response to international rule-making, and in defiance of it. This unsatisfactory state of affairs is further complicated by the troublesome but ever present matter of the American federal system which has resulted, in the area of oil pollution liability, in both state and federal legislation, owing principally to the fact that the United States Supreme Court has upheld the view that the federal legislation had not pre-empted any state from taking its own steps to protect land and navigable waters within its territory.[35] As of 1976, state legislation concerning oil pollution has been enacted in all coastal states.[36] A representative study based on the state statutes of New York and Florida will be considered in Chapter 21, together with a discussion of common law recoveries, where principal developments involve recovery

---

[33] The United States is not a party to Annexes III, IV and V of the 1973 Convention, see United States Treaties in Force, Maritime Treaties. MARPOL 73/78 is analyzed in detail in Chapter 3, above.

[34] The 1969 Intervention Convention is the subject of Chapter 6, above. Vessel manning and navigation standards, which relate to pollution prevention are discussed in Chapter 4, above. There is also some mention of such standards in the context of a general discussion in Chap. 18, below. However, further mention in beyond the scope of this book.

[35] *Askew* v. *The American Waterways Operators, Inc.*, 411 U.S. 325 (1973).

[36] Jarvis, "*Richardson* v. *Foremost Insurance Co.*: A New Opportunity for Industry to End State Regulation of Coastal Oil Pollution," 19 *Gonzaga Law Review* 265 at p. 289 n. 155.

based on common law maritime tort and nuisance principles. Chapter 21 concludes with a discussion of the need for uniformity in the United States oil spill regime, such need having been recognised in legislation presently pending before Congress.

3. CONCLUSION—THE FUTURE

17–14 Despite the reluctance of the United States heretofore to ratify the international statutory scheme for oil pollution liability, it is clear that the present United States framework with its inconsistency, overlap and complexity, is in need of reform and pending legislation is a useful first step. An opportunity for reform has been created by the 1984 Protocol to the CLC and Fund Conventions in which the United States view substantially influenced the final reports. If the United States should adopt the international Conventions, much of the existing legislation could be repealed and the prospect of state legislation becoming pre-empted by international treaty would be enhanced. This would be a welcome respite from the present legal structure which is unduly and unnecessarily complex.

18. Oil Pollution Prevention

SUMMARY

18–01 Discussed in this chapter are the laws of the United States which relate to oil pollution prevention, most importantly, the Port and Waterways Safety Act and the Tank Vessel Act. It will be observed that while the United States in the 1970s unilaterally imposed standards for vessel construction and design for vessels calling at its ports, this resulted in increased consideration of international standards, leading to the 1978 International Conference on Tanker Safety and Pollution Prevention. The proposals adopted at this conference were eventually enacted into law in the United States. Since certain legislation, such as the Intervention on the High Seas Act and the Marpol Act, serve essentially to incorporate international convention into domestic law, they are discussed only briefly here, leaving detailed discussion to Chapters 3 and 6.

1. INTRODUCTION

18–02 Increased United States oil imports in the 1970s stimulated a recognition not only that oil pollution response legislation was necessary, but also that oil pollution prevention legislation was an equally significant factor in planning for the future. In one Congressional report, it was noted that the five-year period 1968–1972 saw a 35 per cent. increase in domestic oil consumption, much of which was transported by tankers.[1] In considering this increase in vessel traffic in the United States and adjoining waters, recent vessel casualty experience was also noted. In 1970, domestic imports were approximately 3.4 million barrels per day. The 1972 Congressional study projected 5.2 million imported barrels per day in 1975 on consumption of 17.8 million barrels per day, 9.9 million imported barrels per day in 1980 based on consumption of 21.8 million barrels per day and 14.9 million imported barrels per day by

---

[1] Sen. Rep. No. 92–724, reported in 1972 U.S. Code Cong. & Admin. News 2767.

1985 based on consumption of 26.5 million barrels per day.[2] Carriage of these incremental quantities would alone require more than 350 tankers, each of 250,000 DWT, and when transferred to the handy-sized vessels capable of calling at most United States ports, the projected increase of tanker traffic was substantial. The report noted that even if the projections were overstated, the increase in waterborne movement of oil which will occur "has grave implications for the quality of the marine environment."[3]

18–03 While increased potential for marine pollution arises out of increased harbour activity, this was by no means the sole concern. Congress recognized that improvement in the design, construction and maintenance of the vessels themselves was also required. In addition, improved operations to minimize the effect of non-casualty discharges were considered to be of vital importance. The Committee received information of 1,416 tanker casualties in 1969 and 1970 which showed that of 269 marine casualty polluting incidents during this period, 30.5 per cent. (81) resulted from collisions, 26 per cent. (70) resulted from groundings, 18.6 per cent. (51) resulted from rammings, 7.4 per cent. (20) resulted from fires, 5.9 per cent. (16) resulted from explosions and 2.7 per cent. (7) resulted from breakdowns and other miscellaneous causes.[4] As a consequence of this study, Congress determined that a"systems approach" to pollution was more appropriate than consideration of one or another aspect alone.

18–04 Apart from its finding of a need for a systems approach to vessel pollution prevention, the Committee also found that new standards for tanker design and construction were required. It noted[5]:

> "[U]nfortunately, tankers have been and are being designed and built exclusively for the economic benefit of their owners and customers, with little thought being given to their impact on the marine environment. Further, as tanker size has increased, the problem has gotten rapidly worse. The testimony received at the committee's hearings overwhelmingly supported the proposition that much can

---

[2] Sen. Rep. No. 92–724, *supra*, p. 2771. Actual domestic consumption and supplemental imports have not been as high as estimated. In 1978, peak consumption was 18.8 million barrels per day; in 1981, consumption was 16 million barrels per day, and, as of September, 1982 (the last figures available), consumption was 15.3 million barrels per day. Temple, Barker & Sloan, Inc., *Cost Benefit Analysis of Possible U.S. Adherence to Two International Conventions on Liability and Consumption for Oil Pollution Damages* (June 30, 1983).

[3] Sen. Rep. No. 92–724, *supra*, at p. 2772.

[4] Sen. Rep. No. 92–724, *supra*, at p. 2775. Pollution incidents from routine vessel operations clearly were not included in the study.

[5] Sen. Rep. No. 92–724, *supra*, at p. 2777.

and must be done to improve tanker design, construction and operation."

18–05 One United States Coast Guard Study, for example, showed that penetration to cargo tanks resulting from groundings could have been avoided in 92 per cent. of the cases analysed had such tankers been constructed with double bottoms of two metres.[6] Another area of concern was over the manoeuvrability of large tankers and the need for continued research leading to improved standards. Finally, the Committee specifically noted that 11 per cent. of total oil pollution resulted from operational discharges based on accepted load-on-top procedures. This was deemed to be excessive and in need of more stringent regulation.

18–06 In 1972, when these Congressional findings were made, existing domestic legislation was sparse. The National Emergency Act of 1917[7] authorised the President to promulgate rules and regulations governing the movement, inspection and guarding of vessels, harbors, ports and waterfront facilities in the United States upon a declaration of national emergency. This Act was passed at the outbreak of World War I and was principally directed to wartime port security. It was also used as the legislative basis for the Coast Guard's port security program during World War II.

18–07 The National Emergency Act was amended in 1950 to become the Magnuson Act.[8] The Act changed the jurisdictional requirement for its invocation to be that "the security of the United States is endangered." Upon such a finding, various protective measures could be instituted to safeguard against destruction, loss or injury to vessels, ports or waterfront facilities from sabotage or from accidents. Under this authority, port security not strictly related to national defence was considered to be within its ambit, yet it was clear that this Act was not broad enough to cover the sweeping prevention regulation which was under consideration.[9]

18–08 As to tanker safety, the Tank Vessel Act of 1936 was enacted to regulate hazards to "life and property" from carriage of flammable or combustible liquid cargoes in bulk.[10] Since "environmental protection" was not specifically mentioned in the 1936 legislation, it was felt that the Act might not be sufficiently broad in

---

[6] Sen. Rep. No. 92–724, *supra*, at p. 2777.
[7] 50 U.S.C. §191.
[8] 50 U.S.C. §191.
[9] Sen. Rep. No. 92–724, *supra*, at p. 2767; H. Rep. 95–1384–Part I, reported in 1978 U.S. Code Cong. & Admin. News 3270–3272.
[10] Formerly, 46 U.S.C. §391(*a*), recodified as 46 U.S.C. §§3701–3718.

its proposed coverage. The United States had also ratified and implemented the International Convention for the Prevention of Pollution of the Sea by Oil, 1954 (the "1954 Convention").[11] In 1966, the United States adopted the 1962 international amendments to the 1954 Convention. The United States subsequently adopted the 1969 and 1971 international amendments in 1973.

18–09 In 1972 there commenced a period of several years in which four major pollution prevention statutes were enacted by Congress. In 1972, the Ports and Waterways Safety Act of 1972 was enacted which vastly expanded Coast Guard authority over vessel activity in the navigable and adjoining waters of the United States and amended the Tank Vessel Act of 1936.[12] In 1974, there was enacted the Intervention on the High Seas Act, which implemented United States ratification of the International Convention Relating to Intervention on the High Seas in Cases of Oil Pollution Casualties, 1969.[13] In 1978, the Port and Tanker Safety Act was passed to amend both the Ports and Waterways Safety Act of 1972 and the Tank Vessel Act.[14] Finally, in 1980, Congress passed the Act to Prevent Pollution from Ships, implementing United States ratification of those aspects of Marpol 73/78 not already implemented by the Ports and Waterways Safety Act and the Tank Vessel Act. These statutes will be discussed in this Chapter.

18–10 The United States has also ratified and implemented the Convention on International Regulations for Preventing Collision at Sea, 1972 ("72 COLREGS").[15] The 72 COLREGS are applicable on the high seas and in all waters connected therewith navigable by seagoing vessels, except that certain "inland waters" may be the subject of special rules. The Inland Navigational Rules Act of 1980 codifies the rules applicable to waters inward of certain established demarcation lines.[16] Another statute relating to vessel

---

[11] The 1954 Convention was implemented in August 1961 by the Oil Pollution Act 1961.

[12] P.L. 92–340 (1972).

[13] P.L. 93–248 (1974), enacted as 33 U.S.C. §§1471–1487 (the "Intervention Act"). The Intervention Act was amended in 1978 to reflect U.S. ratification of the Protocol Relating to Intervention on the High Seas in Cases of Marine Pollution by Substances Other than Oil, 1973. The Protocol came into force for he United States on March 30, 1983.

[14] P.L. 95–474 (1978).

[15] P.L. 95–75, codified as 33 U.S.C. §§1601–1608 (The International Navigation Rules Act of 1977). The 1972 Collision Regulations: Implementing Rules are found at 33 C.F.R. Section 81. The Interpretive Rules are found at 33 C.F.R. Section 82. The 1981 IMO amendments to the 72 COLREGS became effective on June 1, 1983 and are found at 48 F.R. 28634.

[16] P.L. 96–591, codified as 33 U.S.C. §§2001–2038, 2071–2073. The COLREGS Demarcation Lines are published at 33 C.F.R. Section 80. Annexes to the Inland Rules are published at 33 C.F.R. §§84–88. Implementing Rules are published at 33 C.F.R. Section 99 and Interpretative Rules are published at 33 C.F.R. Section 90.

conduct in United States waters is the Vessel Bridge-to-Bridge Radio-Telephone Act, which establishes requirements for the operation and use of bridge radio-telephones for vessels operating within the three-mile territorial limit.[17] Neither the act implementing the 72 COLREGS nor the Vessel Bridge-to-Bridge Radio-Telephone Act are discussed in this Chapter because they are largely beyond its scope.[18]

18–11 It should also be noted that the International Convention for the Safety of Life at Sea, 1974 entered into force for the United States on May 25, 1980.[19] The Protocol of 1978 to 1974 SOLAS entered into force on May 1, 1981.[20] In view of the extensive discussion of these Conventions in Chapter 4, further treatment of them is not required in this Chapter.

2. PORTS AND WATERWAYS SAFETY ACT

18–12 The Ports and Waterways Safety Act ("PWSA") was enacted in 1972 after several years of Congressional investigation.[21] The PWSA was amended in 1978 to strengthen and broaden the Coast Guard's authority in dealing with pollution prevention in United States waters.[22]

18–13 The expressed purpose for the legislation is found in the Statement of Policy prefacing the Act.[23] It notes:

"(a) that navigation and vessel safety and protection of the marine environment are matters of major national importance;

(b) that increased vessel traffic in the Nation's ports and waterways creates a substantial hazard to life, property, and the marine environment;

(c) that increased supervision of vessel and port operations is necessary in order to—

(1) reduce the possibility of vessel or cargo loss, or damage to life, property or the marine environment;

---

[17] P.L. 92–63, codified as 33 U.S.C. §§1201–1208. The regulations may be found at 33 C.F.R. Section 26.

[18] The 72 COLREGS are, however, part of the discussion in paras. 4–25 *et seq.*

[19] Executive Order No. 12234, September 3, 1980, 45 F.R. 58801.

[20] TIAS 10009.

[21] P.L. 92–340, originally codified as 33 U.S.C. §§ 1221–1227. The provisions of the PWSA which amended the Tank Vessel Act of 1936, formerly 46 U.S.C. §391a, are discussed commencing at para. 18–29.

[22] P.L. 95–474, codified as 33 U.S.C. §§1228–1232 (Port Safety and Tank Vessel Safety Act of 1978). That portion of the 1978 Act which amended the Tank Vessel Act of 1936, as amended in 1972, is discussed commencing at para. 18–29.

[23] 33 U.S.C. §1221.

(2) . . . ;

(3) insure that vessels operating in the navigable waters of the United States shall comply with all applicable standards and requirements for vessel construction, equipment, manning and operational procedures;

(4) . . . ; and

(d) that advance planning is critical in determining proper and adequate protective measures for the Nation's ports and waterways and the marine environment. . . . "

A. Vessel Operating Requirements

18–14   The Coast Guard is authorised by the PWSA to establish, operate and maintain vessel traffic services which consist of measures for controlling or supervising vessel traffic or for protecting the marine environment.[24] A vessel traffic system may be instituted in any of several ways: (1) reporting and operating requirements; (2) surveillance and communications systems; (3) routing systems; and (4) fair-ways. All of the foregoing have been utilized in one area or another as appropriate. Once established, the Act requires vessels operating in the service areas to comply with or utilize that service. In furtherance of such compliance, the Act authorizes the Coast Guard to require vessels to install certain equipment for communication and navigation in a service area in the interests of vessel safety.[25]

18–15   In hazardous areas, or under conditions of reduced visibility, adverse weather, vessel congestion or other "hazardous circumstances", the Coast Guard is authorised to control vessel traffic by: (1) specifying times of entry, movement or departure; (2) establishing routing schemes; (3) establishing size, draft and speed limitations; and (4) restricting vessel operation in any other manner consistent with vessel safety. The Coast Guard is also authorised to require receipt of pre-arrival messages to permit advance traffic planning.

18–16   Special powers are granted to the Coast Guard to order any vessel in a port or place subject to the jurisdiction, or within the navigable waters of the United States to operate or anchor in any manner the Captain of the Port directs if he reasonably believes (1) that the vessel does not comply wih any applicable law or treaty; (2) that the vessel does not satisfy the conditions established in the

---

[24] 33 U.S.C. §1223.

[25] Small fishing recreation vessels are exempted from any such requirements. 33 U.S.C. §1223(a)(3).

act for port entry; or (3) that safety requires the directed action to be taken.

18–17 Vessel traffic management requirements have been successfully used in Puget Sound, San Francisco Bay the Houston Ship Channel and Berwick Bay.[26] Successful Vessel Traffic Service (VTS) systems are also now in place in the Rosario Straits,[27] Prince William Sound,[28] the Valdez Narrows[29] and New Orleans.[30] A sophisticated vessel surveillance and radio-telephone system was put into place on an experimental basis for the Port of New York and was the subject of extensive testimony during the Coast Guard Board of Inquiry Hearings in connection with the *M.V. Hoegh Orchid/Staten Island Ferry* Collision in the harbour on May 6, 1981. Apart from utilizing the system for general anchor watch activities, it was never subsequently made fully operational, perhaps in part owing to substantial opposition from the local pilots' associations which contended that the system would remove too much discretion from the vessel bridge.

18–18 In considering the establishment of vessel traffic systems, Congress recognized the greater risk of potential liability of the Federal Government, which risk could be even greater where active controls were utilized.[31] The Committee however, accepted the view of the Coast Guard that the potential benefits far outweighed the potential risks and were, therefore, desirable. There have been no cases reported in which the United States has been sued for negligence in the operation of a VTS system.

B. Vessel Operation Regulations

18–19 Pursuant to the PWSA, the Coast Guard published its "Navigation Safety Regulations."[32] The regulations apply to all self-propelled vessels of 1600 gross tons or more operating on the navigable waters of the United States, except the St. Lawrence Seaway. The regulations also do not apply to vessels which are not destined for or departing from a port or place subject to the jurisdiction of the United States and are in innocent passage through United States territorial waters or are in transit through waters which form an international strait.

---

[26] House Rep. No. 95–1384, reported in 1978 U.S. Code Cong. & Admin. News 3281.
[27] 33 C.F.R. §161.170 *et seq.*
[28] 33 C.F.R. §161.301 *et seq.*
[29] 33 C.F.R. §161.370 *et seq.*
[30] 33 C.F.R. §161.401–161.402.
[31] Sen. Rep. No. 92–724, reported at 1972 U.S. Code Cong. & Admin. News 2792.
[32] 33 C.F.R. Part 164.

18-20 The Navigation Safety Regulations are very broad in scope and specify several requirements for vessel operation. The more important regulations concern:

(1) Wheelhouse manning requirements for vessels underway;
(2) Crew and officer ability to perform appropriate navigation duties;
(3) Performance of propulsion machinery and the engine room in confined or restricted waters and the availability of crew in an emergency in such waters;
(4) Manning and mooring requirements for vessels at anchor;
(5) Tests of steering systems, stand-by generators, alarms, and main propulsion machinery before entering or getting underway;
(6) Requirements for charts and publications[33];
(7) Equipment for all vessels[34];
(8) Requirement for vessels of more than 10,000 GRT carrying oil as cargo or residue to be fitted with an IMO-approved Automatic Radar Plotting Aid;
(9) Requirements for tank vessels or more than 10,000 GRT as to steering gear systems as more specifically set out in the Tank Vessel Act[35];
(10) Requirement for LORAN equipment for vessels calling at a port in the Continental United States.

C. Considerations for Port Entry

18-21 The 1978 amendments to the PWSA[36] included for the first time a prohibition against vessels carrying oil from operating in United States waters if the vessel has a history of accidents, pollution incidents or serious repair problems which creates a reason for the Coast Guard "to believe that such vessel may be unsafe or may create a threat to the maritime environmnent."[37] This broad grant of authority has not been judicially interpreted or made the

---

[33] Special provision is made for charts and publications. Foreign charts and publications are acceptable if they contain similar information to the U.S. Government charts or publications. A vessel bound from a foreign port may have the latest charts and publications available at the previous port of call rather than the general requirement for the latest charts.

[34] Such equipment includes a marine radar system, an illuminated magnetic steering compass, a current compass direction table, augyrocompass, an illuminated rudder angle indicator, appropriately displayed vessel manoeuvering characteristics, an echo sounder and equipment for relative motion plotting. Vessels of 10,000 GRT or more must also have a second independent marine radar system.

[35] 46 U.S.C. §§3701–3718.

[36] P.L. 95–474, known popularly as the Port Safety and Tank Vessel Safety Act of 1978.

[37] 33 U.S.C. §1228(a). International legislation is not as broad in its grant of authority. See, e.g. paras. 4–25 to 4–28, 4–38 to 4–41 and 4–58 to 4–62.

subject of published regulations, but the import is clear: vessels involved in pollution incidents may be deemed unsafe for minor violations even if they do not create a serious pollution threat. Thus, vessels involved in a series of even minor operational spills may likely attract Coast Guard attention and possible severe sanction, including denial of port entry.

18–22 Similarly, vessels which fail to comply with applicable regulations issued under the Tank Vessel Act or under any other applicable law or treaty may be denied entry, as may vessels which discharge oil in violation of applicable law, fail to comply with vessel traffic service requirements, fail to meet required manning levels under the Tank Vessel Act (or as may be proper under the circumstances) or do not have at least one licensed deck officer on the bridge while underway who is capable of clearly understanding English.

18–23 Exceptions may be made for provisional entry of a vessel not in compliance with these provisions if the owner or operator proves to the satisfaction of the Coast Guard that the vessel is not unsafe or a threat to the maritime environment *and* that entry is necessary for the safety of the vessel or persons aboard.[38]

D. Coast Guard Investigatory Powers

18–24 The Coast Guard is authorised to investigate any incident, accident or act involving the loss, damage or destruction of any land or water structure or "which affects or may affect the safety or environmental quality of the ports, harbors or navigable waters of the United States."[39] In carrying out this function, the Coast Guard is given *subpoena* power to require witness attendance and production of documents or other evidence which may be required to conduct an investigation. Although the wording of this section of the Act is vague, a fair reading suggests that the Coast Guard need not wait for an actual polluting incident to occur in order to invoke its investigatory powers, but that there only need be an act which "may affect" the safety or environmental quality of United States waters.

E. Enforcement Provisions

18–25 Any person found in violation of the PWSA or regulations issued thereunder, after a notice and opportunity for a hearing, is

---

[38] 33 U.S.C. §1228(*b*).
[39] 33 U.S.C. §1227.

liable to pay a civil penalty not to exceed $25,000 for each viola-tion.[40] Each day of a continuing violation constitutes a separate violation. In assessing the amount of the penalty, the Coast Guard is directed to take into account, among other things, the nature, cir-cumstances, extent and gravity of the acts and, with respect to the violation, the degree of culpability, history of prior offences and the ability to pay.

18–26 Criminal penalties up to the amount of $150,000 per viola-tion or imprisonment of not more than five years, or both, may be assessed against any person found to have wilfully and knowingly violated any provision of the Act.[41]

18–27 Any vessel subject to the Act may be held liable *in rem* for any civil assessment.

18–28 The PWSA authorises the federal courts to issue orders to restrain violations of its provisions.

3. THE TANK VESSEL ACT AND MARPOL 73/78

A. Introduction: The United States Approach to Regulation of Vessel Design Standards

18–29 The Tank Vessel Act of 1936 was substantially revised by the Port and Waterways Safety Act of 1972.[42] The major amend-ment mandated the preparation of rules and regulation for "protec-tion of the marine environment." The 1936 legislation, while broad in the scope of its coverage, related to protection of life and prop-erty without specified reference to environmental protection. It was the intent of Congress in revising the 1936 Act to emphasize its con-cern with the environment.[43]

18–30 After 1972, maritime traffic in United States waters con-tinued to expand, maritime casualties continued to occur and pol-lution damage increased. In 1972, approximately 35 tankers per

---

[40] 33 U.S.C. §1232(*a*).

[41] Additional criminal penalties, consisting of up to $100,000 and ten years imprisonment may be assessed where knowing and wilful conduct causes bodily injury to an officer auth-orized to take any action under the PWSA.

[42] Formerly 46 U.S.C. §391a. A partial recodification of Title 46 of the United States Code, P.L. 98–89, was enacted on August 23, 1983. The provisions of the Tank Vessel Act, for-merly 46 U.S.C. §391a, have been moved to several recodified sections of Title 46. The principal provisions are now contained in 46 U.S.C. §§3701–3718, which is Chapter 37— Carriage of Liquid Bulk Dangerous Cargoes.

[43] Sen. Rep. No. 92–724, reported in 1972 U.S. Code Cong. & Admin. News 2781.

day entered United States ports. Continued growth in United States importation of foreign oil, if not for immediate use, for implementation of the Government's strategic petroleum reserve programme, continued to concern Congress as to the effectiveness of its pollution prevention legislation.[44] The most noteworthy of a series of casualties around the United States Coast was the *Argo Merchant* in late 1976 in which the Liberian flag vessel went aground 28 miles south-east of Nantucket Island resulting in the discharge of approximately 204,00 barrels of heavy heating oil. Both the Liberian Board of Inquiry and the United States District Court found evidence of mismanagement, improper maintenance and the absence of unworkability of crucial navigation and radiotelegraphy equipment.[45]

18–31 As a result of the *Argo Merchant* spill and others, the Secretary of Transportation established in December, 1976 a special task force to make a comprehensive study of the effectiveness of existing safety regulations in the field of spill prevention. An interim report was published in January, 1977 and resulted in a broader interagency review instituted by the President to examine all tanker-related safety laws concerned with marine protection. This resulted in the President's message to Congress of March 17, 1977 in which he announced a "diverse but interrelated group of measures" to reduce pollution risks of vessel oil transportation. Some of the measures included directing the Secretary of Transportation to prepare new regulations on tanker construction and equipment. The Department of State was directed to take diplomatic incentives seeking to improve international standards and to develop an international system for tanker inspection and certification. A more aggressive policy of boarding tankers for inspection by the Coast Guard was announced. It was proposed that the United States ratify Marpol, 1973. Finally, a study of United States pollution clean-up and removal laws was also directed.

18–32 Following the President's message, the Coast Guard in May, 1977 issued proposed rulemaking relating both to domestic and foreign tanker construction and expanded its boarding programme. As a result of the boarding programme, many tanker vessel deficiencies were found and corrected.[46]

18–33 The President's message emphasised the need for an inter-

---

[44] House Rep. No. 95–1304, reported in 1978 U.S. Code Cong. & Admin. News 3274.
[45] *Argo Merchant Limitation Proceedings*, 486 F.Supp. 436 (S.D.N.Y. 1980).
[46] For the period January 21, 1977 to June 8, 1977, 1,262 foreign vessels were boarded and 4,306 deficiencies were found and corrected. House Rep. No. 95–1384, *ibid*.at p. 3276.

national solution to the problem of tank vessel design, construction and operation. This was not a new approach. Congressional findings in the 1972 legislation were to the same effect. As part of its statement of policy, Congress declared[47]:

> "(C) that existing international standards for inspection and enforcement are incomplete, that those international standards that are in existence are often left unenforced by some flag states, and that there is a need to prevent sub-standard vessels from using any port or place subject to the jurisdiction of the United States for the mitigation of the hazards to life, property, or the marine environment.
>
> . . .
>
> (F) that the United States should continue to actively support and encourage efforts to obtain international agreements concerning navigation and vessel safety and protection of the marine environment."

18–34 It was the conclusion of Congress that international standards were grossly deficient in 1972 and that unilateral efforts were required to protect United States' interests. The Senate Report noted[48]:

> "However, notwithstanding the fact that unilateral imposition of tanker construction standards for protection of the marine environment would not appear to violate any treaty to which the United States is a party, the committee recognized that this has traditionally been an area for international rather than national action. Moreover, international solutions in this area are preferable since the problem of marine pollution is world-wide. This point was raised by the Department of State and the Department of Transportation in testimony before the committee. Similarly, the committee received a communication from the Governments of Belgium, Denmark, Finland, the Federal Republic of Germany, Greece, Italy, Japan, the Netherlands, Norway, Spain, Sweden and the United Kingdom expressing concern about this problem and the belief that international agreement should produce a better solution to these problems than unilateral action.
>
> The committee fully concurs that multilateral action with respect to comprehensive standards for the design, construction, maintenance and operation of tankers for the protection of the marine environment would be far preferable to unilateral imposition of standards. However, standards are slow in coming from the multilateral forums. Information received from the Department of State indicated that the Intergovernment Maritime Consultative Organization (IMCO) has looked at various ways to make large vessels more manoeuvrable such as stopping and backing power, vertical axis pro-

---

[47] 46 U.S.C. §3702.
[48] Sen. Rep. No. 92–724, *ibid.* at p. 2783.

pellors, multiple screws, stern anchors, and side thrusters. However, IMCO has as yet produced no concrete proposals in any of these areas.

As a practical matter, the IMCO record has been rather dismal to date in the area of design and construction standards for protection of the main environment. Material received by the committee indicated that the most expensive standard considered by IMCO, indeed the only construction standard that has received any serious consideration, is that dealing with compartmentation of tanks to limit the out flow of cargo in the event of an accident. Expert shipbuilding estimates indicated that the incremental cost of the originally proposed standard amounts to 2 per cent. on a tanker of 300,000 deadweight tons, and far less on smaller tankers. Nevertheless, when IMCO finally acted on this standard, it was watered down beyond the original proposal. While the committee remains committed to the proposition that multilateral action in this area is preferable, it is not willing to sacrifice the objective of protection of the marine environment on the altar of that principal. Much more rapid and comprehensive action will be required if the United States is to continue to rely on multilateral forums.''

It was mainly as a result of this position that the 1978 Conference on Tanker Safety and Pollution Prevention (the "TSPP Conference") was convened in London to try to persuade the United States not to pursue its unilateralist policy.

18–35 A problem associated with the setting of unilateral standards in the inherently international field of vessel design, construction and operation is the scope of coverage of any such legislation. In this regard, Congress recognised some practical facts, among which were that United States oil importation is principally on foreign hulls[49] and that United States flag vessels would operate at a competitive disadvantage if foreign vessels were not made subject to the same standards. Since it was the considered view of Congress that unilateral imposition of these standards for all vessels calling at ports in the United States violated no international treaty obligation, this position was adopted.[50]

18–36 Shortly after passage of the 1972 amendments of the Tank Vessel Act, Marpol 1973 had been adopted, but by 1977 had not been ratified by a sufficient number of states to come into force.

---

[49] The Senate Report noted that 85% of imported oil was carried on foreign hulls. *Ibid.* at p. 2782.

[50] The Department of State, in a letter to the Chairman of the Committee on Commerce dated September 23, 1971, indicated that there may be a violation of certain certification procedures adopted in SOLAS, 1960, but this view was not accepted by Congress. 1972 U.S. Code Cong. & Admin. News 2799.

The Presidential message of March, 1977 called for United States ratification of Marpol 1973 and appropriate legislation was submitted to the Senate for advice and consent in September, 1977. Hearings on oily waste shoreside reception facilities, among others, delayed final Congressional consideration, and in effect, put further consideration of Marpol 1973 aside until 1980 after adoption of the 1978 Marpol Protocol in February, 1978. Marpol 73/78 was signed by the United States, subject to ratification on June 27, 1978. On July 2, 1980, the Senate gave its advice and consent to the ratification. Implementing legislation, the Act to Prevent Pollution from Ships, was enacted by Congress and signed into law on October 21, 1980.[51] Marpol 73/78 came into force for the United States on October 2, 1983.[52]

18–37 The 1978 amendments to the Tank Vessel Act were enacted in October, 1978 shortly after the 1978 TSPP Conference.[53] The 1978 amendments prescribed certain minimum standards for any self-propelled vessel which were consistent with the standards adopted by the Conference. While the 1978 amendmets were not meant to implement any international agreement and were deemed to represent the independent evaluation of Congress as to standards for vessels operating in United States waters, the Congressional Committee adopted the TSPP international standards to the extent feasible. It expressed an unwillingness to await ratification and implementation of any international agreement then under consideration, referring to Marpol 73/78.[54] As will be discussed, the 1978 amendments went further than the TSPP recommendations in certain respects.

B. Relationship of the Tank Vessel Act and the Act to Prevent Pollution from Ships

18–38 The Tank Vessel Act was amended in minor respects in 1980 and 1982, principally to reflect the enactment of the Act to Prevent Pollution from Ships in 1980. Since the 1978 amendments to the Tank Vessel Act presaged the ratification of Marpol 73/78, the Act to Prevent Pollution from Ships may be considered as

---

[51] P.L. 96–478—"Act to Prevent Pollution from Ships," 33 U.S.C. §§1901–1911.

[52] The Coast Guard has taken the position that the proposed amendments and uniform interpretations to Marpol 73/78 set out in the Marine Environmental Protection Committee's ("MEPC") Circular Letter 97 of April 6, 1982 fully conform to the spirit and letter of Marpol 73/78. The MEPC recommendations were incorporated into the proposed rulemaking for the Marpol regulations, 48 F.R. 30674.

[53] A detailed discussion of that Conference may be found in paras. 3–37 *et seq.*

[54] House Report No. 95–1384, *ibid.* at p. 3289–90.

remedial legislation to implement those aspects of Marpol 73/78 which had not previously been implemented by the 1978 amendments. The two Acts may thus be seen as complementary with the intention of giving Marpol 73/78 full domestic effect.

C. Applicability of the Tank Vessel Act: Scope of Vessel Coverage

18–39   The Tank Vessel Act applies to any vessel regardless of tonnage, size or manner of propulsion, whether or not self-propelled or carrying freight or passengers for hire, which is a vessel of the United States or which operates on or enters the navigable waters of the United States or which transfers oil in any port or place subject to the jurisdiction of the United States, and which is constructed or adopted to carry, or which carries, oil in bulk as cargo or residue.[55] The only exceptions are for public vessels,[56] oil exploration vessels of not more than 500 gross tons used in offshore production operations, certain fishing vessels of not more than 500 tons engaged in the fishing industry off certain Western states, certain fish processing vessels of not more than 5,000 gross tons and any foreign vessel not bound for a United States port in innocent passage in United States navigable waters.[57]

D. Regulatory Authority of the Coast Guard under the Tank Vessel Act

18–40   The Act authorises the Coast Guard to issue regulations for the design, construction, alteration, repair, maintenance, operation, equipping, personnel qualification or manning of vessels subject to the Act, as may be necessary for increased protection against hazards to life and property, for navigation and vessel safety, and for enhanced protection of the marine environment.[58] Such regulations may provide different standards for vessels in domestic trade and the standards established may vary from international standards. The areas for regulation include[59]:
   (i) superstructures, hulls, cargo holds or tanks, fittings, equip-

---

[55] 46 U.S.C. §391a(3).
[56] A "public vessel" is defined as one which is owned or chartered by demise and operated by the United States or a foreign government and which is not engaged in public service; 46 U.S.C. §2101(24).
[57] 46 U.S.C. §3702.
[58] 46 U.S.C. §3703.
[59] 46 U.S.C. §3703(i–vii).

ment, appliances, propulsion machinery, auxiliary machinery, and boilers;

(ii) handling or stowage of cargo, the manner of such handling or stowage of cargo, and the machinery and appliances used in such handling or stowage;

(iii) equipment and appliances for lifesaving, fire protection, and prevention and mitigation of damage to the marine environment;

(iv) manning of such vessels and the duties, qualifications, and training of the officers and crew thereof;

(v) improvements in vessel manoeuvering and stopping ability and other features which reduce the possibility of collision, grounding, or other accidents;

(vi) reduction of cargo loss in the event of a collision, grounding, or other accident: and

(vii) reduction or elimination of discharges during ballasting, deballasting, tank cleaning, cargo handling, or other such activity.

18–41 In issuing such regulations, the Act prescribed certain minimum standards depending on the age, type and size of the vessel.[60] The minimum standards may be generally described as setting out requirements for certain sized crude oil and product tankers relating to segregated ballast tanks, a crude oil washing system, a cargo tank protection system with a fixed deck froth system and a fixed inert gas system. The minimum requirements have gradually come into effect over a period of several years, the smallest size product carrier having become obligated to comply by 1980 and certain existing crude oil tankers being required to comply by June 1, 1986, or by the date the vessel reaches fifteen years of age, whichever is sooner.[61] The detailed regulations mandated by the Act are set forth at Title 33 C.F.R. Part 157 (1984).

18–42 The regulations generally follow the standards approved by the 1978 TSPP Conference except that whereas the TSPP Conference imposed no new requirements on existing crude oil tankers and product carriers of between 20,000 to 40,000 DWT, the Tanker Vessel Regulations do impose minimum requirements on vessels of this size and type which reach fifteen years of age before January 1, 1985.

---

[60] 46 U.S.C. §§3705–3709.

[61] It is clear that the latter requirement was meant to encourage the retirement of such older vessels before June 1, 1968 rather than make the expenditures to comply with the minimum requirements.

E. Certificates of Compliance under the Tank Vessel Act

18–43 No vessel of the United States may carry oil or hazardous materials as cargo or in residue unless it has a certificate of inspection endorsed to indicate that it is in compliance with the Tank Vessel Act and regulations.[62]

18–44 No foreign vessel subject to the Act may operate on or enter the navigable waters of the United States or any port or place under the jurisdiction of the United States unless it has been issued a Coast Guard Certificate of Compliance. A Certificate of Compliance may not be issued unless the vessel has been examined by the Coast Guard. A foreign vessel may be given provisional entry for purposes of having an inspection conducted.

18–45 For purposes of compliance, the Coast Guard may accept a Certificate, endorsement or document issued by a foreign nation pursuant to treaty, convention or agreement to which the United States is a party.

18–46 A vessel which does not have a valid certificate may not carry oil in bulk or in residue unless the Certificate is endorsed to permit such carriage. Certificates may be valid for up to twenty-four months subject to renewal. A Certificate may be revoked if the Coast Guard determines that the vessel is not in compliance with the conditions upon which it was issued.

F. Personnel and Manning Standards Under the Tank Vessel Act

18–47 The Coast Guard is authorized to issue regulations for the manning of vessels of the United States, such regulations to include the duties, qualification and training of all officers and crew.[63] These regulations may include standards relating to instruction in vessel and cargo handling and vessel navigation under normal and emergency situations and licensing qualifications, minimum health standards, retraining and special training for upgrading positions and conditions for certain kinds of certification.

18–48 As to manning of foreign vessels, the Coast Guard is delegated the authority to evaluate periodically the manning, training, qualification and watchkeeping standards of any foreign flag country whose vessels operate in United States navigable waters.[64]

[62] 46 U.S.C. §§3711–3712.
[63] 46 U.S.C. §9102.
[64] 46 U.S.C. §9101.

457

In conducting such evaluation, the Coast Guard shall determine whether the foreign state's standards are comparable to or more stringent than United States or acceptable international standards. In enacting this section, it was clearly contemplated that any foreign vessel whose flag country standards are found to be unacceptable may be denied entry to United States waters or ports under the port entry provisions of the Port and Waterways Safety Act.[65] While this grant of authority has never been used to deny entry to a particular flag, this legislation clearly enables the Coast Guard to take such drastic measures.

18–49 The Act makes special provision for a complement of tankermen aboard United States oil-carrying vessels and foreign oil-carrying vessels operating in United States waters.[66]

G. Prohibited Acts and Enforcement Provision Under the Tank Vessel Act

18–50 Any violation of any provision of the Act, regulations or directives, or any refusal to permit a Coast Guard officer to board a vessel to conduct an inspection, is unlawful.[67] It is also unlawful to operate a vessel in waters within the jurisdiction of the United States which is not in compliance with the Act or its regulations.

18–51 Civil penalties for each violation of the Act or its regulations may be assessed up to $25,000. Civil penalties shall be assessed by the Coast Guard after notice and an opportunity for a hearing. Vessels are liable *in rem* for any civil penalty assessed and sailing clearance may be withheld until a surety for such penalties is furnished.

18–52 Any person who wilfully and knowingly violates the provisions of the Act shall be subject to a fine for each violation of not more than $50,000 or five years imprisonment or both.[68]

H. Vessel Inspections Under the Tank Vessel Act

18–53 At least once each year all vessels subject to the Act shall be examined or inspected.[69] Any vessel over ten years of age must

---

[65] See, *e.g.* PWSA, 33 U.S.C. §1228, discussed in para. 18–21, above.

[66] 46 U.S.C. §§8703, 9101, 7317.

[67] 46 U.S.C. §3713.

[68] The penalty may be increased to $100,000 or ten years imprisonment, or both for any person who, in the wilful and knowing violation of the act uses a dangerous weapon causing bodily injury, or fear thereof, to any officer charged with enforcing the provisions of the Act.

[69] 46 U.S.C. §3714.

undergo a special and detailed inspection of structural strength and hull integrity. A Certificate of Compliance shall be issued to all vessels satisfying the requirements of the Act.

18–54 Foreign inspections by authorized agents of the Coast Guard may be accomplished if a Coast Guard officer is unavailable, but in such circumstances only a temporary certificate may be issued.

I. Tank Washing Prohibition Under the Tank Vessel Act

18–55 The Act prohibits a vessel from conducting transfer operations in United States waters or ports after having cleaned its tanks and discharged the residues on the high seas except, of course, if such discharge is in compliance with Marpol 73/78.[70] Unlawful discharges have been a prevalent practice and this section was instituted to discourage such activities.

J. Marpol Act Applicability

18–56 As discussed, the Act to Prevent Pollution from Ships (the "Marpol Act")[71] is the implementing legislation for that portion of Marpol 73/78 which had not been made a part of domestic law by virtue of the 1978 amendments to the Tank Vessel Act. The Marpol Act directs the Secretary of the Department in which the Coast Guard is operating to administer the Marpol Protocol and issue regulations to carry out this obligation.[72] Marpol 73/78 came into force for the United States on October 2, 1983. The Act implements the articles of the Marpol Protocol as well as Annex I—Regulation for Prevention of Pollution by Oil and Annex II—Regulations for the Control of Pollution by Noxious Liquid Substances in Bulk.[73] The three optional Annexes dealing with packaged harmful substances, sewage and garbage were not implemented.[74]

18–57 The Act applies to: (1) ships[75] of United States registry or operated under other authority of the United States[76]; (2) ships

---

[70] 46 U.S.C. §3716.

[71] P.L. 96–478, October 21, 1980, 33 U.S.C. §§1901–1911.

[72] 33 U.S.C. §1903.

[73] 33 U.S.C. §1901(1–2).

[74] The Act repeals the implementing legislation for the 1954 Oil Pollution Convention and all amendments thereto; 33 U.S.C. §§1001–1015.

[75] A "ship" means a vessel of any type, including hydrofoils, submersibles, floating craft whether self-propelled or not, and fixed or floating platforms; 33 U.S.C. §1901(8).

[76] "Other authority" of the United States refers to fixed platforms, deep seabed mining vessels and other vessels operating under U.S. authority; House Rep. No. 96–1224, reported in 1980 U.S. Code Cong. & Admin. News 4849.

registered in or of the nationality of another Marpol signatory while in the navigable waters of the United States; or (3) ships registered in or of the nationality of a country not a Marpol signatory while in the navigable waters of the United States.[77] With respect to ships of such non-signatory countries, the Coast Guard is directed to pre-scribe regulations to ensure that their treatment is not more favour-able than that accorded vessels of Marpol signatories.

18–58 The Act does not apply to a warship, naval auxiliary, or other ship owned or operated by the United States or any other ship excluded by the Marpol Protocol. Only ocean-going ships are within Marpol regulation.[78]

K. Marpol Certificates and Detention Orders

18–59 The Coast Guard is delegated authority to issue Marpol Protocol Certificates of Vessel Compliance (IOPP Certificates) on behalf of the United States.[79] In recognizing the international con-cept of mutuality of validity, an IOPP certificate issued by another Marpol signatory country has the same validity as an IOPP Certifi-cate issued by the Coast Guard.

18–60 IOPP Certificates are issued by the Captain of the Port (COTP) or Officer in Charge of Marine Inspections (OCMI) at the port of inspection.[80] United States inspected ships may be issued Certificates not to exceed four years validity: United States uninspected vessels may be issued Certificates not to exceed five years validity.

18–61 The Coast Guard is authorized to conduct inspections to determine whether a ship has a valid IOPP Certificate. If clear grounds exist which "reasonably indicate" that the condition of the ship or its equipment does not agree with the particulars of its Cer-tificate, then a further inspection is authorised.

18–62 If a valid certificate is not on board or if an inspection reveals that the vessel's condition does not "substantially agree" with the particulars of the Certificate, the ship "shall be detained" until, in the opinion of the Coast Guard, the ship "can proceed to sea without presenting an unreasonable threat of harm to the mar-ine environment."[81] Detention orders issued by local Coast Guard

---

[77] 33 U.S.C. §1902.
[78] 33 U.S.C. §1903.
[79] 33 U.S.C. §1904.
[80] Proposed Rule 33 C.F.R. 151.19(c), as published in 48 F.R. 30681.
[81] 33 U.S.C. §1904(e). A detention order may authorize a ship to proceed to the nearest appropriate available shipyard.

personnel are reviewable through normal administrative procedures up to the Secretary of Transportation. Review of the Secretary's Determination is available in the Federal Court on limited administrative procedures grounds.[82] A ship unreasonably detained or delayed is entitled to compensation for any loss or damage suffered.[83]

L. Marpol Act Pollution Reception Facilities

18–63 The Act authorises the Secretary of Transportation, in consultation with the Administrator of the Environmental Protection Agency, to establish regulations setting criteria for determining the adequacy of reception facilities required under the Marpol Protocol.[84] The regulations also establish procedures for certification of the adequacy of reception facilities.[85]

18–64 Reception facilities are to be established in accordance with the Marpol Protocol mandate in crude oil loading ports, in ports loading more than 1,000 metric tons of product oil and in ship repair ports. Under the Act, the Secretary of Transportation is to issue reception facility certificates after inspection, and to publish in the Federal Register a list of ports or terminals holding a valid certificate.[86] Once fully implemented, port entry will be denied to seagoing ships required to retain onboard oily residues and mixtures for discharge at a reception facility, if the port of destination is required to have a Marpol reception facility and such port does not hold a valid certificate.[87] It will thus become necessary for shipowners and operators to determine in advance of accepting any discharge port orders whether or not that port is in compliance. The failure to do so could well result in substantial claims for demurrage or deviation.

M. Reporting of Incidents under Marpol

18–65 The Master or other person in charge of a ship must report to the Coast Guard an incident as soon as he acquires knowledge of its occurrence.[88] The Secretary is obliged to make a report

---

[82] 5 U.S.C. §551 *et seq.*

[83] 33 U.S.C. §1904(*h*).

[84] 33 U.S.C. §1905(*a*).

[85] The draft regulations are contained in a Notice of Proposed Rulemaking, 49 F.R. 25196 (June 19, 1984).

[86] 33 U.S.C. §1905(*c–d*).

[87] 33 U.S.C. §1905(*e*).

[88] 33 U.S.C. §1906(*a*). An "Incident" is defined in the same manner as the Marpol Protocol, as an actual or probable discharge of oil in violation of the Marpol Protocol.

to IMO, the flag state of the ship involved, and to any other nation which may be affected, in accordance with Article 8 of the Convention.[89]

N. Marpol Violations, Penalties and Enforcement Provisions

18–66 It is unlawful to act in violation of any provision of the Act, the regulation or the Marpol Protocol.[90] The Secretary is directed to co-operate with other parties to the Convention in the detection of violations and in the enforcement of the Protocol. The Secretary is given broad authority, including the right to issue subpoenas to conduct an investigation of alleged violations. Shipboard investigations may be undertaken while the vessel is at a port or terminal in United States waters to determine whether the ship has discharged a harmful substance in violation of the Protocol or to comply with a request from another party to the Protocol to investigate whether the ship may have discharged a harmful substance anywhere in violation of the Protocol.[91] The latter investigation may be undertaken only when the requesting party has furnished sufficient evidence to allow the Secretary reasonably to believe that a prohibited discharge has occurred as alleged. If the investigation involves a foreign ship and indicates that a violation has occurred, the Secretary shall take the action required by Article 6 of the Convention.

18–67 As with the PWSA, knowing violation of the Protocol, the Marpol Act or any regulation may result in criminal penalties of up to five years imprisonment or not more than $50,000 fine, or both.[92] Civil penalties may be assessed by the Secretary of up to $25,000 per violation, except that the making of a false, fictitious or fraudulent statement or representation permits the Secretary to impose a civil penalty of up to $50,000 for each such statement or representation.[93] Civil penalties may be assessed *in rem* against the vessel and clearance may be withheld until appropriate security is furnished.[94]

18–68 Any person having an interest which is or could be adver-

---

[89] 33 U.S.C. §1906(*b*).

[90] 33 U.S.C. §1907(*a*). Of course, the penalty provisions only apply within the scope of the Marpol Act's jurisdiction, see para. 18–58, above. Thus, for example, a Liberian vessel on the High Seas which is in violation of the Marpol Protocol is not in violation of the Marpol Act's penalty provisions.

[91] 33 U.S.C. §1907(*c*).

[92] 33 U.S.C. §1908(*a*).

[93] 33 U.S.C. §1908(*b*).

[94] 33 U.S.C. §1908(*c–d*).

sely affected may bring an action in the federal court in his own name against any alleged violator or against the Secretary where it is alleged that he failed to perform a non-discretionary act or duty, provided only that the person must give 60 days notice to the alleged violator, the Secretary and the Attorney General of the United States. An action may not be brought if the Secretary has commenced and is diligently conducting an enforcement action with respect to the violation.[95]

O. Exclusivity of the Federal Prevention Statutes

18–69 A recurring problem in the field of United States pollution legislation is the Constitutional reservation afforded the states which permits states to regulate any area not pre-empted by Federal authority.[96] While it is difficult to imagine an area which has greater need for uniformity than that relating to oil tanker requirements for design, construction and operation, the question of state authority to legislate in this field has arisen in our courts.

18–70 The leading case of *Ray* v. *Atlantic Richfield Co.*, 435 U.S. 151 (1978), involved a challenge to the constitutionality of the State of Washington Tanker Law which governed the design, size and movement of oil tankers in Puget Sound. The law: (1) required tankers over 50,000 DWT to carry a licensed state pilot while navigating the Sound; (2) required tankers of between 40,000–125,000 DWT to satisfy certain design and safety standards, or, alternatively, to use tug escorts while operating in the Sound; and (3) barred vessels over 125,000 DWT from operating in the Sound.

18–71 A three-judge district court panel held the statute void and unconstitutional in that the PWSA was designed to insure vessel safety and protection of the navigable waters from oil pollution and the Secretary of Transportation had implemented regulations which pre-empted this state legislation in its entirety. The Supreme Court affirmed the decision in part and reversed the decision in part, finding that certain aspects of the state law were proper since there had not been complete pre-emption of the field by passage of the PWSA or its regulations. In particular, the requirement for a local pilot to be aboard vessels over 50,000 DWT was held to be in direct conflict with federal law as to coastwise vessels, but was per-

---

[95] 33 U.S.C. §1910.
[96] This same issue arises in connection with federal and state regulations concerning recovery of oil spill clean-up and removal expenses. A more complete discussion is presented in paras. 21–73 *et seq.*

missible under the PWSA, and thus was enforceable with respect to foreign vessels.

18–72 As to the imposition of certain design and construction standards for tankers of between 40,000–125,000 DWT, failing which appropriate tug escort was required, the court agreed that the design and construction standards were invalid since it was clear that Congress intended to impose uniform national standards for tanker design and construction which would preclude more stringent state standards. However, the alternative requirement, that appropriate tug escort be provided for tankers over 40,000 DWT was held to be a valid exercise of the state's right to protect its waters. Having agreed that the tug escort provision was valid, the Court next determined that the state was also free to exempt from its provisions vessels meeting certain design and construction standards. Thus, the practical effect of this provision was upheld, although the language of the Act which sought to "impose standards" was stricken.

18–73 As to the prohibition of tankers in excess of 125,000 DWT the court held that the Federal Government had pre-empted the field of tanker size limitations and thus the states were not free to impose such restrictions.

18–74 In *Chevron (United States) Inc.* v. *Hammond*,[97] the Ninth Circuit Court of Appeals, reversing the District Court, upheld the right of the State of Alaska to prohibit oil tankers from discharging ballast from their cargo tanks into the State's territorial waters. The court rejected the shipowner's argument that this state legislation was pre-empted by the PWSA and the Tank Vessel Act which, although prohibiting deballasting within 50 miles of shore, excepted certain clean ballast discharges from a stationary vessel in calm water. The court noted that nothing in the federal prevention scheme precluded states from imposing higher operating standards within their territories. Reconciling *Ray*, it was observed that the Alaska statute did not attempt to legislate in the tanker design and construction field which was there held to be an area for uniform federal legislation.

18–75 It is thus clear that states are free to legislate in the prevention area where a uniform national scheme is neither imposed nor required. Thus, vessel operational requirements seem very much to be an area for state regulation so long as those require-

[97] 726 F.2d 483 (9th Cir. 1984).

ments, even if more stringent, do not conflict with national law or regulations.

18–76 Of course, there may arise a situation where a state imposes onerous operational requirements in a thinly concealed effort to require certain design or construction standards. No such case has arisen, but an argument along these lines was made in *Ray*, which the court found no need to address in view of the reasonableness of the tug escort provision. However, should such a case come before a court, it is likely that clearly onerous operational requirements will be struck down as merely a facade for state regulation in a pre-empted field. The types of situations in which this could occur will have to abide the event.

4. INTERVENTION ON THE HIGH SEAS ACT

18–77 The United States signed the International Convention Relating to Intervention on the High Seas in Cases of Oil Pollution Casualties, 1969 (the "Intervention Convention") on November 29, 1969 and ratified it on September 29, 1971 upon the advice and consent of the Senate. Implementing legislation was passed on February 5, 1974 (the "Intervention Act")[98] and the Convention entered into force on May 6, 1975 following deposit of the fifteenth ratification or accession.

18–78 The Protocol Relating to Intervention on the High Seas in Cases of Marine Pollution by Substances Other than Oil, 1973 (the "Intervention Protocol") was signed by the United States on March 7, 1974 and was subsequently ratified. Implementing legislation amending the Intervention Act was passed on June 26, 1978[99] and the Intervention Protocol came into force for the United States on March 30, 1983.

18–79 The Intervention Convention is discussed in detail in paragraphs 6–14 *et seq.*, and requires no further analysis here. Certain provisions of the Intervention Act, however, should be considered. The Intervention Act authorizes the Secretary of Transportation to determine whether there is a "grave and imminent danger of major harmful consequences to the coastline or related interests of the United States."[1] In making this determination, the Secretary is directed to consider, among others, the

---

[98] 33 U.S.C. §§1471–1487.
[99] P.L. 95–302.
[1] 33 U.S.C. §1473.

effect of the pollution damage on human health, fish, shellfish, and other living marine resources, wildlife, coastal zone and estuarine activities and private shorelines and beaches. The Secretary is required, except in cases of "extreme urgency," to consult with other affected countries, particularly the flag country of the ship involved, before taking action.[2]

18–80   In taking preventive measures under the Intervention Act, the Secretary is admonished to act proportionately to the actual or threatened damage.[3] Such measures "may not go beyond what is reasonably necessary to prevent, mitigate, or eliminate that damage." Such factors as the extent and probability of imminent danger, the likelihood of the success of the measures proposed to be undertaken and the likely damage to be caused by such measures, are specifically required to be examined.

18–81   In taking measures, the Secretary must use his best endeavours to avoid the risk to human life, to render all possible aid to distressed persons and not unnecessarily to interfere with rights and interests of others, including the flag state and other foreign states also threatened by damage.[4]

18–82   The Intervention Act provides both for the recovery of damages caused by the taking of unreasonable measures and a forum for bringing suits against the United States in such event.[5]

18–83   Funding for activities in connection with the Intervention Convention is made available through the revolving fund established under CERCLA,[6] discussed more fully in Chapter 21.

5. CONCLUSION

18–84   As a result of its strong criticism in the 1970s of the lack of international consensus on improvement of vessel design, construction and operation standards, together with its willingness to take unilateral action which would impose its standards on all oil-carrying vessels trading within its waters, the United States was able to bring about substantial changes to the international regime, reflected principally in the 1978 Marpol amendments. Since that

---

[2] 33 U.S.C. §§1475–1476.
[3] 33 U.S.C. §1477.
[4] 33 U.S.C. §1478.
[5] 33 U.S.C. §1479. The United States Court of Claims or any district court of the United States has jurisdiction over these matters.
[6] 33 U.S.C. §1486.

time, the United States has enacted domestic legislation which supports and implements the existing international legislation in the prevention field. The supplementary legislation aimed at providing the Coast Guard with adequate authority to enforce compliance with the standards they are charged to uphold and to co-operate on an international level provides a very strong ground for such activities. It is unfortunate that the United States has not participated as fully in the international regime for oil pollution liability as it has done in the field of oil pollution prevention. The consequences of this non-adherence are fully explained in the remaining chapters of this section.

19. Federal Oil Pollution Control and Clean-up—The Federal Water Pollution Control Act

SUMMARY

19–01 Examined in detail in this chapter is the Federal Water Pollution Control Act ("FWPCA"), which is the most significant water pollution legislation governing vessels in and about United States waters. Unlike the CLC and Fund Conventions, the FWPCA establishes both strict but limited liability for prohibited discharges by all vessels (except public vessels) and a fund for government and discharger clean-up costs. State governments may claim against the fund for certain clean-up costs expended which are consistent with a published National Contingency Plan. The detailed provisions of the National Contingency Plan, which are applicable in the event of a discharge, are discussed in this Chapter.

19–02 It is important to understand the relationship between the FWPCA and other pollution liabilities under United States law. The FWPCA only provides liability for clean-up costs with respect to pollution of the navigable waters of the United States or in connection with certain federally approved activities beyond the territorial sea involving exploitation of the outer Continental Shelf and use of deepwater ports. Thus, private recovery for pollution-related damages is not covered by the FWPCA, nor is state recovery under circumstances in which a state determines that the federal remedy is inadequate for it or for those persons which it protects by its legislation. Also, because of what may be seen as special areas of federal concern, remedies are also provided to the federal government (and to private persons injured by oil pollution) under separate legislation for pollution by vessels involved with outer Continental Shelf Land oil activities or which call at deep-water ports. A separate federal statutory pollution regime dealing with ocean transportation of trans-Alaska pipeline oil must also be considered where appropriate. These activity-related pollution statutes are discussed in Chapter 20. Finally, common law remedies may be available where no statutory remedy exclusively applies. In this perspective, it will be observed that the FWPCA is only one, albeit extremely important, aspect of the United States water pollution law.

19–03 As a result of the continued vitality of the common law which offers a fault-based theory of recovery for oil pollution clean-up costs, also discussed in this chapter is whether or not the FWPCA provides the federal government with its exclusive remedy for clean-up costs and, if so, whether this is so both for dischargers and third parties who may be sole or contributing causes of the discharge.

1. PROHIBITED DISCHARGES

19–04 The teeth of the Federal Water Pollution Control Act,[1] in so far as vessel oil spills are concerned, is section 311, codified as 33 U.S.C. §1321 ("section 311"). Section 311 was added to the FWPCA in 1972 with enactment of the Water Quality Improvement Act,[2] and was amended substantially in 1973,[3] 1977, (The Clean Water Act),[4] 1978,[5] and 1980.[6]

A. Scope

19–05 Section 311, in declaring that "it is the policy of the United States that there should be no discharges of oil . . . into or upon the navigable waters of the United States, adjoining shore-lines, or into or upon the water of the contiguous zone" or in connection with other specified activities, prohibits discharges in such waters and in such quantities as may be harmful.[7] The specified activities involve those in connection with the Outer Continental Shelf Lands Act and the Deepwater Port Act which are discussed in Chapter 20, below. Similarly, discharges are prohibited under circumstances where they may affect natural resources under management authority of the United States except where permitted under MARPOL 73/78.[8]

---

[1] 33 U.S.C. §§ 1251–1376.
[2] 86 Stat. 862.
[3] 87 Stat. 906.
[4] 91 Stat. 1593–1596.
[5] 92 Stat. 2467.
[6] 94 Stat. 2303 and 94 Stat. 3300.
[7] 33 U.S.C. §§ 1321(b)(1)(3).
[8] International Convention for the Prevention of Pollution from Ships, 1973, as modified by the Protocol of 1978 ("MARPOL 73/78"). Ratified and implemented by P.L. 96–478 (1980). MARPOL 73/78 entered into force for the United States on October 2, 1983. For analysis of MARPOL 73/78, see Chap. 3.

19–06 Section 311, like the FWPCA in general, is concerned with water pollution by any substance from any source. Thus, for instance, although covered by section 311, pollution by hazardous substances other than oil is not covered in this book. Oil is defined as oil of any kind or in any form, including, but not limited to, petroleum, fuel oil, sludge, oil refuse and oil mixed with wastes other than dredged oil.[9] Therefore, it covers non-persistent oils, as to which, see paragraphs 10–12 *et seq.*, above. All kinds of classes of vessel are covered by the Act.[10]

19–07 Since the Act only prohibits discharges in such quantities as may be harmful it has been left to courts and administrators to consider the parameters of this mandate.[11] Section 311 itself directs the President to promulgate regulations as to what quantities of oil may be harmful.[12] The so-called "sheen test" was devised to implement this requirement.[13] Included within the meaning of harmful quantities are discharges which "cause a film or sheen upon or discoloration of the surface of the water or adjoining shorelines or cause a sludge or emulsion to be deposited beneath the surface of the water or upon adjoining shorelines."[14] A specific exception is provided for discharges of oil from a properly functioning vessel engine, however, accumulated oil in the bilges not is not excepted.[15] While certain criteria as to what constitutes a harmful discharge are so clear as not to be subject to reasonable dispute, the sheen test has been criticized on the ground that it fails to establish a proper quantitative standard to meet the requirements of the statute. The sheen test has, however, withstood court scrutiny and

[9] 33 U.S.C. § 1321(a)(1).

[10] "Vessel" is defined as "every description of watercraft or other artificial continuance used, or capable of being used as a means of transportation on notes other than a public vessel." 33 U.S.C. § 1321(a)(3). Public vessels, which include vessels owned or bareboat chartered and operated by the United States, state-owned vessels and foreign nation owned vessels are not within the coverage of the Act unless such vessels are engaged in commerce. 33 U.S.C. § 1321(a)(4).

[11] The original language of the Act in 1972 was "harmful quantities" but because the absolute nature of this requirement led certain courts to require some evidence of actual harm, the Act was amended in 1978 to "quantities as may be harmful." An example of one such case was *United States* v. *Chevron Oil Co.*, 583 F. 2d 1357 (5th Cir. 1978). See Helfrich, Problems in Pollution Response Liability Under Federal Law: FWPCA Section 311 and the Superfund, 13 Jl Maritime Law & Commerce 455 (Jul.-Oct., 1982).

[12] 33 U.S.C. § 1321(b)(4). The use of the term "President" in domestic legislation refers to a grant of authority to the President of the United States, who is empowered to delegate this authority to an appropriate administrative agency of the federal government. With respect to vessel oil pollution, the Coast Guard or the Department of Transportation, in which the Coast Guard operates, is the usual recipient of such grants of authority.

[13] 40 C.F.R. §§ 110.3, 110.4.

[14] *Ibid.* "Sheen" is defined as "an iridescent appearance on the surface of the water." 40 C.F.R. § 110.1(1).

[15] 40 C.F.R. 110.6.

remains a cornerstone of government enforcement, especially in relation to smaller spills where other more obvious effects of a spill are not evident.[16]

B. Notification

19–08   Any person "in charge of a vessel" must immediately notify the appropriate governmental authority, in this case the Coast Guard, as soon as he has knowledge of a prohibited discharge from his vessel.[17] The term "in charge of a vessel," which is not defined in the Act, has been judicially held to include a corporation which is the owner or operator of a vessel.[18] The notification procedures to the United States Coast Guard, the agency delegated with authority to handle vessel discharges, are set out in the Federal Regulations.[19] This has been interpreted by one court as extending to masters of vessels and owners who acquire knowledge of a spill.[20] There are no reported cases against pilots, port superintendents or others similarly situated but one case suggests that the purpose of the notification requirement is to facilitate implementation of measures calculated to minimize pollution damage and that such policy must be used as a guideline to determine what class of persons are covered by the requirement.[21] Notification has been held to include the initiation of a chain of events which leads to actual notification to the responsible authorities.[22]

19–09   The failure to notify the appropriate agency is a crime punishable by a fine of not more than $10,000, imprisonment of not more than one year, or both.

2. CIVIL PENALTIES FOR PROHIBITED DISCHARGES

19–10   The government has access under the Act to two methods of collecting a civil penalty as a result of a prohibited discharge: (1) administrative determination, or (2) civil action. The former

---

[16] *United States* v. *Boyd*, 491 F.2d 1163 (9th Cir. 1973); *Ward* v. *Coleman*, 423 F.Supp. 1352 (D.C.Okla. 1976).

[17] 33 U.S.C. § 1321(b)(5).

[18] *Apex Oil Co.* v. *United States*, 530 F.2d 1291 (8th Cir. 1976); *United States* v. *Hougland Barge Lines, Inc.*, 387 F.Supp. 1110 (D.C.Pa. 1974).

[19] 33 C.F.R. Part 153, Subpart B.

[20] See, *e.g. United States* v. *Boyd*, 491 F.2d 1163 (9th Cir. 1973).

[21] *United States* v. *Mobil Oil Corpn.*, 464 F.2d 1125 (5th Cir. 1972).

[22] *United States* v. *T/B CTCD 186–20*, 1974 A.M.C. 1044 (S.D.Tex. 1973) in which it was held that the notification requirement was met where a barge employee asked a shoreside employee to notify the Coast Guard of the spill and actual notice resulted.

involves a maximum fine of $5,000 per violation; the latter includes a maximum fine of $250,000 per violation. As will be seen, the language of this section of the Act creates a confusing and perhaps internally inconsistent remedial scheme.

19–11 An administrative determination, the first method of collecting a civil penalty, is made by the Secretary of Transportation through the Coast Guard. The Secretary is required to fine any owner, operator or person in charge of a vessel from which oil is discharged in violation of the Act a mandatory civil penalty of not more than $5,000 for each violation.[23] The Act requires an opportunity for notice and a hearing and permits the Secretary to compromise the amount assessed. In assessing or compromising penalties, the Secretary is called upon to consider the size of the business of the owner or operator, the effect of the penalty on his ability to continue to do business and the gravity of the violation. With respect to vessels, clearance to sail may be withheld until the penalty is paid or a surety furnished.

19–12 In practice, it is the Hearing Officer appointed by each Coast Guard District Commander who considers and sets the penalty.[24] Appeals from the Hearing Officer's determination are to the Commandant of the Coast Guard whose decision becomes a final determination.[25] The Coast Guard may proceed in the Federal Court if the party refuses to pay the penalty.[26] The Federal Court may review the determination only on such factors as whether the party was granted a fair hearing or whether there was substantial evidence before the Hearing Officer such that the award was neither arbitrary not capricious, but a trial *de novo* on the merits is not permissible.[27]

19–13 The second method involves initiation in the Federal Court of a civil action, in the program administrator's discretion, against any person subject to the above-referred administrative penalties.[28] The court is empowered to impose penalties of up to $50,000 per incident for a prohibited discharge unless the United States can show that such discharge was the result of wilful negligence or wilful misconduct within the privity and knowledge of the owner, operator, or person in charge, in which case a penalty of up

---

[23] 33 U.S.C. § 1321(b)(6)(A).
[24] 33 C.F.R. § 1.07 *et seq.*
[25] 33 C.F.R. § 1.07–75.
[26] 33 C.F.R. § 1.07–85.
[27] *United States* v. *Healy Tibbetts Construction Co.*, 713 F.2d 1469 (9th Cir. 1983).
[28] 33 U.S.C. § (b)(6)(B).

to \$250,000 per incident may be imposed. The civil action option is to be considered by the program administrator taking into account such factors as the gravity of the offence and standard of care exercised by the owner. The criteria which the court is asked to consider, apart from the question of wilful negligence or misconduct and the same business factors as those taken into account in the administrative determination, are the gravity of the violation, and the nature, extent and degree of success of any efforts made by the owner, operator or person in charge to minimize or mitigate the effects of such discharge. The Act provides that the civil action option may not be utilized where an administrative penalty has been assessed.[29] Therefore, the civil action option is clearly meant to be utilized in those cases where the administrator believes the discharge to have been caused by wilful negligence or misconduct or where the discharger has failed to take responsible efforts to minimise or control the spill. There are no reported decisions involving exercise of the civil action option and this may well result from the fact that the cost and time litigating a civil penalty action in the Federal Court, compared with an administrative assessment are so disparate.

3. FWPCA OIL CLEAN-UP AND REMOVAL

A. The National Contingency Plan

19–14   The heart of Section 311 lies with the provision which permits the President to act at any time to remove or arrange for removal of oil which has been discharged into waters covered by the Act or under circumstances in which there is a "substantial threat" of such discharge, "unless he determines that such removal will be done properly by the owner or operator of the vessel."[30] The Act does not establish an affirmative duty upon the discharger to remove oil from a prohibited discharge, although there may be a financial incentive to do so if the discharger believes that he, or his insurer, can accomplish a clean-up more efficiently than the government.

---

[29] One commentator has suggested that since the administrative penalty, with its \$5,000 limit, is mandatory, a discharger against whom a civil action has been commenced might argue that the Coast Guard is required to assess the administrative penalty, thereby depriving it of the right to commence a civil penalty action. Helfrich, above, note 11, at pp. 462–463. This interpretation is most unlikely.

[30] 33 U.S.C. § 1321(c)(1).

19–15 The President is required to prepare and publish a National Contingency Plan to effect the purpose of the Act (The Plan).[31] The Plan, which has been revised from time to time, most recently in 1982, provides for a comprehensive response to be followed to the greatest extent possible in the event of an oil spill. The Act specifies in some detail what the plan is to cover[32]:

"(A) assignment of duties and responsibilities among federal departments and agencies in coordination with state and local agencies, including, but not limited to, water pollution control, conservation, and port authorities;

 (B) identification, procurement, maintenance, and storage of equipment and supplies;

 (C) establishment and designation of a strike force consisting of personnel who shall be trained, prepared, and available to provide necessary services to carry out The Plan, including the establishment at major ports, to be determined by the President, of emergency task forces of trained personnel, adequate oil and hazardous substance pollution control equipment and material, and a detailed oil and hazardous substance pollution prevention and removal plan;

 (D) a system of surveillance and notice designed to insure earliest possible notice of discharges of oil and hazardous substances and imminent threats of such discharges to the appropriate state and federal agencies;

 (E) establishment of a national centre to provide co-ordination and direction for operations in carrying out The Plan;

 (F) procedures and techniques to be employed in identifying, containing, dispersing and removing oil and hazardous substances;

 (G) a schedule, prepared in co-operation with the states, identifying (i) dispersants and other chemicals, if any, that may be used in carrying out The Plan, (ii) the waters in which such dispersants and chemicals may be used, and (iii) the quantities of such dispersants or chemicals which can be used safely in such waters, which schedule shall provide in the case of any dispersant, chemical, or waters not specifically identified in such schedule that the President, or his delegate, may, on a case-by-case basis, identify the dispersants and other chemicals which may be used, the waters in which they may be used, and the quantities which can be used safely in such waters; and

 (H) a system whereby the state or states affected by a discharge of oil or hazardous substance may act where necessary to remove such discharge and such state or states may be reimbursed from the fund established under subsection (k) of this section for the reasonable costs incurred in such removal."

[31] 33 U.S.C. § 1321(c)(2).
[32] 33 U.S.C. § 1321(c)(2).

19–16 The Plan, now known as the National Oil and Hazardous Substances Pollution Contingency Plan, is published in the Code of Federal Regulations.[33] It applies to all federally mandated response powers and responsibilities created by Section 311 and by the Comprehensive Environmental Compensation and Liability Act of 1980 ("CERCLA").[34] It also covers activities in connection with the Outer Continental Shelf Lands Act and the Deepwater Port Act.[35]

19–17 The Plan identifies several important areas of federal activity with respect to vessel oil pollution, among the most significant being[36]:

(1) Diversity of responsibilities among federal, state and local governments and appropriate roles for private entities;
(2) Describing the national response organisation to an oil spill;
(3) Establishing requirements for federal regional and federal local contingency planning; and
(4) Establishing procedures for oil removal operations under section 311.

B. Co-ordination of Federal, State and Local Response

19–18 As with any type of contingency planning, co-ordination is the cornerstone of a successful response. The Plan provides for lateral and vertical co-ordination by designating and assigning tasks to responsible federal agencies and by establishing lines of communication between working groups, federal, state and local governments and non-public organisations.[37]

19–19 On the federal level, the National Response Team ("NRT"), consisting of representatives of all involved federal agencies, is given the task of accomplishing national planning and co-ordination.[38] The representatives of the Environmental Protection Agency and the Coast Guard are designated respectively as

---

[33] 40 C.F.R. Part 300.
[34] CERCLA, 42 U.S.C. §§ 9601–9657, is also known as the Superfund legislation. CERCLA, which is discussed in more detail in Chap. 21, establishes a comprehensive federal regime for liability, response and clean-up in the event of hazardous substance pollution. However, Oil does not come within CERCLA's definition of hazardous substances (42 U.S.C. §9601(14)) and thus CERCLA is not applicable to vessel oil spills.
[35] These acts are discussed in Chap. 20.
[36] 40 C.F.R. § 300.22
[37] 40 C.F.R. §§ 300.22–24.
[38] 40 C.F.R. § 300.32.

Chairman and Vice Chairman of the NRT. The NRT has, among others, the following direct responsibilities[39]:

   (1) maintain national readiness to respond to a major discharge of oil which is beyond regional capabilities.[40]

   (2) develop procedures to ensure coordination among federal, state and local government and private sources of response.

   (3) monitor reports from all Regional Response Teams.

19–20 Regional Response Teams ("RRTs") are the regional planning and preparedness bodies and consist of regional representatives of particular federal agencies and state (or local) governments.[41]

19–21 Finally, the On-Scene Co-ordinator ("OSC") is the federal official designated by the Coast Guard to co-ordinate and direct federal responses under the National Contingency Plan.[42] The OSC also is responsible for developing a local federal contingency plan and for directing federally-financed response.

C. Oil Spill and Substantial Threat Response

19–22 All response operations to oil spills are handled, insofar as possible, according to the National, Regional and, where applicable, Local plans. All plans relevant to any area are published and available in Coast Guard district offices.[43] The OSC directs all federal efforts at the scene of the spill.[44] The OSC is responsible for investigating the discharge and determining, if possible, its sources and cause, the parties responsible, the nature, amount and location of the spills its direction and its potential impact on health, property and natural resources. To assist the OSC, National Strike Force ("NSF") teams have been established by the Coast Guard on the Atlantic, Pacific and Gulf Coasts. The strike force teams are able to supply technical and logistical assistance in the event of a spill. In addition, each OSC manages an emergency task force capable of emergency damage control and containment before

---

[39] 40 C.F.R. § 300.22.

[40] A major discharge is defined in the regulations as involving more than 10,000 gallons of oil to the inland waters or more than 100,000 gallons of oil to the coastal waters. (40 C.F.R. § 300.6).

[41] 40 C.F.R. § 300.32.

[42] 40 C.F.R. § 300.6. Certain inland waters are under the supervision of an OSC appointed by the Environmental Protection Agency.

[43] 40 C.F.R. § 300.41.

[44] 40 C.F.R. § 300.33. In the event of delay, the first official at the scene from an agency involved with the plan is authorized to co-ordinate activities until the OSC arrives.

arrival of the NSF or others who will assume responsibility for the removal and clean-up.[45]

19–23   The Plan identifies four "Phases" for pollution response— Phase I—Discovery and notification; Phase II—Preliminary assessment and initiation of action; Phase III—Containment, counter-measures, clean-up and disposal; and Phase IV—Documentation and cost recovery.[46]

19–24   Phase I refers to the methods of discovery and reporting an oil spill to the Coast Guard and others involved in spill response. Upon receipt of notification of a spill, the OSC for the affected area is directed to proceed with Phases II–IV of the Plan.[47] Hence, the OSC has no powers to commence Phases II–IV until there has been a discharge of oil (except where the casualty constitutes a marine disaster—see paragraph 19–30, below). Pre-spill cases of threatened pollution are, however, within the jurisdiction of the OSC since section 311(c)(1) authorizes the President to remove or arrange for the removal of oil where there is a substantial threat of a prohibited discharge.

19–25   Under Phase II, the OSC is responsible for initiating a preliminary assessment. This phase involves, among other activities, evaluating the pollution threat, assessing the feasibility of removal, determining who are the responsible parties and ensuring that jurisdiction exists to proceed with further response actions.[48] It is at this phase that the OSC is directed to make a reasonable effort, where appropriate, to have the discharger voluntarily and promptly take effective removal action, but, as discussed above, the discharger has no statutory duty to respond. In this regard, the OSC should officially make a determination whether the discharger or other person is properly carrying out removal.[49] The regulations further provide that the OSC shall advise the responsible parties if effective actions are not being taken. The mandatory nature of these regulations suggest that if such a determination is not made and communicated to the responsible parties, the government acts at its own risk and expense in commencing removal and clean-up

---

[45] Under circumstances in which the discharge exceeds the response capability of the OSC or exceeds his geographical jurisdiction, the NRT is obliged to activate the Regional Response Team to co-ordinate regional activities (40 C.F.R. § 300.34). Similarly, the N.R.T. can be activated as an emergency response team when the discharge exceeds regional capabilities or crosses regional boundaries *ibid*.

[46] 40 C.F.R. Part 300 Subpart E.

[47] 40 C.F.R. § 300.51.

[48] 40 C.F.R. § 300.52.

[49] 40 C.F.R. § 300.55.

operations but there has been no case which has reached this result.[50]

19–26 Phase III provides for the actual containment and clean-up operation if the OSC has determined that the owner or operator is not taking effective measures. It is during this phase that sampling of water to determine the source and spread of the oil is accomplished and salvage operations may be commenced.[51] It is also during this phase that oil recovery or mitigation actions should be undertaken. The regulations specifically require that the method chosen to mitigate the effects of, or to recover, the oil should be most consistent with protecting the public health and welfare and the environment. Sinking agents are specifically prohibited.[52] Of course, what is effective may be the subject of extensive disagreement between the Coast Guard and the discharger. In some cases, the most effective measure may be to do nothing. In such instances, the OSC has the clear authority to make such a determination and to the extent he disagrees with the discharger, he may take steps which he considers to be effective.

19–27 The Phase IV operation involves the preparation, gathering and consolidation of documentation to support actions taken to obtain cost recovery under section 311.[53]

19–28 If the discharger does not take appropriate and prompt removal actions, the regulations require the OSC, in directing that oil removal actions be undertaken under his supervision, to ensure that all operations are consistent with reimbursement from one of the funds established for oil spill clean-up reimbursement.[54] The funds, which are discussed in paragraphs 19–73 *et seq.*, below, and in Chapter 20, are the oil pollution fund established under Section 311,[55] the Deepwater Port Act fund,[56] the Outer Continental Shelf Lands Act fund,[57] and the Trans-Alaska Pipeline Authorization Act fund.[58] With the exception of the TAP Fund which bears some resemblance in this regard to the IOPC Fund, the others provide

---

[50] Private parties are entitled to hold governmental agencies to their own legislatively imposed internal regulations, see, *e.g. Chrysler Corp.* v. *Brown*, 441 U.S. 281 (1978); *United States ex rel. Accardi* v. *Shaughnessey*, 347 U.S. 760 (1954), Davis, Administrative Law Treatise, § 78 at 36 (2nd ed., 1979).

[51] 40 C.F.R. § 300.53.

[52] *Ibid.*

[53] 40 C.F.R. § 300.54.

[54] 40 C.F.R. § 300.58.

[55] See 33 C.F.R. Part 153.

[56] See 33 C.F.R. Parts 136 and 150.

[57] See 33 C.F.R. Parts 136 and 150.

[58] See 43 C.F.R. Part 29.

funds from the first dollar of liability to appropriate claimants as there is no initial level payable by the discharger until the funds become available.

19–29 The regulations also implement the provision under Section 311 that states may obtain reimbursement from the FWPCA fund for the reasonable costs incurred by such states in taking steps "where necessary" to remove oil discharges which affect them.[59] As will be discussed more fully in Chapter 21, the availability of federal funds to states participating in oil clean-up does not pre-empt states from exercising other remedies to obtain cost recovery. Specifically, state statutory and common law remedies remain available as a basis for recovery against the discharger or other responsible party so long as they do not provide a duplicate recovery. As a practical matter, however, states will be likely, and many state oil spill statutes require, that they conduct clean-up operations in such a manner that they will be in a position to recover from the federal funds.

D. Special Powers in the Event of Marine Disaster

19–30 Notwithstanding the obligations created in The Plan and section 311 for staffing and funding, there is a general exception from such requirements in the event of a "marine disaster" in or upon the navigable waters of the United States which has created a substantial threat of a pollution hazard . . . because of the discharge, or an imminent discharge, of large quantities of oil from a vessel.[60] This section, which, unlike the general geographical scope of the FWPCA, only pertains to the navigable waters of the United States, was intended to permit the Coast Guard to act outside the regulatory framework in the event of a serious emergency. In the event of a marine disaster, which is not specifically defined, the United States is empowered to co-ordinate and direct all public and private efforts for removal and elimination of the threat and summarily to remove and if necessary destroy a vessel which poses a threat. All such efforts are entitled to be reimbursed as actual costs of removal. Thus, under these provisions there need be no actual discharge of oil in order for the United States to act or to be reimbursed. If there is a substantial threat of discharge of large quantities of oil, summary action is authorized.

---

[59] 40 C.F.R. § 300.50; 33 U.S.C. § 1321(c)(2)(4).
[60] 33 U.S.C. § 1321(d).

4. LIABILITY OF DISCHARGER FOR REMOVAL COSTS

19–31 With limited exceptions, the owner or operator of a vessel which discharges oil into water covered by section 311 is strictly liable for the "actual costs of removal of such oil."[61] Once the United States has proven the fact of a discharge and the relationship of ownership or operation, the burden of proof is shifted to the defence to prove one of the exceptions to strict liability applies. If it does not, the owner or operator has the possibility of limiting his liability unless the United States proves wilful negligence or misconduct. If, in fact, the owner or operator of the vessel which has discharged the oil can show that the sole cause of the discharge was the act or omission of a third party, then that third party will be accorded the same deference as the owner or operator of the vessel.

A. Exceptions to Strict Liability

19–32 There are four exceptions to strict liability for removal costs: (1) Act of God; (2) act of war; (3) negligence on the part of the United States Government; and (4) act or omission of a third party. With respect to these exceptions, the owner or operator is bound to prove that one or a combination of them was the *sole* cause of the discharge. It is not sufficient to escape liability under the Act that the owner or operator proves that one of these causes was merely contributory.

19–33 The Act of God and act of war exceptions have yet to be judicially tested in published court decisions. It may be assumed, however, that courts would be likely to construe these exceptions narrowly as has been done with respect to the third party and government fault exceptions.[62]

19–34 The question of government negligence as sole fault has been considered on three occasions. In *Gaspar* v. *United States*,[63] the owner of a fishing vessel which was lost after striking an unmanned barge which had been anchored by the Coast Guard sued the United States government for the loss. The government counterclaimed under section 311 for oil spill clean-up costs which resulted

---

[61] 33 U.S.C. § 1321(f). Owner or operator is conjunctively defined to include any person owning, operating or chartering by demise a vessel. 33 U.S.C. § 1321(a)(6). Thus, any person coming within the terms of this definition is subject to the provisions of the FWPCA and while not specifically mentioned, liability would appear to be joint and several where more than one person is involved as owner or operator. See, *e.g.* U.S. v. *M/V Big Sam*, 505 F.Supp. 1029 (E.D.La. 1981).

[62] See, *e.g.* U.S. v. *LeBoeouf Brothers Towing Co.*, 621 F.2d 787 (5th Cir. 1980).

[63] 460 F.Supp. 656 (D.C.Mass. 1978).

from the sinking of the fishing vessel. The court held that the Coast Guard, which had taken custody of the drifting barge, failed properly to place strobe lights on the barge and was negligent in anchoring the barge by positioning her such that she extended into the navigation channel. The Court declined to find the crew of the fishing vessel contributorily negligent and assigned sole fault to the Coast Guard, thus exonerating the fishing vessel from liability for clean-up costs.

19–35 In *United States* v. *Bear Marine Services*,[64] the facts of which did not involve the issue of government negligence, the court noted, *in dicta*, that any fault on the part of the discharger created clean-up liability and that the exceptions were to be strictly construed.

19–36 Despite its obvious good sense, *Gaspar* does not represent the mainstream of decisions in this area. Most courts have clearly expressed the view that the exceptions are to be so strictly construed that, apart from anchored vessels, there would seem to be few marine casualties in which some *de minimus* fault cannot be assigned to the vessel. Thus, in *Burgess* v. *M/V Tamano*,[65] the Court of Appeals reversed a decision of the District Court, finding clearly erroneous the conclusion that the pilot of a Norwegian ULCC was not negligent and that the government was solely at fault in the grounding of the vessel out of the channel owing, as it was argued, the government's alleged misplacement of a channel buoy.

19–37 There have been several cases which have considered whether a third party has been solely responsible for a prohibited discharge. As with the government sole fault exception, the third party sole fault exception has been narrowly construed. In *Burgess* v. *M/V Tamano*,[66] the court found that the compulsory pilot who brought the vessel through the channel and was directing her navigation at the time she grounded was not a third party within the meaning of section 311. The court concluded that a true third party was a stranger to the ship, not one hired by the owners, even if that hiring was compulsory. By way of illustration, the court suggested that a vandal who opened a valve resulting in an oil spill would not subject the owner to liability, but that a failure of the valve resulting from improper installation would result in liability.

19–38 In *United States* v. *LeBoeouf Towing Co.*,[67] the court held that a tug boat of an independent oil barge on which a tug crewman

[64] 509 F.Supp. 710 (E.D.La. 1980).
[65] 565 F.2d 964 (1st Cir. 1977), cert. denied, 435 U.S. 941.
[66] 565 F.2d 964 (1st Cir. 1977), cert. denied, 435 U.S. 941.
[67] 621 F.2d 787 (5th Cir. 1980).

accidentally opened a valve leading to discharge of 60 barrels of crude oil into the Mississippi River was not a third party within the meaning of Section 311. The barge owner argued that the barge was under the complete supervision and direct control of the tug and her crew at the time of the incident and thus a third party was the sole cause of the discharge. The court rejected this view and held that LeBoeuf exercised ultimate control by having hired the tug and having specified its itinerary.[68] Recognizing that this constituted a narrow interpretation of the statute's language, the court found such to be consistent with the statute's policy. It noted:

> "The statute's comprehensive scheme for preventing and cleaning up oil spills would be undermined if barge owners like LeBoeuf could escape strict liability merely by hiring out their operations to tugs and independent contractors."

19–39 In *United States* v. *M/V Big Sam*,[69] the United States sought clean-up costs against the owner of a barge which had been the lead barge in a flotilla under tow which was struck by a second tug, causing the barge to discharge some 210,000 barrels of crude oil into the Mississippi River. On the barge owner's motion to dismiss, the court held that because the barge may have had a structural or design defect, or for some other reason, this may have contributed to the size of the spill, in which case the owner would be deprived of the defence of *sole* third party cause. The court refused to dismiss the government's claim and required a trial on these points. Subsequently, the district court in *Big Sam* recognized that the second tug was the sole cause of the spill, thus exonerating the barge owner from clean-up liability.[70]

19–40 Also considered in the second *Big Sam* district court opinion was the liability of the second tug's registered owner under circumstances in which it had been bareboat chartered to the tug's operators, now insolvent.[71] The district court concluded that the registered owner which did not participate with the bareboat charterer in the operation of the vessel, could rely on the third party sole cause defence and exempt itself from liability. The court rejected the government's argument that reference in section 311 to an "owner or operator of a vessel"[72] meant that an owner could not distance itself from an operator in connection with the third party

---

[68] 621 F.2d 787, 790 (5th Cir. 1980).
[69] 454 F.Supp. 1144 (E.D.Ca. 1978).
[70] 505 F.Supp. 1029 (E.D.La. 1981).
[71] The second tug was sued under the provisions of section 311(g) as a third party cause of the discharge. This provision is discussed more fully at paras. 21–44 *et seq.*, below.
[72] 33 U.S.C. § 1321(g).

defence. On appeal, the Fifth Circuit reversed and held that even though the vessel had been bareboat chartered, the statutory definition of "owner or operator" required the court to find that the government had a remedy against either or both even though the tug was in the exclusive control of the bareboat charterer.[73] The appeals court found the district court's disjunctive interpretation of owner or operator to be inconsistent with the inclusive term "any" in the definitional section, and was erroneous and inconsistent with the intent of section 311 to provide an effective remedy for recovery of clean-up costs.

19–41 The consequences of *Big Sam* to the financial community could well be substantial. The bareboat charter being a crucial document for the purpose, in part, of insulating the registered owner from the usual liabilities, banks and their trustees may well find that they have purchased considerably more exposure to liability than they had anticipated. In *Big Sam*, the registered owner argued without success that the regulations pursuant to the Certificate of Financial Responsibility provision of section 311 did not permit the non-operating owner to apply for a certificate and thus prohibited it from insuring against oil pollution liability.[74] The court dismissed this argument, noting that the regulations only permitted one or the other to apply for a certificate but that this did not prevent the registered owner from securing insurance to protect against its potential liability. The court also noted that if the owner is certain that the bareboat charterer has obtained the necessary financial responsibility, then it need not do so.

19–42 Unfortunately, this facile argument is unpersuasive. Given the vagueness in the statutory language and the absence of legislative history concerning the definition of "third party sole cause," it is far from clear that the district court was not correct in finding the phrase "owner or operator" to be disjunctive. Whether or not the Fifth Circuit's interpretation will be followed elsewhere has yet to be tested. If not, it is likely that a contrary result could set up a test case for the United States Supreme Court for a definitive interpretation.[75] Under the circumstances, lenders and equity par-

---

[73] 681 F.2d 432 (5th Cir. 1982).

[74] 46 C.F.R. § 542. The Financial Responsibility provisions are discussed at paras. 21–76 *et seq.*, below.

[75] Under U.S. law, the decision of a Federal Circuit Court of Appeals is binding on all District Courts within its jurisdiction. With respect to courts outside of the circuit, such a decision only constitutes persuasive authority and need not be followed. The Fifth Circuit Court of Appeals, as reconstituted in 1982, covers the District Courts of Texas, Louisiana, Mississippi and, where still applicable, the Canal Zone.

ticipants must assure that the full measure of pollution liability insurance is procured in order to provide a reasonable measure of protection against the substantial clean-up expenses which could arise following a major pollution incident.

B. Discharger Liability

19–43 Where the owner or operator fails to prove that the discharge was caused by one of the four stated exceptions, and where the United States fails to prove that the discharge was the result of wilful negligence or wilful misconduct within the privity and knowledge of the owner the government is entitled to recover its actual costs of removal, in an amount not to exceed (a) for an inland oil barge, $125 per gross ton, or $125,000, whichever is greater, or (b) for any other vessel $150 per gross ton. If, however, a vessel other than an inland oil barge is carrying oil as a cargo, the minimum liability is $250,000.[76] The shifting burden of proof contained in section 311 is thus the framework for the concept of strict but limited liability envisioned by the Act. If the government were to prove wilfulness, the statute provides that the discharger is liable for the full amount of the removal costs. The terms wilful negligence and wilful misconduct are not defined in the Act or the regulations, and thus it has been left to the courts to identify the scope of unlimited recovery.

19–44 As to the issue of wilfulness, the Court of Appeals, in *Steuart Transportation Co.* v. *Allied Towing Corp.*[77] found that "wilful negligence" means reckless disregard for the probable consequences of a voluntary act or omission. Applying this test, section 311 limitation was granted to the owner of an oil barge which sank in heavy weather causing its release of oil. Section 311 limited liability was granted even though it was found that flanges securing a cowling on deck were badly deteriorated and scupper plugs had not been removed on deck, thus allowing sea water to fill the pump-room and accumulate on deck, leading to the loss of the barge and subsequent oil discharge. Although the owner was denied limitation of liability under the 1851 Limitation Act for haphazard inspection practices, the court affirmed the District Court's holding that the owner's conduct did not rise to the level of reckless dis-

---

[76] 33 U.S.C. § 1321(f)(1). The 1977 amendments revised the limits of liability for vessels from $100 per gross ton or $14,000,000, whichever is less. There is now no monetary cap on the limited liability provisions, rather there is a floor setting out the minimum recovery, at least with respect to oil carrying vessels.

[77] 596 F.2d 609 (4th Cir. 1979).

regard for the probable consequences of the act. In this regard, the court noted that the barge was routinely maintained and certified and had undergone normal repair and overhaul.

19–45 In *The Tug Ocean Prince v. United States*,[78] the Second Circuit Court of Appeals was compelled to fashion a definition of wilful misconduct. The court noted that the Warsaw Convention, with respect to aviation liabilities, permitted the air carrier to limit liability unless the damage was caused by wilful misconduct. Adopting the jurisprudence relating to the Warsaw Convention, the court stated the criteria it would apply under section 311[79]:

> "An act, intentionally done, with knowledge that the performance will probably result in injury, or done in such a way as to allow an inference of a reckless disregard of the probable consequences . . . . If the harm results from an omission, the omission must be intentional, and the actor must either know the omission will result in damage or the circumstances surrounding the failure to act must allow an implication of reckless disregard of the probable consequences . . . . The knowledge required for a finding of willful misconduct is that there must be either actual knowledge that the act, or the failure to act, is necessary in order to avoid danger, or if there is no actual knowledge, then the probability of harm must be so great that failure to take the required action constitutes recklessness."

The court denied limitation to the tug under section 311 on several grounds, but specifically because of the owner's failure to inform one of the pilots about the second pilot's unfamiliarity with that section of the river where the spill occurred, the failure to assure that a lookout was appointed and its failure to designate a captain. While no single fault would have been likely to have resulted in a finding of wilful misconduct, it was the belief of the court that the pattern of conduct revealed from an examination of the entire record resulted in such a finding.

19–46 In view of the finding in *Steuart*, it is clear that the standard for breaking normal cost limitation under section 311 (proof of the owners' wilful negligence or misconduct), is far greater than the standard for breaking limitation under the Limitation of Liability Act (negligence within the privity or knowledge of the shipowner). Although the government has sought to collect unlimited clean-up costs in other cases,[80] there have been no recent cases interpreting

---

[78] 584 F.2d 1151 (2d Cir. 1978), cert. denied, 440 U.S. 959.
[79] 584 F.2d 1151, 1163.
[80] See, *e.g. U.S.A.* v. *Allseas Maritime S.A. et al.*, 82 Civ. 7199 (S.D.N.Y. 1982), involving the Burmah Agate/Mimosa collision in Texas in November, 1979, which case was ultimately settled for a sum which was less than the section 311 limitation amount.

the wilful misconduct standard of section 311. In comparing *Steuart* with *Ocean Prince*, it has been suggested that the courts may very well look into a number of mistakes made by the owner rather than the seriousness of the error.[81] Since many judges do not have frequent contact with limitation statutes, this unfortunate possibility has a practical appeal. Even the skilful litigator may find that the judge's laundry list of mistakes far outweighs an argument based on the seriousness of such errors or their relation to the cause of the spill.

C. Third Party Liability

19–47 Where a third party is proved to be the *sole* cause of the prohibited discharge, the owner of the discharging vessel is entitled to a complete defence against payment of clean-up costs and the sole cause third party assumes liability for clean-up costs with the same strict but limited liability of a discharger, *i.e.* subject to the same defence and limited liability provisions.[82] As with discharger liability, a third party may rely upon the same defences to strict liability if it can be demonstrated that one of the four exceptions, *i.e.* Act of God, act of war, government negligence or third party, was the sole cause of the casualty.[83]

19–48 Of course, the possibility exists that the third party sole cause of the discharge is not a vessel. The cause could be, for example, a negligently maintained docking facility or a negligently operated drawbridge. In such a case, section 311 specifies that the strict liability of such third party is subject to the limits which would have been applicable to the vessel from which the discharge actually occurred.

19–49 Section 311 makes no special provision for establishing the liability of a non-discharging *contributing* third party as it does for a discharger and a *sole* cause third party. It is, of course, clear that in the event of a discharge in which there is a joint cause third party, the owner of the discharging vessel remains strictly liable under section 1321(f) and subject to its limitation rights. The Act does, however, preserve remedies against joint cause third parties. Section 1321(h) provides[84]:

"The liabilities established by this section shall in no way affect any

---

[81] Note, Oil spills and Clean-up Bills: Federal Recovery of Oil Spill Clean-up Costs, 93 Harv.L.Rev. 1761 at pp. 1767–1768, n.41 (1980).

[82] 33 U.S.C. § 1321(g).

[83] *U.S.* v. *Bear Marine Services*, 509 F.Supp. 710 (E.D.La. 1980).

[84] 33 U.S.C. § 1321(h).

rights (1) the owner or operator of a vessel or of an onshore facility or offshore facility may have against any third party whose acts may in any way have caused or contributed to such discharge or (2) the United States Government may have against any third party whose actions may in any way have caused or contributed to the discharge of oil or hazardous substances."

Under section 1321(h), the government is entitled to pursue its other common law and statutory remedies against joint cause third parties. Any such action would not be governed by the strict or limited liability provisions of section 311, but would rather involve recovery, as appropriate, under general rules of common law tort, more thoroughly discussed in Chapter 21. Thus, in *United States* v. *Bear Marine Services*,[85] a terminal operator was joined in an action by the government seeking clean-up costs on the ground that the barge struck the terminal's dolphin which had been in poor condition. The terminal sought to dismiss the claim on the ground that section 311 provided the government with its exclusive remedy for clean-up costs. The court rejected this argument, holding that section 1321(h) preserved the government's rights against joint cause third parties.

19–50 A similar case in California reached the same result. In *United States* v. *City of Redwood City*,[86] the government sued the port and a private security company for clean-up costs arising out of the loss of a barge at a berth in the town. The court examined the broad purposes of the FWPCA and held not only that section 311 applied to joint fault third parties, but also that the government was free to sue such third parties in maritime tort.

19–51 The two courts differed in one material respect. In *Bear Marine*, it was held that any government recovery based on maritime tort against joint cause third parties would be governed by the Limitation of Liability Act. In *Redwood City*, the court suggested that the joint fault third party may have available to it the limits of liability set out in section 311. In the recent case of *In Re Berkley Curtis Bay Co.*,[87] the District Court in New York held that the government may proceed in tort against a joint cause third party and that section 1321(g) limitations on recovery were not applicable to such an action. The court also rejected the joint cause third party's argument that the act set out a hierarchy of recovery requiring the government to seek primary recovery from the discharger or

---

[85] 509 F.Supp. 710 (E.D.La. 1980).
[86] 640 F.2d 963 (9th Cir. 1981).
[87] 557 F.Supp. 335 (S.D.N.Y. 1983).

the sole cause third party before proceeding against a joint cause third party.

19–52 It may be anticipated that government claims against joint fault third parties will be reserved for those situations in which strict liability recovery against the discharger proves, or is likely to prove, inadequate to cover the full clean-up costs. An action in maritime tort requires the government to prove fault, proximate cause and other elements of a tort. Clearly, if the government can recover its full costs in strict liability, it would make no sense to seek a recovery in which it is required to prove fault. It is not likely, however, that the strict liability of section 1321(h) will reduce litigation since, as discussed, this provision expressly reserves the right of the discharger to seek contribution or indemnity from a joint cause third party.

D. Exclusivity of section 311 for Government's Recovery of Clean-up Costs

19–53 Closely related to the issue of the government's recovery against joint cause third parties is whether the government is entitled to recover clean-up costs against dischargers or sole cause third parties outside of the section 311 recovery provisions, or whether section 311 is its exclusive remedy against these responsible parties. The significance of the issue is rooted in the fundamental structure of the Act which affords the discharger, or sole cause third party, the benefit of limited liability in return for which the government is not required to prove fault. Pursuant to this statutory framework, the limited section 311 recovery may prove to be inadequate for recovery of all clean-up costs, particularly in cases of discharge from small ships. In such situations, the government has taken the position that section 311 does not provide it with its exclusive remedy and that it is entitled to recover against dischargers and sole fault third parties on other legal principles.[88]

19–54 The language of the Act itself is vague on the issue of exclusivity. Section 1321(f), which is concerned with discharger liability, and section 1321(g), which is concerned with sole fault third party liability, states that owners or operators who violate the provisions of the Act, shall, notwithstanding any other provision of

---

[88] See, *e.g. In Re Oswego Barge Corp.*, 664 F.2d 327 (2d Cir. 1981); *U.S.* v. *Dixie Carriers*, 462 F.Supp. 1126 (E.D.La. 1978), *aff'd*, 627 F.2d 736 (5th Cir. 1980); *Steuart Transportation Co.* v. *Allied Towing Corp.*, 596 F.2d 609 (4th Cir. 1979); *U.S.* v. *M/V Big Sam*, 681 F.2d 432 (5th Cir. 1982).

law, be liable to the United States Government for actual removal costs. One court has declared[89]:

> "The statute is not a model of clarity. In the absence of clarifying case law or legislative history in point, one can only speculate as to the meaning of the "notwithstanding any other provision of law" clause."

19–55   The language could have several possible meanings:
   (1) It refers to any statute or common law rule creating liability and thus only recognizes that other federal remedies exist to recover clean-up costs;
   (2) It refers to other statutes and common law rules which section 311 now replaces;
   (3) It refers to other statutes which limit a shipowner's liability, specifically the 1851 Limitation of Liability Act, which section 311 precludes from application to clean-up costs;
   (4) It refers to laws which both create and limit liability; or
   (5) It refers to any other law which might purport to limit clean-up cost recovery specifically under Section 311, but not to any law concerning limitation outside Section 311.

19–56   Courts and commentators have not agreed.[90] The result is that certain parties may be liable for clean-up costs beyond section 311 depending upon the jurisdiction in which suit is commenced. A binding determination on all federal circuits awaits a decision by the United States Supreme Court.

19–57   The Fourth Circuit[91] held in *Steuart Transportation Co.* v. *Allied Towing Corp.*,[92] that the United States could not recover in excess of section 311 limits on a maritime tort claim and that section 311 was the government's exclusive remedy. The court in *Steuart* recognized that section 311 was drafted so as to accommodate both the interests of the government and the interests of the vessel owners, and to permit recovery outside of the framework of the Act could defeat that dual purpose. The court noted[93]:

> "Nothing in the legislative history suggests that Congress included

---

[89] *The Tug Ocean Prince* v. *U.S.* 584 F.2d 1151, 1162 (2d Cir. 1978.)
[90] See, generally, Comment, Federal Water Pollution Control Act—The Federal Government's Exclusive Remedy for Recoupment of Oil Spill Clean-up Costs, 53 Tul.L.Rev. 1421 (1979): Note, Oil Spills and Clean-up Bills; Federal Recovery of Oil and Spill Clean-up Costs, 93 Harv.L.Rev. 1761 (1980): Guss, "Interaction of the Federal Water Pollution Control Act With The Limitation of Liability Act And The General Maritime Law," 6 *The Maritime Lawyer 199* (1981).
[91] Covering Maryland, Virginia, North Carolina, South Carolina and West Virginia.
[92] 596 F.2d 609 (4th Cir. 1979).
[93] 596 F.2d 609, 617–18.

the compromise limitation provision simply as a means of guarantee-
ing minimum cost recoveries by requiring shipowners to file evidence
of their financial responsibility. Congress could have required insurance up to a fixed amount, while setting a still higher figure as the
liability limit . . . . Thus, Congress meant the limitation of liability to
put a ceiling on Federal removal cost claims. Allowing Federal
removal cost recoveries beyond the Pollution Act's limitation would
have economic consequences for oil carriers significantly different
from those envisioned by Congress.''

19–58   In *United States* v. *Dixie Carriers, Inc.*,[94] the Fifth Circuit
upheld the decision of the District Court that section 311 constituted the federal government's sole remedy against dischargers
under section 1321(f).[95] In reaching its conclusion the court indicated that any other result could produce inconsistency in the area
of federal recovery. It was noted that if the government could
recover in maritime tort it could obtain unlimited recovery upon
proof of mere negligence within the privity and knowledge of the
shipowner whereas section 311 requires proof of wilful misconduct
in order to secure unlimited recovery.

19–59   The Second Circuit adopted the exclusivity position with
respect to dischargers in *In Re Oswego Barge Corporation*.[96] The court
focused on the question of federal pre-emption, holding that section
311 was the government's exclusive remedy with respect to dischargers.

19–60   The question of exclusivity becomes more problematic,
however, with respect to non-discharger liability. As has been discussed in paragraph 19–49, above, a joint fault third party is not
covered by any provision of section 311 and, thus, federal clean-up
recovery by way of tort is clearly available to the government in
addition to its section 311 remedy against the owner or operator,
and so at least in that sense section 311 does not provide an exclusive remedy. In addition, there is a line of recent decisions concerning sole fault third parties establishing a trend that both section 311
and recovery by way of tort or other statute is available to the
government, and that section 311 is not the government's exclusive
remedy against them. In *United States* v. *City of Redwood City*,[97] the

---

[94] 462 F.2d 736 (5th Cir. 1980).

[95] In 1981, the Fifth Circuit was divided into two circuits, the Fifth and the new Eleventh
Circuits. The first order of business of the new Eleventh Circuit was to adopt the jurispru-
dence of the Fifth Circuit. The coastal states covered by the two circuits comprise Texas,
Louisiana, Mississippi, Alabama, Florida, Georgia and the Canal Zone.

[96] 664 F.2d 327 (2d Cir. 1981).

[97] 640 F.2d 963 (9th Cir. 1981). The Ninth Circuit covers the Coastal States of California,
Oregon, Washington and Alaska.

court held that section 311 was not the exclusive remedy against sole fault third parties. Noting the language of section 1321(h)(2) which states that section 311 does not affect any rights against third parties, the court found that exclusivity would be inconsistent with this section.

19–61 A similar result was reached by the Fifth Circuit in *United States* v. *M/V Big Sam*,[98] on reasoning similar to *Redwood City*. The court could find no logical reason why the government should be permitted to sue in maritime tort to recover clean-up expenses from a joint cause third party and be prevented from a similar action against a sole cause third party. It reasoned:

> "We can see no functional reason, nor does the legislative history afford any suggestion of such a purpose, that the government's remedy against the negligent third party vessel should be less because that negligence was the sole instead of merely a contributing cause of the accident."

The recent decision of the Fifth Circuit in *United States* v. *T/B Arcadian 95*,[99] allowed that court to reconfirm its view. It also noted that the apparent discrepancy between *Dixie Carriers*, upholding exclusivity with respect to dischargers, and *Big Sam*, denying exclusivity to sole cause third parties, resulted from a strict adherence to the language of the Act, which in section 1321(h) preserved the government's remedies against third parties not dischargers.

19–62 The remaining coastal circuits have not ruled on the exclusivity question. One district court in Massachusetts has, however, ruled that section 311 is the government's exclusive remedy against dischargers and left open the question as to sole fault third parties.[1]

19–63 Unless the Supreme Court intervenes, the question of exclusivity, at least with respect to sole cause third parties, will very much depend upon the jurisdiction in which the spill occurs.[2]

19–64 A related issue is whether a discharger may claim the benefit of the 1851 Limitation of Liability Act in connection with a claim brought by the federal government for strict liability for

---

[98] 681 F.2d 432 (5th Cir. 1982).

[99] 714 F.2d 470 (5th Cir. 1983).

[1] *Frederick E. Bouchard, Inc. et al.* v. *U.S.*, 583 F.Supp. 477 (D.Mass. 1984).

[2] While the Second Circuit in *In Re Oswego Barge Corpn.* did not have to rule on the question of exclusivity of third party liability under the Act, it conceded that the Government stood on stronger ground in its argument for non-exclusivity based on section 1321(h). 664 F.2d 327, 341. The court noted that a decision upholding or denying sole fault third party exclusivity results in an anomaly, suggesting that comprehensive legislative action was preferable to contrary and conflicting court interpretations. *Ibid.*

clean-up costs under section 311. One court has considered this issue and has held that a shipowner may not claim the benefit of the 1851 Act's limitation provisions. In *Re Hokkaido Fisheries Co.*,[3] the District Court in Alaska had before it two limitation petitions brought under the 1851 Act arising out of the collision of the M/V Lee Wang Zim and the M/V Ryuyo Maru in which large quantities of oil were discharged into Alaskan waters by both vessels. Federal clean-up costs amounted to $550,000 for oil discharged from the Ryuyo Maru and $2,238,000 for oil discharged from the Lee Wang Zim. Under the strict liability provisions of section 311, the government could recover its full costs, while it could recover only approximately $47,000, assuming limitation under the 1851 Act. The court examined the purpose and policies of the two Acts and the grounds for proving claims thereunder and held that the government could recover its full costs under section 311 notwithstanding the 1851 Act.

5. RECOVERY OF CLEAN-UP COSTS BY DISCHARGER

19–65   In an attempt to provide an incentive to dischargers to act promptly to effect clean-up operations, §1321(i)(1) states[4]:

> "In any case where an owner or operator of a vessel . . . from which oil . . . is discharged in violation of subsection (b)(3) of this section acts to remove such oil or substance in accordance with regulations promulgated pursuant to this section, such owner or operator shall be entitled to recover the reasonable costs incurred in such removal upon establishing, in a suit which may be brought against the United States Government in the United States Court of Claims, that such discharge was caused solely by (a) an Act of God, (b) an act of war, (c) negligence on the part of the United States Government or (d) an act or omission of a third party, without regard to whether such act or omission was or was not negligent, or of any combination of the foregoing causes."

19–66   Under this section, in order to recover its reasonable clean-up costs, the discharger must bring an action in the United States Court of Claims and prove that the discharge was caused by one of the four exceptions to strict liability. Even with the narrow interpretation courts have given to the scope of these four exceptions, discussed above, it is hard to imagine that this section provides, in most cases, any true incentive for dischargers to take

[3] 506 F.Supp. 631 (D.Alaska 1981).
[4] 33 U.S.C. § 1321(i)(1).

effective, prompt steps to clean up after a spill; and, in fact, judicial interpretation of this provision itself supports that view. In *Cities Service Pipeline Co.* v. *United States*,[5] a pipeline owner sought recovery of clean-up costs expended when a pipeline break caused a spill.[6]

The owner asserted that the break was caused either by an unspecified Act of God or an act of third parties. The court rejected the application for cost recovery on the ground that the owner failed to prove that the cause of the spill was one of the excepted causes and, moreover, that the spill could have been prevented through due care by the owner. The court noted that the pipeline had been damaged on previous occasions simply because of what it perceived to be continual operational stress and that, because of this, the owner should have undertaken preventive testing to determine weak points along the pipeline.

19–67 Recovery has been permitted, however, where the owners of a newly purchased onshore facility exercised reasonable measures to prevent vandalism and despite such measures vandals entered the facility and opened valves resulting in a spill.[7]

19–68 Section 1321(i)(1) only permits clean-up cost recovery from the United States by the discharger. A third party is not entitled to recovery. In one ingenious effort evidently to circumvent this requirement, a sole cause third party's insurer paid the discharger the cost of clean-up. Thereafter, suit was brought in the name of the discharger under section 1321(i) to recover these costs from the United States.[8] The court rejected recovery on the ground that the discharger has already been reimbursed and the third party or his insurer had no cause of action under section 311, even as the discharger's subrogee. This decision must remove all incentive on an insurer to co-operate with the discharger in such a case.

19–69 Section 1321(i) has been urged to be the basis for a set-off in actions brought for spill recovery against dischargers. In *United States* v. *Dixie Carriers, Inc.*,[9] the most recent appeal of this long-running case, the vessel owner argued that the $108,000 it spent in clean-up operations before initiation of action by the government should be recognized as an offset to the $121,600 being sought by

---

[5] 4 Ct. Cl. 207 (1983).
[6] There have been no reported cases of vessel dischargers seeking recovery of clean-up costs but the same principles would be applied.
[7] *Chicago, Minn. St. Paul P.R.R. Co.* v. *U.S.*, 575 F.2d 839 (Ct. Cl. 1978).
[8] *Amoco Oil Co.* v. *U.S.*, 3 Cl.Ct. 785 (1983).
[9] No. 83–3321 (5th Cir. July 12, 1984).

the government. The Court of Appeals affirmed the District Court's decision that the statute did not recognize such an offset and that any recovery under this section would require the discharger to file its suit for recovery in the Court of Claims.[10] In any such action brought in the Court of Claims, prior litigation in a suit establishing the fault of the discharger may collaterally estop efforts by the discharger to recover for clean-up costs.[11]

6. ACTUAL COSTS OF REMOVAL UNDER SECTION 311

19–70 In contradistinction to section 1321(i), which provides for the recovery of reasonable costs by the discharger against the United States, both section 1321(f) and section 1321(g) provide for recovery by the United States of the "actual costs incurred . . . for the removal of such oil" against the discharger. The Act does not specify what costs are to be included, but at least three courts have adopted the literal meaning of the Act and held that such actual costs need not be reasonable in order to secure government recovery.[12] Another court has suggested, without judicial support, that such actual costs must be fair and reasonable.[13] Further limiting government recovery, in *In Re Oswego Barge*,[14] the court rejected an effort by the government to include as actual costs amounts paid to Canada as reimbursement for a portion of Canadian clean-up costs pursuant to the Joint US/Canada Contingency Plan. The actual costs to which the Act refers only include federal costs, not state costs independent of those recoverable under the National Contingency Plan and thus it has been held that states are free to recover their separate, non-reimbursible clean-up costs under state law.[15] Hence, the limit of liability of an owner or operator in respect of any one incident could exceed the section 311 limit, because he is also liable to state and local interests under state law and to private interests under the common law of tort.

19–71 While "actual costs" is nowhere defined and there has

---

[10] A similar result denying a right of offset was reached in *Steuart Transportation Co.* v. *Allied Towing Corpn.*, 596 F.2d 609 (4th Cir. 1979).

[11] *Tanker Hygrade No. 18* v. *U.S.*, 1976 AMC 840 (Ct.Cl. 1975).

[12] *U.S.* v. *Beatty Inc.*, 401 F.Supp. 1040 (D.C.Ky. 1975); *Comm. of Puerto Rico* v. *S.S. Zoe Colocotroni*, 456 F.Supp. 1327 (D.C.P.R. 1978), rev'd on other grounds, 628 F.2d 652 (1st Cir. 1980); *Burgess* v. *M/V Tamano*, 564 F.2d 964 (1st Cir. 1977).

[13] *U.S.* v. *Malitovsky Cooperage Co.*, 472 F.Supp. 452 (D.C.Pa. 1978).

[14] 664 F.2d 327 (2d Cir. 1981).

[15] *Complaint of Allied Towing Corp.*, 478 F.Supp. 398 (D.C.Va. 1978).

been no judicial analysis of this point,[16] the Act and regulations define "remove" or "removal" as not only the removal of oil from the waters and shorelines but also "the taking of such other actions as may be necessary to minimize or mitigate damage to the public health or welfare, including, but not limited to, fish, shellfish, wildlife and public and private property, shorelines, and beaches."[17] Some guidance can be taken from the regulations as to what are actual costs of removal within this broad scope of activities[18]:

(a) The following costs incurred during performance of a Phase III or IV activity, defined by the National Contingency Plan, as authorized by the appropriate OSC under the authority of section 311(c) of the Act and of the provisions of the National Contingency Plan, or during the removal or elimination of threats of pollution hazards from discharges, or imminent discharges, of oil or hazardous substances, and the removal and destruction of vessels, as authorized by the appropriate AC [*sic*: presumably OSC] under the authority of section 311(d) of the Act are reimbursible to federal and state agencies:

(1) Costs found to be reasonable by the Coast Guard incurred by government industrial type facilities, including charges for overhead in accordance with the agency's industrial accounting system.

(2) Actual costs for which an agency is required or authorized by any law to obtain full reimbursement.

(3) Costs found to be reasonable by the Coast Guard incurred as a result of removal activity that are not ordinarily funded by an agency's regular appropriations and that are not incurred during normal operations. These costs include, but are not limited to, the following:

(i) Travel (transportation and per diem) specifically requested of the agency by the On-Scene Coordinator.

(ii) Overtime for civilian personnel specifically requested of the agency by the On-Scene Coordinator.

(iii) Incremental operating costs for vessels, aircraft, vehicles, and equipment incurred in connection with the removal activity.

(iv) Supplies, materials, and equipment procured for the

---

[16] One court noted that salary and other costs incurred by the Coast Guard were recoverable, but the decision does not involve an analysis of what constitutes actual costs of removal. *United States* v. *Hollywood Marine, Inc.*, 519 F.Supp. 688 (D.C.Tex. 1981). Another court has noted that actual costs include not only the costs of outside contractors but also the Coast Guard's internal costs for personnel, equipment and overhead. *U.S.* v. *Morania Barge*, 1983 A.M.C. 2761 (E.D.N.Car. 1982).

[17] 33 U.S.C. § 1321(a)(8).

[18] 33 C.F.R. § 153.407.

specific removal activity and fully expended during the removal activity.

(v) Lease or rental of equipment for the specific removal activity.

(vi) Contract costs for the specific removal activity (covering marine disasters).

(4) Claims payable under Part 25, Subpart H of this title (covering marine disasters).

(b) The District Commander may authorize the direct payment of the costs found to be reasonable under paragraph (a)(3) of this Section.

(c) The Pollution Fund is not available to pay any foreign, federal, state or local government or agency for the payment or reimbursement of its costs incurred in the removal of oil or hazardous substances discharged from a vessel or facility that it owns or operates.

In general, then, it would seem that items in the category of fixed costs would not, under these guidelines, qualify as actual costs. Further, it should be noted that these costs are keyed to Phase III and IV of the National Contingency Plan discussed at paragraphs 19–14 *et seq.* This raises the question of whether or not the government is limited to recovery of actual costs only during Phase III and Phase IV. The government would be likely to argue that a restrictive interpretation is contrary to the expansive phrase "or the taking of such other action as may be necessary to minimize or mitigate damage to the public health or welfare," within the definition of "remove."[19]

19–72 Whether or not removal costs include removal of a vessel before any actual discharge of oil has taken place is open to question. It would seem that the expansive definition of "removal" set out in the Act, together with the grant of authority to remove a substantial threat of a discharge, combined with the public policy of the Act to eliminate all discharges of oil leads to the conclusion that the costs of such pre-emptive action would probably be recoverable. In recognition of this likely result, the National Contingency Plan defines "discharge" to include a substantial threat of discharge.[20]

---

[19] A more extensive discussion of what constitutes actual costs of removal may be found in United States Dept. of Justice, *Methods and Procedures For Implementing A Uniform Law Providing Liability For Clean-Up Costs and Damages Caused by Oil Spills From Ocean Related Sources* (Comm. Print. 1975).

[20] 40 C.F.R. § 300.6.

7. REVOLVING FUND FOR CLEAN-UP ACTIVITIES

19–73   The Act authorizes a federal appropriation in the United States Treasury for a revolving fund of $35 million to carry out clean-up and removal activities.[21] The fund is available to reimburse federal and state authorities for clean-up costs expended under authority of the FWPCA and under the National Contingency Plan, to reimburse owners and operators who act to remove oil following a prohibited discharge from their vessel and to federal authorities acting under authority of section 311(d) covering maritime disasters.[22] In addition to federal funds, which are taken out of general government revenues, any collections derived from payment by dischargers and third parties and from civil penalties under the Act are also credited to the revolving fund. As a result of the unreimbursed expenditures, should the fund have an unobligated balance of less than $12 million, the Secretary of Transportation is called upon to notify Congress, with a request for a supplemental appropriation.

19–74   The existence of the $35 million revolving fund raises the question whether the government could continue with clean-up operations in the event of a depletion of the fund. While this is a theoretical possibility, the likelihood of a termination or delay of vital clean-up activities because of a depletion of the fund is extremely unlikely. Emergency governmental appropriations, either by way of a supplemental appropriation to the revolving fund (more likely long before the fund was fully depleted) or use of funds from other sources until the supplemental appropriation was authorized, would be far more likely.

19–75   A fundamental question arises whether financing clean-up activities, or even advancing clean-up funds to be recovered in whole or part against responsible parties, should be a burden on the general public as it is with the use of general appropriations to constitute the revolving fund. The argument has been advanced that the cost of such activities should be laid at the doorstep of

---

[21] 33 U.S.C. § 1321(K)(1). As of March, 1983, the section 311 Fund had total obligations from 1971–1983 of $124,749,595. What portion of this expense resulted from tanker discharges is not known. Congressional appropriation during this same period totalled a net of $91,000,000 and collections from fines and recoveries was $50,999,165. The unobligated balance as of January 31, 1983 was $17,249,570. Temple, Barker & Sloane, Inc., *Cost Benefit Analysis Of Possible U.S. Adherence To Two International Conventions on Liability and Compensation For Oil Pollution Damages* (Prepared for Commandant, United States Coast Guard, June 1983).

[22] As to state recovery, see 33 U.S.C. § 1321(c)(2)(H); as to owner or operator recovery, see 33 U.S.C. § 1321(i)(3); and as to maritime disasters, see 33 U.S.C. § 1321(d).

those most closely associated with the entrepreneurial exercise and the opportunity to protect against damage, the oil companies and the shipowners.[23] In response, it has been asserted that the general public benefits from these inherently dangerous activities and that turning a blind eye to this reality is to avoid a sincere confrontation of the problem.[24] While the FWPCA has not been amended to reflect changing trends, it will be seen in Chapter 20, with respect to the activity-related oil spill statutes, that funds contributed by involved industries in amounts considerably greater than $35 million have supplanted the use of general treasury appropriations to finance clean-up expense. As discussed in Chapter 21, recent legislation introduced in Congress seeks to remedy this problem by abolishing the section 311 Fund and replacing it with a Fund created from levies on oil imports and exports.

8. VESSEL COMPULSORY INSURANCE OR EVIDENCE OF FINANCIAL RESPONSIBILITY

19–76 By section 1321(p) any vessel or barge over 300 gross registered tons, excluding non-self propelled barges which do not carry oil or hazardous substances as cargo or fuel, which uses any port or place in the United States or the navigable waters of the United States, is required to establish and maintain evidence of financial responsibility to meet its potential liability under section 311.[25] For vessels, financial responsibility must be established for the amount of their section 311 limitation funds, *i.e.* $150 per gross registered ton, or $250,000, whichever is greater. A certificate is provided to all qualifying vessels which must be carried on the vessel and displayed on request.

19–77 Evidence of financial responsibility may be established by one or a combination of the following methods: (a) evidence of insurance issued by an insurer acceptable to the Coast Guard; (b) surety bonds issued by a surety company acceptable to the Coast Guard; (c) qualification as a self-insurer; or (d) other acceptable evidence of financial responsibility. As to the self-insurance qualification, the owner must maintain working capital and net worth in the United States in the section 311 limitation amount of the largest

---

[23] Note, Oil Spills and Clean-up Bills: Federal Recovery of Oil Spill Clean-up Costs, 93 Harv. L. Rev. 1761 (1980). This same argument has also been advanced with respect to the issues of limitation of vessel and facility liability in the event of a spill.

[24] *Ibid.*

[25] 33 U.S.C. § 1321(p)(1).

vessel in the fleet to be insured.[26] Evidence of net worth or working capital is demonstrated by the filing of certified financial statements and other information relevant to making a proper analysis of value.[27] The filings must be regularly updated to demonstrate continued compliance.[28]

19–78 To alleviate some of the burden which would follow from requiring financial responsibility of all vessels owned or operated by a single owner or operator, and presumably in recognition of the improbability of requiring the use of such security for more than one vessel in the fleet on a single occasion, the Act permits the filing of financial responsibility in the amount only of the largest vessel in the fleet.[29]

19–79 Any underwriter or other person providing evidence of financial responsibility may be sued directly for the amount of the vessel's section 311 liability to the United States up to the amount guaranteed.[30] In any such action, the underwriter or other such person may invoke only the defences of the owner or operator under section 311. The underwriter may not, therefore, invoke defences based on the owners' or operators' breach of the insurance agreement.

19–80 Provision is made for the issue of Master Certificates in favour of contractors or other persons responsible for vessels as builders, repairers, scrappers or sellers in lieu of a certificate for each vessel.[31] The Master Certificate covers all of such applicant's vessels to the extent that covered vessels are held only for purposes of construction, repair, scrapping or sale and not for any other commercial or business purpose. A Master Certificate must cover the maximum section 311 limitation amount of the largest vessel of those held by such applicant.

19–81 Failure to maintain evidence of financial responsibility or to comply with the regulations related thereto is punishable by a fine of not more than $10,000. In addition, the secretary may refuse to grant customs clearance to any vessel lacking a certificate. The

---

[26] 33 C.F.R. § 130.8. Working Capital is defined as the amount of current assets located in the United States less all liabilities. Net Worth is defined as the amount of all assets located in the United States less all liabilities.

[27] 33 C.F.R. § 130.8(b)(3)(i).

[28] 33 C.F.R. § 130.8(b)(3)(ii).

[29] 33 U.S.C. § 1321(p)(1).

[30] 33 U.S.C. § 1321(p)(3). The guarantor must accept United States federal court jurisdiction and designate an agent for service of process: 33 U.S.C. § 1321(w) and 33 C.F.R. § 130.15.

[31] 33 C.F.R. § 130.11.

Coast Guard may deny entry to any port in the United States or the navigable waters of the United States and may detain any vessel which does not produce a valid certificate.[32]

9. CONCLUSION

19–82 The fundamental problem with the FWPCA regime is not so much associated with the drafting inconsistencies contained in the legislation or the general scheme for clean-up cost recovery which the Act establishes, but it is the fact that it represents only a partial solution to a problem which most observers agree must be dealt with comprehensively. This stems in large part from the United States federal system in which the Constitution reserves rights to the states which are not within the sphere of exclusive federal regulation, about which a more extensive discussion may be found in Chapter 21. However, the failure of the FWPCA to legislate in the area of private recovery for oil spill damages leaves open a substantial area for protracted litigation in which recovery is by no means assured.

19–83 Perhaps the greatest defect in the legislation with respect to the area with which it is involved is the manner in which it allocates the costs of clean-up. Assuming the owner, who is the sole discharger, satisfies the very minimal standard to achieve the right to limit its liability, the balance of federal clean-up costs above the strict liability limitation are borne by the general public. At a time when the trend of pollution legislation has moved in the direction of allocating the burden of these costs to the industries which are most closely associated with the entrepreneurial effort, the FWPCA is unique in its failure to follow suit. As will be observed in Chapters 20 and 21, the activity-related statutes have moved clearly in the proper direction in this regard and recent legislation goes even further.

---

[32] 33 U.S.C. § 1231(p)(4)(5).

20. Activity-Related Federal Pollution Liability Statutes

SUMMARY

20–01 Apart from the Federal Water Pollution Control Act, there are three additional federal statutes which relate to vessel oil pollution liability in or about United States waters. The three statutes are the Trans-Alaska Pipeline Authorization Act,[1] the Deepwater Port Act[2] and the Outer Continental Shelf Lands Act Amendments.[3] Unlike the FWPCA, these three statutes are activity-related and, as will be discussed, legislate with respect to vessel pollution liability within the framework only of those specific activities. In certain instances, there is jurisdictional overlap with the FWPCA and the provisions concerning vessel pollution liability are inconsistent or irreconcilable. Such inconsistency leaves the shipowner simply to scratch his head and ponder his potential liabilities and obligations until resolved by the courts. Because of the relatively sparse vessel activity in the areas covered by the legislation, there has been little judicial interpretation of the interaction among these laws and thus much remains to speculation and scholarly pronouncement. Discussed in this chapter are the three activity-related pollution statutes. It will be noted, in particular, the introduction into these statutes of industry-supported liability funds which are available to cover private damages as well as government clean-up costs.

1. TRANS-ALASKA PIPELINE AUTHORIZATION ACT

A. Introduction
20–02 The Trans-Alaska Pipeline Authorization Act ("TAP")[4] was enacted by Congress in 1973 to promote exploitation of the oil resources of Alaska's North Slope for domestic use.

---

[1] 43 U.S.C. §§ 1651–1655.
[2] 33 U.S.C. §§ 1501–1524.
[3] 43 U.S.C. §§ 1801–1824.
[4] 43 U.S.C. §§ 1651–1655.

20–03 Apart from providing authorization for construction of the pipeline itself and establishing strict liability upon the holder of the pipeline right-of-way for damages arising out of operation of the pipeline, the Act also provides for strict joint and several liability upon owners and operators of vessels "operating between the terminal facilities of the pipeline and ports under the jurisdiction of the United States."[5] The Act, however, only applies to vessels carrying oil that has been transported through the trans-Alaska pipeline[6] and strict liability ceases "when the oil has first been brought ashore at a port under the jurisdiction of the United States.[7] Under these requirements, smaller trans-shipment vessels upon which oil has been loaded at an offshore transfer point remain subject to the Act's strict liability provisions. In fact, the regulations implemented pursuant to TAP require that all such transfer vessels have a Certificate of Financial Responsibility.[8] Because of the requirement that North Slope oil be used for domestic purposes only, American Flag vessels must be utilized for carriage of oil between Alaska and the lower forty-eight states. TAP, therefore, has application limited to vessels of the United States Flag which operate in this service.

B. Oil Spill Liability

20–04 TAP provides for strict joint and several liability on the owners and operators of vessels subject to the Act and, where applicable, on the Fund, for all damages, including cleanup costs, sustained by any person or entity, public or private, including residents of Canada, as the result of discharges of oil from the vessel.[9] Thus, whereas section 311 of the FWPCA only involves federal clean-up and removal costs, TAP more comprehensively involves all damages arising out of an incident and covers all persons so damaged. The scope of this provision, "all damages," has not been judicially defined, but regulations issued pursuant to the Act provide a broad construction of the term[10]:

---

[5] 43 U.S.C. § 1653(c)(7).
[6] 43 U.S.C. § 1653(c)(1).
[7] 43 U.S.C. § 1653(c)(7).
[8] 33 C.F.R. § 131.3.
[9] 43 U.S.C. § 1653(c)(1). The phrase "including residents of Canada" is redundant in view of the fact that "person" means "an individual, a corporation, a partnership, an association, a joint-stock company, a business trust or an unincorporated organization.': 44 U.S.C. § 1653(c)(12). While Canadian residents would be among those foreign residents likely to be most immediately effected by a TAP oil spill, it is unlikely that this phrase would be interpreted to exclude from recovery other foreign nationals, such as foreign flag vessel owners in waters where a TAP spill has occurred.
[10] 43 C.F.R. § 29.1(d).

"Damage or damages means any economic loss, arising out of or directly resulting from an incident, including but not limited to:

(1) Removal costs.
(2) Injury to, or destruction of, real or personal property.
(3) Loss of use of real or personal property.
(4) Injury to, or destruction of, natural resources.
(5) Loss of use of natural resources.
(6) Loss of profits or impairment of earning capacity due to injury or destruction of real or personal property or natural resources, including loss of subsistence hunting, fishing and gathering opportunities.
(7) Loss of tax revenue for a period of one year due to injury to real or personal property.

There is no geographical limitation to recoverability, thus a vessel carrying TAP oil through the Panama Canal, for example, could be liable under TAP for damages sustained in that jurisdiction resulting from an oil spill.

20–05 The only defences to strict liability on owner, operator and Fund under TAP arise if the damages were caused by an act of war or by negligence of the United States or one of its agencies.[11] The Act of God and acts of third parties defences contained in FWPCA, section 311 are not included. Moreover, strict liability will not be imposed if it can be demonstrated that the damage was caused by the party claiming damages.

C. Trans-Alaska Pipeline Liability Fund and Subrogation

20–06 The Act is the first domestic pollution legislation to establish a special fund for the benefit of damage claimants. The Fund, named the Trans-Alaska Pipeline Liability Fund ("TAP Fund"), is a non-profit making corporation administered by the holders of the pipeline right of way[12]; with a Board of Trustees comprised of the participating oil company representatives, United States Government representatives appointed by the Secretary of the Interior and one representative appointed by the Governor of Alaska.[13] The TAP Fund is derived from a fee, collected from the owners of the oil at the time of loading, of five cents per barrel.[14] Collections cease when the Fund reaches $100,000,000 and resume when the Fund falls below such amount.[15]

[11] 43 U.S.C. § 1653(c)(2).
[12] 43 USC § 1653(c)(4); 43 C.F.R. § 29.2.
[13] 43 C.F.R. § 29.2.
[14] 43 U.S.C. § 1653(c)(5).
[15] Income from investment of the amounts collected are added to the principal amount of the Fund and all costs of administration are payable out of the Fund: 43 C.F.R. § 29.4.

20–07 While liability is strict upon the owners and operators of vessels and the Fund is not limited in amount as in section 311 of the FWPCA, the operative provision of the Act state as follows[16]:

> "Strict liability for all claims arising out of any one incident shall not exceed $100,000,000. The owner and operator of the vessel shall be jointly and severally liable for the first $14,000,000 of such claims that are allowed . . . .The Fund shall be liable for the balance of the claims that are allowed up to $100,000,000. If the total claims allowed exceed $100,000,000 they shall be reduced proportionately. The unpaid portion of any claim may be asserted and adjudicated under other applicable Federal or state law."

Under this section, while the shipowner and operator have joint and several strict liability, secured by a Certificate of Financial Responsibility, of up to the first $14,000,000, they remain liable to damage claimants for any amounts not compensated after the TAP Fund has paid up its $86 million maximum. This residual liability of the shipowner and operator would not be governed by principles of strict liability under TAP, but would be governed by other applicable state or federal law, such as negligence.

D. Subrogation

20–08 Nor is the shipowner or operator insulated from liability for the maximum $86,000,000 paid out of the TAP Fund. In fact, if the TAP Fund pays claimants, in most circumstances it becomes subrogated to those claims and is entitled to seek recovery from the shipowner or affiliates of the shipowner and operators or against third parties. Section 1653(c) (8) provides:

> "In any cases were liability without regard to fault is imposed pursuant to this subsection and the damages involved were caused by the unseaworthiness of the vessel or by negligence, the owner and operator of the vessel, and the Fund, as the case may be shall be subrogated under applicable State or Federal laws to the rights under said laws of any person entitled to recovery hereunder. If any subrogee brings an action based on unseaworthiness of the vessel or negligence of its owner or operator, it may recover from any affiliate of the owner or operator, if the respective owner or operator fails to satisfy any claims by the subrogee allowed under this paragraph."

20–09 An "affiliate" is defined for purposes of this subsection as "(A) any person owned or effectively controlled by the vessel owner or operator; or (B) any person that effectively controls or has the

---

[16] 43 U.S.C. § 1653(c)(3).

power effectively to control the vessel owner or operator by (i) stock interest, or (ii) representation on a board of directors or similar body, or (iii) contract or other agreement with other stockholders, or (iv) otherwise; or (C) any person which is under common ownership or control with the vessel owner or operator." Under this provision not only the vessel owner and operator, but also any of their affiliates, as well as third parties, may become liable to the TAP Fund, as subrogee, for any amounts paid under the Act. However, as with residual liability to damage claimants, the TAP Fund and others claiming as subrogees would be required to prove unseaworthiness or negligence and could not recover on a strict liability basis.

E. Notification and Claims Procedure

20–10 Although the vessel is responsible for the first $14,000,000 of liability which is guaranteed by evidence of financial responsibility, the TAP claims procedure holds open a further possible remedy to claimants for the vessel's liability, as shall be discussed below.

20–11 As soon as he has knowledge of an incident involving possible Fund liability, the person in charge of the vessel must notify the Fund "using the fastest available communication."[17] Where possible, the Fund must designate the source or sources of oil pollution and notify the owner or operator of such designation.[18]

20–12 Where there is a likelihood of damages from an incident, the Fund must advertise the designation and state the procedures by which claims may be presented.[19] All claims are initially presented to the Fund.[20]

20–13 Where a designation is accepted by a vessel owner or operator, all claims presented to the Fund shall be promptly transmitted to the owner or operator. If the owner or operator for any reason denies liability for the claim, or if the claim is not settled within 90 days, the claimant may make an irrevocable election to sue the owner or operator in court or to present the claim to the

---

[17] 43 C.F.R. § 29.8(a).
[18] 43 C.F.R. § 29.8(b).
[19] 43 C.F.R. § 29.8(c). Advertisement, which must run for not less than 30 days, must commence no later than 15 days after the designation and in any event no later than 20 days from the date the Fund learns of the incident.
[20] 43 C.F.R. § 29.9(b)(1).

Fund.[21] In view of the Fund claims procedure, and the extension of pollution compensation beyond the shipowner's limited exposure of $14,000,000, it is unlikely that the court action election would ever be utilized in a major spill situation.

20–14 Where a claim is not satisfied by the shipowner either because it exceeds the $14,000,000 limitation or because the owner, operator and guarantor are financially incapable of meeting their obligations, a claim for the uncompensated amount may be presented to the Fund.[22] If the Fund, for any reason, pays any part of this first $14,000,000, it may seek reimbursement, by court action or otherwise, from the owner, operator or guarantor.[23] There is no suggestion, however, that the Fund is *required* to pay any portion of the first $14,000,000.

20–15 If the source of the discharge cannot be designated the Fund shall nevertheless "process and adjudicate" all claims, but this does not mean that the Fund is required to satisfy the first $14,000,000 in claims, nor does it mean that the Fund necessarily will accept liability under circumstances in which it cannot be demonstrated that the Fund is liable. Since there has been no practice in this area, it is uncertain what position the Fund would take should this occur.

20–16 Any properly documented claim presented to the Fund must be settled and paid within 120 days of presentation or the date upon which advertising, as described in paragraph 20–12, above, was commenced, whichever is later.[24]

20–17 Disputed claims submitted to the vessel interests or to the Fund may be submitted to arbitration upon agreement of all parties.

20–18 Any claim not submitted to the Fund or made the basis of a court action is time-barred after three years from the date of discovery of the damages, or within six years of the date of the incident, whichever is earlier.[25]

---

[21] 43 C.F.R. § 29.9(c).
[22] 43 C.F.R. § 29.9(e). Presumably this refers to a single or an aggregate of claims which cause the limitation amount to be exceeded.
[23] 43 C.F.R. § 29.9(f). If the refusal or denial by the owner, operator and guarantor is "without any reasonable basis for doing so," the Fund may request the revocation of the Certificate of Financial Responsibility and prohibit the guarantor from providing evidence of financial responsibility in the future. *Ibid.*
[24] 43 C.F.R. § 29.9(g).
[25] 43 C.F.R. § 29.9(i).

F. Financial Responsibility

20–19 A vessel owner or operator within TAP jurisdiction is required to obtain a Certificate of Financial Responsibility, generally similar in its rights and obligations to the FWPCA certificate, from the Coast Guard before TAP oil is loaded.[26] The Certificate is specifically for Alaska Pipeline financial responsibility and must be in the amount of $14,000,000. Unlike section 311 of Financial Responsibility, the Alaska Pipeline Certificate is not tonnage related.

20–20 The legislative history of the Act provides a justification for the strict liability and financial responsibility of a fixed $14,000,000 as follows[27]:

> "Strict liability is primarily a question of insurance. The fundamental reason for the limit placed on liability in the Federal Water Quality Improvement Act stemmed from the availability, or non-availability, of marine insurance. Without a readily available commercial source of insurance, liability without a dollar limitation would be meaningless and many independent owners could not operate their vessels. Since the world-wide maritime insurance industry claimed $14 million was the limit of the risk they would assume, this was the limit provided for in the Federal Water Quality Improvement Act. There has been no indication that this level has since increased. Accordingly, the Conference adopted a liability plan which would make the owner or operator strictly liable for all claims (for both clean-up costs and damages to public and private parties) up to $14 million. This limit would provide an incentive to the owner or operator to operate the vessel with due care and would not create too heavy an insurance burden for independent vessel owners lacking the means to self-insure."

It is interesting to note that the $14,000,000 limit contained in TAP in 1973 was modelled on the $14,000,000 limit contained in section 311 of FWPCA prior to that Act's amendment in 1977.[28] The present maximum limit of liability under section 311 is, however, based

---

[26] By Executive Order 12418, dated May 5, 1983, the President transferred responsibility for vessel financial certification from the Federal Maritime Commission to the Secretary of Transportation. The Secretary redelegated this authority, effective October 11, 1983, to the Coast Guard.

[27] Conference Report No. 93–924, reported in 1973 U.S. Code Cong. & Admin. News at pp. 2530–31.

[28] See Chap. 19, n.76.

507

on gross vessel tonnage and can rise above \$14,000,000. Evidently, by 1977 it was determined that the insurance industry could and would accept higher strict liability limits but this concept was not incorporated into the TAP legislation.[29]

20–21 One must also be circumspect in considering the suggestion contained in the Conference Report to the effect that the strict liability limits contained in the Act "provide an incentive" to vessel owners. Careful consideration of the Act suggests very little by way of incentives to vessel owners. Since strict liability is indeed onerous, limits imposed on strict liability may be regarded by some as "incentives," but this is a cavalier manner of using that phrase. Rather, the pollution liability provisions of TAP provide to shipowners only a catalogue of difficulties: (1) they are strictly liable for the first \$14,000,000 rather than held to a negligence standard; (2) there are meagre defences to strict liability and fewer defences than under section 311; (3) the Act prescribes no limit to fault-based liability[30]; and (4) liability is statutorily extended to "affiliates" of owners and operators of vessels. While the provision on affiliates has not been judicially tested, the language of the Act suggests that the Fund has available to it a method in this specific area of piercing corporate veils which eliminates most of the fine points in common law pleading and proof in the jurisprudence of that area, making it far easier to reach assets of one entity to satisfy the debts of another.

2. DEEPWATER PORT ACT

A. Introduction

20–22 The Deepwater Port Act ("DPA"), enacted in 1975, is the second of the three United States activity-related oil spill schemes to become effective.[31] The Act establishes procedures for the construction, licensing and operation of deepwater ports outside of the three-mile territorial sea of the United States. Since such facilities

---

[29] At present tankers generally in the 250,000 DWT range are frequent carriers of Alaska crude oil. Such tankers as the M/V Brooklyn have a weight of approximately 104,000 gross tons. Under the existing statutory framework, the owner of the M/V Brooklyn would have strict section 311 clean-up liability to the government limited to \$15,600,000, whereas strict liability for all damages under TAP would be only \$14,000,000.

[30] One can only assume that fault-based liability would be subject to the 1851 Limitation of Liability Act to limit the shipowner's or demise charterer's liability. In most cases, the operator would fall within the protection of the 1851 Limitation Act and thus would not be protected by it.

[31] 33 U.S.C. §§ 1501–1524.

are outside the navigable waters of the United States, a legal regime consistent with the use of the High Seas was also established. In connection with this legal regime, the DPA, modelled in relevant part on the Trans-Alaska Pipeline Act liability provisions, establishes a scheme for liability, clean-up and damages recovery as a result of an oil spill at or around a deepwater port.[32]

20–23 The DPA was originally considered by Congress prior to the time of the 1973–1974 Arab oil embargo, when it was believed that greater reliance on foreign oil was inevitable.[33] It was felt that the ULCC and VLCC would represent an increasingly greater segment of the world tankship fleet, yet vessels of such size and draft could not safely enter United States ports. Except for two West Coast ports, loaded vessels in excess of 80,000 deadweight tons could not be accommodated. While port dredging was considered possible, the preferred alternative was development of the deepwater port concept, typically single point mooring buoys anchored in deep water with submarine hose connections ashore. The legislative committees which examined the bill prior to enactment were candid to observe that circumstances had changed since the deepwater port concept was first proposed. Specifically, as a result of the 1973–1974 oil embargo, a high national priority became the reduction of foreign oil dependence.[34] Nevertheless, it was recognized that a high level of importation would be foreseeable for at least a decade and thus development of the plant was not abandoned. Of course, as the market price of crude oil and other petroleum products continues to fluctuate, imports continue to occupy a significant role in domestic consumption.[35]

20–24 It was also candidly recognized that the DPA would constitute interim legislation until a comprehensive ocean oil spill regime was adopted.[36] However, more than ten years after the DPA legislation was passed, there is a continued, although reduced, use of foreign oil in the United States and despite the des-

---

[32] 33 U.S.C. § 1517.

[33] Senate Report No. 93–1217, reported in 1974 U.S. Code Cong. & Admin. News at pp. 7533–7535.

[34] Senate Report No. 93–1217, reported in 1974 U.S. Code Cong. & Admin. News at p. 7534.

[35] In 1982, 3,461,000 barrels per day of crude oil and 758,000 barrels per day of residual oil were imported into the United States. Temple, Barker & Sloane, Inc., *Cost-Benefit Analysis of Possible U.S. Adherence to Two International Conventions on Liability and Compensation For Oil Pollution Damages at V-33* (Prepared for Commandant, United States Coast Guard, June, 1983).

[36] Senate Report No. 93–1217, reported in 1974 U.S. Code Cong. & Admin. News at p. 7543.

perate need for it, there is still no comprehensive ocean oil spill legislation. Theoretically, the need for the DPA is as strong today as it was in 1974. In practice, of the three deepwater ports under consideration in 1974, only one is in use[37] and its economic viability is questionable.[38]

20–25 The DPA establishes a licensing procedure for the ownership, construction and operation of deepwater port facilities off the United States. Such licensing calls upon the Secretary of Transportation to confirm, *inter alia,* that the applicant is financially responsible, that he is willing to comply with all applicable laws, that the operation of the facility is in the national interest, that it will not unreasonably interfere with navigation and other uses of the high seas, that it is environmentally safe, that the Governors of adjacent states approve of the facility and that the ownership and use does not violate the anti-trust laws of the United States.[39] All deepwater ports and related storage facilities are subject to regulation as common carriers under the Interstate Commerce Act.[40]

20–26 With respect to vessel operations, "subject to recognized principles of international law," the Secretary of Transportation is required to prescribe and enforce regulations governing vessel movement, loading and unloading procedures, designation and marking of anchorage areas, maintenance, law enforcement and training in pollution avoidance, and control in the operation of deepwater ports.[41] To help ensure operational safety, the Secretary is called upon to designate a "safety zone" of appropriate size around each deepwater port.[42] In such zones, uses inconsistent with the operation of the deepwater port are prohibited. To ensure compliance by foreign vessels with the DPA together with its rules and regulations, the Secretary of State is required to notify the flags state of each vessel which may call at a deepwater port that, absent the flag state's objection, such vessels will be subject to United States jurisdiction when calling at or utilizing a deepwater port or when within a safety zone of a deepwater port in activities connected, associated or potentially interfering with the use and operation of a deepwater port. The vessel owner must appoint an agent

---

[37] The Louisiana Offshore Oil Port (LOOP) facility located 19 miles off the coast of Louisiana in the Gulf of Mexico.

[38] See, *Legislative History for the Deepwater Port Act Amendments of 1984,* published in 1984 U.S. Code Cong. & Admin. News, at p. 2729.

[39] 33 U.S.C. § 1503(c).

[40] 33 U.S.C. § 1507.

[41] 33 U.S.C. § 1509(a). The regulations are published in 33 C.F.R. Subchapter NN.

[42] 33 U.S.C. § 1509(c). The only safety zone thus far established is in connection with the LOOP facility. 33 C.F.R. § 150, Annex A.

for service of process in the United States in the event of legal proceedings in connection with such operations. Where an objection is made, except in cases of *force majeure*, a licensee may not permit a vessel of such objecting flag state to call at the deepwater port.[43]

20–27 The DPA was envisioned as establishing a procedure to reduce the general level of vessel source oil pollution in and around the United States, but also as increasing the risk of a large spill.[44]

> "The construction and operation of deepwater ports off the coast of the United States promises to reduce oil pollution damage to the marine environment. Tanker traffic in congested harbours and ports should be reduced and the need to lighten supertankers at offshore locations should be almost eliminated. As a result, the risk of collision and the number of cargo transfer and other chronic spills should be minimized.
>
> In spite of these environmental advantages the Committees recognize that increasing the number of supertankers operating off U.S. shores also increases the risk of a catastrophic super spill."

20–28 The liability provisions of the Act thus took on special significance in view of the perceived increase in the risk of a large spill. Three major objectives were addressed[45]:
  (1) to provide the fullest and most expeditious compensation possible;
  (2) to distribute the burden of risk equitably among deepwater port licensees, vessel owners and operators and consumers who will ultimately benefit from the use of these facilities; and
  (3) to impose standards of liability that will induce maximum effort to prevent prohibited discharge without imposing such onerous financial conditions as would impair competition for use of deepwater ports.

B. Oil Spill Liability

20–29 Section 1517 of the Act[46] is the oil spill liability provision. It prohibits the discharge of oil into the marine environment from any vessel within a safety zone which has received oil from another

---

[43] 33 U.S.C. § 1518. These provisions were added through enactment of the Deepwater Port Act Amendments of 1984, P.L. 98–419, effective September 25, 1984 ("DPAA–1984"). Prior to DPAA–1984, the licensee was required to obtain such jurisdictional recognition by the flag state.

[44] Senate Report No. 93–1217, reported in 1974 U.S. Code Cong. & Admin. News at p. 7542.

[45] *Ibid.* at 7543.

[46] 33 U.S.C. § 1517.

vessel at a deepwater port or which has itself received oil from a deepwater port.[47]

20–30 As in TAP and FWPCA, liability under the DPA is strict but limited, with exceptions. There are, however, substantial differences. Under the DPA, the owner and operator of a vessel: (1) which discharges oil within a safety zone, or (2) which discharges oil after it has received oil from another vessel at a deepwater port (except when moored at the port) are jointly and severally liable for clean-up costs and damages resulting therefrom.[48] Joint and several liability shall not exceed the smaller of $150 per gross ton or $20,000,000, unless it can be shown that the discharge was the result of gross negligence or wilful misconduct within the privity and knowledge of the owner or operator, in which case both are jointly and severally liable for the full amount of all clean-up costs and damages. Similarly, licensees of deepwater ports are strictly liable for clean-up costs and damages that result from a discharge of oil from the deepwater port or from a moored vessel, except that their strict liability is limited to $50,000,000. "Clean-up costs" include, but are not limited to, "all actual costs by Federal, state or local governments, foreign governments and each of their contractors or subcontractors incurred in removing or attempting to remove or taking other measures to reduce or mitigate damage from a prohibited discharge of oil."[49] "Damages" include damages suffered by any person or involving real or personal property, natural resources of the marine or coastal environment, including "damages claimed without regard to ownership of any affected lands, structures, fish, wildlife, or biotic or natural resources."[50] "Damages" do not include "clean-up costs" which are, under the Act, a separate category of recovery.

20–31 There are only two defences to strict liability. If the owner, operator or licensee can show that the discharge was caused *solely* by (1) an act of war, or (2) negligence on the part of the federal

---

[47] 33 U.S.C. § 1517(a)(1). "Oil," as defined in the DPA, means "petroleum, crude oil, or any substance refined from petroleum or crude oil." 33 U.S.C. § 1502(14).

[48] 33 U.S.C. § 1517(d). Discharge of oil from a moored vessel at a deepwater port subjects the deepwater port licensee to strict liabililty, and thus to provide focus to the recovery scheme, moored vessels are not subject to strict liability. The licensee's liability is up to $50,000,000 except that full liability will result from gross negligence or wilful misconduct within the privity and knowledge of the licensee. 33 U.S.C. § 1517(c).

[49] 33 U.S.C. § 1517(m)(1).

[50] 33 U.S.C. § 1517(m)(2). The Act also provides for the Attorney General to act on behalf of any group of damaged citizens if he determines that a "class action" would be appropriate. If the Attorney General fails to act within 90 days of a discharge of oil, any aggrieved party may bring a class action on behalf of the class: 33 U.S.C. § 1517(i).

government in maintaining aids to navigation, there shall be no liability. In addition, if the owner, operator or licensee (or the Fund, as discussed in paragraphs 20–33 *et seq.*, below), can show that the damages claimed by a particular party were caused *solely* by that party's negligence, there will also be no liability.[51]

20–32 The Secretary of Transportation is authorized to remove or arrange for the removal of oil discharged from a vessel within a safety zone or from a vessel which has received oil from another vessel at a deepwater port unless he determines that such removal will be done properly and expeditiously by the vessel owner or by the deepwater port licensee.[52] Such removal, whether by the vessel owner, the licensee or the Secretary, shall "to the greatest extent possible" be in accordance with the National Contingency Plan established pursuant to FWPCA, section 311.[53] In acting to remove or minimize the effect of an oil spill, the Secretary is authorized to draw upon available money in the Deepwater Port Liability Fund established by the Act.[54]

C. The Deepwater Port Liability Fund

20–33 The DPA establishes as a non-profit-making corporate entity the Deepwater Port Liability Fund (the "DPA Fund") administered by the Secretary of Transportation or his designee.[55] The Secretary has designated the Chief of the Pollution Liability Funds Management Staff within the Office of Marine Environment and Systems at the United States Coast Guard as administrator of the DPA Fund.[56]

20–34 The DPA Fund, created to assure full and expeditious recovery for oil spill clean-up costs and damages, is strictly liable for all clean-up costs and all damages in excess of those "actually compensated" by vessel owners, operators and licensees.[57] This differs from TAP in that the TAP Fund's liability is operative only where damages exceed $14,000,000 and the limit of the TAP Fund's liability is $100,000,000.[58] Apart from the defence that the party claiming damages was solely responsible for the loss, the Fund has no other defences to strict liability. Any damages not

---

[51] 33 U.S.C. § 1517(g).
[52] 33 U.S.C. § 1517(c)(1).
[53] The National Contingency Plan is discussed in detail at paras. 19–14 to 19–29, above.
[54] 33 U.S.C. § 1517(c)(3).
[55] 33 U.S.C. § 1517(f)(1).
[56] 33 C.F.R. § 137.101.
[57] 33 U.S.C. § 1517(f).
[58] See discussion at para. 20–07, above.

"actually compensated" by the vessel owner or operator will be paid by the DPA Fund.

20–35 The DPA Fund was established by the payment by deep-water port licensees of a fee of two cents per barrel on any oil loaded or unloaded at the port,[59] except that no fees were payable on bunker and fuel oil to be used by a vessel or on oil which has been transported through the trans-Alaska pipeline. Collections in the original Act were to cease when the Fund reached $100,000,000, unless there were adjudicated claims against the Fund to be satisfied. The Fund may borrow from the Treasury of the United States any amounts necessary to pay clean-up costs and damages for which it is liable even if such liability exceeds the $100,000,000 maximum amount of the Fund.[60] The 1984 amendments to the DPA suspended the barrel fee as of the date of the law, September 25, 1984. On that date, the Fund had assets of $4,000,000 and apart from administrative fees, there had been no payments out of the Fund. The 1984 amendments provide that collection of barrel fees will be renewed only when the Fund balance falls below $4,000,000.[61]

20–36 In making claim for recovery from the Fund for clean-up costs and damages, the revised regulations adopt the regulations applicable for recovery of clean-up costs and damages from the Fund established pursuant to the Outer Continental Shelf Lands Act Amendments.[62] While a more complete discussion of that procedure appears in paragraphs 20–61 *et seq.*, the procedures call for claimants to proceed initially against vessel owners and operators and deepwater port licensees who accept the obligation to pay under the strict liability provisions of the Act.

D. Certificates of Financial Responsibility

20–37 The Secretary of Transportation shall require any owner or operator of a vessel using a deepwater port or a deepwater port licensee to carry insurance or provide evidence of appropriate financial responsibility to meet the liabilities imposed by the Act.[63]

---

[59] 33 U.S.C. § 1517(f)(3).

[60] *Ibid.*

[61] P.L. 98–419, § 2 (September 25, 1984). The 1984 Amendments were enacted, in part, in an effort to make use of deepwater ports more competitive and it was for this reason that the user fees were suspended.

[62] 33 C.F.R. § 137.503.

[63] 33 U.S.C. § 1517(l).

E. Notification and Penalties

20–38   As with TAP, any individual in charge of a vessel from which a prohibited discharge of oil is made is required to notify the authorities as soon as he has knowledge of the discharge. Failure to notify "immediately" is a crime punishable by a fine of up to $10,000, imprisonment of not more than one year, or both.[64]

20–39   There are also civil penalties of up to $10,000 for each day of violation for each prohibited discharge. The Secretary of Transportation may withhold clearances of any vessel subject to a civil penalty which fails to furnish satisfactory surety.

F. Subrogation and Third Party Liability

20–40   Unlike section 311, the DPA does not permit a shipowner or port licensee to avoid strict liability for third party sole fault (except an act of war or federal government negligence as specified in paragraph 20–31, above) if the vessel or part is the source of the discharge. It is therefore logical and appropriate that the Act provides that a shipowner or operator subject to strict liability, who is compelled to pay claims even though the discharge was the result of the port licensee's negligence, should be subrogated to the rights of any person entitled to recovery against such licensee.[65] Similarly, where the licensee is forced under the strict liability provisions to pay claims even though the discharge was the result of vessel unseaworthiness or negligence of the owner or operator, the licensee is subrogated to the rights of any person entitled to recovery against the shipowner or operator.[66] It is unclear from the words of this subsection or from the legislative history whether the rights which are subrogated are common law rights or the rights set out in the DPA itself, including strict liability, or both. Since the language is broad, it suggests that the shipowner or the licensee is subrogated to all such rights, including those created by the Act. In such a case, the licensee may well find that his common law remedies against the owner and operator are limited by the 1851 Limitation Act and that his DPA rights are limited by the Act's own limitation provisions. In such a case, while the port's liability is $50 million,[67]

---

[64] 33 U.S.C. § 1517(b).
[65] 33 U.S.C. § 1517(h)(1).
[66] 33 U.S.C. § 1517(h)(2).
[67] 33 U.S.C. § 1517(e).

the vessel's liability will not exceed $20 million and possibly much less with respect to vessels under 250,000 DWT.[68]

20–41 Should the DPA Fund pay any compensation for "damages," it should be subrogated to all rights of claimants so paid. Since the definition of "damages" contained in the Act excludes "clean-up costs"[69] the plain language of the Act means that the Fund would not be subrogated to the rights of claimants to recover for clean-up costs. Thus, in large spill situations where clean-up costs exceed the port and/or vessel limits of liability or where the rare circumstances exist in which the vessel or port are not strictly liable, the Fund may find that it must pay clean-up claims without any right of subrogation. If this was merely poor draftmanship and not a purposeful omission, congressional amendment would be required to avoid such a result.

3. OUTER CONTINENTAL SHELF LANDS ACT AMENDMENTS

A. Introduction

20–42 Congress passed the Outer Continental Shelf Lands Act[70] in 1953 which, together with the Submerged Lands Acts of 1953,[71] was meant to clarify disputed questions of state and federal jurisdiction in the area of natural resources exploitation and control of the Continental Shelf of the United States. The focus of the dispute prior to 1953 was whether it was the states or the federal government which had the right to control development of these valuable resources.[72] The so-called Truman Proclamation issued in 1945 stated[73]:

> "The Government of the United States regards the natural resources of the subsoil and seabed of the continental shelf beneath the high

---

[68] 33 U.S.C. § 1517(d). DPAA 1001 amends the DPA to clarify the point that when the licensee or the Fund are strictly liable as a result of the negligence of the owner or operator of a vessel or the unseaworthiness of the vessel, then the owner and operator of the vessel are jointly and severally liable to the licensee or the Fund to the same extent as if they were themselves liable to the claimants under section 1517(d). See paras. 22–29 to 22–31, above.

[69] 33 U.S.C. § 1517(i)(2).

[70] 43 U.S.C. §§ 1331–1343.

[71] 43 U.S.C. §§ 1301–1315.

[72] See, generally, 1978 U.S. Code Cong. & Admin. News at pp. 1462–4: Comment "A Broad Overview of the Outer Continental Shelf Lands Act Amendment of 1978," 4 *The Maritime Lawyer* 108.

[73] Proclamation No. 2667, 10 Fed. Reg. 12303 (1948).

seas but contiguous to the coasts of the United States as appertaining to the United States [and] subject to its jurisdiction and control."

Several states challenged the claim of federal sovereignty announced by the Truman Proclamation but the United States Supreme Court upheld the federal government's position that these areas were subject to federal control.[74]

20–43 The dispute was resolved with the passage of the Submerged Lands Act and the Outer Continental Shelf Lands Act in 1953. Under the former, that portion of the Continental Shelf within the three-mile limit was released and relinquished to the states. Under the latter, the federal government reasserted its jurisdiction to the Continental Shelf outside of that area relinquished to the states under the Submerged Lands Act.[75] Neither Act made substantive provision for pollution, but provided a structure for the assertion of jurisdiction over these lands. The Outer Continental Shelf Lands Act also authorized the Secretary of the Interior to manage mineral resources in the federal territory and to lease lands for exploitation.[76]

20–44 In response to a complex bureaucratic administration of the Outer Continental Shelf which lead to disincentives for exploration, the Act was comprehensively revised in 1978.[77] That revision is contained in the Outer Continental Shelf Lands Act Amendments of 1978 ("OSCLA–1978"). OSCLA–1978 extensively revised the mineral resources management scheme and specified a procedure for granting oil and gas leases on the Outer Continental Shelf, among other provisions, for extracting minerals from the seabed and subsoil. Most pertinent to vessel oil pollution, OCSLA–1978 established the Offshore Oil Spill Pollution Fund.[78] In addition to this fund, the Act provides for strict but limited oil

---

[74] *U.S.* v. *Texas*, 339 U.S. 707 (1950); *U.S.* v. *Louisiana*, 339 U.S. 669 (1950); *U.S.* v. *California*, 332 U.S. 19 (1947).

[75] Subsequent to the 1953 legislation, the United States became a signatory to the 1958 Geneva Convention on the Continental Shelf, 15 U.S.T. 471, which defines the Continental Shelf as: "The seabed and subsoil of the submarine areas adjacent to the coast but outside the area of the territorial sea, to a depth of 200 meters or, beyond that limit, to where the depth of the superadjacent waters admits of the exploration of the natural resources of the said areas." The act, as amended, defines "Outer Continental Shelf" as "All submerged lands lying seaward and outside of the area of lands beneath navigable waters as defined in the Submerged Lands Act and of which the subsoil and seabed appertain to the United States are are subject to its jurisdiction and control."

[76] 43 U.S.C. § 1337.

[77] See generally, Comment, 4 *The Maritime Lawyer* 108; Coulter, "The Outer Continental Shelf Lands Act—Its Adequacies and Limitations," 4 *Nat. Resource Law* 725 (1971).

[78] 43 U.S.C. §§ 1801–1824.

pollution damage liability for owners and operators of vessels and offshore facilities operating on the Outer Continental Shelf.

B. Oil Spill Liability

20–45   With respect to vessels, the Act provides that the owner and operator of a non-public vessel, "which is the source of oil pollution, or poses a threat of oil pollution," shall be jointly, severally and strictly liable for loss to defined classes of persons under the provisions of the Act.[79] A "source" is not defined, and while clearly this refers to vessels from which oil is discharged into the water, it may be anticipated that courts will one day be confronted with the argument that "source" also refers to the "cause" of oil pollution, not merely its immediate source.[80] For purposes of restricting the application of this section, the Act defines a vessel as one which is operating in the waters above the Outer Continental Shelf or which is operating within the territory covered by the Submerged Lands Act[81] and which is transporting oil directly from an offshore facility.[82] Hence, the Act would not apply, for instance, to a ship on her ballast voyage to an offshore installation from a terminal not used in connection with the offshore facility. "Oil pollution" is defined as: (1) the presence of oil in an unlawful quantity or which has been discharged at an unlawful rate in or on waters above Submerged Lands or on the waters of the Contiguous Zone[83]; or (2) the "presence" of oil on the high seas outside the territorial limit of the United States when discharged in connection with Outer Continental Shelf Lands activities or which causes injury to or loss of natural resources under United States jurisdiction; or (3) the presence of oil in or on the territorial sea, navigable or internal waters,

---

[79] 43 U.S.C. § 1814(a). A "loss," within the meaning of OCSLA–1978, must be considered with reference to 43 U.S.C. § 1813, which sets out classes of "economic loss' recoverable as a result of oil pollution. Economic loss under the Act is discussed at paras. 20–51 *et seq.* An "owner" means any person holding title or other indicia of ownership, except that it does not include "a person who, without partipating in the management or operation of a vessel . . . holds indicia of ownership primarily to protect his security interest in the vessel: . . . " 43 U.S.C. § 1811(19). An "operator" means "a charterer by demise or any other person, except the owner, who is responsible for the operation, manning, victualing and supplying of the vessel": 43 U.S.C. § 1811(20).

[80] Some support for the argument is found in comparing the language of this subsection with subsection § 1814(d), which makes specific reference to vessels "from which the discharge occurred."

[81] See discussion at paras. 20–42 to 20–44, above.

[82] 43 U.S.C. § 1811(5). An "offshore facility" under the Act is any oil refinery, drilling structure, oil storage or transfer terminal, or pipeline used in connection with Outer Continental Shelf Oil production operations. 43 U.S.C. § 1811(8).

[83] Established under Article 24 of the 1958 Convention on the Territorial Sea and the Contiguous Zone, 15 U.S.T. 1606.

or adjacent shorelines of a foreign country when the pollution occurred in connection with Outer Continental Shelf Lands activities.[84]

20–46 The owner and operator of a vessel are entitled to limit their liability in the same fund to the greater of $250,000 or $300 per gross ton,[85]; unless the incident is caused "primarily by wilful misconduct or gross negligence within the privity or knowledge of the owner or operator" or primarily by a violation of safety, operation or construction standards of the federal government.[86] Also, presumably as a reaction to such pollution incidents as the *Tamano* and *Ocean Eagle*,[87] the owner or operator of a vessel which fails or refuses to provide "all reasonable co-operation and assistance requested by the responsible Federal Officials in furtherance of clean-up activities may be deprived of limitation."[88] The owner and operator of a vessel are also liable for interest at the commercial rate without benefit of limitation, on all amounts ultimately paid in satisfaction of a claim from the date of presentation of the claim to the date of payment.[89]

20–47 The Act excepts the vessel from liability: (1) "if the incident is caused solely by an act of war, hostilities, civil war or insurrection," or by natural *force majeure* circumstances "which could not have been prevented by the exercise of due care or foresight"; or (2) "if the incident is caused solely by the negligent or intentional act of the damaged party or any third party (including any governmental entity)."[90]

20–48 Notwithstanding the exceptions and limitation from liability contained in the Act, the owner and operator of a vessel subject to the jurisdiction of the Act from which oil is discharged, are jointly and severally strictly liable without limitation or exception

---

[84] 44 U.S.C. § 1811(9). While not specifically stated, an "unlawful quantity" clearly refers to the FWPCA proscription against discharges "in such quantities as may be harmful," as to which, see para. 19–07, above. An "unlawful rate" must refer to the requirements of MARPOL 73/78, as to which, see paras. 3–52 *et seq.*, above.

[85] 43 U.S.C. § 1814(e) requires the Secretary to report to Congress on the desirability of adjusting these limits. It was specifically envisioned that Congress would make inflation adjustments to these limits. See 1978 U.S. Code Cong. & Admin. News at p. 1589.

[86] 43 U.S.C. § 1813(b). The legislative history suggests that this latter exception was intended to cover offshore facilities, not vessels. See 1978 U.S. Code Cong. & Admin. News at p. 1589.

[87] See *Ocean Eagle-Limitation Proceedings*, 1974 A.M.C. 1629 (D.P.R. 1974).

[88] *Ibid.*

[89] Interest will not, however, run for the period in which the owner has offered to pay a disputed claim and it is finally determined that the offer is equal to or greater than the award.

[90] 43 U.S.C. § 1814(c).

for all costs of removal incurred by the federal government or any state or local agency.[91]

C. The Offshore Oil Pollution Compensation Fund

20–49 The Act establishes a fund, the Offshore Oil Pollution Compensation Fund ("the OPC Fund"), in the Treasury of the United States in an amount not to exceed $200,000,000.[92] The OPC Fund is given legal status such that it may sue and be sued in its own name. The requisite funding is through a levy by the Secretary of Transportation of a fee of three cents per barrel on oil obtained from the Outer Continental Shelf to be paid by the owner of such oil when produced.[93] Additional funding is through monies collected or recovered on behalf of the OPC Fund against dischargers and others as provided in the Act. As shall be discussed below, the OPC Fund is liable to pay certain claims resulting from oil pollution. In the event that the OPC Fund's liabilities exceed monies available in the Fund, the Secretary of Transportation is authorized to have issued notes and obligations for purchase by the Secretary of the Treasury to form part of the public debt. Thus, subject to the solvency of the United States Treasury, liquidity of the OPC Fund to pay its obligations is assured.

20–50 With limited exceptions, the OPC Fund is liable, without benefit of limitation, to pay for all losses which may be compensated for under the Act to the extent that such losses are not otherwise compensated.[94] The only exceptions are that the OPC Fund shall not be liable for losses over and above removal costs where a particular claimant's gross negligence or wilful misconduct was the cause of the loss or, as to a particular claimant, to the extent that the incident or consequent loss was caused by that person's own negligence.

D. OCSLA–1978 Claims for Oil Pollution

20–51 OCSLA–1978 is the first federal statute to *specify* that claims for so-called economic loss, in addition to removal costs, are

---

[91] "Removal Costs" are defined in OCSLA–1978 as costs incurred under the oil removal cost provision of the Federal Water Pollution Control Act, section 1321(c), (d) or (e). These costs, generally, refer to actual removal costs. See paras. 19–70 *et seq.*, above.

[92] 43 U.S.C. § 1812. The OPC Fund may, however, be increased to permit recoveries and collections to be paid into the Fund.

[93] 43 U.S.C. § 1812(d).

[94] 43 U.S.C. § 1814(f).

recoverable. The classes of recoverable economic claims under § 1813(a) include[95]:

"(1) removal costs; and
(2) damages, including—
    (A) injury to or destruction of, real or personal property;
    (B) loss of use of real or personal property;
    (C) injury to, or destruction of, natural resources;
    (D) loss of use of natural resources;
    (E) loss of profits or impairment of earning capacity due to injury to, or destruction of, real or personal property or natural resources; and
    (F) loss of tax revenue for a period of one year due to injury to real or personal property."

These classes of recovery are considerably broader than recovery based on the common law. However, given the broad scope of damages, the Act limits the availability of certain classes of damages to specific classes of claimants.[96] As will be discussed in Chapter 21, enumeration of these types of damage does not solve the problems of quantification and proof of loss.

20–52 Any claimant may recover provable removal costs, including a vessel owner or operator who can show that he is entitled to the benefit of one of the exceptions to liability contained in the Act, or if not so entitled, that he is entitled to limit his liability, in which case he can recover costs in excess of the limitation amount from the OPC Fund.[97] Included within the meaning of "removal costs" are "clean-up costs" which are defined as "costs of reasonable measures taken, after an incident has occurred, to prevent, minimize, or mitigate oil pollution from such incident."[98]

20–53 A United States claimant who owns or leases property or who "utilizes" the natural resource, may assert a claim for its injury or destruction and its loss of use resulting from an oil discharge. Although not defined, loss of use presumably refers to costs associated with such damage items as temporary replacement of the property pending replacement or repair and related out-of-pocket expenses. A United States claimant who derives at least 25 per cent. of his earnings from activities which "utilize" the property or natural resource may also recover for loss of profits or

---

[95] 43 U.S.C. § 1813(a).
[96] 43 U.S.C. § 1813(b).
[97] 43 U.S.C. § 1813(b)(1).
[98] 43 U.S.C. §§ 1811(14) (22). An "incident" means "any occurrence or series of related occurrences, involving one or more offshore facilities or vessels, or a combination thereof, which causes or poses an imminent threat of oil pollution." 43 U.S.C. § 1811(4).

impairment of earning capacity resulting from its loss or injury. As appears from the legislative history of this section, the provision permitting recovery of lost profits does not depend upon ownership of the property, but simply "utilization."[99] Thus, claimants such as fishermen whose potential catch is destroyed or polluted, or hotel and other resort facility owners, together with their employees, may recover for lost profits and earnings as a result of lost business from an oil spill. This section is not intended, however, to modify traditional common law principles of proof of damages. In making these classes of damages recoverable, it remains for the court or other tribunal to examine the proof that the damages claimed were likely consequences of the oil spill. A primer on damages proof is well beyond the scope of this book, but it must be noted that damages must be proven with reasonable certainty and that there must exist a proximate connection, without intervening cause, between the oil spill and the loss claimed.

20–54 The President, as trustee for natural resources of the federal government, or the states, as trustees for natural resources within their boundaries, can seek recovery for injury or destruction to natural resources.[1] This includes, but is not limited to costs of restoring, replacing, rehabilitating or acquiring the equivalent of such natural resources.[2] Recovery would include restoration or replacement of property such as public beaches, marshlands, wetlands, fisheries, flora, fauna, wildlife and other natural resources.[3] The federal government and any state or sub-division may seek recovery for loss of tax revenue resulting from injury to real or personal property but any such recovery is limited to a period of one year.[4]

20–55 A foreign claimant can recover for any damage item set out in section 1813(a)(2) to the same extent as a United States claimant if: (i) the oil pollution occurred in the waters or on the shoreline of the foreign country of which the claimant is a resident; (ii) the claimant is not otherwise compensated; (iii) the oil was discharged from a vessel in connection with Outer Continental Shelf

---

[99] 1978 U.S. Code Cong. & Admin. News at p. 1588.

[1] 44 U.S.C. § 1813(b)(3).

[2] *Ibid.* The issue of recovery for acquiring the equivalent of destroyed or damaged natural resources is extremely complex in view of how a court can quantify the loss. Since there is a natural regeneration of certain resources, immediate replacement may be imprudent and improper in some instances. Clearly the statute opens an area for extended litigation. See, *e.g. Comm. of Puerto Rico* v. *M/V Zoe Colocotroni*, 528 F.2d 652 (1st Cir. 1980), where certain of these problems were raised. A discussion of the case appears in Chap. 23.

[3] 1978 U.S. Code Cong. & Admin. News at p. 1588.

[4] 43 U.S.C. § 1813(b)(5).

Lands activities; and (iv) a United States claimant under similar circumstances could recover in the foreign country.[5]

20–56 The Act authorizes the Attorney General of the United States to maintain a class action on behalf of United States claimants if such action would more adequately represent their interests.[6] Should the Attorney General fail to act within a specified time, any member of the group affected by the oil spill may bring a class action on behalf of the group and the failure of the Attorney General to act has no bearing on the maintenance of the class.[7]

E. Financial Responsibility

20–57 As with the Federal Water Pollution Control Act, the Trans-Alaska Pipeline Authorization Act and the Deepwater Port Act, OCSLA–1978 requires the owner or operator of a vessel which uses an offshore facility to establish and maintain evidence of financial responsibility to the Coast Guard up to the vessel's maximum exposure were it entitled to limited liability.[8] Financial responsibility for Outer Continental Shelf Lands activity is separate and distinct from financial responsibility for other vessel-related activities.[9]

20–58 In cases in which an owner or operator owns, operates or charters more than one vessel, evidence of financial responsibility under OCSLA–1978 need only be established in the amount of the largest such vessel, as with the FWPCA.[10]

20–59 Failure to comply with these provisions requires the Secretary to deny entry of a vessel into any port or the navigable waters of the United States, and to detain any vessel which does not comply.[11]

20–60 The Act permits any authorized claim to be asserted directly against the person providing financial responsibility for the vessel.[12] In the event of a direct action by a claimant, the guarantor may invoke all of the shipowners' or operators' rights and defences available under the Act. Moreover, the guarantor may raise as a

---

[5] 43 U.S.C. § 1813(b)(6).
[6] 43 U.S.C. § 1813(b)(7).
[7] 43 U.S.C. § 1813(c).
[8] 43 U.S.C. § 1815. The Secretary of the Department in which the Coast Guard is operating was delegated this function, which was primarily administered by the Federal Maritime Commission, effective May 5, 1983 by Ex. Ord. No. 12123.
[9] 43 C.F.R. § 132.8(2).
[10] 43 U.S.C. § 1815(a)(1).
[11] 43 U.S.C. 1815(a)(2).
[12] 43 U.S.C. § 1815(c).

defence to its obligation to pay that the incident was cased by the wilful misconduct of such owner or operator, but no other defences, such as failure to pay premiums or unseaworthiness, may be asserted to avoid liability.[13] This raises the interesting possibility that the guarantor could seek to prove the misconduct of its insured in order to escape liability whereas the claimant may be put in the position of unholding the bona fides of the shipowner in order to prevent the guarantor from avoiding its liability. Of course, the standard of wilful misconduct is such a severe departure from normal conduct that it is anticipated that it will be a rare case in which such conduct is demonstrated. The standard has been met, however, in other oil spill cases under section 311 under circumstances in which the test is "willful negligence or willful misconduct."[14]

F. Notification and Procedures for Claims Settlement

20–61 The procedure for reviewing, assessing and paying claims, which is similar in form to the TAP procedure, begins with compulsory notification to the Coast Guard by the person in charge of a vessel involved in an oil spill.[15] Upon receipt of such notification, the Coast Guard is required to designate, where possible, the source or sources of the oil pollution, and to notify the owner, operator and guarantor of such designation.[16] Within five days of notification, unless the owner, operator or guarantor denies the designation, it must "advertise" the designation and the procedures by which to present claims to it.[17] Where the designation is denied, where the Coast Guard is unable to make a designation or where the source of discharge is a public vessel, the Coast Guard is required to advertise and make notification as appropriate.[18]

20–62 The Act makes the owner, operator and guarantor responsible in the first instance for paying claims.[19] Assuming the owner accepts responsibility as the source of the spill, it must establish procedures to receive and pay claims in accordance with its obligation under the Act. In a zealous effort to assure prompt payment, the Act virtually assures OPC Fund participation by requiring that

---

[13] *Ibid.*

[14] *Tug Ocean Prince* v. *United States*, 584 F.2d 1151 (2d Cir. 1978). A discussion of the case appears at paras. 19–45 *et seq.*, above.

[15] 43 U.S.C. § 1816(a). Failure to give notification is a crime punishable by a fine of not more than $10,000, imprisonment for not more than one year, or both. 43 U.S.C. § 1822(b).

[16] 43 U.S.C. § 1816(b)(1).

[17] 43 U.S.C. § 1816(b)(2). If the owner, operator or guarantor fails to advertise, the Coast Guard shall do so at owner's expense. Advertisements must commence within 15 days of the designation and continue for at least 30 days: 43 U.S.C. § 1816(d).

[18] 43 U.S.C. § 1816(c).

[19] 43 U.S.C. 1817.

all claims be settled within 60 days of advertisement or presentation, whichever is later.[20] Failing this, the claimant may make an irrevocable election to sue the owner, operator or guarantor,[21] or present the claim to the OPC Fund.[22] Since there are no incentives to litigate against the owner, operator or guarantor and, in fact, in most circumstances there are disincentives to do so, the likelihood of OPC Fund intervention is very strong. As a practical matter, claims simply cannot be reviewed and approved within sixty days. Experience suggests that with complex claims, such as those which may be compensated for under the Act, review (assuming an admission of liability) requires a minimum of six months to in excess of one year. Moreover, in advising claimants, one would be well advised to proceed against the OPC Fund at the first opportunity. First, suit against the owner or operator is uncertain in terms of full recovery because of the possibility in some cases of exceeding the vessel limit. Secondly, the guarantor has a defence available to it, wilful misconduct, which the Fund does not have available. Although there is a residual clause in the Act which allows a claimant to proceed against the OPC Fund for uncompensated damages,[23] this is statutorily limited to where recovery is prevented by the limitation of liability provisions or the insolvency of the owner, operator or guarantor. Strictly speaking, this provision does not apply where the wilful misconduct defence is proved. Given this uncertainty, and the obvious delay involved in suing the owner, it is evident that the Fund will most likely be involved in virtually all instances in which claim is made. In summary, it is simply more likely that the OPC Fund will pay claimants more quickly than will the owner, operator or guarantor. This provides ample incentive to proceed against the Fund as soon as possible.

20–63 If the vessel denies liability after designation, the claimant must irrevocably elect either to sue the owner, operator or guarantor or to file a claim with the Fund.[24] Claims are also filed with the Fund when the source of discharge is a public vessel or when the

---

[20] 43 U.S.C. § 1817(c)(2). Of course, the same criticism can be made of the TAP Fund claims procedure, described at paras. 20–10 *et seq.*, above.

[21] The United States District Courts are granted original and exclusive jurisdiction for all controversies arising under the Act. The venue is where the injury occurred, or where the defendant resides or has his principal place of business: 43 U.S.C. § 1819. Query as to venue involving a foreign-owned vessel which causes damage outside of any district. The legislative history suggests, without support, that the nearest district would then be appropriate: 1978 U.S. Code Cong. & Admin. News at p. 1597.

[22] 43 U.S.C. § 1817(d).

[23] 43 U.S.C. § 1817(d).

[24] 43 U.S.C. § 1817(c). If the Fund is notified of the commencement of an action against the vessel interests, it will be bound by any action as a matter of right: 43 U.S.C. § 1817(j).

Coast Guard is unable to designate the source of the oil pollution. The manner and procedures for presenting a claim are set out in detailed regulations.[25]

20–64 Like the vessel owner, the OPC Fund must settle such claims within the latter of 60 days of presentation or advertisement, failing which the claimant must again make an irrevocable election either to sue the OPC Fund or submit the dispute for resolution to the Secretary of Transportation for decision.[26] The Secretary of Transportation may elect either to submit the dispute for resolution to an administrative law judge or to a panel of three specialists in analysis of lost claims. Their decision constitutes a final determination of the Secretary of Transportation and is reviewable in the federal courts only on limited grounds.

20–65 Claims for economic loss will be barred unless they are either presented, or an action commenced, within three years after the date of discovery of the economic loss, or within six years of the date of the incident, whichever is later.[27]

G. Subrogation

20–66 Any person or governmental entity, including the Fund, which makes payment for an economic loss is subrogated to the claimant's rights under the Act. The Fund is authorized to commence an action through the Attorney General of the United States to recover any such payments. In any such action, the OPC Fund is entitled to full recovery of any amounts paid, subject to all of the rights and exceptions to which the owner and operator are entitled under the Act. There are complex provisions in the Act which specify how and under what circumstances interest is payable on any judgment in favour of the OPC Fund. These provisions are drafted to create an inducement to owners and operators to settle proper claims. The Act further provides for the award of processing and investigation costs, court costs and attorney's fees.[28]

4. RELATIONSHIP BETWEEN THE ACTIVITY-RELATED STATUTES AND THE FWPCA

20–67 As noted in Chapter 19, the FWPCA applies specifically to activities in connection with DPA and OCSLA–1978, not

[25] 33 C.F.R. Part 136.
[26] 43 U.S.C. § 1817(f).
[27] 43 U.S.C. § 1817(l).
[28] 43 U.S.C. § 1817(c).

TAP.[29] Therefore any vessel which comes within the mandate of DPA or OCSLA–1978 also comes within the mandate of FWPCA and all FWPCA rules and regulations must also be satisfied.

20–68   Although TAP does not specifically incorporate FWPCA, any vessel discharging oil in or on the navigable waters of the United States or the contiguous zone, within the jurisdiction of FWPCA, even when carrying TAP oil, is also subject to FWPCA and its requirements. A vessel carrying TAP oil which pollutes waters not within FWPCA's jurisdiction is not, of course, subject to FWPCA.

20–69   In those circumstances in which the rules and regulations of FWPCA overlap with the rules and regulations of any of the activity-related statutes, and where there is inconsistency between the Acts, the fundamental question of which law governs must be considered. In this regard, three rules of legislative interpretation are relevant:
   (1) The provisions of a statute which are more specifically related to a particular activity prevails over a less specific statute;
   (2) The provisions of a more recent statute prevail over an older statute; and
   (3) A statute may specifically mandate in what manner inconsistencies shall be reconciled.

Inconsistency may, for example, be found in the provisions concerning reimbursement of clean-up costs to the federal government under FWPCA and under OCSLA–1978. Under FWPCA, federal clean-up costs are subject to rights of limitation of liability under section 311(f)(1) and the exceptions to liability under section 311(f)(2),[30] whereas under OCSLA–1978, all costs of removal are borne by the discharging vessel without limitation or exception under 43 U.S.C. 1814(d).[31] Clearly, a vessel within a safety zone carrying oil from an OCSLA facility which discharges oil into such waters would be subject to both Acts. OCSLA–1978 specifies what provisions control in the event of inconsistencies. Section 1814(i) states that to the extent they are in conflict with any other provision of law relating to liability or limitation thereof, OCSLA–1978 shall supersede any other provision of law. Thus, in the example given, OCSLA–1978 controls because of the statutory mandate. However, even if this were not the case, since OCSLA–1978 is both more specific and more recent than FWPCA, the OCSLA–1978 pro-

---

[29] See para. 19–05, above.
[30] See paras. 19–31 to 19–46 above.
[31] See para. 20–48, above.

visions would nevertheless prevail and the vessel interests would be subject to unlimited liability for removal costs without exception. Neither TAP nor DPA contain a provision similar to that of OCSLA–1978 section 1814(i) and thus two general rules of interpretation stated above must be applied, leading in each case to the result that the activity-related statutes will control in most cases of inconsistency with FWPCA.

20–70   Only one case, not involving a vessel, has considered the relationship of FWPCA with an activity-related statute. In *Alyeska Pipeline Service Company* v. *United States*,[32] the company ("Alyeska") which services the Trans-Alaska Pipeline System sued to recover the costs of clean-up of oil discharged from the pipeline into the navigable waters. Alyeska invoked section 311(i) of FWPCA which provides, in pertinent part, that a discharger who cleans up its oil discharge can recover the reasonable costs of the operation if it establishes that the sole cause of the discharge was one of the four exceptions contained in FWPCA.[33] Alyeska alleged that the oil discharge was caused by an unknown saboteur who detonated an explosive charge which breached the pipeline and resulted in the discharge. The company cleaned up the spill at a cost of $1,169,035.51 for which it sought reimbursement from the United States under section 311(i).

20–71   The Government, on the other hand, argued that section 1653(b) of TAP makes the owner or operator of the pipeline liable for all removal costs without the right of reimbursement. In relevant part, section 1653(b) states that "control and total removal of the pollutant shall be at the expense of" the pipeline owners and operators.

20–72   The court recognized that these two provisions were inconsistent on their facts. In resolving this inconsistency against Alyeska, the court stated the appropriate rule of construction as follows[34]:

> "Under normal principles of statutory interpretation, the later enacted Pipeline Act prevails over the earlier enacted Water Pollution Control Act. See *Regional Rail Reorganizational Act Cases (Blanchette* v. *Connecticut General Insurance Corp.)* 419 U.S. 102 (1974). This is particularly true where, as here, the later legislation is a special statute addressed to the specific case while the earlier one is a more general law. *Morton* v. *Mancare*, 417 U.S. 535 (1974)."

[32] 649 F.2d 831 (Ct. Cl. 1981).
[33] This section is discussed with respect to vessels in paras. 19–65, *et seq.*, above.
[34] 649 F.2d, 831, 833.

The court carefully examined the legislative history of TAP and found within it an intention to make the pipeline owners and operators liable for these expenses.

20–73 Although *Alyeska* involved an onshore facility and a specific provision concerning pipeline responsibility for oil spills, the court's reasoning provides a proper basis to reconcile conflicting language contained in FWPCA and in any of the activity-related statutes. Utilising this analysis, TAP provisions will always prevail over inconsistent FWPCA provisions because of the specific statutory mandate of TAP. Also, in most cases, DPA and OCSLA–1978 will prevail over inconsistent FWPCA provisions in view of the rules of interpretation discussed in *Alyeska*.

5. CONCLUSION

20–74 The activity-related statutes represent a substantial departure from, and expansion of, the FWPCA scheme which covered only clean-up and removal costs. One of the significant features of all three Acts was the establishment of industry sponsored funds available for relatively prompt reimbursement to claimants of broad classes of damages. The activity-related statutes are, however, jurisdictionally limited and, it is suggested, most vessel activity in or about the waters of the United States involving oil spills does not come within the jurisdiction of any of these three federal statutes. In that sense, while theoretically important for the concepts introduced, they are insignificant to the development of a comprehensive pollution scheme covering most potential oil spills in the United States.

20–75 The recent amendment of the Deepwater Port Act reveals yet another problem. Because of intense competitive pressures, the 1984 amendments of the DPA suspended collection of the barrel fees for use of the port. Clearly such user fees, contained in all three statutes, impose a competitive disadvantage for these facilities and products as compared to facilities and products which do not impose such fees. If it is accepted that industry sponsored pollution liability funds are a desirable method of compensating claimants for pollution damage, a system which evenly spreads the costs among competitors is clearly more desirable than one that does not. As we have seen, the activity-related statutes do not evenly spread the costs and, as in the case of deepwater ports, were at least partly responsible for the economic failure of LOOP. For this reason alone, the activity-related statutes are inherently defective in the broader economic sense. This provides yet another compelling reason to abandon the existing scheme in favour of the approach of the IOPC Fund Convention which would spread the costs among virtually all net importing oil companies within the United States without regard to activity or source. In such a manner, the harmful competitive effect of assessing the costs against one facility or product, as opposed to another competitor, is eliminated.

20–76   The basic defect of the activity-related statutes is that they are too limited in scope and do not address the problem of oil pollution in the comprehensive manner which this field requires. As recognized specifically in the DPA legislation, these statutes must be viewed only as interim measures until a comprehensive scheme is implemented.

21. State and Private Recovery for Oil Spills

SUMMARY

21–01   Supplementing the major federal legislation in the field of oil pollution are a bevy of state laws, enacted principally after the *Torrey Canyon* oil spill in 1967.[1] Currently, nineteen coastal states and five Great Lakes states have oil pollution legislation on their books.[2] These state laws are generally neither entirely consistent with federal laws nor are they uniform in their coverage. The consequences of oil pollution under these laws range from civil or criminal penalties or recovery by the state for clean-up costs to establishment of private causes of action for economic loss. Most provide for damages similar to those covered by the Federal Water Pollution Control Act, *i.e.* state costs of clean-up and removal.[3] Several states supplement these basic elements with statutory recovery for replacement of natural resources. The Oregon statute,

---

[1] See, generally, Wallace & Ratcliffe, "Water Pollution Laws: Can They Be Cleaned Up?" 57 Tul.L.Rev. 1343 (1983); Jarvis, "*Richardson* v. *Foremost Ins. Co.*, A new Opportunity for Indemnity to End State Regulation of Oil Pollution" 19 Gonzaga L. Rev. 265 (1984); Dep't of Justice Methods and Procedures For Implementing A Uniform Law Providing Liability For Clean Up Costs and Damages Caused by Oil Spills From Ocean Related Sources (Comm. Print 1975).

[2] Alabama Code §22–22–9(p) (1975 & Supp. 1982); Alaska Stat. §46.04 (1977 & Supp. 1982); California [Water] Code §13440 (Deering 1977); Connecticut Gen. Stat. Ann. §25–54 (West 1975); Delaware Code Ann. tit. 7, §§6207–6209, 6211 (1974 & Supp. 1982); Florida Stat. Ann. §§376. 011–376. 21 (West 1974 Supp. 1983); Georgia Code Ann., tit. 12–5.20 to 53 (1982 & Supp. 1982); Maine Rev. Stat. Ann. tit. 38 §§41–571 (1978 & Supp. 1982); Maryland [Nat. Res.] Code Ann. §§8–1405(3), 1407, 1408 1411 (1983); Massachusetts Ann. Laws ch. 21, §§27(14), 50 (Michie/Law. Coop. 1980 & Supp. 1982); New Hampshire Rev. Stat. Ann. §146.A (1977 & Supp. 1981); New Jersey Stat. Ann. §§48: 10–23.11a to–23.11z (West 1982); New York [Nav.] Law §180 (McKinnery Supp. 1982); North Carolina Gen. Stat. §§143–215.87, –215.93 (1978 & Supp. 1983); Oregon Rev. Stat. §§468.785–790 (1983); Texas (Water) Code Ann. 1982); Virginia Code §§62.1–44.34 : 1 to –44.34 : 7 (1982); Washington Rev. Code Ann. §§90.48.336, §26.265 (Vernon Supp. 338, 390, 400 (Supp. 1981). Additionally, the five states surrounding the Great Lakes have also enacted pollution statutes: Illinois Ann. Stat. ch. 85, §§1701–1706 (Smith-Hurd Supp. 1982); Indiana Code Ann. §=13–1–3–1 to 4–4 (Burns 1981 & Supp. 1982); Michigan Stat. Ann. §§3.521–3.533 (Callahan 1977 & Supp. 1982); Minnesota Stat. §115 (1976 & Supp. 1982); Ohio Rev. Code Ann. §§ 3767.14, 3767.99, 6111.04 (Supp. 1980).

[3] See, *e.g.* statutes of the following states: Alabama, Alaska, Connecticut, Delaware, Florida, Massachusetts, New Hampshire, New Jersey, New York, North Carolina, Oregon, Texas, Virginia and Washington. For FWPCA, see Chapter 21 above.

for example, provides for recovery of "any amount reasonably necessary to restock or replace . . . fish or wildlife and to restore natural fish or wildlife production in the affected waters."[4] Other states provide for recovery of provable damages caused by the spill[5] and thus sanction private civil suits against those responsible for pollution damage.

21–02   Many states provide for unlimited liability for clean-up costs.[6] A few states have limited liability similar to the Federal Water Pollution Control Act, but specific limits vary.[7] Thus, depending on the situs of the casualty, a vessel might be limited in its liability for federal clean-up costs under section 311 but subject to unlimited state clean-up costs.

21–03   Examined herein are two representative state oil spill statutes—New York's[8] and Florida's[9]—to provide some understanding of a state level approach. These two statutes are particularly interesting in view of the scope of liability which they permit and the use of a fund as the primary vehicle for recovery by injured parties.

21–04   Also examined in this Chapter is the law relating to private non-governmental recovery for oil spill damages under doctrines other than those discussed in Chapter 20: as discussed in Chapter 20, the activity-related federal acts do provide for such recovery. However, with respect to spills which occur outside the jurisdiction of such statutes, or outside the jurisdiction of state statutes providing for private recovery, the courts have been left to consider whether maritime or other common law tort remedies are available, and if so, what limits of liability are available.

21–05   Finally, this Chapter considers the fundamental issue of uniformity of federal and state legislation. In particular, the overlap, in some instances, of federal and state laws has raised the question whether or not federal legislation has pre-empted the oil spill recovery field and, in so doing, deprived the states of the right to enact oil pollution laws. A related issue is whether these laws have supplanted the common law recoveries in certain instances. The

---

[4] Ore.Rev.Stat. §468.745 (1974). Several states have similar provisions, *e.g.* Alabama, Georgia, Maryland, Massachusetts, Mississippi, North Carolina, South Carolina, Virginia and Washington.

[5] Alabama, Alaska, Delaware, Florida, New Jersey, New York and South Carolina.

[6] Alabama, Alaska, California, Connecticut, Florida, Georgia, Maine, Massachusetts, New Hampshire, New York, Oregon, South Carolina, Virginia and Washington.

[7] Delaware, Florida, New Jersey, New York, North Carolina and Texas.

[8] N.Y. Navigation Law, §172–202 (McKinney's Supp. 1982).

[9] Fla. Stat. Ann., §376.011 *et seq.* (West 1974, Supp. 1982).

chapter concludes with a discussion of the benefits to the United States of adherence to the international regime.

1. NEW YORK OIL SPILL PREVENTION, CONTROL AND COMPENSATION LAW

21–06   The New York oil spill liability law was enacted in 1977 and is codified as a part of its Navigation Law.[10] There is established within the state treasury a New York Environmental Protection and Spill Compensation Fund, a non-lapsing, revolving fund which has allotted to it all license fees from oil facilities, the proceeds of penalties from oil spills and collections, together with interest thereon, from liabilities arising from prohibited discharges of oil.[11] Like the federal activity-related pollution statutes, the New York Act prohibits oil discharges, sets out the obligation to remove any such prohibited discharges, establishes methods of compensation for clean-up costs and identifies liability therefor.

A. Prohibited Discharges and Notification Thereof

21–07   The Act prohibits the discharge of petroleum except where permitted under state or federal permit.[12] "Discharge" is defined as "any intentional or unintentional action or omission resulting in the releasing, spilling, leaking, pumping, pouring, emitting, emptying or dumping of petroleum into the navigable waters of the state or onto lands from which it might flow or drain into said waters, or into waters outside the jurisdiction of the state when damage may result to the lands, waters or natural resources within the jurisdiction of the state."[13]

21–08   Responsibility for immediate reporting of a prohibited discharge is upon "any person responsible" for causing a discharge.[14] "Immediate" requires that a report be made in any case no more than two hours after the discharge. Anyone failing to make proper notification to the state is liable to pay a penalty of up to $25,000 per violation per day.[15]

---

[10] Navigation Law, §170–204.
[11] Navigation Law, §179.
[12] Navigation Law, §173.
[13] Navigation Law, §172(8).
[14] Navigation Law, §175.
[15] Navigation Law, §§175, 192.

B. Removal of Prohibited Discharges

21–09 Unlike the federal statutes, the New York Act places upon the discharger a legal obligation to undertake immediate action to contain the discharge.[16] The State may, in its discretion, undertake the removal of such discharge and hire agents or contractors for such purpose. The ultimate responsibility for clean-up is upon the State, which "shall respond promptly to clean-up and remove the discharge in accord with environmental priorities."[17] In this regard, the State Department of Environmental Conservation is charged with the responsibility of developing such priorities and providing technical expertise to ensure prompt and correct clean-up.[18]

21–10 Spills affecting New York waters from an unknown or unexplained source are to be removed by the State. Any expense incurred, except that incurred by the party causing the discharge, shall be paid promptly from the New York Environmental Protection and Spill Compensation Fund established by the Act.[19] Therefore, unlike the IOPC Fund discussed in Chapter 11, the New York Fund will reimburse expenses even if the source of discharge is unknown. Any clean-up and removal or actions to minimize damage are to be conducted, to the greatest extent possible, in accordance with the National Contingency Plan, discussed in detail in Chapter 20.[20] Since the failure to so act could prejudice State recovery from the FWPCA fund, discussed in paragraph 19–29, a discharger might well argue that such a violation by the State of its own law should not subject him to greater liability for clean-up costs under State law.

21–11 Any person threatened with injury resulting from an oil spill is not precluded from taking clean-up and removal actions provided that, as soon as reasonably possible, that person co-ordinates and obtains approval for such actions with State and federal authorities.[21] As an incentive to communal self-help, any person who renders assistance in clean-up and removal shall not be liable in civil damages to third parties resulting solely from his or her acts

---

[16] Navigation Law, §176. The failure to comply with any duty created by the Act gives rise to liability in an action brought by the State for penalties of not more than $25,000 for such violation. Each day of a continuing violation constitutes a separate offence. Navigation Law, §192.

[17] Navigation Law, §176(2).

[18] *Ibid.*

[19] Navigation Law, §176(3).

[20] Navigation Law, §176(4).

[21] Navigation Law, §176(7).

or omissions unless that person is grossly negligent or guilty of wilful misconduct.[22]

C. Discharger Liability and the New York Environmental Protection and Spill Compensation Fund and Recoverable Damages

21–12   Any person who has discharged petroleum in violation of the act is strictly liable, without regard to fault, for all clean-up and removal costs and all direct and indirect damages.[23] The only defences available to the owner or operator of a vessel or facility is if the act or omission was caused *solely* by war, sabotage or governmental negligence. It would appear that third party sole cause is not a defence to the vessel owner's strict liablity, as it is in the FWPCA.[24] "Clean-up and removal" is defined as "the (a) containment or attempted containment of a discharge, (b) removal or attempted removal of a discharge or (c) taking of reasonable measures to prevent or mitigate damages to the public health, safety or welfare, including but not limited to, public and private property, shorelines, beaches, surface waters, water columns and bottom sediments, soils and other affected property, including wildlife and other natural resources."[25] "Clean-up and removal costs" are defined as "all costs associated with clean-up and removal" incurred by the state, its subdivisions, its agents or any person authorized to act by the State.[26]

21–13   The New York Environmental Protection and Spill Compensation Fund (the "New York Fund") is a non-lapsing, revolving fund established in the State Department of Audit and Control.[27] The New York Fund is credited with annual licence fees from the operation of oil refineries, storage or transfer terminals, pipelines, deepwater ports, drilling platforms or related facilities used in the

---

[22] *Ibid.*

[23] Navigation Law, §181(1). See paras. 21–47 *et seq.*, for a discussion of damages. The Act as originally enacted provided for the filing of evidence of financial responsibility to the State on the part of facilities and vessels, in the amount of $50 million and $300 per gross registered ton, respectively: Navigation Law, §174(9)(e) (10). These sections were repealed in 1978 partly under pressure from the International Group of P & I Clubs and there is presently no requirement for providing evidence of financial responsibility to the State. Curiously, there remains a section of the Act permitting direct action against the insurer or any other person providing evidence of financial responsibility, but to what or whom this applies is questionable. Navigation Law, §190.

[24] See paras. 19–47 *et seq.*

[25] Navigation Law, §172(4).

[26] Navigation Law, §172(5).

[27] Navigation Law, §179.

handling or processing of oil within the State.[28] The licence fee is set at one cent per barrel transferred through the facilitiy until the Fund equals or exceeds $25 million.[29] Provision is made to impose a greater fee in certain instances, such as when the existing balance of the Fund is exceeded by pending claims. No monies are allocated to the Fund from general revenues.

21–14 The New York Fund is also strictly liable, without regard to fault, for all clean-up and removal costs and all direct and indirect damages, whether or not the source of the spill has been identified.[30] The Act, like the Outer Continental Shelf Lands Act Amendments (see paragraphs 20–42 *et seq.*), provides for a very wide range of damages to be recoverable, and attempts a categorization of some of them. It provides that, without limitation of the generality of the phrase, the following are included as recoverable items: (1) the cost of restoring, repairing or replacing any real or personal property damaged or destroyed by a discharge; (2) any lost income from the time of injury to the time of replacement; (3) any reduction in value of such property; (4) the cost of restoration and replacement, where possible, of any damaged or destroyed natural resource; (5) loss of income or impairment of earning capacity due to damage to real or personal property, including natural resources, provided that the loss exceeds at the time ten per cent. of the claimant's earnings for the relevant period; (6) loss of tax revenue by the state and local governments for a period of up to one year; and (7) interest on loans and other obligations incurred for the purpose of "ameliorating the adverse effects of a discharge pending payment of a claim in full."[31] Clearly other classes of damages are not excluded by reference to these particular claims. As with any other statute, or for that matter the common law, while the Act permits recovery for these classes of damage, the award of damages must still be based on proof of a loss and proper quantification of that loss.

21–15 The Act provides that damages recoverable by or from any person other than the owner or operator of a vessel or facility shall be limited to those authorized by common or statutory law.[32] Since the Act also expressly disclaims any interpretation of its provisions which preclude any other civil or injunctive remedy,[33] it

---

[28] Navigation Law, §§179, 172(11).
[29] Navigation Law, §1174(4).
[30] Navigation Law, §181(2).
[31] Navigation Law, §1181(2)(a)–(e).
[32] Navigation Law, §181(3).
[33] *Ibid.*

clearly envisions recovery against third parties (such as ship-builders, repairers and classification societies) by any available method.

D. Vessel Limitation of Liability

21–16 The liability of the owner or operator of a vessel is strict but limited to $300 per gross ton ($50 million for a facility) unless it can be shown that the discharge was the result of (a) gross negligence or wilful misconduct within the privity or knowledge of the owner, operator or person in charge,or (b) a gross or wilful violation of applicable safety, construction or operating standards or regulations, in which case liability is unlimited.[34]

E. Claims Against the Fund and Subrogation

21–17 Injured parties may file a claim with the Fund administrator within three years after the date of discovery of the damage but not later than ten years after the date of the incident.[35] The administrator is required to attempt to promote and arrange a settlement of the claim between the claimant and the person responsible for the discharge.[36] If the source of the discharge is known and liability is conceded, any such settlement is final and binding upon the parties and is a waiver of any recourse against the Fund.[37] Where the source of discharge is not known, the Fund administrator will attempt to arrange a settlement between the claimant and the Fund.[38]

21–18 Although not specifically stated, the structure of the New York Act clearly channels all claims for pollution damage through the Fund in the first instance, not through the discharger or other responsible party. The Fund assumes strict, unlimited liability to injured parties and does not pay only where liability of the responsible party is admitted and damages are paid. Such protection to injured parties is broader than any federal statute or the Fund Convention.

21–19 In the event of a dispute over the validity or amount of damages claimed, the person alleged to be responsible for the injury, the claimant or the Fund administrator may convene a

---

[34] Navigation Law, §192.
[35] Navigation Law, §182.
[36] Navigation Law, §183.
[37] *Ibid.*
[38] Navigation Law, §184.

board of arbitration to adjudicate the dispute. Where the source of discharge is unknown, any person may dispute the claims presented to the Fund.[39]

21–20 The board of arbitators may consist either of one neutral arbitrator or three arbitrators, one selected by the discharger or alleged discharger and one selected by the first two chosen by the parties. If the source of discharge is unknown, then the American Arbitration Association may be asked to select an arbitrator in place of the unknown discharger.[40] Representation by any person on the board does not constitute an admission of liability. One board may be selected to adjudicate all claims arising out of a common discharge.[41] The costs and expenses of the arbitrators are payable from the Fund. All decisions are final subject to state rules on judicial review of administrative proceedings.[42]

21–21 All awards ae payable within 60 days unless judicial review is sought.[43] If total awards for a specific occurrence exceed the current balance of the Fund, then immediate payments will be pro rata and claimants share in further proceeds on the same basis.[44] This procedure may be varied for cases of "extreme hardship."[45]

21–22 Payment of any claims for clean-up or damages by the Fund shall be conditional upon the Fund acquiring by subrogation all of the rights of the claimant against the discharger or other responsible party.[46] The Fund administrator is required to recover for the Fund monies disbursed for clean-up costs when the discharger had failed promptly to perform satisfactory clean-up activities.[47] The administrator is also called upon to recover all penalties and all direct and indirect damages paid by the Fund.[48] For this purpose, the administrator is required to make claim against the discharger or other responsible party in the state court if reimbursement is not forthcoming. In any such action, the administrator need only prove that there has been an unlawful discharge which is the responsibility of the defendant.[49] Presumably,

---

[39] Navigation Law, §185(2).
[40] Navigation Law, §185(2).
[41] Navigation Law, §185(3).
[42] Navigation Law, §185(7).
[43] Navigation Law, §185(8).
[44] Navigation Law, §189.
[45] *Ibid.*
[46] Navigation Law, §188.
[47] Navigation Law, §187.
[48] *Ibid.*
[49] Navigation Law, §188.

since the defendant will have had an opportunity to contest the damages in the administrative phase of the proceedings, it is likely that any protest based on the validity or quantum of such damages will be collaterally estopped.

F. Conclusion

21–23 The New York Act provides the State and private parties with a broad basis for recovery of damages resulting from prohibited discharges. The New York Fund assumes an unusual and extremely important role in the New York reimbursement scheme which is different from that of the federal funds or the IOPC Fund in that it pays all claims in the first instance, even for unknown source spills, and only thereafter seeks recovery from the responsible parties. The potential cost to the oil industry, which provides the funding for this broad legislation, could be quite high. If it be assumed that a small vessel which is the sole cause of the spill is entitled to limitation under the Act following a major spill, the Fund faces unlimited liability without recourse above the vessel's limitation up to the amount of all provable damages. No such situation has yet arisen.

2. FLORIDA POLLUTANT DISCHARGE PREVENTION AND REMOVAL LAW

21–24 The Florida Pollutant Spill Prevention and Control Act[50] was extensively amended in 1974 following the 1973 Supreme Court decision in *Askew* v. *American Waterways Operators, Inc.*,[51] which, while upholding the constitutionality of the statute, provided an opportunity for the Florida legislature to reconcile certain aspects with existing federal legislation.

21–25 The Florida statute must be considered in light of the declared legislative intent in enacting the law. Among its findings, the Act states the following general guidelines[52]:
  (1) the highest and best use of the seacoast of the state is as a source of public and private recreation;
  (2) transfer of "pollutants," including oil, between and among vessels and shoreside facilities is considered a "hazardous undertaking";

---

[50] Fla. Stat. Ann. §376.
[51] 411 U.S. 325 (1973). The *Askew* case is more fully discussed at paras. 21–76 *et seq.*
[52] Fla. Stat. Ann. §376.021.

(3) spills, discharges and escapes of pollutants during transfer, storage and transportation pose threats of great danger and damage to the environment of the state, to owners and users of shore front property, and to citizens and other interests deriving level land from marine-related activities and to the beauty of the Florida coast;

(4) it is the intent of the Act to "support and complement" the Federal Water Pollution Control Act, specifically, the National Contingency Plan for removal of pollutants.

A. Prohibited Discharges

21–26   In recognition of this declaration of intent, the Act prohibits the discharge of pollutants into or upon any coastal waters, estuaries, tidal flats, beaches and lands adjoining the seacoast of the State.[53] "Pollutants" is defined to include oil of any kind and in any form[54] and "pollution" means presence in the water of any quantities of pollutants "which are or may be potentially harmful or injurious to human health or welfare, animal or plant life, or property or which may unreasonably interfere with the enjoyment of life or property, including outdoor recreation."[55]

B. Clean-up Responsibility

21–27   The Florida Act requires any person discharging pollutants immediately to undertake to contain, remove and abate the discharge to the satisfaction of State officials.[56] Should the person causing the discharge fail to take appropriate action, the State may do so or arrange for such action. If the prohibited discharge was into or upon the navigable waters of the United States, the removal activities are to be accomplished in accordance with the National Contingency Plan in order that the State may obtain reimbursement under section 311. It is specifically provided that "federal funds provided under said act shall be used to the maximum extent possible prior to the expenditure of state funds."[57]

21–28   If the source of the discharge is unknown, then any local

---

[53] Fla. Stat. Ann. §376.041.

[54] Fla. Stat. Ann. §376.031(12).

[55] Fla. Stat. Ann. §376.031(13).

[56] Fla. Stat. Ann. §376.09(1). Even if the discharger does accept this undertaking, the Act permits the state also to undertake removal operations. This differs from the FWPCA in the crucial respect that the federal statute only provides an incentive to the discharger to clean-up, it does not impose a requirement to do so. Compare with 33 U.S.C. §1321(C)(1), discussed at paras. 19–14 et seq.

[57] Fla. Stat. Ann. §376.09(2).

discharge clean-up organisation shall, at the State's request, immediately contain and remove the discharge.[58] Any clean-up action taken may not be construed as an admission of liability and as in the New York Act, no person engaged in clean-up operations, voluntarily or at the State's request, shall be liable for civil damages to third parties in the course of rendering such assistance, except for acts amounting to gross negligence or wilful misconduct.[59]

C. Penalties

21–29　The Act provides that any violation of its provisions or of any rule, regulation or order made thereunder, is punishable by a civil penalty of up to $50,000 per day to be assessed by the Florida Department of National Resources.[60] As a further incentive to prompt and effective reporting and removal by the discharger, the penalty provisions are not applicable to any discharge promptly reported and removed by the terminal facility or by the vessel consistent with the rules, regulations and orders of the Department.[61]

D. Florida Coastal Protection Trust Fund

21–30　In order to make funds immediately available for "clean-up and rehabilitation" following a prohibited discharge, to prevent further pollution damage and to pay for damages, the Florida Coastal Protection Trust Fund, a non-lapsing revolving fund, was established.[62] Credited to the Fund are all excise taxes, registration fees, penalties, judgments and other collections related to the Act. The principal source of revenues for the Fund is a two cent per barrel excise tax on oil transferred to or from facilities registered to operate within the State. The registrants of the facilities are responsible for the payment. Such a levy is to be made until the Fund equals or exceeds $30,000,000. Provision is made for increasing the levy up to ten cents per barrel in the event of a discharge of "catastrophic proportions" and for a levy of up to four cents per barrel in the event of shortfalls in the Fund's reimbursement obligations. The Fund is liable to pay all proven claims against it even if the aggregate of such claims exceed either the amount then in the Fund or the maximum $30,000,000. In this event, proven claims are paid

---

[58] Fla. Stat. Ann. §376.09(3).
[59] *Ibid.*
[60] Fla. Stat. Ann. §376.16(1).
[61] Fla. Stat. Ann. §376.16(3).
[62] Fla. Stat. Ann. §376.11.

and supplemented on a pro rata basis as monies are made available.[63]

E. Claims against the Fund and the Discharger

21–31 Like the New York Act, the Florida Act is structured such that the Fund is the primary mechanism for the State and third parties to recover damages from an oil spill. Hence, the Fund is made strictly liable for all damages and clean-up costs without benefit of limitation of liability and beyond the maximum size of the Fund. While not specifically mentioned, and the point has not yet been litigated, the Act would appear to make the Fund liable for damages where the source of the discharge is not known or proved.

21–32 Although the Act expressly provides that proceeding against the Fund is not a condition precedent to any other available remedies and that an aggrieved party may pursue such remedies in court, the clearly preferable manner would be to proceed against the Fund. If, for instance, an injured party were to sue the shipowner under the principles outlined in paragraphs 21–24, below or otherwise, the Act does not specify what limitation of liability would apply. Since the Florida Act specifically states that the only defences of the discharger in such an action are the same sole fault exceptions as FWPCA, section 311,[64] one could anticipate a persuasive argument that the 1851 Limitation Act applies.[65] In any event, an injured party might well prefer to proceed against the Fund where he can recover 100 per cent. of his provable damages rather than risk his chances in court where he may find little or no recovery because of the shipowners' limitation rights or insolvency. It should be noted, however, that the remedies in the Act are deemed to be cumulative and not exclusive.[66] Thus, in the event of less than full recovery in a suit against the shipowner, the injured party could elect to file its claim against the Fund, but any such action would only serve to accomplish what could have been accomplished by originally filing a claim against the Fund. Moreover, all claims against the Fund must be filed within 180 days of the discharge.[67] Given the unlikelihood of any lawsuit against the shipowner being concluded within 180 days, the injured party

---

[63] Fla. Stat. Ann. §376.11.

[64] Fla. Stat. Ann. §376.205. See paras. 19–32 *et seq.*

[65] The Florida Act's limitation applies only to claims against the vessel by the Fund: Fla. Stat. Ann. §376.12(1).

[66] Fla. Stat. Ann. §376.205.

[67] Fla. Stat. Ann. §376.12(2).

would be well advised to proceed against the Fund with its more assured method of payment.[68]

21–33 In paying claims, a procedure is established by which the Fund administrator establishes the amount of the damage award and certifies that amount to the State Treasurer who will pay the award from the Fund.[69] In view of the pro rating provisions in the event of large claims likely to exceed the amount of the Fund, it would appear that no claim should be paid until all provable amounts are certified. If the claimant disagrees with the amount certified, he may request a hearing and review by the Governor or his designee, such decision to be final.

F. Vessel Liability and Limitation of Liability

21–34 The Act provides for strict but limited clean-up cost liability upon the vessel as a result of a prohibited discharge subject to certain defences discussed in paragraph 21–37, below. Recovery for private damages is, however, not so limited by the Act. In the event of a prohibited discharge, the vessel is liable to the Fund by way or subrogation, for all costs of clean-up or abatement up to $14,000,000 or $100 per gross registered ton of the vessel, whichever is less.[70] If the state can show that the discharge was the result of wilful or gross negligence or wilful misconduct within the privity or knowledge of the owner, operator or agent thereof, then the owner or operator is liable to the Fund for the full amount of clean-up cost expended.[71]

21–35 With respect to private damages suffered by any person apart from clean-up and abatement costs, the Act makes no provision for limited liability of the vessel to the Fund by way of subrogation, or to private persons directly. Therefore, subject to the strict liability exceptions discussed in paragraph 21–34, above, there is no State law limitation of the vessel's liability for private damage. However, as discussed in paragraph 21–32, above, the shipowner or bareboat charterer might be in a position to have the benefit of the 1851 Limitation of Liability Act which effectively limits his liability to the vessel's value at the conclusion of the voyage plus pending freight.[72]

---

[68] The limitation period may be waived for good cause shown: Fla. Stat. Ann. §376.12(2).

[69] Fla. Stat. Ann. §376.12(2).

[70] Fla. Stat. Ann. §376.12(1).

[71] *Ibid.*

[72] Since vessel "operator" is not defined in the Act, it is likely that a court would apply the definition in section 311 of the FWPCA, which includes a demise charterer.

21–36 Upon payment of clean-up and removal expenses together with proven damages of private persons, the Fund is subrogated to all such rights and claims.[73] The Fund is directed diligently to pursue reimbursement from the offending parties. In any suit for reimbursement, the Fund is entitled to prove the amount of its subrogated claims by submitting a written report to the court of all amounts paid or owed by the Fund to claimants. The report is deemed by the statute to be admissible evidence and the amounts paid from or owed by the Fund to the claimants stated therein constitute rebuttable evidence of the Fund's damages.[74]

21–37 In any suit brought by a private individual or by the Fund, against the discharger, there is no need to plead or prove negligence, but only that a prohibited discharge occurred.[75] The only defences of one found to be responsible for the discharge are basically the same as those found in section 311 of the FWPCA; namely, that the occurrence was *solely* the result of any one or a combination of the following:

(1) act of war;

(2) act of government, either state, federal or municipal;

(3) act of God; or

(4) act or omission of a third party, whether or not negligent.

Unlike section 311 of the FWCPA, the Florida Act does not seem to discriminate between sole fault dischargers, sole fault third parties and joint fault third parties. It would appear that any of these categories of persons are strictly liable under the Florida Act.

21–38 An interesting prospect arises under the Florida statute. If a vessel engages in a prohibited discharge in Florida causing, as might be anticipated, the expenditure of clean-up costs as well as damage to private persons which are paid by the Fund, the shipowner may well be compelled to proceed in the Federal Court with a Limitation of Liability petition under the 1851 Act. The limitation petition does not affect the owner's liability for state clean-up costs under the Florida Act,[76] and thus liability for clean-up costs will be limited under the State Act unless the state can prove gross negligence or wilful misconduct. The subrogated damages claim

---

[73] Fla. Stat. Ann. §376.12(2)(d).

[74] Fla. Stat. Ann. §136.12(3)(d). *Quaere* whether this provision would be applied by a Federal Court sitting by virtue of its admiralty or its diversity jurisdiction. Under federal law, a Federal Court will apply the substantive law of the state in which it sits, but will apply its own procedural law. Typically, an evidentiary question in the Federal Court is considered to be within the ambit of federal, not state, law.

[75] F.a. Stat. Ann. §376.12(3).

[76] See *In re Hokkaido Fisheries Co.*, 506 F.Supp. 631 (D. Alaska 1981), discussed in para. 19–64.

brought by the Fund subsequent to its payment to third parties is not subject to the Florida Act's limitation provisions and would therefore be controlled by the 1851 Limitation Act. Under that law, the shipowner must prove that the damage was caused without his privity and knowledge. Thus, in the same claim, there would be involved two different limitation funds on the same vessel, two different burdens of proof and two different applicable legal standards. No such case has yet arisen and, fairly put, it is the rare judge who will be anxious to hear the trial of such a case.[77]

G. Financial Responsibility

21–39   Perhaps the most significant change in the Florida law in 1974 concerned financial responsibility. Under the 1970 law, vessel owners or operators using Florida ports were required to file with the State evidence of financial responsibility for their vessel's oil spill liabilities. This meant that in addition to the section 311 FWPCA Certificate of Financial Responsibility,[78] to satisfy federal clean-up liability, which at the time was based on a maximum exposure of $14,000,000, the shipowner also had to establish financial responsibility in the State of Florida for a like amount.[79] While the economic and administrative consequences of the imposition of such a requirement by Florida could be substantial, it was not insurmountable. However, if other states had also imposed financial responsibility obligations, the result could have made it financially and administratively unfeasible in some instances to continue trading in the United States.

21–40   As a result of extended discussions with shipowners and underwriters, the Florida legislature amended this provision in 1974 to allow financial responsibility for vessels to be demonstrated by compliance and possession of a Federal Certificate of Financial Responsibility.[80] While it may appear that the State is unsecured in recovery of its clean-up costs, this is not entirely correct. As discussed earlier, the State's plan requires that those administering clean-up operations follow the National Contingency Plan and federal funding will be used "to the maximum extent possible."[81] The federal evidence of financial responsibility thus provides a

---

[77] A similar situation could also occur under any of the federal schemes, as the FWPCA, where there is statutory limitation for clean-up costs, but not for private damages.

[78] See paras. 21–07 *et seq*.

[79] As discussed in para. 21–43, the maximum liability figure has been deleted, and maximum liability is now based on gross tonnage.

[80] Fla. Stat. Ann. §376.14.

[81] Fla. Stat. Ann. §376.09(2).

measure of security for State clean-up costs and the duty to follow the National Contingency Plan provides some degree of continuity to clean-up operations on all administrative levels.

H. Conclusion

21–41 The Florida Act, like New York's, provides those falling within its protection with a reimbursement mechanism which is both complete and prompt with respect to properly recoverable damages from an oil spill. Unlike the New York Act, however, the Florida Act removes all limitations for strict liability of the vessel for reimbursement of private damages paid by the Fund to injured persons. This is considerably more onerous than existing federal statutes or than the international conventions. There have as yet to be any reported decisions in which the Florida Act has been applied but clearly, injured persons protected by the Act may utilize a procedure which assures them recovery of their provable damages without having to engage in costly fault-based litigation.

3. ASPECTS OF COMMON AND STATE LAW BASED RECOVERIES

A. Introduction

21–42 The discussion in Chapters 19 and 20 focused on federal statutory clean-up and damage remedies and the discussion thus far in Chapter 21 has analysed two representative coastal state statutes concerning oil spill recoveries. The discussion which follows examines court cases in which either substantive state or common law, or both, has been used to effect recovery of clean-up costs and damages. The discussion in the section thereafter will consider the issue of federal pre-emption and the need for uniformity in the United States oil spill regime. While logically a discussion of federal pre-emption should precede examination of those cases in which non-pre-emption has been decided or assumed, the issue is closely related to the pressing problem of uniformity of legislation which provides a suitable topic to summarize the status of existing oil spill legislation and the need for adherence to the international regime.

B. Scope of Common and State Law Recoveries

21–43 One area not covered by FWPCA, section 311 concerns certain state and all private recovery for oil spill damage. Several states provide for private recovery, such as the Florida and New York statutes and the activity-related statutes discussed herein, but in most cases, such statutes specifically permit common law recovery in addition to recovery under the statute, provided that there be no multiple recovery for the same damage or expense.[82] Also not covered by section 311, as it is more fully discussed in Chapter 19,[83] is joint cause third party liability and, in some jurisdictions, sole cause third party liability for clean-up and removal costs. It is with respect to these gaps in the legislation that recovery for oil pollution damage is still permitted based on common law and other statutory grounds, discussed below.

C. Maritime Tort

21–44 The common law theory most often advanced for private or governmental recovery of damages and costs is that of common law or maritime tort. A tort generally becomes "maritime" when the tortious activity (1) occurs in a place subject to admiralty and maritime jurisdiction of the court or causes an injury within such jurisdiction, or (2) occurs under circumstances in which the status of the parties or their relationship or the activity itself relates to maritime service or the navigation or operation of a vessel.[84] Maritime tort sounding in negligence, nuisance or intentional activity has been upheld consistently as a basis to proceed against those responsible for oil spill damage.[85]

D. Refuse Act

21–45 Another asserted basis for recovery has been expressed as a right deriving from the Federal Refuse Act.[86] The Refuse Act, however, only prohibits discharge of refuse matter into the navigable waters of the United States but does not provide for recovery

---

[82] See, for example, the New York Oil Spill statute, Navigation Law §193, which states: "Nothing in this article shall be deemed to preclude the pursuit of any other civil or injunctive remedy by any person. The remedies provided in this article are in addition to those provided by existing statutory or common law, but no person who receives compensation for damages or clean-up and removal csts pursuant to any other state or federal law shall be permitted to receive compensation for the same damages or clean-up and removal costs under this article."

[83] See paras. 19–49 et seq.

[84] See, generally, 2 Benedict on Admiralty § (1975); State Department of Fish and Game v. Bournemouth, 307 F.Supp. 922, 929 (D.C. Cal. 1969).

[85] See, e.g. United States v. M/V Big Sam, 681 F.2d 432 (5th Cir. 1982); Re Oswego Barge Corporation, 664 F.2d 327 (2d Cir. 1981); United States v. City of Redwood City, 640 F.2d 963 (9th Cir. 1981); State Department of Fish and Game v. S.S. Bournemouth, 307 F.Supp. 922 (D.C. Cal. 1969).

[86] 33 U.S.C. §407. See Chap. 17 for a brief discussion of the statute.

of clean-up costs. Although United States courts have recognized a right to fashion cost remedies for the United States where none statutorily exist,[87] they have held that section 311 has pre-empted government cost recovery under the Refuse Act.[88] The government usually seeks Refuse Act recovery as a separate count in complaints against dischargers seeking section 311 cost recovery, but such a basis for recovery has been denied in every case.

E. 1851 Limitation Act and Common Law Recovery

21–46   Common law recoveries are, of course, subject to the ship-owner's limitation rights under the 1851 Limitation Act. Although recovery is not based on strict liability, as are most state and federal statutes, the prospect of securing a full recovery is substantially enhanced since the 1851 Limitation Act denies the right to limit upon proof of simple negligence within the privity and knowledge of the owner, not wilful negligence or gross negligence as the oil spill statutes generally require. Of course, in return, fault must be proved, failing which, the limitation amount may be quite small.

F. Damages Recoverable

21–47   Another significant difference between statutory and common law recovery relates to the damages recoverable under each. For example, the activity-related oil spill statutes and most of the state statutes provide for certain types of private recovery not permitted at common law. For example, even the more limited recovery provided in section 311 calls for the recovery of "actual" clean-up costs, which some courts have suggested do not have to be "reasonable" as they would at common law.[89] Another example is the permitted recovery in OSCLA—1978 and other statutes of tax revenues by state and local aurhorities lost as a result of a prohibited discharge.

21–48   Recovery of damages under the common law clearly includes costs for repair or replacement of real and personal property damaged by an oil spill.[90] More difficult issues are presented

---

[87] See, *e.g. Wyandotte Transportation Co.* v. *United States*, 389 U.S. 191 (1967), recognizing the right of the government to recover wreck removal costs under the Wreck Act, 33 U.S.C. §409.

[88] See, *e.g. Re Oswego Barge Corpn.*, above, applying the Supreme Court test set out in *City of Milwaukee* v. *Illinois*, 451 U.S. 304 (1981) to determine the pre-emptive effect of a federal statute.

[89] 33 U.S.C. §1321(f). See discussion at paras. 19–70 *et seq.*

[90] Dep't of Justice Report, above, note 1, at p. 23.

with respect to state recovery for environmental damages and recovery by private persons who utilize living resources and the environment as a source of income.[91]

21–49 The question of common law recovery by such classes of persons as fishermen, tourist hoteliers, non-riparian landowners, and pleasure boat owners are all addressed to whether the damages resulting from an oil spill are specifically related to that plaintiff or whether such damages are non-specifically suffered by the public at large. The failure to establish special and particular injury has been the basis for denial of recovery in several instances.

21–50 A spate of litigation arose out of the 1969 Santa Barbara Channel oil spill in which an oil platform maintained by Union Oil Company off the California coast on the Outer Continental Shelf was said to have been the cause. One case was *Oppen* v. *Aetna Insurance Co.*[92] *Oppen* involved consolidated actions brought by representative owners of pleasure boats which had sustained physical damage from contact with the oil slick which ensued following the escape of oil from beneath the platform. As a result of the spill, the boats were also rendered unusable for a certain period of time. The court held that since the injuries "took effect" in the navigable waters of the United States and since the injuries complained of bore a "significant relationship to traditional maritime activities," the claims for physical damage and interference with their right of navigation sounded in maritime tort.[93] While the physical loss caused by the contact with oil was held to be recoverable by way of compensation, the Court reaffirmed the long-standing principle of maritime law that the loss of use of a private pleasure boat is not. Although the Court denied the plaintiff's arguments that California statutory nuisance law applied to their claims since the injury occurred on navigable waters, it noted that even under that law, mere deprivation of navigational rights not the result of physical

---

[91] Recognizing these problems, federal and state legislation makes specific provision for areas of questionable recovery at common law. Even section 311, as amended in 1977 by the Clean Water Act amendments, provides for federal or state recovery, as trustees of public property, for costs of replacing or restoring natural resources: 33 U.S.C. §1321(f) (4) (5).

[92] 485 F.2d 252 (9th Cir. 1973).

[93] The test of whether a claim is maritime was examined by the Supreme Court in *Executive Jet Aviation, Inc.* v. *City of Cleveland*, 409 U.S. 249 (1972). More recently, the Supreme Court expanded the scope of federal admiralty jurisdiction by making it applicable to a collision between two pleasure boats *Richardson* v. *Foremost Insurance Company*, 102 S.Ct. 2654 (1982). See generally, Jarvis, "*Richardson* v. *Foremost Insurance Company:* A new Opportunity for Industry to End State Regulation of Coastal Oil Pollution," 19 Gonzaga L. Rev. 265 (1984).

damage is not compensable under the law of public nuisance.[94] Summarizing the law of damages arising out of a nuisance, the Court stated:

> "The plaintiff's physical damages are recoverable in negligence and probably also constitute such special injury as to present them with a cause of action or these damages in nuisance. But the damage suffered on account of their loss of navigation right in the Santa Barbara Channel and harbor is no different in kind from that suffered by the public generally. . . . These plaintiffs were deprived of no more than their 'occasional Sunday piscatorial pleasure.' For this deprivation there is no recovery either under California law or general maritime law."

21–51   In yet another case arising out of the Santa Barbara spill, the plaintiffs, commercial fishermen, sued for lost profits and other injuries as a result of reduction of fishing potential in the Channel.[95] The court held that the alleged diminution of sea life resulted in a right to compensation for the commercial fishermen. The court refused to accept the defendants' argument that the fishermen had no claim because they had no proprietary interest in the sea life of the Channel. It concluded that recovery on a theory of negligence required only a showing that these defendants engaged in a kind of activity which created a duty to act in a manner so as not to injure others. In essence, it was held that those engaged in oil exploration in navigable waters owed a duty to local commercial fishermen not to injure their economic interests, which injuries were special and particular, not general. While this case did not involve vessel source oil pollution, it may be anticipated that similar arguments could be successfully advanced by commercial fishermen in such a case.

21–52   In a similar case, the Federal Court in Maine held that commercial fishermen and clammers were entitled to recover for their economic losses arising out of diminution of sea life caused by the grounding of the *M/V Tamano* which resulted in an oil spill in the bay from which such activities were conducted.[96] The court rejected the defendants' argument that since the plaintiffs had no proprietary interest in the coastal waters and marine life, they had no right to compensation.

---

[94] At least one case has suggested that a single act of pollution does not amount to a common law nuisance, thus the State of Maryland could not maintain an action grounded in nuisance to abte an oil spill resulting from a transfer line rupture on one occasion: *Md. Dept. of Natural Resources* v. *Amerada Hess Corp.*, 350 F. Supp. 1060 (D.Md. 1972).

[95] *Union Oil Co.* v. *Oppen*, 501 F.2d 558 (9th Cir. 1974).

[96] *Burgess* v. *M/V Tamano*, 370 F.Supp. 247 (D.;Me. 1973).

21–53   Another class of claimants which has been recognised as having rights to compensation for economic loss is riparian and littoral proprietors. Thus in *Burgess* v. *M/V Tamano*,[97] the claimants were owners of a variety of businesses, including motels, trailer parks, campgrounds, restaurants and groceries in and around the area of the spill, who claimed economic loss from the diminution of the tourist trade as a result of the spill. Recognizing that a private person can recover for a tort for invasion of a "public right" (such as use of the waters within the state) only if there has been a special and particular injury, the court dismissed the claims of all businessmen except those who owned property along the shore which had been physically injured. It noted that the "loss of customers indirectly resulting from alleged pollution of the coastal waters and beaches" was derivative from the public right at large and not directly special or particular.[98]

G. The *Zoe Colocotroni* case—Damage to Publicly Held Natural Resources

21–54   Recovery by a state for damage to natural resources and other environmental damage was considered in *Commonwealth of Puerto Rico* v. *M/V Zoe Colocotroni*.[99] The tanker *M/V Zoe Colocotroni* ran aground on a reef off the coast of Puerto Rico in March 1973, releasing more than 5,000 tons of crude oil from the refloating operation. A substantial portion of the oil came to rest on the southwestern tip of the island at Bahia Sucia Bay. Among the claims filed against the shipowner, the vessel *in rem*, and the vessel's P & I Club were those of the Commonwealth of Puerto Rico and the Commonwealth Environmental Quality Board for damages arising out of harm to the coastal environment.

21–55   The oil spill damage was extensive, particularly to certain areas of Mangrove forest on the east and west sides of the bay. The suit by the Commonwealth and its adjunct, the Environmental Quality Board, sought recovery for damage to the natural resources and marine species in the area of the spill as "trustee of the public trust in these resources" and in its capacity as *parens patriae*. On appeal, the Court of Appeals upheld the plaintiff's standing to sue for economic loss to its own property which the defendants did not

---

[97] *Ibid.*

[98] Other cases have permitted recovery by riparian hotel owners and owners of shoreside fishing and swimming facilities: *Kirwin* v. *Mexican Petroleum Co.*, 267 F.460 (D.R.I. 1920); *Maine* v. *M/V Tamano*, 357 F.Supp. 1097 (D.Me. 1970); *Petition of New Jersey Barging Corpn.* 168 F.Supp. 925 (S.D.N.Y. 1958).

[99] 628 F.2d 652 (1st Cir. 1980).

seriously contest. The defendants did contest, however, the plaintiff's right to recover "on behalf of its people for the loss of living natural resources on the land such as trees and animals."[1] The focus of the defendant's argument was that the state, claiming a right as public trustee, lacked a sufficient proprietary interest in the resource actually damaged. The Court refused to reach this issue and instead relied on a Commonwealth statute which authorized the Environmental Quality Board to protect the environment by maintaining actions such as the instant one. The Court found no Constitutional or common law prohibition against such exercise of the Commonwealth's law-making power and thus upheld this exercise of authority.

21–56 An equally significant issue considered by the Court of Appeals in *Zoe Colocotroni* was the question of how to measure the loss, and this is relevant in varying degrees to all claims for damage to real property, particularly undeveloped real property, whether the claim is brought under statute or in tort. The district court awarded damages based on the plaintiff's estimate of the replacement cost of the living organisms and trees found to be destroyed. Specifically, the court found that 92,109,720 invertebrate marine organisms were destroyed by the spill, based on expert testimony reporting on comparison of sampling of the 20 damaged acres of Mangrove forest with sampling of control acreage. The expert estimated that the difference between the density of organisms between the damaged and the control areas was "1,138 creatures per square meter" or 4,605,486 creatures per acre. Applying the lowest estimated replacement cost of $0.06 per animal, if purchased from biological supply laboratories, and ordering replacement of animals lost only in the East Mangrove area, the Court awarded $5,526,583.20 for this loss. As to the mangrove plantings, the district court found that approximately 23 acres required replanting and monitoring for a five-year period for a total cost of $559,500. Finally, the court awarded $78,108.89 for the Commonwealth's reimbursible clean-up costs already undertaken. Total damages were therefore awarded in the principal amount of $6,164,192.09.

21–57 The defendants argued on appeal that the district court should have applied the common law diminution in value rule to

---

[1] The discussion specifically did not cover the issue of state recovery for damage to privately owned land, since all of the land which was the basis for the suit was owned by the Commonwealth. Also, the living natural resources which were the basis of the suit were non-transitory, such as trees and non-migratory animals. The court expressly excluded from its decision whether damage or destruction of birds, fish and the like would entitle the state to any recovery.

calculate damages, in which damages are assessed on the basis of difference in market or commercial value of the property as a result of the injury, unless full restoration can be accomplished for a lesser amount. The defendants argued that comparable property was valued on the market and had been sold for $5,000 per acre and thus recovery for the 40 damaged acres could not exceed $200,000 for a full depletion in value.

21–58   The Court of Appeals rejected the defendants' argument, finding that the language of the statute authorizing recovery for "the total value of the damages caused to the environment and/or natural resources" implied a determination not to restrict recovery to mere diminution of value. It noted[2]:

> "Many unspoiled natural areas of considerable ecological value have little or no commercial or market value. Indeed, to the extent such areas have a commercial value, it is logical to assume they will not long remain unspoiled absent some governmental or philanthropic protection. A strict application of the diminution in value rule would deny the damages for harm to such areas, and would frustrate appropriate measures to restore or rehabilitate the environment."

After rejecting the diminution in value rule, the court was left to consider whether the replacement value analysis applied by the district court was supportable. The primary standard which the court applied was the "cost reasonably to be incurred by the sovereign or its designated agency to restore or rehabilitate the environment in the affected area to its pre-existing condition, or as close thereto as is feasible without grossly disproportionate expenditures."[3] In applying this standard, the court admonished that the focus should be on the reasonable and prudent steps a sovereign would take to mitigate the harm, including such factors as "technical feasibility, harmful side effects, compatibility with or duplication of natural regeneration" and disproportionability of the expense.[4] The court admitted that such a test lacked the certainty of damages assessable in connection with breach of a commercial contract, but observed that it had more inherent accuracy and certainty than damages typically awarded for pain and suffering or mental anguish in a personal injury or death case.

21–59   In applying this analysis, the court overruled the district court's assessment of damages in view of the fact that replacement

---

[2] 628 F.2d 652, 673.
[3] 628 F.2d 652, 675.
[4] 628 F.2d 652, 675.

of the "small valueless creatures" said to have been lost was neither practical nor probable, and even if replaced, they could not survive in the altered environment. In fact, the Commonwealth had disclaimed any intention to replace these creatures, which the court noted would probably replenish themselves with the restoration of the environment. As to the replacement of the mangrove trees, the court found conflicting evidence of need. The court remanded the case to the district court to examine more reasonable proposals for restoring the land held in public trust. The court suggested the possibility of requiring the defendants to pay for the purchase of comparable lands or relocation of the land at a similar proximate site. Following remand, the case was settled out of court and the final disposition is not reported.

21–60   The *Zoe Colocotroni* case is but an extreme example of the difficulty in any environmental injury case of proving the quantum of damages. The court recognized that it was indeed "venturing far into uncharted waters."[5] No case since *Zoe Colocotroni* has been called upon to articulate such a precise standard for assessing environmental damages and it may be assumed that this constitutes a landmark decision in this area. Unfortunately, the Court of Appeals left as many unanswered or unsatisfactorily answered questions as were raised before it. For instance, in dealing with cost of restoration as a basis for damage to marine animals which would not be replaced, the test enunciated is vague. It appears to allow for "disproportionate" expenditure (which will not be made) but not "grossly" disproportionate expenditure. It does not state what is the basis for establishing whether a hypothetical expenditure is "disproportionate"—to have the concept of disproportion, it is necessary to compare two things. In the court's view, courts must compare hypothetical, unspent "expenditure" with—something it failed to specify. The guidelines on application sound good, until it is appreciated that, for instance, technical feasibility is a paper concept until the restoration is actually tried. The grounds for rejection of diminution in value as a basis for damages is unclear and, one might suspect, seem to be made simply because this principle would have led to what the court might have felt to be an unacceptably low figure. It is to be hoped that next time the issue is before the court, a greater precision in formulating legal policy will be adopted. One approach would be to allow reasonable restoration costs and assess them *after* the restoration action has been taken. If the evidence is that prior thereto the balance of probability is that

[5] 628 F.2d 652, 678.

the cheapest reasonable method is to leave nature to restore herself, then the argument is reduced to a claim for loss of use while that is happening.

H. The *Amoco Cadiz* Case—Corporate Liability

21–61   The litigation arising from the *Amoco Cadiz* disaster in which a fully loaded VLCC ran aground off the Brittany coast of France in March 1978, spilling her entire cargo of 65 million gallons of crude oil on the water and adjoining shoreline, has raised issues of great significance, which extend outside the United States. Suits for clean-up costs and damages were commenced in the United States by the Government of France, the administrative departments of Finistere and Conseil Côtes du Nord and certain French municipalities and private businesses against the registered owner of the vessel, Amoco Transport Company ("Transport"), several affiliated Amoco companies and the parent, Standard Oil Company (Indiana) as operators and corporate *alter egos* of the owner, Astilleros Espanoles, S.A., the builder, Bugsier Reederei und Bergungs, A.G., the salvor, and the American Bureau of Shipping, the Classification Society. Eventually, all suits were consolidated by the United States Multidistrict Litigation Panel for discovery and trial before United States Judge Frank J. McGarr in the Federal Court in Chicago.[6] Astilleros Espanoles elected not to participate on the merits after it was held that they were subject to the personal jurisdiction of the court and a default was subsequently entered against them. All of the Amoco parties filed for limitation under the 1851 Act, but the court dismissed the petitions of all but the registered owner, Amoco Transport Company, on the ground that these other parties had no standing.[7]

21–62   The claims against the Amoco parties were grounded in maritime tort alleging negligence in the design, construction, maintenance and operation of the vessel. It was also alleged that the relationship between the Amoco parties was such that each affiliate and subsidiary was merely a corporate shell of the other and, as such, were corporate *alter egos* making each liable for the debts of the other.

21–63   For its part, the Amoco parties asserted counter and third-

---

[6] *In Re Oil Spill By The Amoco Cadiz Off the Coast of France on March 16, 1978*, M.D.L. No. 376 (N.D. Ill.)

[7] *In Re Amoco Transport Co.*, 1979 A.M.C. 1017 (N.D. Ill. 1979).

party claims against the French claimants for failure to prevent or contain the oil spill.[8]

21–64   The court issued its decision on liability on April 18, 1984, holding that the Amoco parties, through individual acts of negligence in the design, construction, maintenance and operation of the *Amoco Cadiz*, and as corporate *alter egos* of the registered owner, were liable to the French claimants for all provable damages and clean-up costs.[9] The court denied the registered owner's right to limit its liability under the 1851 Limitation Act on grounds of the above mentioned negligence. The French claimants' actions against Bugsier were, however, denied on the ground of failure to prove that any actionable negligence on the salvor's part caused oil pollution damage, the standard of proof being causative gross negligence or wilful misconduct.[10] Astilleros was held liable to Amoco for indemnity or contribution in an amount to be later determined.[11]

21–65   In reaching its decision, the court noted that since the damage was sustained in French territorial waters, French substantive law would govern. However, since no effort was made to prove that French law was different from United States law, and since several claimants stipulated without objection that United States law applied, the claims were decided in accordance with United States law.[12]

21–66   As to the liability of the Amoco parties, the court con-

---

[8] The court denied the French parties' motion to dismiss the Amoco claims on the ground of sovereign immunity and the Act of State doctrine. The court held that such claims by Amoco were properly brought and that the French parties' claims of immunity were waived by the initiation of suit: *In Re Oil Spill by the Amoco Cadiz Off The Coast of France on March 16, 1978*, 491 F.Supp. 161 (N.D. Ill. 1979).

[9] *In Re Oil Spill By The Amoco Cadiz off the Coast of France on March 16, 1978*, M.D.L. No. 376 (N.D.Ill. 1984, not officially reported); [1984] 2 Lloyd's Law Rep. 304. An appeal was not taken of the decision. Since this was a judgment on liability only, it was not a final judgment for purposes of an appeal because it did not determine damages: 9 Moore's Federal Practice 110.08[1] (1983). While interlocutory appeal of liability determinations are available in admiralty actions, 28 U.S.C. §1292(a) (3), this is not mandatory and a party may elect to await a final judgment to make an appeal on all matters adjudicated by the District Court: 9 Moore's Federal Practice 110.19[3] (1983).

[10] Conclusions of Law, Nos. 3–9. This statement of the law by the court is correct, but incomplete. Under United States law, a salvor is liable only for wilful or gross negligence in connection with "non-distinguishable" or "non-independent" injuries. One type of "non-distinguishable" injury covers "errors that made salvage ineffectual." *The Noah's Ark* v. *Bentley & Felton Corpn.*, 292 F.2d 437 (5th Cir. 1961). A "distinguishable" injury, for which liability may be for simple negligence, is "damage sustained by the salved vessel other than that which she would have suffered had not salvage efforts been undertaken." *Ibid.* The Court's implicit finding of fact must have been that this was a "non-distinguishable" injury.

[11] *Ibid.*, Conclusions of Law, No. 53.

[12] *Ibid.*, Conclusions of Law, No. 2.

sidered in detail the corporate interrelationships of these companies in addition to their active roles in the construction and operation of the *Amoco Cadiz*. As to Amoco International Oil Company ("AIOC"), the court found that it "exercised complete control over the operation, maintenance and repair of the Amoco Cadiz and the selection and training of its crew." It held that AIOC negligently performed its duty to ensure seaworthiness since it was privy to many of the vessel problems which made her unseaworthy, some of which were scored rams, slipping belts, poor maintenance practices and steering gear problems. Also, the crewing of the vessel was held to have been negligent. Each of these acts was held to be a proximate cause of the grounding and resultant pollution damage.[13]

21–67 Amoco Transport Company, the registered owner, was held to have had actual knowledge of the unseaworthiness of the vessel and failed to remedy the deficiencies. As the party which had a non-delegable duty to control and supervise the maintenance, repair and crewing of the vessel, it failed in its exercise of care in these respects and thus was held liable to the French claimants for all damages without benefit of limitation.[14]

21–68 In perhaps the most controversial part of its decision, Standard Oil Company (Indiana), the corporate parent, was held by the court to be liable for the tortious acts of its wholly owned subsidiaries AIOC and Transport as a result of its control over these companies, making them "mere instrumentalities of the parent."[15] Also, less controversially, it was found that Standard was itself initially involved in and controlled the design, construction, operation and management of the *Amoco Cadiz* and was thus itself liable to the French claimants for its own negligence.[16]

21–69 The Amoco Cadiz decision is the first in which a United States court was faced with the questions whether the Civil Liab-

---

[13] *Ibid*, Conclusions of Law, Nos. 22 33. The court also noted, apparently to protect the record on appeal, that AIOC's acts were such that it could not limit its liability if the 1851 Limitation Act or other limitation statutes were to apply to it. *Ibid*. Conclusions of Law, No. 34.

[14] *Ibid.*, Conclusions of Law, Nos. 35–42.

[15] A discussion of United States corporate *alter ego* law is beyond the scope of this book, but it should be observed that the issue of corporate integrity is generally a matter of state law. The court did not indicate what law it was applying or what were the elements of proof, but presumably it applied the law either of Indiana, the place of incorporation of Standard, or Illinois, its principal place of business. It should also be noted that significant choice of law problems may arise in considering corporate *alter ego* issues since generally many companies incorporated and doing business in several jurisdictions may be involved and the situs of the injury or the activity leading to the injury may differ from the location of the companies involved.

[16] *Ibid.*, Conclusions of Law, Nos. 43–46.

ility Convention would be applied to limit the Owner's liability and whether the plaintiffs had standing to sue. The Court rejected application of the Convention to this dispute since it was the law of France, not of the United States, and it was United States law which applied to whether AIOC and Standard could be sued. As to AIOC and Standard Oil Company (Indiana), the parent, the Court noted that even if the Convention were to have applied, neither AIOC nor the parent were mandatories or preposes (agents and servants) of the owners and thus not subject to the Convention's proscription against suit for these classes of persons. The legislative history of the Convention, according to Judge McGarr, was meant to refer to and immunise the master and crew of the vessel; major operators of vessels were not meant to be immunised against suit by these provisions of the Convention, according to the Court.[17]

21–70 The Court dismissed the Amoco counterclaims and third party claims against France and certain of the communes for their alleged negligence in failing to prevent or contain the oil spill. It was found that the French parties had no duty which could accrue to Amoco's benefit and thus there was no cause of action in favour of Amoco for failure of the French parties to plan for clean-up or for ineffectual clean-up efforts. It was noted, however, that Amoco could not be liable for damages resulting from "inept clean-up efforts which exacerbated the harm," and resolution of this issue was left for the damages trial.[18]

21–71 Since the trial was bifurcated between liability and damages issues, there remain possibly years of further discovery addressed to damages. Such damages issues will undoubtedly be resolved by analysis and refinement of the prnciples of recovery set out in detail in the *Zoe Colocotroni* case. There is, however, an important distinction between the two cases. In *Zoe Colocotroni*, the restoration of the environment was never undertaken and it was conceded that most of the work for which damages were being sught would never be performed. In *Amoco Cadiz*, the French Government and the several communes did take action to restore the environment and thus there is no speculation as to what amounts might be spent. In fact, most of the damaged shoreline was restored before 1980 when the *Tanio* casualty caused damage in the same area. In *Zoe Colocotroni*, the Court was clearly influenced by the theoretical nature òf the damage calculations since much of

---

[17] *Ibid.*, Conclusions of Law, Nos. 10–21.
[18] Conclusions of Law, Nos. 49–52.

the work had not been performed or even contemplated in the future. Such considerations are not generally present in the *Amoco Cadiz* case. Nevertheless, the question of the reasonableness of the restoration in terms of need and appropriateness, as discussed in *Zoe Colocotroni*, remains as an important damages issue in *Amoco Cadiz*.

21–72 The *Amoco Cadiz* liability decision is significant not merely because it opens up exposure to an extremely large damages claim which is expected to be broadly contested, but also because it provides an expression of the law in areas which are very relevant in the vessel pollution field. In particular, the court's views on the conflicts of laws issues with respect to applicability of the Civil Liability Convention, while predictable, offer guidance as to how courts in general can be expected to resolve this issue. Also, the decision on denial of 1851 Act limitation to each of the Amoco parties tends to confirm the view within the United States admiralty bar that limitation continues to lose its vitality as an effective remedy for shipowners.

4. FEDERAL PRE-EMPTION AND THE NEED FOR UNIFORMITY

21–73 It does not require superior perceptive abilities to discern that the present vessel oil pollution regime in the United States is burdened with excessive layers of laws and regulations which can present a formidable obstacle to compliance. No small part of the problem arises from the nature of the United States federal legal system in which the Constitution reserves to the states all powers not expressly delegated to the federal government. What has resulted is a regime of at least two layers of regulations, state and federal. The problem for the shipowner, the insurer and the oil industry is that there are not one, but fifty states, each of which has the power to legislate in areas of local concern not pre-empted by federal legislation. To exacerbate the problem, regulation on the federal level has been piecemeal and re-active to specific pressures at different times, leading to enactment and frequent amendment of four different and occasionally contradictory oil pollution schemes.

21–74 The maritime field is particularly well suited for a uniform legal scheme for vessel oil pollution. Stating the obvious, vessels are not fixed facilities which can be constructed and operated according to the laws which exist in a particular place. By design they are movable and by purpose they transport goods from one jurisdiction to another. By imposing different standards, different liability limits and different levels and filings of financial responsibility, among others, the burden on the maritime industry is increased. The legal regime becomes unnecessarily complex, when a single, and much simpler, scheme is available for United States participation which provides for full and adequate recovery of oil pollution damage in all but the most disastrous of cases. While the worldwide public may pay indirectly for all, or a portion of, the expense of these varying standards, it is nevertheless an extravagant expense when a properly considered uniform standard would be likely to protect all of those interested in vessel oil pollution.

21–75 In the context of United States legislation, it is to be recalled that significant oil spill laws were not enacted on the federal level until the 1970 Water Quality Improvement Act, following the *Torrey Canyon* disaster. Simply, coastal states, too, began to recognize that they had a legitimate interest in the protection of their waters and many enacted oil spill legislation of their own, some providing for more onerous standards than the federal laws and some requiring separate evidence of financial responsibility.[19] It was in the light of these varying federal and state standards that the question of uniformity in the maritime oil spill field first arose as a constitutional challenge to the Florida Oil Pollution Act, followed by a constitutional challenge to the Maine Oil Pollution Act.[20] In both instances, the courts declined the invitation to mandate uniformity.

21–76 In *Askew* v. *American Waterways Operators, Inc.*, the Florida Act was challenged by shipowners using Florida ports, P&I Associations (the insurers of shipowner's oil pollution and other third party liabilities), world shipping groups and others involved in seaborne trade in Florida. The Florida Act, passed in 1970, was a precursor to the present Florida Act discussed earlier in this chapter.[21] Unlike the FWPCA, as amended by the Water Quality Improvement Act, the Florida Act imposed strict and unlimited liability on any vessel which discharged oil while proceeding to or departing from a Florida port and required separate evidence of financial responsibility. The Act also required that vessels transiting Florida waters be equipped with special containment gear and a crew trained in its use. To enforce these provisions, every vessel was sub-

---

[19] See, paras. 21–01 *et seq.*

[20] *Askew* v. *The American Waterways Operators, Inc., 411 U.S. 325 (1973); Portland Pipe Line Corpn.* v. *Environmental Improvement Commission*, 307 A.2d 1 (Me. 1973).

[21] 1970 Fla. Laws Ch. 70–244.

ject to state inspection before entering a Florida port. The liabilities and requirements of the FWPCA were concededly less stringent than the Florida provisions.

21–77 The district court three-judge panel in Florida granted declaratory judgment to the plaintiffs, holding that the doctrine of federal pre-emption in the maritime field rendered the Florida statute unconstitutional.[22] The court, in a lengthy discusssion, perceptively articulated the need for uniformity in maritime affairs, which need had long been recognized:

> "It is well settled that state legislation is invalid where it is in contravention with general admiralty rules or congressional enactments in the maritime field. In the landmark *Jensen* case, the Supreme Court, in holding that the New York State Workmen's Compensation Statute could not constitutionally be applied where an accidental death occurred on a vessel afloat in navigable waters within New York's boundaries, said:
>
> 'And plainly, we think, no such legislation is valid if it contravenes the essential purpose expressed by act of Congress or works material prejudice to the characteristic features of the general maritime law or interferes with the proper harmony and uniformity of that law in its international and interstate relations. This limitation, at the least, is essential to the effective operation of the fundamental purposes for which such law was incorporated into our national laws by the Constitution itself. If New York can subject foreign ships coming into her ports to such obligations as those imposed by her Compensation Statute, other States may do likewise. The necessary consequence would be destruction of the very uniformity in respect to maritime matters which the Constitution was designed to establish; and freedom of navigation between the States and with foreign countries would be seriously hampered and impeded. . . . The legislature exceeded its authority in attempting to extend the statute under consideration to conditions like those here disclosed. So applied, it conflicts with the Constitution and to that extent is invalid.'
>
> The Florida Act here constitutes a far greater intrusion into the federal maritime domain than the New York statute in the *Jensen* case. If applied to the plaintiffs and intervenors in this case, the Florida Act would effect—in the words of *Jensen*—the 'destruction of the very uniformity in respect to maritime matters which the Constitution was designed to establish; and freedom of navigation between the states with foreign countries, would be seriously hampered and impeded.'
>
> This is not a situation in which a state legislature has sought to act in an area of purely local concern and its enactment is no real encroachment of federal interests. Rather, this is a case where the

---

[22] *American Waterways Operators, Inc.* v. *Askew*, 335 F.Supp. 1241 (M.D. Fla. 1971.

State purports to impose upon shipping and related industries duties which under the federal law they do not bear. It can hardly be said that Florida is not seeking to regulate conduct in the federal maritime jurisdiction. We need not belabor the point that to permit the states severally to regulate these industries as Florida seeks to do would sound the death knell to the principle of uniformity."

21–78 The *Askew* case reached the United States Supreme Court on direct appeal in which the Supreme Court unanimously reversed the lower court decision. It held that states may enact laws to protect property within their territories as a legitimate exercise of policy power as long as the federal government has not enacted laws which have pre-empted the field or which are inconsistent with the state law. In reviewing the Florida Act, the court decided that its coverage was not inconsistent with the Water Quality Improvement Act amendments to the FWPCA and that existing federal legislation did not pre-empt the field. In particular, it was held that the Florida Act's requirements concerning clean-up of state waters and state and private recovery for clean-up costs and damages did not conflict with the federal legislation which concerned federal clean-up and recovery.

21–79 A similar case followed closely on the tail of *Askew* in the state courts of Maine. In *Portland Pipe Line Corpn.* v. *Environmental Improvement Commission*,[23] several companies doing business in Maine sought to overturn the Maine Oil Discharge Prevention and Pollution Control Act. The Maine Act was similar in form to the present New York Act discussed earlier in this chapter in that a pollution fund was established by the collection of one half-cent per barrel charge on oil transfers and a claims procedure called for arbitration of disputed claims. The plaintiffs asserted that this statute was unconstitutional in that it violated the Admiralty Clause in the United States Constitution, the due process clause, the equal protection clause and the commerce clause. They also argued that the Act was pre-empted by the FWPCA. The Supreme Court of Maine upheld the Act and, with respect to pre-emption, noted that the language of FWPCA, section 1321 (o) was a broad waiver of pre-emption.[24] The United States Supreme Court declined to review the case, thus impliedly indicating that it did not believe that *Askew* required further explanation.[25]

---

[23] 307 A.2d 1 (Me. 1973).
[24] Section 1321(o) states, in relevant part: "Nothing in this section shall be construed as pre-empting any state or political subdivision thereof from imposing any requirement or liability with respect to the discharge of oil or hazardous substance into any waters within such state."
[25] 414 U.S. 1035 (1935).

21–80   With efforts having failed to persuade the courts to impose uniformity on the states, instances of inconsistent state legislation escalated.[26] Political lobbying supplanted courtroom activity, but movement toward uniformity has been extremely slow.[27]

5. COMPREHENSIVE ENVIRONMENTAL RESPONSE, COMPENSATION AND LIABILITY ACT OF 1980

21–81   One moderate effort in the direction of uniformity came with passage of the Deepwater Port Act in 1974. In that Act, a detailed study by the Attorney General of the United States was ordered to examine methods and procedures for implementing a uniform law providing liability for clean-up costs from oil spills.[28] That report was delivered to Congress in 1975 by Attorney General Edward Levi and became known as the Levi Report.[29] The Levi Report, which reviewed existing federal and state legislation and examined issues of damages recovery, was candidly critical of the overlap and inconsistency of existing law and recommended development of uniform legislation which would take into account industry as well as environmental needs.[30]

21–82   The Levi Report specifically endorsed the "Superfund" concept which would, in effect, comprise a comprehensive federal, as opposed to international, pollution scheme which would take into account the concerns of the states. Superfund bills were introduced in Congress,[31] which continued to think in terms of going its own way, rather than seeking the solution of the Civil Liability Convention and the IOPC Fund; but comprehensive legislation was not passed and enacted until 1980. In that year, Congress enacted the Comprehensive Environmental Response, Compensation and Liability Act of 1980 (CERCLA).[32] CERCLA establishes a comprehensive federal scheme for response, compensation and liability with respect to "hazardous substances" released into the environment. CERCLA, however, does not deal with oil which is excluded from the definition of "hazardous substances."[33]

---

[26] See Jarvis, above, note 1; Wallace & Ratcliffe, above, note 1.
[27] The politics of oil pollution regulation are beyond the scope of this book. An excellent coverage of the topic can be found in M'Gonigle and Zacker, *Pollution, Politics and International Law*, Univ. of California Press (1979).
[28] 33 U.S.C. §1517(n).
[29] Dep't of Justice Report, above, note 1.
[30] *Ibid.* at p. 60.
[31] H.R. 6803 (95th Cong.); S.2083 (95th Cong.).
[32] 42 U.S.C. §§9601–9657.
[33] 42 U.S.C. §9601(14).

21–83 While CERCLA is not applicable to vessel oil spills, the CERCLA legislation was being considered in Congress at the same time as oil Superfund legislation. It has been suggested that CER-CLA may well serve as a model for such legislation in the future, and therefore a brief look at its pertinent provisions is in point here.[34]

21–84 CERCLA is modelled upon the activity-related federal statutes in that a "Hazardous Substance Response Trust Fund" is established for expeditious removal and response under the Intervention on the High Seas Act, under FWPCA, and damages to natural resources.[35] The fund has received a yearly $44,000,000 appropriation. In addition, the Fund is credited with all penalties and amounts recovered from responsible parties for prohibited discharges. The fund is not liable for claims beyond the amounts available in the fund.[36]

21–85 Liability under CERCLA is strict but limited for vessel owners to: (1) all costs of removal or remedial actions "not inconsistent with the National Contingency Plan" incurred by federal or state authorities; (2) any necessary costs of response incurred by any other person consistent with the National Contingency Plan; and (3) damages for injury to, destruction, or loss of natural resources.[37] The only available defences are those in which the discharge was caused soley by an Act of God, act of war, act or omission of a third party who is neither an employee or agent of the defendant nor one with a "contractual relationship" to the defendant, or a combination of these.[38] Liability is limited to $300 per gross ton or $5,000,000, whichever is greater, for vessels carrying hazardous substances as cargo or residue ($300 per gross ton or $500,000 for any other vessel) unless the defendant is guilty of wilful misconduct or wilful negligence or unless the defendant has failed to render all reasonable co-operation and assistance in response requested by a responsible public official.[39] In addition, if the discharger fails without sufficient cause properly to provide removal or remedial action, then punitive damages of up to three times the clean-up costs may be claimed in any civil action which may be commenced.[40]

---

[34] See, *e.g.* Wallace and Ratcliffe, above, note 1, at p. 1351.
[35] 42 U.S.C. §9611.
[36] 42 U.S.C. §9611(e).
[37] 42 U.S.C. §9608(a).
[38] 42 U.S.C. §9607(b).
[39] 42 U.S.C. §9607(c).
[40] 42 U.S.C. §9607(c) (3).

21–86 The issue of applicability of the 1851 Limitation of Liability Act to vessel liability under existing federal and state statutes remains without a definitive answer. As discussed in paragraph 21–64, above, one district court case has thus far ruled on the subject in relation to the FWPCA, finding that the strict liability and limits of the FWPCA could not be diminished by the 1851 Act. While this would appear to be the anticipated result in consideration of the relationship between the 1851 Act and other federal and state strict liability laws,[41] the result is by no means clear or well-settled. CERCLA responds to this problem by specifically making the 1851 Act inapplicable to claims under CERCLA.[42]

21–87 As with other oil spill federal legislation, under CERCLA evidence of financial responsibility is required to be demonstrated up to the vessel's strict liability limitation amount.[43] The guarantor, who may be sued directly, may invoke the defences of the owner and operator and may also assert the defence that the incident was caused by the owners' or operators' wilful misconduct. All other defences under the policy are unavailable to the guarantor in a direct recourse action.[44]

21–88 The claims procedure requires that each claim be submitted to the owner, operator or guarantor of the known discharging vessel in the first instance. If payment is not made within 60 days of presentation, the claimant may elect to commence suit against the discharger or guarantor or to present the claim to the Fund for payment.[45] If the discharger is unknown, claims may be presented to the Fund. If not settled, the claim may be arbitrated.[46] All decisions are final and may not be overturned for any reason except for "arbitrary or capricious abuse of an arbitrator's discretion."[47]

21–89 All claims paid by the Fund are subject to the Fund's rights of subrogation. The fund may seek compensation from any responsible person without regard to limitation of liability and may also seek costs and fees arising out of the responsible person's failure to make payment in the first instance.[48]

---

[41] See, Sisson, "Oil Pollution Law and the Limitation of Liability Act. A Murky Sea for Claimants Against Vessels," 9 Jl. Mar. Law and Comm. 285 (1978).
[42] 42 U.S.C. §9607(i).
[43] 42 U.S.C. §9608(a).
[44] 42 U.S.C. §9608(c).
[45] 42 U.S.C. §9612(a).
[46] 42 U.S.C. §9612(b) (4) (A).
[47] 42 U.S.C. §9612(b) (4) (G).
[48] 42 U.S.C. §9612(c).

21–90   Although endorsed as a Superfund (in the sense of an all-embracing fund), CERCLA is hardly such a statute. It remedies, for the non-oil pollution area, certain of the non-uniformity problems associated with existing legislation in the oil pollution area, such as the relationship between it and the 1851 Limitation Act, but it leaves unresolved, of course, the problems of overlapping state, federal and common law oil pollution law in the United States. Specifically, CERCLA states that no other statutory or common law remedies are deemed to be waived by asserting a claim against the Fund.[49] The Act also states that state and common law remedies are not meant to be pre-empted.[50] CERCLA is a useful beginning, but it is hardly the comprehensive Superfund legislation so badly needed in the United States.

6. THE COMPREHENSIVE OIL SPILL BILL OF 1985— AN APPROACH TO UNIFICATION

21–91   On February 21, 1985, H.R. 1232 was introduced to the 99th Congress, a modification of legislation which passed the House of Representatives but not the Senate in the 98th Congress. H.R. 1232 is a bill seeking passage of the Comprehensive Oil Pollution and Compensation Act which, like its predecessors, is expected to pass in the House of Representatives whereupon, in 1986 it will be transmitted to the Senate for consideration. There is cautious optimism that this bill will be enacted because there is Administration support for its passage even though oil company support does not appear to be unanimous and further extensive lobbying against the bill is anticipated. Because, unlike CERCLA, H.R. 1232 would cover oil spill liability from vessels and would supplant some of the now-existing oil spill legislation, an examination of its provisions is useful.

A. Relationship with Existing Oil Spill Statutes

21–92   The Bill, if enacted, would not repeal any of the four major oil spill statutes, but would modify each of them. It would, for example, replace the four oil spill funds created by FWPCA and the three activity-related statutes with a single federal fund financed by premiums on imported and exported oil.[51–53] The Bill would also

[49] 42 U.S.C. §9612(e).
[50] 42 U.S.C. §9604.
[51–53] The DPA and OCSLA Funds would be transferred to a new Fund created by the Bill, the TAP Fund would be repealed and the section 311 Revolving Fund from general appropriations would be abolished. See para. 21–101, below, for elaboration of this point.

alter to a uniform amount the varying limits of strict liability contained in the several statutes and would standardise the degree of culpability required to break limitation of liability.[54] Existing provisions, such as those relating to financial responsibility would, of course, also be superceded.

B. Scope of Coverage for Oil Pollution—Strict Liability

21–93 The Bill covers oil pollution from vessels (other than public vessels and facilities) in or upon United States waters or in connection with OCSLA, DPA or TAP activities.[55] The owner of any vessel "that is the source of oil pollution or poses a substantial threat of oil pollution" in circumstances that justify the incurrence of "oil removal costs," would be "jointly, severally and strictly liable for all damages for which a claim may be asserted" under the Bill. The class of claims assertable would be very broad, much as are the broad range of assertable claims set out in the activity-related statutes discussed in Chapter 20.

21–94 Under the Bill, liability would be strict but limited, with certain exceptions as are described below. The limits of liability would be, for a vessel other than one carrying oil in bulk as cargo or residue, the greater of $500,000 or $300 per gross ton; for a "ship," which is defined as a vessel carrying oil in bulk as cargo or residue, the greater of $3,000,000 or $420 per gross ton (up to $60,000,000); and for an inland oil barge, the greater of $150,000 or $150 per gross ton. These limits are roughly equal to those of the 1984 Protocol to the Liability Convention: see paragraphs 10–141 *et seq.*

21–95 Limited liability would not be permitted (1) where the

---

[54] The Bill provides that, to the extent there is any inconsistency or conflict between its provisions and any other law relating to liability or the limitation thereof, this Bill would supercede such other law: s.104(h).

[55] s.104. A "vessel" is broadly defined as every description of watercraft used or capable of being used as a means of transportation: s.101(3). A "facility" means a structure licensed under the Deepwater Port Act or located on the outer Continental Shelf used in the exploration or transportation of oil from the outer Continental Shelf: s.101(5).

"Oil Pollution in United States waters" is defined as the presence of oil in or on the navigable waters of the United States, on immediately adjacent land, or in or on the waters of the contiguous zone, which has been discharged in harmful quantities from a vessel or facility. "Oil Pollution in or on waters outside United States territorial limits or within a foreign country" is defined to include the presence of oil discharged in connection with OCSLA or DPA activities, as to which see Chapter 20, and which causes injury or loss to natural resources, or has been aboard a TAP oil vessel before discharge to a United States port. Also, "Oil Pollution" includes the presence of oil in the waters of a foreign country which has been discharged from a vessel in United States navigable waters, in connection with DPA and OSCLA activities or a TAP oil vessel before discharge to a United States port. "Oil" means petroleum including crude oil, or any fraction or residue therefrom: s.101(b).

polluting incident was "caused primarily by willful misconduct or gross negligence within the privity or knowledge of a responsible party"; (2) where the incident was caused primarily by a violation, within the responsible party's privity or knowledge, of applicable safety, construction or operating standards or regulations of the United States; or (3) where a responsible party failed or refused to report the incident or to provide requested co-operation and assistance to Federal officials in furtherance of clean-up and removal activities.[56]

21–96 There would be complete and partial defences to strict liability. A complete defence would be available if the responsible party proved that the incident (1) resulted from an act of war or hostilities or from a *force majeure* event, or (2) was wholly caused by an act or omission of a person other than a responsible party, his agent or employee or one whose act or omission occurred in connection with a contractual relationship with a responsible party. Partial defences would be available as to particular claimants where or to the extent that the incident or loss was caused, in whole or in part, by the gross negligence or wilful misconduct of that claimant.

C. Assertable Claims

21–97 Like the activity-related statutes, the Bill sets out a listing of damages "arising out of or directly resulting from oil pollution or the substantial threat of oil pollution" which may be asserted. These are:
  (1) removal costs;
  (2) injury to, or destruction of, real or personal property;
  (3) reasonable costs incurred in assessing injury to or destruction of natural resources, preparing a restoration plan and restoring or acquiring equivalent resources;
  (4) loss of substantive use of natural resources;
  (5) loss of profits or impairment of earning capacity for up to two years due to injury or destruction of real or personal property or natural resources; and
  (6) loss of tax revenues for up to one year.

21–98 As with the activity-related statutes, there would be qualifications on what class of injured party could recover the above-

---

[56] s.101(17)(A). The Bill adopts the operational description of "harmful quantities" as found in FWPCA, s.311, meaning adoption of the so-called sheen test, as to which see para. 19–07.

listed damages. In general, a claim could be asserted for such damages by any person, with certain exceptions, as follows:

(1) An owner or operator of an offending vessel could recover its costs associated with oil removal only where he would be entitled to a defence against liability or to a limitation of liability. If he were entitled to limit liability, he would only recover any excess of such costs over the limitation amount;

(2) Items (2) and (4) above could be recovered by a United States claimant only if the property involved were owned or leased, or if the natural resource were utilised by him;

(3) Item (3) above could be recovered by the President as trustee for natural resources controlled by the United States or by the Governor of any State in which the natural resource was controlled by that State or local authority;

(4) Item (5) above could be recovered by a United States claimant only if he derived at least 25 per cent. of his earnings from activities which utilise the natural resource or the property;

(5) Item (6) above could be recovered only by a State of the United States or political subdivision thereof;

(6) Foreign claimants could recover for items (2)–(5) above to the same extent as a United States claimant if (a) he were not otherwise compensated for his loss, and (b) recovery was authorised by a treaty or agreement between the United States and the foreign country and the Secretary of State certified that a comparable remedy would be available in that country to a United States citizen.

D. Financial Responsibility

21–99 Vessels over 300 gross tons which use a facility or the navigable waters of the United States would be required to maintain evidence of financial responsibility sufficient to satisfy the maximum liability to which the owner or operator would be exposed if he were entitled to limited liability. As with existing legislation, fleet certificates for the largest vessel would be acceptable. Penalties for failure to meet this provision would include withholding sailing clearance and denying entry or otherwise detaining the vessel.

21–100 The guarantor providing financial responsibility would be directly liable to satisfy any claims authorised by the Bill. Such guarantor would be able to invoke all rights and defences of the vessel and, as with existing legislation, could assert that the incident was caused by the wilful misconduct of the responsible party. No other defences would be acceptable.

E. Marine Oil Pollution Insurance Corporation and Compensation Fund

21–101 The Bill would establish the Marine Oil Pollution Insurance Corporation ("MOPIC") as a wholly-owned Government corporation supervised and directed by the Secretary of Transportation. MOPIC would be managed by an Administrator, appointed by the President upon the advice and consent of the Senate, for a term of seven years. The principal function of MOPIC would be to hold and administer the Marine Oil Pollution Compensation Fund ("MOPC Fund"). The MOPC Fund would be comprised of: (1) premiums, in the amount of 1.3 cents per barrel (subject to prohibition against double collection) on (a) crude oil or other petroleum products entered into the United States for consumption, use or warehousing, payable by the person entering such products; (b) crude oil received at a United States refinery, payable by the refinery; and (c) crude oil exported from the United States, payable by the person exporting such products. Such premiums would be imposed only when the amount in the MOPC Fund was less than $200,000,000; (2) amounts collected on behalf of the MOPC Fund from parties responsible for oil pollution damage and for which the Fund would be required to make payment; (3) amounts transferred to the MOPC fund from the then defunct Deepwater Port Act and OCSLA Funds, for which the MOPC Fund would then fulfil the same responsibilities; (4) income earned on investments; and (5) amounts borrowed in order to make payments authorised by the Bill.

21–102 Unlike the OPC Fund discussed in Chapter 20, neither the MOPC Fund nor MOPIC would bear unlimited liability for claims. All claims would be paid only out of the Fund and liability with respect to any one incident would not exceed $200,000,000.

21–103 Except for removal costs, no claim would be paid out of the Fund if such payment would reduce the outstanding balance to less than $30,000,000. If the Fund were unable to make a payment for this reason, claims would be paid in full in the sequential order in which they were finalised. The Fund would be authorised to borrow to pay claims in that order. Should claims for any one incident exceed the $200,000,000 limit, all claims would be proportionately reduced. Under this system, it would seem that when the prospect of claims in excess of fund limits was evident, final payments would have to be delayed until the appropriate proportion could be calculated. Of course, interim payments would be a fair manner of compensation prior to a final calculation.

21–104 Subject to the above-described limitations, MOPIC would be strictly liable for damages for all claims assertable under the Bill to the extent that damages would not otherwise be compensated. The only defences of the Fund to liability, except for removal costs for which there would be no defence, would be: (1) where the incident was caused wholly by an act of war, hostilities, civil war or insurrection; (2) as to a particular claimant, where the incident or loss is caused in whole or in part by its own gross negligence or wilful misconduct, or (3) as to a particular claimant, to the extent the incident or economic loss is caused by its own negligence.

F. Claims Procedure

21–105 Each of the four vessel pollution statutes which would be affected by the Bill provide for notification to the Secretary of Transportation or his designee in the event of an oil spill. Those provisions would not be changed by this Bill. The Bill provides, upon such notification, that the Secretary must designate and notify the "source" of the oil pollution, where possible. Thereafter, it would become the duty of such source, and its guarantor, to advertise the designation and establish a procedure for presentation of claims to them. As with the existing statutes, there appears to be no incentive in the Bill for a discharger or its guarantor to accept the designation and pay claims in the first instance. Although the Bill makes a step in the right direction by depriving a responsible party of his right to limit liability if he "fails or refuses to report the incident where required by law or to provide all reasonable co-operation and assistance requested by the responsible Federal official in furtherance of clean-up and removal activities," this would not appear to cover a refusal to accept a designation.

21–106 If the designation were to be refused, the Secretary would be called upon to advertise and establish a claims procedure. In such a case, claims would be presented to MOPIC.

21–107 The procedures for reviewing and settling claims would be virtually identical to those established under the activity-related statutes. Significantly, in circumstances where claims would be submitted to a designated responsible party who, by virtue of limited liability, incapacity or otherwise, fails to pay the full amount of a proved assertable claim, the uncompensated balance would then be submitted to MOPIC for full reimbursement. Of course, full reimbursement from MOPIC would be subject to the

571

$200,000,000 per incident limit, but it may be anticipated that this amount would cover all but the most catastrophic events.

21–108 Under circumstances in which the responsible party accepts a designation but thereafter denies liability for the claim or fails to settle a claim within the later of 180 days after presentation or advertisement, the claimant would be entitled irrevocably and exclusively to elect to sue the responsible party or his guarantor or to present the claim to MOPIC. As with the activity-related statutes, there appears to be no strong incentive in this Bill to elect the former option and given the short periods involved, it would be likely that most claims would proceed to settlement through MOPIC.

21–109 If, however, suit were to be commenced against the responsible party, MOPIC would be granted the right to be given notice of the suit and to intervene. MOPIC would be conclusively bound by any judgment thereafter rendered. Failure to notify MOPIC of the suit and serve upon it copies of the pleadings could have dire consequences for both parties: the vessel owner would not be entitled to the benefit of limited liability and the claimant would not be entitled to collect any outstanding balances from MOPIC. Attention to procedural detail would be extremely important under such circumstances.

21–110 As with the existing statutes, claims would be barred unless they were presented, or a suit commenced, within the lesser of three years of the discovery of the loss or six years of the incident.

G. Subrogation

21–111 Any person who compensates a claimant would be subrogated to all rights of that claimant. There would be special rules regarding subrogation rights of MOPIC concerning, *inter alia*, how interest would be computed based on whether the responsible party has accepted a designation and whether he has paid or offered a proper settlement of the claim. Although such distinctions are made in the Bill, the rewards for accepting a designation or paying a liability do not appear to be strong enough to encourage such actions by the responsible party.

H. Co-ordination With International Oil Pollution Conventions

21–112 The Bill specifically provides that during any period in which the 1969 Liability and 1971 Fund Conventions, as amended in 1984, are in force for the United States, the provisions of the Bill would not apply to the extent that compensation is available under the Conventions. While the Bill lacks clarity, it would appear that the intention of the drafters was to substitute the liabilities, defences and compensation scheme contained in the International Conventions, when in force, for those contained in the Bill. Despite the use of the ambiguous phrase "to the extent that compensation is available under such [International] Conventions" it does not appear that the MOPC Fund was meant to provide a supplemental or further source of recovery over and above the IOPC Fund, nor are the defences contained in the Bill meant to provide a different basis for liability than that found in the 1984 Liability Convention.

21–113 In anticipation of United States ratification of the Conventions, the Bill sets out implementing legislation to give them full force and effect. Contributions payable to the IOPC Fund under Article 10 would be paid by the MOPC Fund.

I. Analysis of H.R. 1232

21–114 H.R. 1232 is the sixth effort since 1975 to streamline existing United States oil pollution law. It is an energetic effort to begin what will undoubtedly be a long process of disengaging existing laws and substituting in their place legislation leading to full implementation of the International Conventions. As pointed out by Congressman Gary Studds in his statement introducing the Bill, enactment would improve upon the present law in several important respects:
  (1) It would ensure all claimants of prompt and fair compensation for oil spill damage, rather than being required to prove negligence in the typical case of a spill in United States waters;
  (2) It would impose strict liability on producers and transporters of oil which could encourage a high standard of care and would allocate the costs of clean-up and loss to those with the most direct entrepreneurial interest;
  (3) It would reduce the cost to the taxpayers of clean-up and removal by creating an industry-financed fund;
  (4) It would simplify and reduce federal administrative expense by eliminating four funds and merging them into one comprehensive fund; and
  (5) It would standardise the regulatory and legal framework thereby making the regime more predictable.

21–115 Each of these points addresses important problems with existing oil pollution legislation in the United States. By proposing a method of reconciling this Bill with the International Conventions and by proposing implementing legislation to give domestic effect to these International Conventions, this legislation, if enacted, would significantly advance efforts to secure a uniform international oil pollution regime.

7. CONCLUSION

21–116 Confronted with an extremely complex and unnecessarily duplicitive body of federal, state and local laws in the vessel pollution area, it is appropriate for the Government of the United States to take strong affirmative action to remedy the situation. H.R. 1232 is a good first step. Clearly, through such decisions as *Askew* and *Portland Pipe*, the courts have acknowledged their limited role in making the system more comprehensible, and this unification is at the doorstep of Congress.

21–117 The United States participated actively in the 1984 IMO Conference which revised the 1969 and 1971 Conventions and the views of the United States delegation were largely adopted by the Conference, particularly on the crucial issue of liability limits. There have been many calls for the United States to ratify the 1969 and 1971 Conventions if the concern over liability limits could be satisfied.[57] Now that such concerns appear to have been resolved, the only impediment to immediate ratification is the development of an interim scheme to supplement the lower limits of the 1969/ 1971 Conventions until the 1984 limits come into force. One simple method of achieving this could involve passage of H.R. 1232 or amendment of CERCLA to include oil discharges and thus making its higher limits available until the 1984 limits on the Civil Liability and Fund Conventions come into force. In enacting the 1969/ 1971 Conventions, there would need to be an explicit Congressional recognition of the pre-emptive nature of these laws, thereby rendering state legislation void. Whether or not such Congressional pronouncements would carry the day in a confrontation with the states is an open question, but given the fact that the 1969/

---

[57] See, *e.g.* Dep't of Justice, Report, 1 above, note at pp. 97–102; Temple, Barker & Sloane, Inc., *Cost-Benefit Analysis of Possible U.S. Adherence to Two International Conventions on Liability and Compensation for Oil Pollution Damages*, pp. 1–28–31 (Prepared for Commandant, United States Coast Guard, June, 1983); Wallace & Ratcliffe, above, note. 2.

1971 Conventions are far broader in terms of coverage than the FWPCA, upon which prior court tests of pre-emption were based, it is believed that a pre-emption argument would be likely to succeed.

21–118 Immediately following the 1984 Diplomatic Conference on the revision of the 1969/1971 Convention, the Director of the IOPC Fund, Dr. Reinhard H. Ganten, prepared remarks for presentation to the House Subcommittee on Coast Guard and Navigation wherein he stated:

> "Mr. Chairman, let me conclude my statement by expressing the hope that, after careful consideration of the interests of the United States, this nation comes to the conclusion that it is in its best interest to join the CLC and the Fund Convention and that by depositing the two instruments with IMO this great nation improves its poor reputation regarding ratification of liability conventions in the field of transport law. The ratification of the two Conventions by the United States would, in my opinion, be to the benefit of the United States; this view is supported by the excellent study of Temple, Barker and Sloane. An early ratification would also greatly encourage the ratification of these two Conventions by other States, and enable their early entry into force. Last, but by no means least, such a change in attitude towards ratification of liability conventions in transport law would remarkably increase the negotiating position of U.S. delegations at future Diplomatic Conferences. The US delegation at the London Diplomatic Conference succeeded in convincing delegates of the good prospects of a U.S. ratification of the two Conventions and, as a result of this, considerable concessions were made to the United States delegation and many of the U.S. objectives at the Conference were largely achieved. There is now a great expectation for positive steps towards U.S. ratification of the CLC and Fund Convention. It would be a great achievement if these steps could be taken in the 'Year of the Ocean,' just declared by President Reagan. Should the United States decide not to ratify the two new CLC and Fund Conventions, the enormous amount of goodwill built up by this very remarkable delegation, would not only be lost but could have the contrary effect of delegations becoming unwilling ever again to make efforts to accommodate wishes of United States delegations at Diplomatic Conferences on private transport law issues."

Mr. Ganten's comments are echoed by many others in the United States who support ratification of the 1984 Protocols.[58] The present

---

[58] See, Paulsen, "Ratification of 1984 Protocols," 20 The Forum 164 (1984). The Maritime Law Association of the United States has long advocated ratification of the CLC and the Fund Convention and recently endorsed ratification of the 1984 Protocols.

United States scheme is so clearly inefficient and cost-burdensome that a welcome improvement would be the acceptance of the 1969/1971 Conventions as amended by the 1984 Protocols, together with a purposeful winding down of the existing legislation.

INDEX

ACCIDENTAL DISCHARGES, 4–01 *et seq.*
  amount of spillage, estimation of,
    4–06
  annual statistics, 4–64, 4–65
  casualties, arising from, 4–04
  complications inherent in, 1–06 *et seq.*
  design of ships, 4–18
  government enforcement standards,
    poor, 4–03
  grounding, 4–05, 4–06, 4–07
  high seas, on,
    flag state jurisdiction, 5–24
  impact of, factors affecting, 4–06
  Liability Convention, covered by,
    10–14
  occasional, tolerated by customary
    international law, 9–05
  operational procedures, associated
    with, 4–04
  personnel standards, 4–10, 4–12,
    4–16
  prevention of,
    role of ship management, 4–65
  routine operations, 4–10
  shipowners, poor management by,
    4–03
  statistics,
    limited use of, 4–06, 4–07, 4–08
  trespass, action for, 15–12
  1954 Convention, exempt from
    provisions of, 9–14
ACTION *IN PERSONAM, IN REM. See IN*
  PERSONAM; IN REM.
AMOCO CADIZ CASE, 21–61 *et seq.*
  U.K. government response to,
    14–11
ANGLO-FRENCH SAFETY OF NAVIGATION
  GROUP, 14–08
ARCHIPELAGIC STATES, 5–54, n. 47
ARCHIPELAGIC WATERS, 5–57
  definition, 5–75
  duties of ships in, 5–77
  enforcement of standards in, 5–78
  right of passage through, 5–76
    transit passage, treated as, 5–77
  sea lanes through, 5–76
  sovereignty over, 5–75

ARREST OF SHIP, 5–19, 5–24, 16–11
  issue of writ, 16–11
  release from arrest, 16–13
  security for claim, as, 16–07, 16–10
  substantive action, part of, 16–12
  whether possible where ship entitled
    to transit passage, 5–72
ARREST CONVENTION, 5–33
  arrest, definition of, 5–33
  existing state powers preserved, 5–33,
    n. 34
  jurisdiction of states,
    geographical scope not defined,
      5–34
    grounds for, 5–35
    merits of claim, on,
      provision for, 16–17, n. 22
  maritime claim, definition of, 5–33
  national law, in relation to, 5–36
  oil pollution damage covered by, 5–33
  ship not responsible for damage,
    arrest of, 16–38
  United Kingdom, ratification by,
    16–12

BALLAST,
  clean. *See* DEDICATED.
  dedicated, 3–65 *et seq.*, 3–72
  oily,
    oil water separator for, 3–09
    reception facilities, and, 3–32
    separator, discharge through, 3–84
  segregated. *See* SEGREGATED BALLAST
    TANKERS.
BALTIC SEA CONVENTION, 7–26
  regular surveillance patrols, 7–26
  undertaking to develop detailed
    rules, 9–24
BALTIC SEA,
  discharge standards in, 14–07
BILGES,
  oily,
    discharge standards for, 3–79 *et seq.*
    oil/water separator for, 3–08
    reception facilities, discharge to,
      3–08
    reception facilities, for, 3–32

UNITED KINGDOM—*cont.*

Oil Pollution Prevention Certificate, 14–27

oil record books, 13–12

oil tankers, requirements for, 13–06 *et seq.*

pollution prevention certificates, 13–11

pollution prevention legislation, 13–01 *et seq.*

port and harbour authorities, role of, 14–13

public nuisance, 13–25 ʼ

Oil Pollution Prevention Certificate, validity of, 14–27 and n. 49

prosecution for pollution offences, 13–15, 14–52

initiation of, 14–45

number of, 14–45

See also CRIMINAL PENALTIES.

reception facilities, 13–10

regional water authorities, role of, 14–37

Report published 1978, 14–11

action taken following, 14–11

reporting requirements,

application of, 14–23

appropriate authorities, 14–24

co-ordination of, 14–50

harbours, transfer of oil in, 14–25

Master's role, 14–24

organisation of, 14–09

response to pollution. *See also* CONTINGENCY PLANS.

Royal Commission on Environmental Pollution, Eighth Report on, 14–11

action following, 14–11

salvor, immunity of, 15–05, n. 2

Scotland, 14–14

river purification boards, responsibilities of, 14–37

Secretary of State for Transport, advisory committee, 14–15

prosecution by, 14–45

ships registered in,

annual survey of, 14–27

certification as to prevention equipment, 14–27

reporting requirements for, 14–23

sightings of oil, reporting system for, 14–12

special areas, discharges prohibited in, 13–04

UNITED KINGDOM—*cont.*

Standing Committee on Pollution at Sea, 14–15

statutory provisions, 13–02

statutory duty, special action for breach of, 15–05, n. 1

survey of ships,

annual, 14–27

unilateral measures by states, opposition to, 14–07

Welsh Water Authority, responsibility of, 14–37

See also ENGLAND, PROCEEDINGS IN.

UNITED NATIONS,

coordination of oil pollution activities, 1–02

UNITED NATIONS CONFERENCE ON THE HUMAN ENVIRONMENT,

duty of states to neighbouring territories, 7–09

UNITED NATIONS CONFERENCE ON THE LAW OF THE SEA, THIRD. *See* LAW OF THE SEA CONFERENCE.

UNITED NATIONS CONVENTION ON THE LAW OF THE SEA. *See* LAW OF THE SEA CONVENTION.

UNITED NATIONS ENVIRONMENT PROGRAMME, 7–06, 7–17

action plans, 7–06

Regional Seas Programme,

Kuwait Convention, 7–34

Mediterranean Convention, 7–27

work of, 1–02

UNITED STATES,

Act to Prevent Pollution from Ships, 18–38

implementation of MARPOL 73/78, 18–38

See also MARPOL ACT, 18–56

Alaska,

oil resources of, 20–02

pipeline, 20–03

alteration of, proposals for, 21–91 *et seq.*

Amoco Cadiz case,

counterclaims in, 21–70

decision of the court, 21–64

vessel, state of, 21–66

Zoe Colocotroni case, compared with, 21–71

Argo Merchant spill, effect of, 18–31

casualties, statistics on, 18–03

CERCLA,

claims procedure, 21–88